Biology, Conservation and Sustainable Development of Sturgeons

Ramón Carmona • Alberto Domezain
Manuel García-Gallego • José Antonio Hernando
Fernando Rodríguez • Manuel Ruiz-Rejón
Editors

Biology, Conservation and Sustainable Development of Sturgeons

Springer

Editors
Ramón Carmona
University of Granada
Spain

José Antonio Hernando
University of Cadiz
Spain

Alberto Domezain
Department of Research
and Development
Sierra Nevada Fish Farm
Spain

Fernando Rodríguez
University of Sevilla
Spain

Manuel García-Gallego
University of Granada
Spain

Manuel Ruiz-Rejón
University of Granada
Spain

ISBN 978-1-4020-8436-2 e-ISBN 978-1-4020-8437-9

Library of Congress Control Number: 2008933626

Printed on acid-free paper

9 8 7 6 5 4 3 2 1

springer.com

Volume Foreword

This volume complements the recent book on North American Paddlefish and Sturgeons (LeBreton et al. 2004) in this series. Together these two volumes bring together our knowledge of these remarkable fishes. There are notable contrasts in the volumes, as well as obvious similarities. Sturgeons have influenced our lives for millennia, not just in fisheries but also in art, economics and mythology. Perhaps we would think first of sturgeon caviar, and its incredible history (Saffron 2002, Vecsei 2005). That story certainly brings together much of what sturgeon species have in common. All sturgeon species have a long biological history. We are still deciphering that history, with a combination of techniques as traditional as morphology and as modern as molecular genetics. Sturgeons also have a long, and mostly sad, history where humans are concerned. All that history, and much more, is presented in this volume. Of course all sturgeons appear very similar, even to the trained eye. Every textbook of ichthyology shows the requisite line drawing and lists the common morphological features of sturgeons. Their obvious similarities also include the history of exploitation, habitat degradation and seriously depleted native populations of virtually every species. This volume goes well beyond a simple compilation of those similarities and concerns.

This volume originated from a meeting of researchers from Asia, Europe and America, held in Spain in 2005 to address the fundamental concerns of conservation and restoration of sturgeons. Those researchers, with some other invited authors, have brought together a wealth of information on subjects as diverse as forensic DNA in the caviar trade to the production of sturgeon in conservation hatcheries. The volume carries a wealth of information for anyone concerned with sturgeon anywhere. Europeans have led the way in the culture and production of sturgeons, whether for conservation and restoration stocking, or for commercial harvest. Chapters related to aquaculture include studies of early development, feed requirements, and management for conservation stocking. A set of chapters deal with the evolution of sturgeons, their relationship to the critical habitat, and large-scale efforts directed to conserve or restore native species. Recovery projects in Italy, Spain, Russia, France and Germany are detailed and documented. Taxonomy, systematics and evolution are critical to conservation and recovery programs and all those subjects are considered as well. In particular, the resolution of long-standing taxonomic and systematic problems with sturgeons in southern Europe is one of the most important contributions of this volume.

Together with the wealth of information on artificial production of sturgeons and the widespread conservation efforts, this gives a positive conclusion to what might easily be considered yet another litany of threatened and vanishing species. It will take more effort by dedicated researchers and managers before we can go beyond what might be a faint positive outlook for sturgeons. This volume certainly shows the way.

Dr. David L. G. Noakes
Series Editor, Fish and Fisheries Series
Professor of Fisheries and Wildlife
Oregon State University
Senior Scientist, Oregon Hatchery Research Center
Corvallis, Oregon

References

LeBreton, G. T. O., F. W. H. Beamish & R. S. McKinley. 2004. North American sturgeons and paddlefish. Fish & Fisheries, Springer, New York.
Saffron, I. 2002. Caviar. The strange history and uncertain future of the world's most coveted delicacy. Broadway Books, New York. 270 pp.
Vecsei, P. 2005. Gastronome 101: How capitalism killed the sturgeons. Environmental Biology of Fishes 73: 111–116.

Foreword

Sturgeons can be considered living fossils having many morphologic and biological features of ancient fish; in fact, sturgeon fossils date from about 300 million years ago, and they have essentially maintained their present features for more than 200 million years. Thus, this phylogenetic group presents unique characteristics from a biological point of view, representing a real challenge for researchers. Furthermore, sturgeons are of the utmost interest also from an applied, economic perspective. In this regard, these species are in great demand, not only for the preparation of caviar, which reaches exceptionally high prices in the international markets, but also for the meat.

There are about 27 sturgeon species living in the seas and rivers of the Northern hemisphere. However, sturgeons are virtually unknown from a scientific point of view and, most importantly, they might soon become a part of the story of evolution; over-fishing, and the loss of their natural habitats might soon make it impossibile to learn about them. In fact, the wild populations of the majority of sturgeon species are actually in great danger of extinction all over the world. Being a diadromous species, that is, migrating from the sea to rivers and lakes for spawning and also staying there during the early juvenile years, and maturing slowly to the reproduction age, make them particularly vulnerable to excessive capture for commercial purposes; just as important is the degradation of their natural habitats due to human establishment and industry. In this situation, it is important to develop strategies for their conservation in natural habitats and also for their culture in fish farms, not only for providing meat and caviar for the market, but also to raise adequate reproducer sturgeons, with the three main objectives of conservation, repopulation of their original habitats and aquaculture.

Rational recovery of the wild sturgeon species requires a sound knowledge of their taxonomy, biogeographical distribution, and biological features. Furthermore, information concerning previous efforts for the conservation and recovery of this group in Europe, Asia and America is also needed.

The story of the recovery of sturgeons began in Russia around 1850, when attempts at artificial breeding were made to enhance the wild populations. In Western Europe, the recovery of autochthonous species was initiated in France, after a training period with an allochthonous species, *Acipenser baerii* from Russia. Soon, groups from several European countries followed the French initiative and worked with *A. sturio* and *A. naccarii*. *A. sturio* is the most common of the European sturgeons, but is actually in danger of extinction and is difficult to reproduce in captivity. In fact, only four successful breedings in captivity have been obtained by the CEMAGREF, France, the last one very recently (summer, 2007). *A. naccarii* is a species sympatric of *A. sturio* all over Southern Europe. *A. naccarii* was virtually unknown until systematic studies were recently initiated, as it is usually confused with *A. sturio*, or, when identified as a different species, as exclusive to the Adriatic Sea and thus known as the Adriatic sturgeon. The research concerning *A. naccarii* began in Italy where, in contrast to what happened with *A. sturio* in other European countries, the studies were conducted not by public organizations but by a private company, the Azienda Agricola V.I.P., owned by Giacinto Giovannini, in Brescia. This researcher, initially without public financial support, obtained the last *A. naccarii* specimens captured alive in the Po and other rivers of Northern Italy (in the Veneto and Lombardia regions), and has been able to reproduce them in captivity. In fact, in 1988, with the collaboration of Italian researchers and public organizations, Giacinto Giovannini succeeded (for the first time and in his own aquafarm), in reproducing in captivity an autochthonous sturgeon outside Russia. The efforts at aquaculture and the recovery of *A. naccarii* soon extended to Spain, when the Piscifactoría de Sierra Nevada, in Granada, also a private company, imported Giovannini's F1 specimens. The research department of this fish farm obtained immediate support from public organizations and was quickly joined by research groups from several Andalusian Universities. The joint efforts of researchers with a multidisciplinary background enabled the development of the largest stock of a European autochthonous sturgeon species.

Taking into account all these developments, the members of the Spanish research groups decided to organize a meeting of experts from several American, European, and Asian countries, aimed at sharing the experiences of the recovery and conservation of sturgeons, to facilitate and promote optimal strategies for the recovery of European sturgeons. With this purpose, the workshop BIORESTURGEONS was organized in Granada, Spain, in November 2005. The main conclusions of this meeting were as follows:

1. Two sympatric sturgeon species have lived in southern Europe: *A. naccarii* and *A. sturio*. This fact could be of great interest for conservation and recovery.
2. The genetic studies have proved useful for the identification of sturgeon species in general and for caviar in particular.
3. The physical environments where the recovery plans will be or are being carried out are very similar to those that existed when sturgeons were abundant.
4. Notable advance has been made in the farming of sturgeons, especially regarding the techniques of nutrition, maintenance, sanitary control, etc., particularly in the species *A. naccarii*, leading to the availability of an ample stock of sturgeons

of different ages with appropriate health controls, thus providing an adequate base for carrying out recovery programmes.

5. The analysis of the recovery experiences of the sturgeon populations in Russia, Italy, and Germany, as well as the projects proposed for Spain and France, has helped to identify problems and propose possible solutions.
6. The farming and recovery of sturgeons can become a valid tool in the field of sustainable development. Therefore, it is urgent that the different governments involved (Andalusian regional government, the Spanish national government, European governments, and the government of the European Union) take action and develop projects for the farming and recovery of sturgeons in Southern Europe.
7. The Andalusian regional government, in particular, is urged to initiate the recovery programme for *A. naccarii* in the Guadalquivir River, taking into consideration the model proposed in this workshop.

This book includes the papers presented at this meeting, as well as additional papers considered to be of utmost interest for the objectives of this endeavour. The contents of this book are organized in three sections: Taxonomy and Biogeography, comprising six chapters; Biology and Aquaculture, with seven chapters; and Recovery and Conservation, with thirteen chapters. Section I includes the morphological and genetic analyses that clarify the taxonomy and phylogeny of sturgeons, concentrating on those from Southern Europe. In Section II, several aspects of developmental biology (digestive tract, heart and brain), feeding, nutrition and reproduction are considered in relation to the improvement of sturgeon farming. Finally, in Section III, the various recovery and research activities developed in several countries and the problems related to the trade of caviar and restoration of the ecology of the rivers are analysed.

R. Carmona, A. Domezain, M. García-Gallego, J.A. Hernando, F. Rodríguez & M. Ruiz-Rejón (Editors)

To the Pioneers in the Recovery of Sturgeons

The editors of this volume are aware that neither the contributions presented here, nor our work, would have been possible without the previous efforts of all the pioneers who devoted their life to the study of sturgeons, first in Russia and later in the United States and Western Europe. We wish to thank all the scientists and technicians who have dedicated their life to work for and with the different species of sturgeons. They are many, and some of them have graciously accepted our invitation to contribute to this volume.

Although it would be impossible to name all of them, we must mention the Russian pioneers, at least in general, who, as early as 1850, began their work for maintaining and restoring the stocks of wild sturgeons, stabilized the technology for reproduction and guaranteed the survival of so many species, and played an essential role in the production of caviar. They developed the techniques for processing the sturgeon roe to unprecedented levels and shared their knowledge with the Persians—now Iranians—first, and soon after, with the rest of the world. Professor Igor Burtsev, from VNIRO, who has devoted his life to the study of sturgeons, has explained how the ancient Cossaks protected the sturgeons, even avoiding the ringing of bells during the spawning season. For this, we also thank the Cossaks of today.

As this volume is mainly concerned with the sturgeons of Southern Europe, we wish to honour Giacinto Giovannini, the person directly responsible for preventing the extinction of *A. naccarii,* the only species whose distribution area is exclusively the South of Europe. When no one paid any attention to this species, he took up its conservation and recovery. It was in the 1970s, when he owned a small aquaculture farm dedicated exclusively to the trout, that he realized that the capture of sturgeons had dramatically decreased in the rivers of Southern Europe, even in his own region (the Northwestern Adriatic) which had been their last refuge. And he took a fundamental decision to buy all the specimens that the fishermen occasionally captured. These sturgeons were usually juvenile *A. naccarii* captured in the rivers flowing into the Adriatic Sea.

Although no one could provide him data concerning the culture of *A. naccarii* as it had never been tried before, he housed the specimens in his aquafarm in Orzinuovi (Brescia, Italy), where he began to take care of them, and, even risking being sentimental, we dare to say that he began to love them. He has devoted his life to sturgeons, and, after many years of solitary effort, of learning from his sturgeons and understanding their needs, he succeeded in 1988: he was able to reproduce them in captivity. This was the first reproduction of autochthonous sturgeons outside the former USSR.

This was a major success that generated hopes and opened up new possibilities for the conservation and recovery of sturgeons, initially in Italy and soon in other European countries (although most of these plans have not yet crossed the proposal stage).

As mentioned, Giacinto Giovannini worked alone during the 1980s, until several groups, headed for example, by Drs. Arlati, Bronzi, Colombo, Fontana and Rossi, gave a strong impulse to the recovery of the Southern European sturgeon *A. naccarii.* The public administration also recognized Giovannini's essential contribution and began to support his efforts.

Even today, Giacinto Giovannini is the most important scientist responsible for the recovery of *A. naccarii*, not only in Italy, but all over Southern Europe. Fortunately, he has recently obtained strong support from scientists in different countries who share his objectives, and also from public institutions, as well as his own family.

We also wish to mention the recent achievement of Dr. Patrick Williot and his colleagues from the CEMAGREF of artificial reproduction of the other important

European species, *A. sturio*, which makes the recovery of this species in the European rivers possible.

We sincerely thank Giacinto Giovannini, Patrick Williot, and all the pioneers for their contributions to the recovery of every sturgeon species.

The editors

Acknowledgements

This book, and BIORESTURGEONS, the conference that inspired it, were possible due to the generous support of the following organizations and institutions:

Sponsor

Ministry of Agriculture and Fisheries, Andalusian Government

Collaborators

Ministry of Environment, Andalusian Government
Ministry of Innovation, Science and Enterprise, Andalusian Government
Ministry of Tourism, Sports and Commerce, Andalusian Government
Ministry of Education and Science of the Spanish Government
Andalusian Institute for Research and Training in Agriculture, Food, Fisheries and for Ecological Production (IFAPA)
Agency for Innovation and Development of Andalusia (IDEA)
Agency wildlife photographer Antonio Sabater/Enfoque10
Sierra Nevada Fish Farm Ltd
Sectorial Association of Organic Aquaculture (ASAE)
University of Granada
University of Sevilla
University of Cádiz
University of Malaga

Contents

Part II Biology and Aquaculture

Part III Recovery and Conservation

Part I
Taxonomy and Biogeography

Chapter 1
The Regression of Sturgeons in Southern Europe

J.A. Hernando, A. Domezain, C. Zabala, R. Cabrera, J. Domezain and M.C. Soriguer

Abstract Analysing historical citations and the specimens preserved in museums, as well as zoological collections in France, Portugal, Spain, and Italy, the historical and current distributions of sturgeons in southern Europe have been studied, using morphological and genetic techniques. Errors were found in the classification of specimens of non-Adriatic origin, and 13 specimens were found to be classified as *Acipenser sturio* when they were in fact *Acipenser naccarii*. Of the 86 specimens considered valid, 44 had been captured in the Adriatic and 42 from other parts of the study area (Spain, France, Mediterranean Italy, Portugal, and Sweden). *A. sturio* is considered to have been present throughout Europe and currently is an endangered population in the Gironde-Garonne-Dordogne basin (France), while *A. naccarii* is considered endemic to the Adriatic. However, the data gathered show that it had a much broader distribution than the current one, extending at least from the French Atlantic to the Adriatic Sea.

Keywords *Acipenser naccarii*, sturgeon, autochthonous, taxonomy, biogeography, species in decline

1.1 Introduction

Sturgeons (Osteicthyes, Actinopterigia, Condrosts, Ascipenseriformes) appeared on Earth some 200–250 million years ago and established regular and abundant populations in all the great rivers of the Northern Hemisphere. Today, they are all

J.A. Hernando, C. Zabala, R. Cabrera, and M.C. Soriguer
Departamento de Biología, Facultad de Ciencias del Mar y Ambientales,
Universidad de Cádiz, 11510 Puerto Real, Cádiz, Spain

A. Domezain, and J. Domezain
Departamento de I + D, Piscifactoría de Sierra Nevada S.l., 18313 Riofrío,
Granada, Spain

in danger of extinction (Holcik et al., 1989a; IUCN, 1996; Williot et al., 1997; Vasil'eva, 1999; Billard, 2002; Domezain, 2003).

Although they have a long history, have been very abundant until 100–150 years ago (i.e. for 250,000,000 years), have survived glaciations and cataclysms, and have adapted to many different living conditions (from the cold waters of Canada and Siberia to the warm waters of Italy and Andalusia), all the species of sturgeons share the same fate: in the last 150 years, when human civilization took full command over nature, all the species of sturgeons, which were abundant and common, are in grave danger of extinction.

In the last decades of the twentieth century, the only sturgeon populations that maintained good numbers were in the Soviet Union. Though they were exploited intensely, their numbers were regulated and supported by effective repopulation measures. This has demonstrated that, with adequate management on the principles of aquaculture, it is possible to maintain an economically profitable exploitation level without placing the species in danger of extinction. In fact, when the Soviet Union lost control of the fisheries, all the sturgeon species with which they had worked moved towards the brink of extinction. Fishing continued, but without a consistent repopulation policy. As a result, the number of captured sturgeons fell more than 20-fold in 11 years, going from 15,000 tonnes in 1991 to 700 tonnes in 2002 (Domezain, 2003).

Sturgeons are becoming extinct not because of accidents or because they cannot adapt to natural changes, but because bad management by humans is driving them to extinction, especially the construction of dams to prevent water flowing to the sea, which impedes sturgeon migrations. The little water that remains is contaminated by agricultural biocides (pesticides, herbicides, fungicides, etc.) and industrial wastes. Above all, fishing depletes the number of adults, without allowing them to reproduce, and repopulation is not undertaken. In short, humans are driving the sturgeon to extinction and therefore their repopulation is our responsibility.

The aim of the present study is to examine the minimum historical zoographical distribution area of two species: *Acipenser naccarii* (Bonaparte 1836) and *Acipenser sturio* (Linnaeus 1758). In addition, an analysis (at least a summary one) is made of our refusal to understand the situation of *A. naccarii*, our motives for doing so, and their implications in the management of the natural environment.

1.2 Background

The case of *A. naccarii* is not an exception, and its situation is worsened by the serious and strange lack of knowledge that up to now has surrounded its authentic taxonomic status, which has impeded its recovery, despite this being technically possible and socially necessary. That is, until recently, *A. naccarii* was erroneously considered endemic to the Adriatic (Holcik et al., 1989a), without acknowledging or wishing to validate the numerous old studies of many renowned naturalists (Capello 1869, 1880; Osorio, 1894; Seabra, 1911; Nobre, 1931, 1935; Gonçalves,

1942; Helling, 1943; Albuquerque, 1956), as well as other more recent references (Bauchot, 1987) who cited *A. naccarii* in the Iberian Peninsula as well.

Meanwhile, the Atlantic European sturgeon *A. sturio* was considered to be present in Eastern Europe (Magnin, 1962; Holcik et al., 1989b) and has been in regression in France since the nineteenth century (Williot et al., 2008). In the Iberian Peninsula, this species was considered to be widely distributed in the large basins of the Miño, Duero, Tagus, Guadiana, Guadalquivir, Júcar, Turia, and Ebro (Lozano Rey, 1935). Of these, the Guadalquivir, where sturgeon processing commenced in 1932, was most thoroughly studied until the factory was shut down on 18 June 1970 (Algarin, 2002). In a detailed study of the specimens captured for this factory, Classen (1944) recorded the variations in length and weight of the specimens caught between 1932 and 1966.

The capture grew from 1932 to 1935 but fell from 1935 to 1939, possibly due to various biological reasons, but later rose again until 1946. From that time on, the capture gradually declined, except in 1951, when 281 specimens were caught (179 females and 102 males), until in 1966 only four females were captured (Figs. 1.1 and 1.2). The mean weight of the 2987 females captured between 1932 and 1966 was 45.37 kg, ranging from 40.9 to 54.95 kg, while the 1027 males captured during the same period weighed an average of 22.28 kg, the smallest weighing 18 kg and the largest 29.09 (Fig. 1.2).

In terms of the frequency of the size of the fish, three size intervals can be considered for the period 1932–1942: 18.28% (n = 230) of the females measured 141–175 cm, 62.96% (n = 792) 176–200 cm, and the remaining 18.76% (n = 236) 201–250 cm, with the shortest length being 148 cm and the longest 248 cm. During the period 1943–1966, 25.38% (n = 267) fit the first interval, 65.78% (n = 692) the

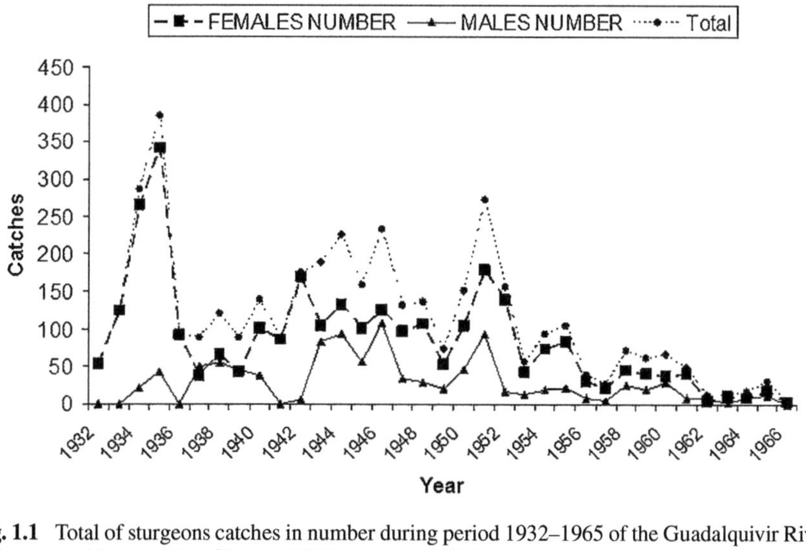

Fig. 1.1 Total of sturgeons catches in number during period 1932–1965 of the Guadalquivir River and separated by sex (data Classen, 1944 in Algarin, 2002)

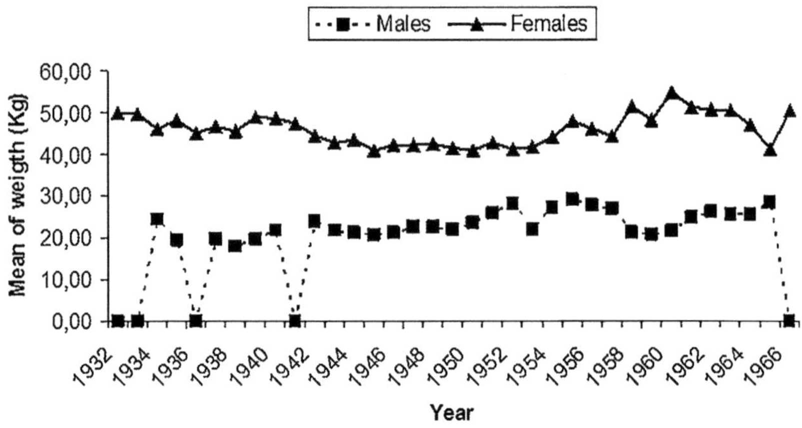

Fig. 1.2 Annual variation of the average weight of the sturgeons captured in the Guadalquivir River during period 1932–1965 separated by sexes (data Classen, 1944 in Algarin, 2002)

second, and 8.84% (n = 93) the third, with sizes ranging from 143 to 284 cm. For the entire period of active fishing from 1932 to 1966, 21.52% of the females captured (n = 497) were in the first interval, 65.78% (n = 1484) in the second, and 14.24% (n = 329) in the third, with lengths ranging from 143 to 248 cm (Fig. 1.3).

The length of the males during the period 1932–1944 ranged from a minimum of 107.5 cm to a maximum of 187 cm. The first interval (105–130 cm) accounted for 8.23% (n = 20), the second interval (131–155 cm) for 69.7% (n = 161), the third interval (156–180 cm) for 27.78% (n = 48), and the fourth interval (181–205 cm) for 0.87% (n = 2). Between 1943 and 1966, the first interval accounted for 1.59% (n = 4), the second interval for 59.52% (n = 150), the third for 29.76% (n = 75), and the fourth for 9.13% (n = 23). For the entire period of 1932–1966, of the 483 males captured, 4.97% (n = 24) measured 105–130 cm, 64.39% (n = 311) were 131–155 cm, 25.47% (n = 123) were 156–180 cm, and only 5.18% (n = 25) were 181–205 cm (Fig. 1.4).

Taking into account that the mean length of *A. naccarii* was 100–180 cm and that this fish could exceed 200 cm, while *A. sturio* measures 130–220 cm and can reach more than 300 cm (CITES, 2001), it is no surprise that Magnin (1962) disqualified the biostatistical study of the sturgeons of the Guadalquivir (Classen, 1944), as the data of Classen actually included a population formed by specimens of two species (*A. sturio* and *A. naccarii*). But this we know now, after many morphological and genetic studies; Classen did not have this advantage.

1.3 Sturgeons in Museums

To denote the different collections (Table 1.1), the nomenclature of Leviton et al. (1985) was used whenever possible. For those having no acronym assigned by Leviton, the following denomination was applied using the same style: EBD

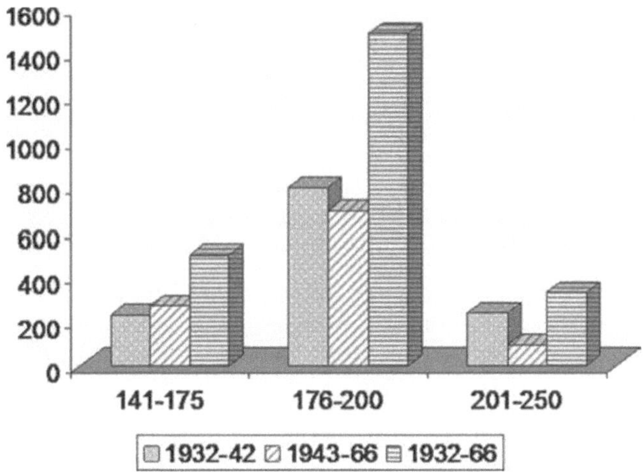

Fig. 1.3 Length frequency of sturgeons female of the Guadalquivir River between 1932 and 1966

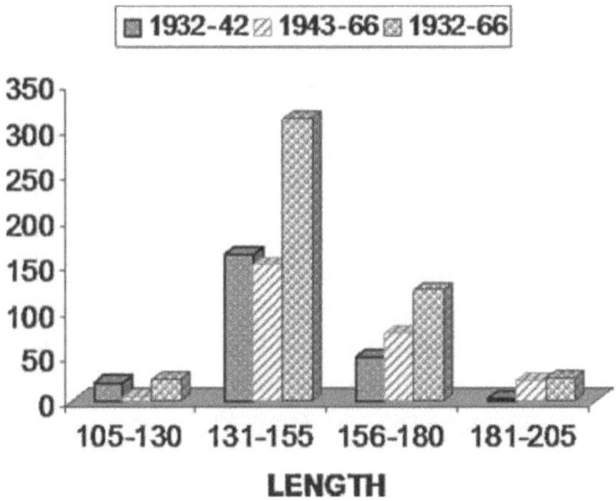

Fig. 1.4 Length frequency of sturgeons male of the Guadalquivir River between 1932 and 1966

(Estación Biológica de Doñana, Seville, Spain), USEDBA (University of Seville, Department of Animal Biology, Spain), MCNAE (Museum of Natural Sciences Aguilar y Eslava, Cabra, Spain), MZB (Museum of Zoology of Barcelona, Spain), MOPA (Museum of Oceanography, Portinho da Arrabida, Portugal), MHNT (Museum of Natural History, Toulouse, France).

For the taxonomic classification, two technical blocks were used. For the morphological classification, the application combined the keys most generally accepted for these species on a world scale: Soljan (1975), Svetovidov (1984), Sokolov (1989), Tortonese (1989) and Holcik et al. (1989a,b).

Table 1.1 Museums and zoological collections in which we found specimens of the study area, in suitable conditions to be able to be analysed and included in this work

EDB	Biological Station of Coto Doñana, Spain
USEDBA	University of Seville, Department of Animal Biology, Spain
MCNAE	Aguilar and Eslava Natural Sciences Museum, Spain
MNCN	Spanish National Museum of Natural Sciences
MZB	Zoological Museum of Barcelona, Spain
AVG	Vasco da Gama Aquarium, Portugal
MZC	Zoological Museum, Department of Zoology, University of Coimbra, Portugal
MAN	Augusto Nobre Museum, University of Porto, Portugal
MOPA	Portinho da Arrabida Oceanographic Museum, Portugal
MSNG	Giacomo Doria Civic Natural History Museum, Italy
MZUF	Zoological Museum Observatory, University of Florence, Italy
MOM	Monaco Oceanographic Museum
MG	Guimet Museum, France
ISTPM	Scientific and Technical Institute for Marine Fish, France
MHNT	Natural History Museum, Toulouse, France
MNHN	National Museum of Natural History, Paris, France

The total length (TL) was measured together with four biometric variables described by Holcik et al. (1989a) and one by Soljan (1975), and three indices were calculated. Following the nomenclature of Holcik et al. (1989a) and Soljan (1975), the six variables studied were: total length (TL), distance from the snout to the base of the barbells (s – b), snout width at the barbells (lab), distance from the base of the barbels to the cartilaginous arc of the mouth (b – mc), distance from the tip of the snout to the cartilaginous arc of the mouth (s – mc), and internal width of the mouth (IWM). Three indices were defined to determine first, the relative position of the barbels (CA), calculating the difference between b – mc and s – b: CA = {(b – mc) – (s – b)}; second, the proportion of the length (s – mc) to snout width (lab) as FB = {(s – mc): (lab)}; and third, the number of times that the mouth width fits into the snout length, referred to as the Soljan index (SOLJ) = (s – mc): lab. The statistical analyses were made with STATGRAPHICS PLUS 5.0.

For the genetic classification, cytogenetic assays (erythrocyte cytometry) and molecular genetics (satellite DNA as a differential marker) were used. The methodology has been described by Garrido-Ramos et al. (1997), Hernando et al. (1999b,c), Robles et al. (2003), Robles (2003), De la Herrán et al. (2004). For this, the research area was expanded and more specimens were studied, in addition to those used in the studies mentioned above, in numerous museums in Spain, Portugal, France, and parts of Mediterranean Italy, in an effort to increase our knowledge of the historical distribution as well as its spatiotemporal evolution.

Table 1.1 lists the museums in which the zoological collections included in this study were examined. A study was made of all the specimens of *A. naccarii* and *A. sturio* which could be verified as being captured in the study area, were preserved well enough to permit accurate classification, and had reliable and adequate dating (at least the date and place of capture).

Of the total number of specimens (Table 1.2) studied, 86 were found to fulfil the necessary criteria. Of these specimens, 44 had been captured in the Adriatic Sea

Table 1.2 Specimens of Acipenser naccarii and A. sturio, with validated data, held by the collection studied

Collection/ museum	Town/country	Identification	Old classification	New classifica-tion	Year of catch	Location
EBD	Seville/Spain	EBD8173	*A. sturio*	*A. naccarii*	1974	Guadalquivir River (Spain/Atlantic)
EBD	Seville/Spain	EBD8174	*A. sturio*	*A. naccarii*	1975	Guadalquivir River (Spain/Atlantic)
EBD	Seville/Spain	EBD8401	*A. sturio*	*A. sturio*	1981	Guadalquivir River (Spain/Atlantic)
USEDBA	Seville/Spain	SE-1	*A. sturio*	*A. sturio*	End of nineteenth century	Guadalquivir River (Spain/Atlantic)
USEDBA	Seville/Spain	SE-2	*A. sturio*	*A. sturio*	End of nineteenth century	Guadalquivir River (Spain/Atlantic)
USEDBA	Seville/Spain	SE-3	*A. sturio*	*A. sturio*	End of nineteenth century	Guadalquivir River (Spain/Atlantic)
MCNAE	Cabra/Spain	AEC-1	*A. sturio*	*A. sturio*	End of nineteenth century	Guadalquivir River (Spain/Atlantic)
MNCN	Madrid/Spain	MNCZ1582	*A. sturio*	*A. sturio*	Between 1936 and 1944	Guadalquivir River (Spain/Atlantic)
MZB	Barcelona/Spain	B-82-5340	*A. sturio*	*A. sturio*	1983?	Ebro River (Spain/Atlantic)
MZB	Barcelona/Spain	B-82-5342	*A. sturio*	*A. sturio*	1983?	Ebro River (Spain/Atlantic)
AVG	Lisbon/Portugal	AVG-1	*A. sturio*	*A. sturio*	1961	Miño River (Portugal/Atlantic)
AVG	Lisbon/Portugal	AVG-2	*A. sturio*	*A. sturio*	1965	Guadiana River (Portugal/Atlantic)
MZC	Coimbra/Portugal	MUC1[a]	*A. sturio*	*A. naccarii*	1890	Tajo River (Portugal/Atlantic)
MZC	Coimbra/Portugal	MUC46B[a]	*A. sturio*	*A. naccarii*	1897	Mondego River (Portugal/Atlantic)
MOPA	Arrabida/Portugal	457	*A. sturio*	*A. sturio*	1920	Arrabida (Portugal/Atlantic)
MSNG	Genoa/Italy	34547	*A. naccarii*	*A. naccarii*	1868	Adriatic Sea
MSNG	Genoa/Italy	34553	*A. sturio*	*A. sturio*	1920	Adriatic Sea
MSNG	Genoa/Italy	40364	*A. naccarii*	*A. naccarii*	1861	Gulf of Genoa (Mediterranean)
MSNG	Genoa/Italy	41000[b]	*A. sturio*	*A. naccarii*	1967	Ticino River (Italy/Adriatic)
MSNG	Genoa/Italy	41001	*A. naccarii*	*A. naccarii*	1967	Ticino River (Italy/Adriatic)
MSNG	Genoa/Italy	47931	*A. naccarii*	*A. naccarii*	1987	Po River (Italy/Adriatic)
MZUF	Florence/Italy	2488	*A. naccarii*	*A. naccarii*	1899	Cremona (Italy/Adriatic)
MZUF	Florence/Italy	3035	*A. sturio*	*A. sturio*	1873	Kattegat (Sweden/Atlantic)
MZUF	Florence/Italy	5700	*A. naccarii*	*A. naccarii*	1882	Chioggia (Italy/Adriatic)
MZUF	Florence/Italy	5701	*A. naccarii*	*A. naccarii*	1872	Venezia (Italy/Adriatic)
MZUF	Florence/Italy	5702	*A. naccarii*	*A. naccarii*	1908	Cremona (Italy/Adriatic)

(continued)

Table 1.2 (continued)

Collection/ museum	Town/country	Identification	Old classification	New classifica-tion	Year of catch	Location
MZUF	Florence/Italy	5704[c]	*A. naccarii*[c]	*A. naccarii*	1879	Livorno (Italy/Mediterranean)
MZUF	Florence/Italy	5705	*A. naccarii*	*A. naccarii*	1882	Chioggia ((Italy/Adriatic))
MZUF	Florence/Italy	5706	*A. naccarii*	*A. naccarii*	1882	Chioggia ((Italy/Adriatic))
MZUF	Florence/Italy	5707	*A. sturio*	*A. sturio*	1872	Venezia (Italy/Adriatic)
MZUF	Florence/Italy	5708	*A. sturio*	*A. sturio*	1878	Tevere River (Italy/Mediterranean)
MZUF	Florence/Italy	5709	*A. naccarii*	*A. naccarii*	1899	Cremona (Italy/Adriatic)
MZUF	Florence/Italy	5710	*A. naccarii*	*A. naccarii*	1885	Venezia (Italy/Adriatic)
MZUF	Florence/Italy	5711	*A. naccarii*	*A. naccarii*	1899	Po River (Italy/Adriatic)
MZUF	Florence/Italy	5714	*A. sturio*	*A. sturio*	1886	Salermo (Italy/Mediterranean)
MZUF	Florence/Italy	5716	*A. sturio*	*A. sturio*	1908	Livorno (Italy/Mediterranean)
MZUF	Florence/Italy	5717	*A. sturio*	*A. sturio*	1880	Chioggia (Italy/Adriatic)
MZUF	Florence/Italy	5718	*A. sturio*	*A. sturio*	1877	Rimini (Italy/Adriatic)
MZUF	Florence/Italy	5719	*A. sturio*	*A. sturio*	1898	Cremona (Italy/Adriatic)
MZUF	Florence/Italy	6401	*A. sturio*	*A. sturio*	1908	Livorno (Italy/Mediterranean)
MZUF	Florence/Italy	6472	*A. naccarii*	*A. naccarii*	1882	Chioggia (Italy/Adriatic)
MZUF	Florence/Italy	6473	*A. naccarii*	*A. naccarii*	1882	Chioggia (Italy/Adriatic)
MZUF	Florence/Italy	6475	*A. sturio*	*A. sturio*	1886	Salermo (Italy/Mediterranean)
MZUF	Florence/Italy	6476	*A. naccarii*	*A. naccarii*	1908	Po River (Italy/ Adriatic)
MZUF	Florence/Italy	6477	*A. naccarii*	*A. naccarii*	1908	Po River (Italy/Adriatic)
MZUF	Florence/Italy	6478	*A. naccarii*	*A. naccarii*	1909	Po River (Italy/Adriatic)
MZUF	Florence/Italy	6479	*A. naccarii*	*A. naccarii*	1909	Po River (Italy/Adriatic)
MZUF	Florence/Italy	6480	*A. naccarii*	*A. naccarii*	1909	Po River (Italy/Adriatic)
MZUF	Florence/Italy	6486	*A. naccarii*	*A. naccarii*	1884	Venezia (Italy/Adriatic)
MZUF	Florence/Italy	6486	*A. sturio*	*A. sturio*	1872	Venezia (Italy/Adriatic)
MZUF	Florence/Italy	6486	*A. sturio*	*A. sturio*	1872	Venezia (Italy/Adriatic)
MZUF	Florence/Italy	6486	*A. sturio*	*A. sturio*	1872	Venezia (Italy/Adriatic)
MZUF	Florence/Italy	6487	*A. naccarii*	*A. naccarii*	1882	Chioggia (Italy/Adriatic)
MZUF	Florence/Italy	6489	*A. naccarii*	*A. naccarii*	1899	Cremona (Italy/Adriatic)
MZUF	Florence/Italy	6490	*A. naccarii*	*A. naccarii*	1899	Cremona (Italy/Adriatic)

MZUF	Florence/Italy	6491	*A. naccarii*	*A. naccarii*	1899	Cremona (Italy/Adriatic)
MZUF	Florence/Italy	6492	*A. naccarii*	*A. naccarii*	1899	Cremona (Italy/Adriatic)
MZUF	Florence/Italy	6493	*A. naccarii*	*A. naccarii*	1899	Cremona (Italy/Adriatic)
MZUF	Florence/Italy	6494	*A. sturio*	*A. sturio*	1908	Cremona (Italy/Adriatic)
MZUF	Florence/Italy	6496	*A. sturio*	*A. sturio*	1881	Chioggia (Italy/Adriatic)
MZUF	Florence/Italy	6497	*A. sturio*	*A. sturio*	1884	Chioggia (Italy/Adriatic)
MZUF	Florence/Italy	6498	*A. sturio*	*A. sturio*	1885	Chioggia (Italy/Adriatic)
MZUF	Florence/Italy	6499	*A. sturio*	*A. sturio*	1878	Tevere River (Italy/Mediterranean)
MZUF	Florence/Italy	6500	*A. sturio*	*A. sturio*	1877	Rimini (Italy/Adriatic)
MZUF	Florence/Italy	6501	*A. naccarii*	*A. naccarii*	1882	Chioggia (Italy/Adriatic)
MZUF	Florence/Italy	6502	*A. naccarii*	*A. naccarii*	1882	Chioggia (Italy/Adriatic)
MZUF	Florence/Italy	6995	*A. sturio*	*A. sturio*	1909	Cremona (Adriatic)
MOM	Monaco/Monaco	POI4591	*A. sturio*	*A. sturio*	1938	La Rochelle (France/Atlantic)
MOM[b]	Monaco/Monaco	POI0001	*A. sturio*	*A. naccarii*	1906	Marseille (France/Mediterranean)
MOM	Monaco/Monaco	POI	*A. sturio*	*A. sturio*	1958	Gironde (France/Atlantic)
MG	Lyons/France	42006250	*A. sturio*	*A. sturio*	1879	Nice (France/Maditerraneo)
ISTPM	Nantes/France	019558	*A. sturio*	*A. sturio*	1880	Loire (France/Atlantic)
MHNT[b]	Toulouse/France	ICHT1995-380	*A. sturio*	*A. naccarii*	Nineteenth century	Garonne River (France/Atlantic)
MHNT[b]	Toulouse/France	ICHT1995-381	*A. sturio*	*A. naccarii*	Nineteenth century	Garonne River (France/Atlantic)
MNHN	Paris V/France	0000 2610	*A. sturio*	*A. sturio*	1850	Gironde (France/Atlantic)
MNHN	Paris V/France	0000 3108	*A. sturio*	*A. sturio*	1851	Seine (France/Atlantic)
MNHN	Paris V/France	0000 3115	*A. sturio*	*A. sturio*	1870	Seine (France/Atlantic)
MNHN	Paris V/France	0000 5158	*A. sturio*	*A. sturio*	1870	Elbe (Germany/Atlantic)
MNHN	Paris V/France	0000 5159	*A. sturio*	*A. sturio*	1870	Gironde (France/Atlantic)
MNHN	Paris V/France	4898 1257	*A. naccarii*	*A. naccarii*	1876	Nice (France/Mediterranean)
MNHN	Paris V/France	0000 9531	*A. naccarii*	*A. naccarii*	End of nineteenth century	Po River (Italy/Adriatic)

(continued)

Table 1.2 (continued)

Collection/ museum	Town/country	Identification	Old classification	New classification	Year of catch	Location
CLASSEN	Classen (1944)	CM-1	*A. sturio*	*A. naccarii*	1940	Guadalquivir River (Spain/Atlantic)
CLASSEN	Classen (1944)	CM-2	*A. sturio*	*A. naccarii*	1940	Guadalquivir River (Spain/Atlantic)
CLASSEN	Classen (1944)	CM-3	*A. sturio*	*A. naccarii*	1940	Guadalquivir River (Spain/Atlantic)
CLASSEN	Classen (1944)	CM-4	*A. sturio*	*A. naccarii*	1940	Guadalquivir River (Spain/Atlantic)
CLASSEN	Classen (1944)	CM-5	*A. sturio*	*A. naccarii*	1940	Guadalquivir River (Spain/Atlantic)

The first column gives the name of the collection, as designated by Leviton et al. (1985). In the case of a collection not classified by these authors, it has been assigned a code name according to the system used by these authors, this also confirming that the collection was not covered by these authors.

The key to code names is as follows: EBD, Biological Station of Cota Doñana, Spain; USEDBA, University of Seville, Department of Animal Biology, Spain; MCNAE, Aguilar and Eslava Natural Sciences Museum, Spain; MNCN, Spanish National Museum of Natural Sciences; MZB, Zoological Museum of Barcelona, Spain; AVG, Vasco da Gama Aquarium, Portugal; MZC, Zoological Museum, Department of Zoology, University of Coimbra, Portugal; MOPA, Portinho da Arrabida Oceanographic Museum, Portugal; MSNG, Giacomo Doria Civic Natural History Museum, Italy; MZUF, Zoological Museum Observatory, University of Florence, Italy; MOM, Monaco Oceanographic Museum; MG, Guimet Museum, France; ISTPM, Scientific and Technical Institute for Marine Fish, France; MHNT, Natural History Museum, Toulouse, France; MNHN, National Museum of Natural History, Paris, France. The last five examples listed correspond to those referenced in Classen (1944).

[a]Hernando et al. (1999c).

[b]Hernando et al. (in preparation).

[c]Bernini and Vanni (1995).

(and its tributary basin) and 42 in other parts of the study area (Spain, France, Mediterranean Italy, Portugal, and Sweden). Of the non-Adriatic specimens, 3 were already correctly classified as *A. naccarii* in their respective museums, while 13 specimens were labelled *A. sturio* when they were in fact *A. naccarii*. Of the 44 Adriatic specimens, 40 dated prior to 1910. Of these, 27 were *A. naccarii* and 13 *A. sturio*. Of the four that dated from 1910 onwards, three were *A. naccarii* and one was *A. sturio* (Fig. 1.5). Of the 42 non-Adriatic specimens, 26 predated 1910, of which 8 were *A. naccarii* and 18 were *A. sturio*. Of the 16 dated 1910 and later, 8 were *A. naccarii* and 8 *A. sturio* (Fig. 1.6).

All the *A. naccarii* captured outside the Adriatic after 1910 that we were able to use in the study had been captured in the Guadalquivir river.

For studying TL with respect to the rest of the variables analysed, a simple linear-regression analysis was applied, as shown in Table 1.3. After it was determined that SOLJ and FB did not correlate with TL, and that the correlation of CA with TL was not significant, a forward stepwise discriminate analysis (SDA; Fig. 1.5) was applied to test statistically the changes that the morphological and genetic tools,

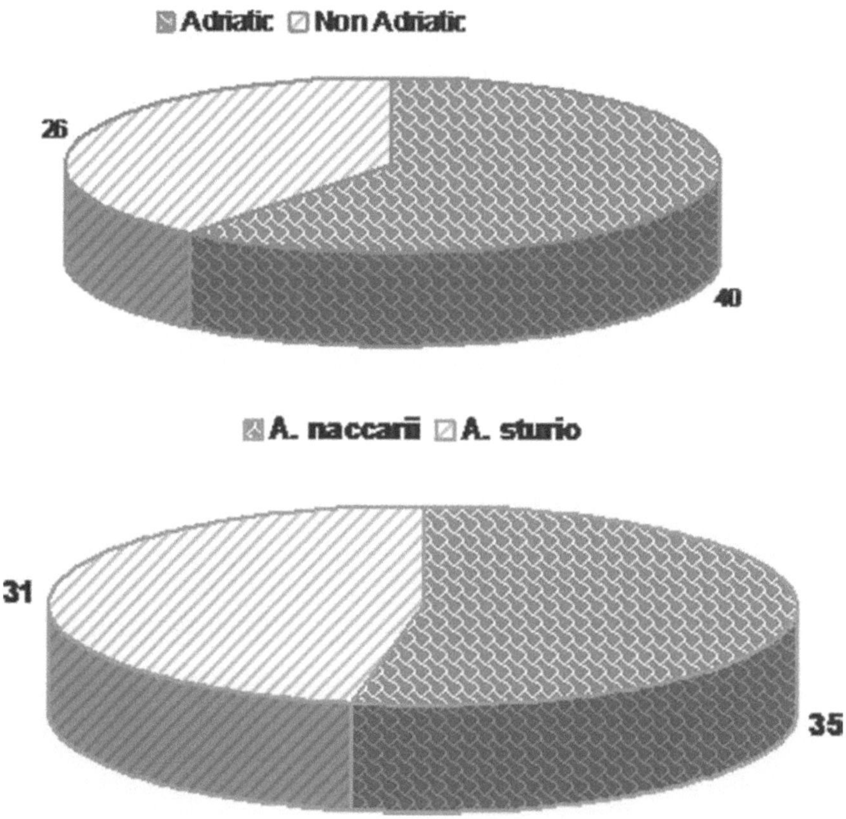

Fig. 1.5 Sturgeons of before 1910 in the European collections studied

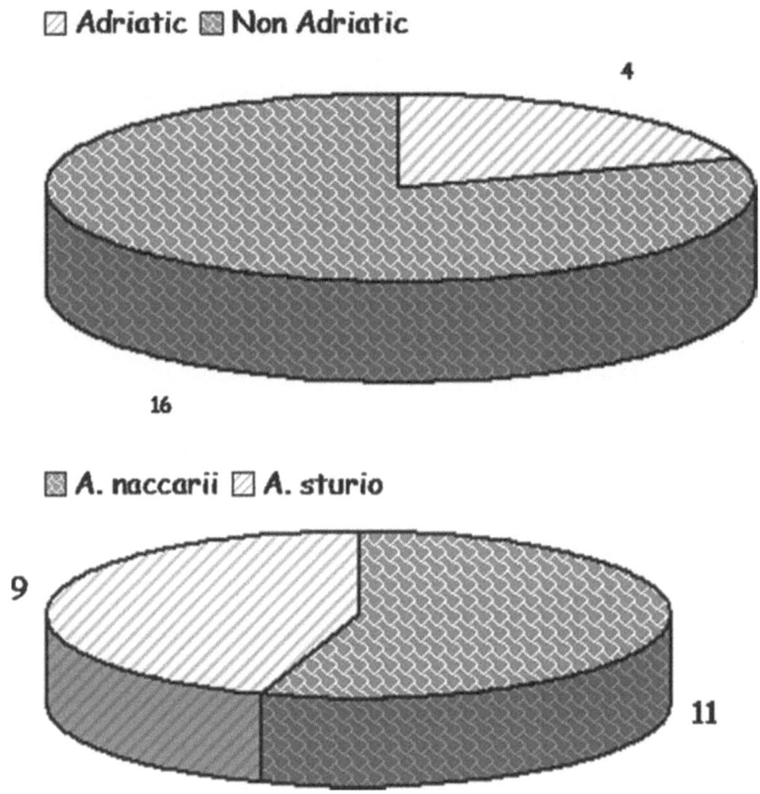

Fig. 1.6 Sturgeons later to 1910 presents in the studied collections

Table 1.3 Mean and median for both groups (AN: *Acipenser naccarii*; AS: *Acipenser sturio*)

	Mean				Median			
Index	AN	AS	*t*	*p* Value	AN	AS	*W*	*p* Value
FB	1.44 ± 0.078	2.25 ± 0.142	−10.18	0.0	23.43	61.02	1641.0	1.25×10^{-12}
SOLJ	1.64 ± 0.069	2.77 ± 0.18	−11.56	0.0	1.64	2.7	1470	0.0
CA	0.39 ± 0.273	−0.56 ± 0.14	6.14	1.32×10^{-7}	0.065	−0.63	242.5	1.76×10^{-8}

applied individually, caused in the taxonomic review of the 86 specimens, using AS (*A. sturio*) and AN (*A. naccarii*) as input groups.

The SDA confirmed the existence of groups separated on the basis of SOLJ ($\lambda = 0.359$; $F = 130.34$; $p = 0.0$) as the first canonical variable, and FB ($\lambda = 0.289$; $F = 88.32$; $p = 0.0$), CA ($\lambda = 0.25$; $F = 70.87$; $p = 0.0$), and TL ($\lambda = 0.187$; $F = 76.04$; $p = 0.0$) as the second, third, and fourth canonical variables, respectively. The first standardized discriminate function with the p value <0.05 was 0.53 SOLJ + 0.72 FB − 0.72 CA + 0.69 TL, which confirmed the new classification of the 86 specimens with 98.67% of the cases correctly classified (Fig. 1.5).

The third step was to conduct an analysis to compare the variation of the means and medians (Mann-Whitney W test) of both the groups for the three indices (SOLJ, FB, and CA). This analysis confirmed that there were significant statistical differences (Table 1.3).

1.4 Historic Distribution

The taxonomic keys used in this study are universally accepted and, furthermore, their validity has been expressly verified for the species and areas studied (Bernini and Nardi, 1989; Garrido Ramos et al., 1997; De la Herrán, 1998; Hernando et al., 1999b,c; Ruiz-Rejón et al., 2000; Domezain, 2003; Robles, 2003, Robles et al., 2003; De la Herrán et al., 2004; among others).

The possible influence of the allometry on the morphological characteristics considered is nominal and known, affecting only the small specimens, specifically in the relative position of the barbels. This has been described in detail for *A. sturio* (Holcik et al., 1989b) as well as for *A. naccarii* (Hernando et al., 1999a) and does not affect any of the cases in this study.

Another taxonomic trait, the number of gill rakers of the first branchial arc, could not be used in many cases, given that we were working with specimens from old collections in which the branchia/gills were extirpated in the preservation process. For the specimens in which this character remained visible, the analysis in all cases coincided with the other traits analysed, confirming the assigned classification.

The three statistical analyses showed that the indices used (SOLJ, FB, and CA) definitely differentiated the two species. After reviewing the classification of the 86 specimens individually, only one, which was listed in the museum as *A. sturio* and which in the morphological analysis presented various characteristics intermediate between the two species, was left under this classification (given the doubt involved) and was assigned by SDA to the group *A. naccarii*, though the attribution was not entirely conclusive. This result occurred because of the values of the SOLJ and FB indices, which were closer to the lowest limit of the range corresponding to *A. sturio*. Therefore, after identifying this specimen, we felt that it should remain classified as *A. sturio*, reserving changes in classification for those cases that were clearly evident. This case could be a hybrid between the two species. Recently, a similar case has been identified among the specimens found in the Guadalquivir river (Robles, 2003; Robles et al., 2003; De la Herrán et al., 2004).

The suitability of the DNA satellite *Hind*III used as a marker to differentiate *A. naccarii* and *A. sturio* (always present in *A. naccarii* but never in *A. sturio*) has been definitively demonstrated in previous studies (Garrido-Ramos et al., 1997; De la Herrán, 1998; Ruiz-Rejón et al., 2000; De la Herrán et al., 2001; Lanfredi et al., 2001; Fontana, 2002). In subsequent genetic studies, several of the samples analysed by us were verified, by using up to five different genetic markers and forensic techniques and, in all cases, the classification was upheld (Robles, 2003; Robles et al., 2003; De la Herrán et al., 2004). The results of these studies can be summarized as follows:

With respect to the mitDNA analysis based on Cytb and 12S, Almodóvar et al. (2000), Gasent-Ramirez et al. (2001), and De la Herrán et al. (2004) found that the specimen EBD8174 (Table 1.2) was *A. sturio*. Almodóvar et al. (2000) and Gasent-Ramirez et al. (2001) failed to extract mitDNA from specimens EBD8173 and EBD8041 (Table 1.2) whereas De la Herrán et al. (2004) were successful and concluded that the specimens were *A. naccarii*. The studies on satellite DNA with *Hind*III were also conclusive as Almodóvar et al. (2000) and Gasent-Ramirez et al. (2001) did not analyse the material, while Garrido-Ramos et al. (1997) and De la Herrán et al. (2004) found that the three specimens studied were *A. naccarii*. Therefore, the conclusion was that the specimens EBD8173 and EBD8401 were *A. naccarii* whereas EBD8174 was a hybrid *A. sturio* (ADNmit) × *A. naccarii* (*Hind*III).

Focussing on the study of the 42 specimens identified as not belonging to the Adriatic (given that the sympatry between *A. naccarii* and *A. sturio* in the Adriatic is not questioned), we find that of these specimens, 38.1% are *A. naccarii* and 61.9% are *A. sturio*. When the analysis is made as a function of the two periods that we established above, we find that, of those predating 1910, 31% were *A. naccarii* and 69% *A. sturio*; of those after this date, 50% were *A. naccarii* and 50% *A. sturio*. Given that these specimens comprise the totality of well-preserved material in all the main museums of the study area, the same should be considered extremely significant (Figs. 1.6 and 1.7).

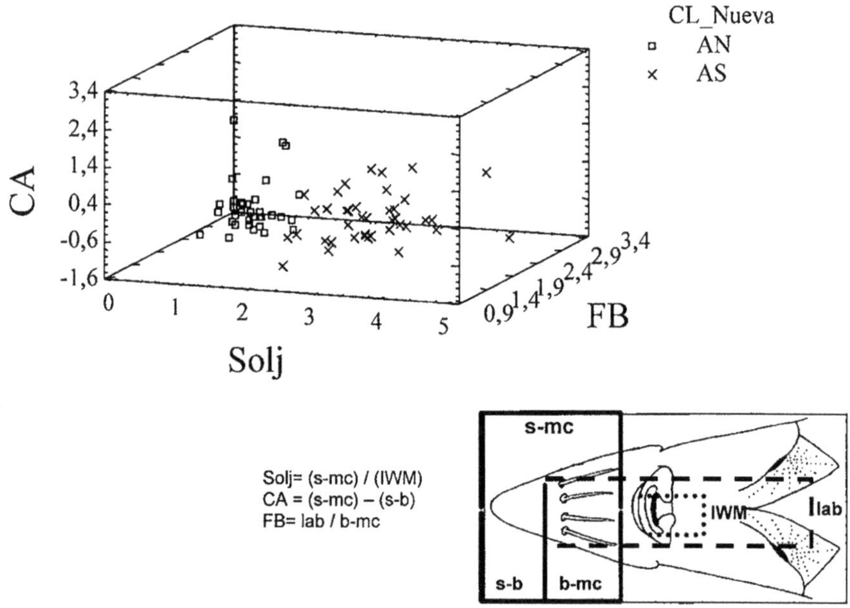

Fig. 1.7 Forward stepwise discriminate analysis: three-dimensional scatterplot of the specimens of the two species (AN: *Acipenser naccarii* and AS: *Acipenser sturio*) resulting from the new classification of the specimens given in Table 1.2. SOLJ, FB and CA are the indices used to calculate the canonical functions

It can be seen that before 1910, *A. naccarii* had a proportionally abundant and homogeneous distribution from the Atlantic to the Adriatic; however, the analysis of the information after 1910 reveals that its only distribution area outside the Adriatic was the Guadalquivir. The species disappeared from the rest of the area.

Combining this information with earlier Portuguese citations (Capello, 1869, 1880; Osorio, 1894; Seabra, 1911; Nobre, 1931, 1935; Gonçalves, 1942; Helling, 1943; Albuquerque, 1956), as proposed by Hernando et al. (1997) and Domezain et al. (1999), we see that *A. naccarii* was continuously autochthonous at least from the French Atlantic to the Adriatic. Its presence throughout this area was abundant until the end of the nineteenth century, after which it underwent a regression similar to that of *A. sturio*, whereupon the only areas of resistance of *A. naccarii* were the Guadiana–Guadalquivir and the northern Adriatic.

We should emphasize that, on the basis of our great discovery that *A. naccarii* was autochthonous to the Iberian Peninsula, we have confirmed what classical authors (mentioned above) had established much earlier. In fact, in this study we found that important museums such as the Natural History Museum of Paris, Giacomo Doria of Genoa, and the Zoological Museum Observatory of Florence have sturgeons from the area of Rodano/Gulf of Genoa correctly classified as *A. naccarii* and even published as such (Bernini and Vanni, 1995). In short, what we discovered in the Iberian Peninsula is nothing new; others had reached the same conclusions long before us, but their data had been forgotten by science. Will the same happen now?

From the analysis of these results, we find that historically not only was *A. naccarii* sympatric with *A. sturio* in the Adriatic, as believed for some time, but also that the distribution proposed by Hernando et al. (1999b,c) has been confirmed and expanded. *A. naccarii* should therefore be considered autochthonous, at least from the French Atlantic (Gironde) to the Adriatic, without any interruption in continuity (Fig. 1.8).

Also, we find an inflexion point in the abundance of the specimens, which we can date around 1910, and therefore we divide the study into two periods—before and after this date.

1.5 Current State

1.5.1 A. naccarii

With regard to its state of conservation, *A. naccarii* is considered 'vulnerable' (A1ac Sturgeon Specialist Group 1996. *A. sturio*. In: IUCN 2007. *2007 IUCN Red List of Threatened Species*. < www.iucnredlist.org>). All the foregoing data showing that *A. naccarii* is autochthonous to southern Europe are, strangely enough, doubted by some authors in the Iberian Peninsula, who claim (and surprisingly continue to claim) that the only species present in this area is *A. sturio* (Elvira and Almodóvar, 1993; Elvira et al., 1991; among others). Previously, a Portuguese work (Almaça, 1988) did not take into account the works of the renowned naturalists who cited *A. naccarii* in the waters of the Iberian Peninsula, although the author admits

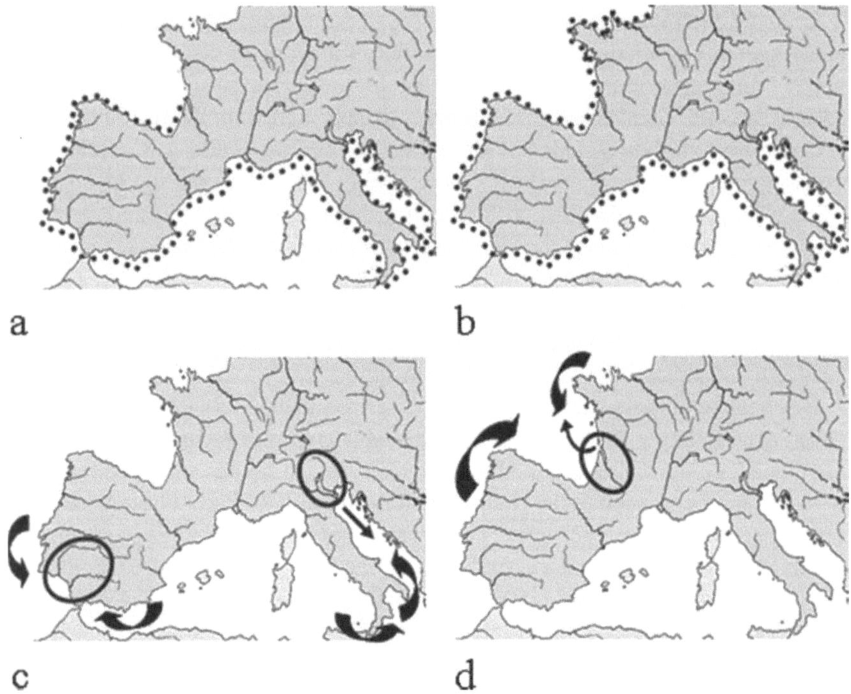

Fig. 1.8 Historical distribution of *Acipenser naccarii* (A) and of *Acipenser sturio* (B), indicated by points, in the study area. Direction of the regression (heavy arrows); areas of resistance (circles); and direction of re-colonization (fine arrows) by means of recovery programmes for *A. naccarii* (C) and for *A. sturio* (D)

in his own work that he draws his conclusions without personally analysing the specimens studied by the other authors, but rather on the basis of the museum identification cards (some were lost in a fire).

A. naccarii, which this author considered endemic to the Adriatic, is in reality autochthonous to the whole of southern Europe. Its historical distribution area reaches at least from the French Atlantic (Gironde) to the Adriatic (Mediterranean), without a break in continuity. It is a species in regression, abundant and widespread till the beginning of the twentieth century, when it became rare or extinct in many areas.

Its last refuge is the Adriatic, where its recovery has begun by repopulation from aquacultural specimens, and the zone of the Guadalquivir–Guadiana, where it cannot be considered extinct as there have been verified citations in the last 50 years (the 1970s and 1980s); these are strong recommendations for its repopulation to be undertaken (De la Herrán et al., 2004).

The errors committed previously in the taxonomy and management of sturgeons in the Iberian Peninsula (shown in the present work) have impeded the recovery of this species in the wild. We would like to stress the importance of scientific accuracy before the works of old naturalists are dismissed, even before they have been correctly analysed.

1.5.2 A. sturio

The state of preservation of *A. sturio* is considered critical CR (A2d) (Sturgeon Specialist Group 1996. *A. sturio*. In: IUCN 2007. *2007 IUCN Red List of Threatened Species*. < www.iucnredlist.org>), this species being found currently only in France (Williot et al., 2002a,b), with a threatened population in the basins of the Gironde, Garonne and Dordogne Rivers (Williot et al., this volume). However, its recovery is being actively taken up. Since the 1990s studies have been undertaken to reintroduce this species in Germany (Kirschbaum et al., 2000), and more recently in the Rhone River in the French Mediterranean, where investigations are under way for the possible reintroduction of *A. sturio* and *A. naccarii* (Brosse et al., 2008). The Spanish authorities consider, for legal reasons, that it has been sighted in diverse water courses such as the Urumea, Miño, Duero, Guadiana, Guadalquivir, Júcar, Turia and Ebro. However, in recent years, it has been found only in the Duero, Guadiana, and Guadalquivir River Basin (Ministerio de Medio Ambiente, 2007).

1.6 Conclusion

The status of sturgeons in southern Europe:

A. sturio is autochthonous to the whole of southern Europe while *A. naccarii* is autochthonous at least from the French Atlantic to the Adriatic, and both are sympatric throughout the distribution area of the latter. This implies a distribution more in accord with the biology of these species and with the information on the biogeographical distribution of the different sturgeon species, such as the proposal by Choudhury and Dick (1998) that would explain why the Adriatic (which is probably the limit of the distribution area of this species and not the centre) is where this sturgeon currently survives in relatively greater abundance.

The current distribution of sturgeons in southern Europe:

Both *A. sturio* and *A. naccarii* have undergone drastic regression, which has left them in their current zones of resistance. Only the decisive action of supporting their recovery by repopulation with the help of aquacultural specimens will provide the opportunity for the recovery of these species in their old distribution area and prevent their total disappearance.

The status of sturgeons in the Iberian Peninsula and implications for management:

All the scientific data show beyond any reasonable doubt that *A. naccarii* is autochthonous to the waters of the Iberian Peninsula and sympatric with *A. sturio*. Therefore, when doubts about these facts are expressed, the question arises: 'Why is there such interest in covering up such a clear fact?'

Our hypothesis is that the real problem is not a refusal to recognize that *A. naccarii* is autochthonous, but rather that success has been achieved in recovering the species in captivity to the level that would guarantee success with a hypothetical attempt to recover the species in the natural environment (Domezain, 2003; Domezain et al., 2008). The opposition is also perhaps to the recovery of the sturgeon in the Guadalquivir, regardless of the species. Opposition to recognize *A. naccarii* as autochthonous is not only a logical tactic as the recovery of *A. sturio* is not viable, since it is not available and unfortunately it appears that it will not be for a long time. If the artificial scientific dispute ceased and it were recognized that *A. naccarii* is autochthonous, there would be specimens of this species available in the quantity necessary to undertake its recovery.

Acknowledgements We wish to express our appreciation to all the caretakers of the museums and zoological collections we visited, for their warm reception and their help during our stay at those facilities. Also, we thank the Instituto de Fomento de Andalucía (Actualmente Agencia IDEA, Junta de Andalucía), Centro para el Desarrollo Tecnológico Industrial (CDTI), PROFIT (MCyT), University of Cádiz, and the Dirección General de Pesca y Acuicultura (Consejería de Agricultura y Pesca, Junta de Andalucía), as well as Observatorio de Pesca Andalucía-Marruecos (INTERREG IIIA) for their collaboration in financing a part of this project.

References

Albuquerque, R.M. 1956. Peixes do Portugal e ilhas adjacentes. Chaves para a sua determinaçao. *Port. Acta Biol.* 5B, 1164 pp (p. 195).

Algarin, S. 2002. La historia última de los esturiones del Guadalquivir. *Azotea No.* 13–14, 19–88.

Almaça, C. 1988. On the sturgeons, *Acipenser sturio*, in the Portuguese rivers and sea. *Folia Zool.*, 37, 183–191.

Almodóvar, A., Machardom, A., Suárez, J. 2000. Preliminary results from characterization of the Iberian Peninsula sturgeon based on analysis of the mtDNA cytochrome b. In: Symposium on Conservation of the Atlantic Sturgeon *Acipenser sturio* L., 1758 in Europe. *Bol. Inst. Esp. Oceanogr.*, 16, 17–27.

Bauchot, M.-L. 1987. *Poissons osseux.* In: W. Fischer, M.L. Bauchot and M. Schneider (eds.), *Fiches FAO d'identification pour les besoins de la pêche* (rev. 1). *Méditerranée et mer Noire. Zone de pêche 37*, vol. II. Commission des Communautés Européennes and FAO, Rome, pp. 891–1421.

Bernini, F., Nardi, P.A. 1989. Caractei morfomrtrici e meristici del genere *Acipenser* L. (Osteichthyes, Acipenseridae) nel tratto pavese dei Fiumi Po e Ticino. *Boll. Mus. Reg. Scien. Nat. Torino* 7, 321–340.

Bernini, F., Vanni, S. 1995. Cataloghi del Museo di Storia Naturale dell'Universitá di Firenze Sesione di Zoologia "La Specola" XIV: Ochtheichtyes Acipenseriformes. *Atti. Soc. Tosc. Sci. Nat. Mem. Serie B.* 102, 1–2.

Billard, R. 2002. *Esturgeons et caviar.* Lavoisier, Paris, 298 pp.

Brosse, L., Berrebi, P., Desse-Berset, M., Lepage, M. (2008). Sturgeon recovery plan in the Rhône River (France). Preliminary results on species determination and habitat suitability. In: Carmona et al (edits.). *Biology, Conservation and Sustainable Development of Sturgeons.* Springer Science + Business Media B.V., pp (to be completed by the publishers).

Capello, F.B. 1869. Catalogo dos peixes do Portugal que existem do Museu de Lisboa. *Jorn. Sci. Math Phys. Nat.* 1a serie, 2, 131–193.

Capello, F.B. 1880. Catalogo do peixes do Portugal. *Mem. Acad. R. Sc. Lisboa*, 6, 1–78.

Choudhury, A., Dick, T.A. 1998. The historical biogeography of sturgeons (Osteichthyes: Acipenseridae): a synthesis of phylogenetics, palaeontology and palaeogeography. *J. Biogeogr.*, 25, 623–640.

CITES, 2001. *CITES Identification Guide: Sturgeons and Paddlefish.* Minister of Supply and Service, Canada, 62 pp.

Classen, T.E.A. 1944. *Estudio bio-estadístico del esturión o sollo del Guadalquivir (Acpenser sturio* L.). Trabajos del Instituto Español de Oceanografía n° 14. Ministerio de Marina, Madrid, 112 pags + 27 láminas.

De la Herrán, R. 1998. Utilidad del ADN satelite y del ADN ribosomico en la filogenia, taxonomia y diagnosis de enfermedades de peces y almejas. Tesis doctoral, Universidad de Granada, Granada, España.

De la Herrán, R., Fontana, F., Lanfredi, M., Congiu, L., Leis, M., Rossi, R., Ruiz Rejón, C., Ruiz Rejón, M., Garrido-Ramos, M.A. 2001. Slow rates of evolution and sequence homogenization of an ancient satellite DNA family of sturgeons. *Mol. Biol. Evol.* 18(1), 432–436.

De la Herrán, R., Robles, F., Martínez-Espín, E., Lorente, J.A., Ruiz-Rejón, C., Garrido-Ramos, M., Ruiz-Rejón, M. 2004. Genetic identification of western Mediterranean sturgeons and its implication for conservation. *Conser. Genet.* 5: 545–551.

Domezain A. 2003. La acuicultura como herramienta para la recuperación de especies: el esturión autóctono de la Península Ibérica *Acipenser naccarii* Bonaparte 1836. Tesis doctoral, Universidad de Granada, Granada, España.

Domezain, A. 2008. Reflections, main steps and proposals on program for recovery plan of sturgeon in the Guadalquivir River, Spain. In: Carmona et al (edits.). *Biology, Conservation and Sustainable Development of Sturgeons*. Springer Science + Business Media B.V., pp (to be completed by the publishers).

Domezain, A., Soriguer, M.C., Ruíz Rejón, M., Hernando, J.A. 1999. The sturgeon as a valid tool for sustainable development: application in the river Guadalquivir (Andalusia, Spain). *J. Appl. Ichth.* 15(4–5), 299–301.

Elvira, B., Almodóvar, A. 1993. Notice about the survival of sturgeon (*Acipenser sturio* L., 1.758) in the Guadalquivir estuary (SW Spain). *Arch. Hydrobiol.*, 129, 253–255.

Elvira, B., Almodóvar, A., Lobón-Cerviá, J. 1991. Sturgeon (*Acipenser sturio* L., 1758) in Spain. The population of the river Guadalquivir: a case history and claim for a restoration programme. In: P. Willot (ed.), Actes du premier colloque international sur l'sturgeon, CEMAGREF, Bordeaux, France, pp. 337–347.

Fontana, F. 2002. A cytogenetic approach to the study of taxonomy and evolution in sturgeons. *J. Appl. Ichthyol.*, 18, 226–233.

Garrido-Ramos, M.A., Soriguer, M.C., De la Herránz, R., Jamilena, M., Ruiz Rejón, C., Domezain, A., Hernando, J.A., Ruiz Rejón, M. 1997. Morphometric and genetic analysis as proof for the existence of two sturgeon species in the Guadalquivir river. *Mar. Biol.*, 129, 33–39.

Gasent-Ramírez, J.M., Godoy, J.A., Jordano, P. 2001. Identificación de esturiones procedentes del Guadalquivir mediante análisis de ADN en especímenes de museo. *Medio Ambiente. Publicaciones de la Consejería de Medio Ambiente de la Junta de Andalucía*, 36, 44–49.

Gonçalves, B.C. 1942. Colecçao ocenográfica de D. Carlos I.-Peixes. *Trav. Stat. Biol. Mar. Lisbonne*, 46: 1–108.

Helling, H. 1943. *Novo catalogo dos Peixes do Portugal em colecçao no Museu de Zoologia da Universidade de Coimbra*. Mem. Est. Mus. Zool. Univ. Coimbra. 149: 1–110.

Hernando, J.A., Ruiz-Rejón, M., Soriguer, M.C., Domezain, A. 1997. Actualización de la distribución histórica de *Huso huso* (Linnaeus 1.758) y *Acipenser naccarii* Bonaparte 1.836: Nuevos casos de regresión. *II Congreso Iberico de Biologos Ambientalistas. Badajoz. (España)*

Hernando, J.A., Arlati, G., Domezain, A., Soriguer, M.C., Poliakova-Belysceva, L.A., Domezain, J., Vallespín, C., Bravo, R. 1999a. Morphometric study of *Acipenser naccarii* (Bonaparte, 1836) in fish farm individuals. *J. Appl. Ichth.* 15(4–5), 46–50.

Hernando, J.A., Vasil'eva, E.D., Arlati, G., Vasil'ev, V.P., Santiago, J.A., Belysceva-Poliakova, L., Domezain, A., Soriguer, M.C. 1999b. A new proof for the historical presence of two European sturgeons in the Iberian Peninsula: *Huso huso* (Linnaeus, 1758) and *Acipenser naccarii* Bonaparte, 1836. *J. Appl. Ichth.* 15(4–5), 280–281.

Hernando, J.A., Vasil'eva, E.D., Arlati, G., Vasil'ev, V.P., Santiago, J.A., Belysceva-Poliakova. L., Domezain, A., Soriguer, M.C. 1999c. New evidence for a wider historical area of two European sturgeons: *Acipenser naccarii* and *Huso huso* (Acipenseridae). *J. Ichthyol.*, 39, 841–845.

Hernando, J.A., Domezain A, Cabrera R., Domezain J., Soriguer M.C. New evidence for historical distribution of *Acipenser naccarii* in the Mediterranean sea (in preparation).

Holcik, J., Banarescu, P., Evans, D. 1989a. A general introduction to fishes. In: J. Holcik (ed.), *The Freshwater Fishes of Europe*, vol. 1(II). Aula-Verlag, Wiesbaden, pp. 18–147.

Holcik, J., Kinzelback, R., Sokolov, L.I., Vasil'ev, V.P. 1989b. *Acipenser sturio* Linnaeus., 1758. In: J. Holcik (ed.), *The Freshwater Fishes of Europe*, vol. 1(II). Aula-Verlag, Wiesbaden, pp. 367–394.

IUCN, 1996. *Red List of Threatened Animals*. IUCN, Gland, Switzerland.

IUCN, 2007. *2007 IUCN Red List of Threatened Species*. IUCN, Gland, Switzerland. Available at: www.iucnredlist.org.

Kirschbaum F., Gessner J., Wiliot P. 2000. Restoration of *Acipenser sturio* L., 1758 in Germany: Growth characteristics of juvenile fish reared under exprimental indoor conditions. Bol. Inst. Esp. Oceanogr. 16: 157–165

Lanfredi, M., Congiu, L., Garrido-Ramos, M.A., De la Herrán, R., Leis, M., Chicca, M., Rossi, R., Tagliavini, J., Ruiz Rejón, C., Ruiz Rejón, M., Fontana, F. 2001. Chromosomal location and evolution of a satellite DNA family in seven sturgeons species. *Chromosome Res.*, 9, 47–52.

Leviton, A.L., Gibbs Jr. R.H., Heal, E., Dawson, C.E. 1985. Standards in herpetology and ichthyology. Part I. Standard symbolic codes for Institutional resource collections in herpetology and ichthyology. *Copeia* 1985(3), 802–832.

Lozano Rey, L. 1935. *Los peces fluviales de España*. Memorias de la Real Academia de las Ciencias Exactas, Físicas y Naturales, Madrid, 345 pp.

Magnin, E. 1962. Recherches sur la systématique et la biologie des Acipensérides. *Ann. Sta. Centr. Hydrobiol. Appl.*, 9, 7–242.

Ministerio de Medio Ambiente, 2007. Available at: http://www.programaagua.com/portal/secciones/biodiversidad/inventarios/inb/atlas_Peces/Acipenser_sturio.htm.

Nobre, A. 1931. Peixes das aguas doces de Portugal. *Bolm. Min. Agric.*, 13(2), 73–112.

Nobre, A. 1935. *Fauna Marinha de Portugal*. Vertebrados, Porto, 579 pp.

Osorio, B. 1894. D'algunas especies a juntar ao Catalogo dos peixes do Portugal de Capello. *Jorn. Sc. Math. Phys. Nat.*, 3, 186–188.

Robles, F. 2003. Análisis de la evolución concertada, filogenia e identificación de especimenes en el Orden Acipenseriformes. Tesis doctoral, Universidad de Granada, Granada, España.

Robles, F., De la Herrán, R., Garrido-Ramos, M.A., Ruiz-Rejón, C., Martínez-Espín, E., Lorente, J.A., Ruiz-Rejón, M. 2003. Identificación de los esturiones del Guadalquivir utilizando cinco marcadores moleculares y técnicas forenses. IX Congreso Nac Acuicultura, Cadiz, España, Mayo, 2003.

Ruíz-Rejón, M., De la Herrán, R., Ruíz-Rejón, C., Garrido Ramos, M.A. 2000. Genetic characterization of *Acipenser sturio* in relation to other sturgeon species using satellite DNA. *Bol. Inst. Esp. Oceanogr.*, 16(1–4), 231–236.

Seabra, A.F. 1911. Catalogue systematique des vertebros du Portugal. V. Poissons. *Bull. Soc. Port. Sc. Nac.* 5, 129–224.

Sokolov, L.I. 1989. *Acipenser* Linnaeus, 1758. In: J. Holcik (ed.), *The Freshwater Fishes of Europe*, vol. 1(II). Aula-Verlag, Wiesbaden, pp. 201–205.

Soljan, T. 1975. *Il pesci dell'Adriatico*. Mondori, Verona, 522 pp.

Svetovidov, A.N. 1984. Acipenseridae. In: P.J.P. Whitehead, M.L. Bauchot, J.C. Hureau, J. Nielsen and E. Tortonese (eds.), *Fishes of the North-eastern Atlantic and the Mediterranean*, 2nd edn. UNESCO, Paris, pp. 220–225.

Tortonese, E. 1989. *Acipenser naccarii* Bonaparte, 1836. In: J. Holcik (ed.), *The Freshwater Fishes of Europe*, vol. 1(II). Aula-Verlag, Wiesbaden, pp. 285–293.

Vasil'eva, E.D. 1999. Some morphological characteristics of Acipenserid fishes: considerations of their variability and utility in taxonomy. *J. Appl. Ichth.*, 15(4–5), 32–35.

Williot, P., Rochard, E., Castelnaud, G., Ronault, Th., Brun, R., Lapage y, M., Ellie, P. 1997. Biological characteristics of European Atlantic sturgeon, *Acipenser sturio*, as the basis for restoration program in France. *Environ. Biol. Fish.*, 48, 359–370.

Williot, P., Rouault, T., Brun, R., Pelard, M., Mercier, D. 2002a. Status of caught wild spawners and propagation of the endangered sturgeon *Acipenser sturio* in France: a synthesis. *Int. Rev Hydrobiol.*, 87, 515–524.

Williot, P., Arlati, G., Chebanov, M., Gulyas, T., Kasimov, R., Kirschbaum, F., Patriche, N, Pavlovskaya, L., Poliakova, L., Pourkazemi, M., Kim, Y., Zhuang, P., Zholdasova, I.M. 2002b. Status and management of Eurasian sturgeon: an overview. *Int. Rev. Hydrobiol.*, 87, 483–506.

Williot, P., Rochard, E., Rouault, T., Kirschbaum, F. (This volume). *Acipenser sturio* recovery research actions in France. In: Carmona et al (edits.). *Biology, Conservation and Sustainable Development of Sturgeons*. Springer Science + Business Media B.V., pp (to be completed by the publishers).

Chapter 2
Analysis of Mitochondrial and Nuclear DNA Markers in Old Museum Sturgeons Yield Insights About the Species Existing in Western Europe: *A. sturio, A. naccarii* and *A. oxyrinchus*

Manuel A. Garrido-Ramos, Francisca Robles, Roberto de la Herrán, Esther Martínez-Espín, José A. Lorente, Carmelo Ruiz-Rejón, and Manuel Ruiz-Rejón

Abstract Today, with all the sturgeon species almost disappearing all over the world, it is necessary to undertake their recovery under the programs for the conservation of genetic resources. The complete absence of these fish from most rivers increases the difficulties in carrying out such programs, hampering the genetic identification of specimens and their correct species assignment. However, with the development of new and reliable molecular genetic techniques, many studies such as this are yielding insights concerning the sturgeon species that inhabited European rivers in the past. In the last few years, our group has developed forensic techniques to isolate DNA from ancient sturgeon specimens preserved in museums. These DNA samples have been the subject of various analyses conducted on several nuclear and mitochondrial DNA markers. The combined use of both types of markers has provided accurate genetic identification of these specimens and has overcome the problem of misinterpretation caused by hybridization and introgression. Here, we show that, in addition to *Acipenser sturio* (the only species previously believed to inhabit the rivers of Western Europe), two other species, *Acipenser naccarii* and *Acipenser oxyrinchus* lived in these rivers. Thus, we have found evidence for the presence of *A. naccarii* in the Guadalquivir river in the Iberian Peninsula and in some rivers in Italy from the Tyrrhenian/Ligurian side, as well as for the presence of *A. oxyrinchus* in the Ebro river in Spain. Our studies clarify the distribution of sturgeon species in the Western Mediterranean and open new perspectives for recovery plans.

M. A. Garrido-Ramos, R. De la Herrán, C.R. Rejón, and M.R. Rejón
Departamento de Genética (Facultad de Ciencias), Universidad de Granada, 18071 Granada, Spain

E. Martínez-Espín, J.A. Lorente
Departamento de Medicina Legal y Forense (Facultad de Medicina), Universidad de Granada, 18071 Granada, Spain

F. Robles,
Present address: Institute for Zoo and Wildlife Research, Department of Evolutionary Genetics, 10252 Berlin, Germany

R. Carmona et al. (eds.) *Biology, Conservation and Sustainable Development of Sturgeons*,

Keywords *A. naccarii*, *A. sturio*, *A. oxyrinchus*, forensic techniques, genetic identification, molecular markers, conservation genetics, Western Europe Rivers

2.1 Introduction

Condrostean sturgeons are very primitive bony fish that have inhabited the earth for about 200 million years. Sturgeons are distributed exclusively throughout the Northern Hemisphere, occupying large rivers, lakes, coastal waters, and interior seas in China, Kazakhstan, Turkmenistan, the Islamic Republic of Iran, Azerbaijan, the Russian Federation, Ukraine, Turkey, Romania, Bulgaria, and countries of Western Europe as well as of North America. Most sturgeon species have an amphibiotic behaviour, being anadromous, given that males and females, when sexually mature, return from the sea to the river where they were born. These fish are highly appreciated not only because of the quality of their meat but more especially because they produce caviar, an expensive and highly nutritive delicacy.

The superorder Condrostean is composed of 17 orders, of which 16 are extinct, the only extant order being Acipenseriformes. This order has two families—that of the paddlefishes (family Polyodontidae), with two extant representative species of the genera *Polyodon* (*Polyodon spathula*) and *Psephurus* (*Psephurus gladius*), and that of the sturgeons (family Acipenseridae). Within the family Acipenseridae, there are 25 species included in four genera (*Acipenser* with 17 species, *Huso* with 2 species, *Scaphirhynchus* with 3 species and *Pseudoscaphirhynchus* with 3 species). The genera *Acipenser* and *Huso* comprise the subfamily Acipenserinae and the genera *Scaphirhynchus* and *Pseudoscaphirhynchus* constitute the subfamily Scaphirhynchinae.

However, this classification is still under discussion (Artyukhin, 1995; Birstein and DeSalle, 1998; Ludwig et al., 2001; Birstein et al., 2002). According to new data gathered from mitochondrial DNA sequence alignments used as an input database for phylogenetic analysis, within the Acipenseridae, sturgeon can be assigned to four phylobiogeographic clades (Ludwig et al., 2001): one clade comprised of the species of the subfamily Scaphirhynchinae (*Scaphirhynchus* and *Pseudoscaphirhynchus*), one clade composed of the sturgeon *A. sturio* and *A. oxyrinchus*, one made up of the Pacific species of *Acipenser* and *Huso* (*A. medirostris*, *A. mikadoi*, *A. schrenckii*, *A. transmontanus*, *A. sinensis* and *H. dauricus*) and another clade containing the Atlantic species of *Acipenser* and *Huso* (*A. gueldenstaedtii*, *A. naccarii*, *A. persicus*, *A. baerii*, *A. brevirostrum*, *A. fulvescens*, *A. nuvidentris*, *A. ruthenus*, *A. stellatus* and *H. huso*). However, Birstein et al. (2002), also using mitochondrial DNA sequences, proposed the paraphyly of the genera *Scaphirhynchus* and *Pseudoscaphirhynchus,* as the latter genus should be embedded within *Acipenser* clustering with *A. stellatus*.

According to Choudhury and Dick (1998), sturgeons are distributed throughout four broad geographic areas which resemble the phylobiogeography of the group (Fig. 2.1). Thus, species such as *A. oxyrinchus*, *A. brevirostrum*, *A. fulvescens* and those

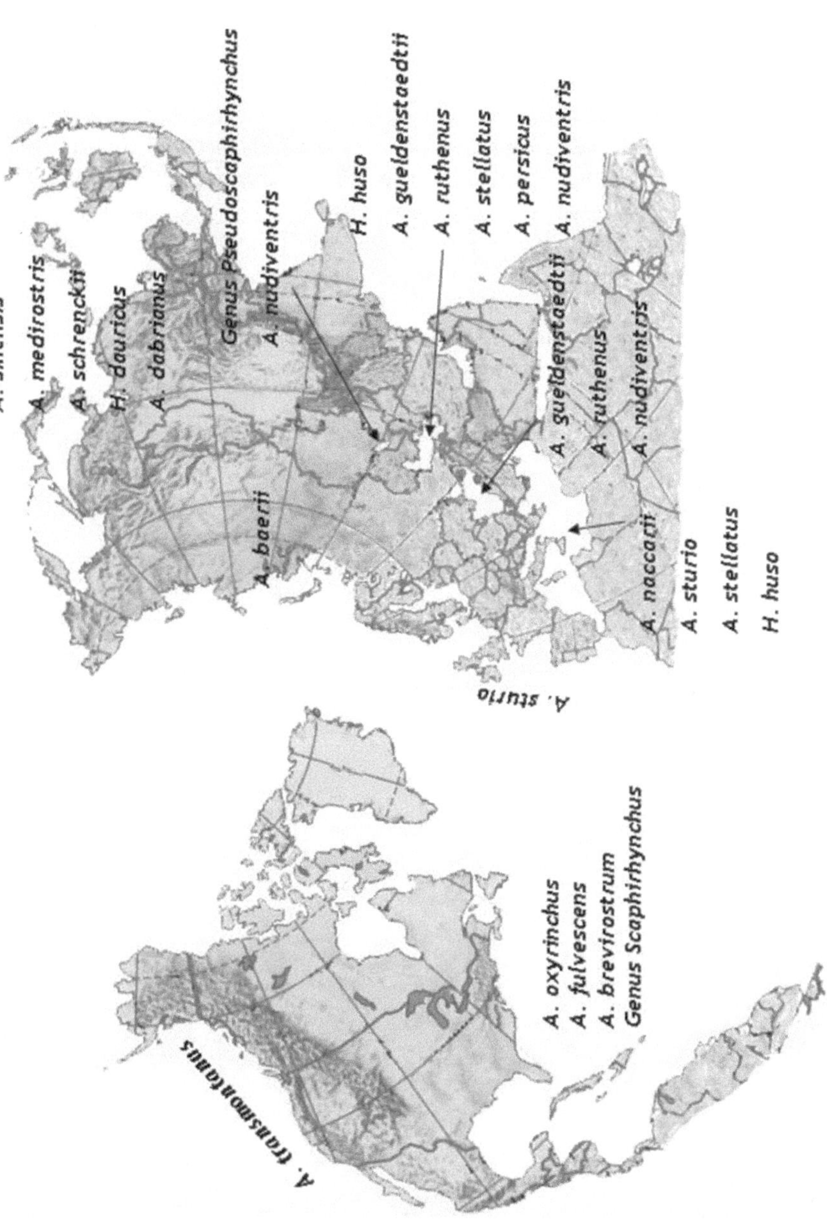

Fig. 2.1 Geographic distribution of sturgeon species

of the genus *Scaphirhynchus* are found on the Atlantic side of North America. The area comprising the Mediterranean, Black, Caspian and Aral Seas (Mediterranean-Ponto-Caspian area) is the area with the greatest species diversity. Species of the Mediterranean area are *H. huso*, *A. stellatus*, *A. sturio* and *A. naccarii*, although this latter is considered endemic to the Adriatic Sea. The species *A. nudiventris*, *A. gueldenstaedtii* and *A. ruthenus* are found in the Black Sea. The Caspian Sea has species such as *A. stellatus, H. huso, A. nudiventris, A. gueldenstaedtii*, *A. ruthenus* and *A. persicus*, with the last found mostly on the Iranian coast. Species of the genus *Pseudoscaphirhynchus* live in the Rivers Syr-Darya and Amu-Darya, which flow into the Aral Sea, where *A. nudiventris* lives. *A. baerii* is distributed in the rivers of Siberia in Asia. *A. transmontanus* and *A. medirostris* are species of the Pacific coast of North America, while *A. schrenckii*, *H. dauricus*, *A. mikadoi*, *A. dabrianus* and *A. sinensis* are distributed along the Pacific coast of Asia.

Among the sturgeon species, the Atlantic sturgeon, *A. sturio* was the species with the broadest distribution, inhabiting Western Europe from the North Sea to the Atlantic Ocean and from the Mediterranean Sea to the Black Sea. The distribution of sturgeon species in Western Europe might be considered, however, somewhat

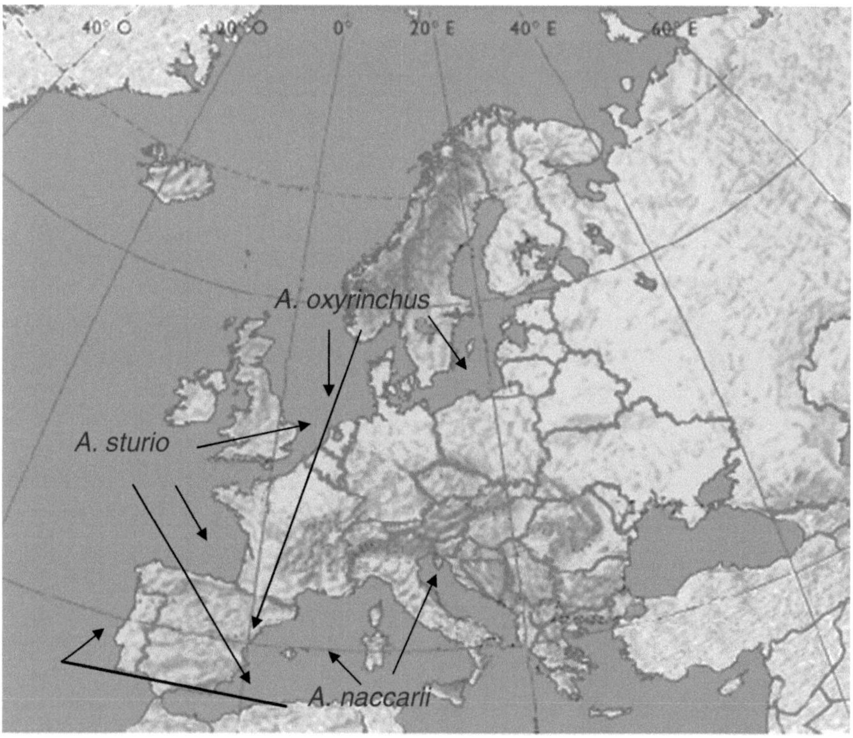

Fig. 2.2 Distribution of *A. sturio*, *A. naccarii* and *A. oxyrinchus* according to Capello (1869); Gonçalves (1942); Garrido-Ramos et al. (1997); Ludwig et al. (2002)

controversial as the only recognized native species has traditionally been *A. sturio*. However, there are observations supporting its sympatry with *A. naccarii*, held to be endemic to the Adriatic region, in south-western Europe (Capello, 1869; Gonçalves, 1942) and some centuries ago with *A. oxyrinchus*, a species from the North American side of the Atlantic Ocean, in Northern Europe (Ludwig et al., 2002) (Fig. 2.2).

Sturgeons are 'living fossil' fish threatened with extinction throughout the world (McDowall, 1999). Since the wild populations of sturgeons are currently at critically low levels, recovery efforts focus on sturgeons raised in fish farms. If sturgeons from fish farms are to be released into the wild, it is important to have accurate identification of the sturgeon species already inhabiting the river. In the rivers of Western Europe, the only recognized native species, *A. sturio*, is currently restricted to the Gironde and is difficult to rear in captivity. However, another sturgeon species, *A. oxyrinchus*, might be used for recovery plans of sturgeons in the rivers of Germany (see the chapter by Kirschbaum et al., in this volume) and Poland as its presence has been demonstrated in these rivers together with *A. sturio* (Ludwig et al., 2002). Also, *A. naccarii* is being reared successfully in fish farms in Italy and Spain. This perspective opens up the possibility of the recovery of this species in European rivers which *A. naccarii* inhabited in the past, such as the rivers of the Iberian Peninsula, where the species have been reported (Capello, 1869; Gonçalves, 1942; Garrido-Ramos et al., 1997). Unfortunately, there are no sturgeons currently found in those rivers. For example, in the Guadalquivir, only four specimens have been captured since 1974, of which three are preserved at the Biological Station of Doñana in Seville (Spain). No captures have been reported from the Ebro since the 1960s (Porres and Farnós, 1999). Thus, there is a need for a clear genetic identification of the species used for any recovery plan, as identification by morphological grounds is difficult, and sometimes disputed, for sturgeon species. For example, two species such as *A. naccarii* and *A. sturio* differ only in the position of the barbels and the size and shape of the snout (Svetovidov, 1989), so that the application of these meristic characteristics is controversial (Garrido-Ramos et al., 1997; Rincón, 2000). There is no less controversy concerning the morphological differentiation between *A. sturio* and *A. oxyrinchus* specimens. In 1963, Magnin and Beaulieu, on the basis of morphometric differences of the head, considered them two different species (Magnin and Beaulieu, 1963). Furthermore, the problem is accentuated in each study to be undertaken because only a few museum specimens are available for reference.

There are two main problems to be solved while looking for accurate techniques for genetic identification of sturgeon museum specimens. The first is the obvious necessity of using forensic techniques to obtain good-quality, authentic DNA to avoid PCR contamination and hence misidentification; the second is to find a reliable genetic marker to ensure an accurate genetic identification and to overcome misidentifications caused by hybridization and introgression. The first problem will be examined in depth in a separate chapter in this volume (Martínez-Espín et al., 2008). In this chapter, we take up the second by analysing available DNA markers for sturgeon genetic identification and their usefulness for assigning the species of the museum specimens.

2.2 The Value of Nuclear Repetitive DNA as a Genetic Identification Marker in Sturgeon

The usefulness of mitochondrial DNA as a robust genetic marker for genetic identification of sturgeons even when they are museum specimens, is widely accepted. In fact, there are many publications on this topic (Birstein and DeSalle, 1998; Ludwig and Kirschbaum, 1998; Almodóvar et al., 2000; Birstein and Doukakis, 2000; Krieger et al., 2000; Ludwig et al., 2000, 2001, 2002; Gasent-Ramírez et al., 2001). Notably, several nucleotide positions in the cytochrome *b* and the 12S ribosomal RNA (sRNA) genes of the mitochondrion are diagnostic and can be used with certainty to ascertain the species of the sturgeons. Several sets of primers, such as those developed by Birstein and DeSalle (1998) or by Ludwig and Kirschbaum (1998), which are specific for the PCR amplification of several regions of these two genes, have aided in the development of PCR genetic testing in sturgeons. Of course, in addition, these types of PCR reactions have led to the availability of a set of sequence alignments as the basis for phylogenetic inferences (Birstein and DeSalle, 1998; Birstein et al., 1998; Krieger et al., 2000; Ludwig et al., 2001). In this sense, the robustness of mitochondrial DNA as a phylogenetic marker has proved successful in shedding light in several studies on the evolutionary pathways leading to current extant sturgeon species.

However, the use of mitochondrial sequences can be problematic when specimens are F_1 hybrid (maternal mitochondria and maternal/paternal nucleus) or the result of a process of genetic introgression (maternal mitochondria and paternal nucleus). In this sense, there are many claims of having developed nuclear molecular markers for sturgeon identification (Fontana, 2002; Ludwig et al., 2003).

The development of newer and more powerful nuclear markers is necessary for both evolutionary and taxonomic genetic analysis in sturgeons. However, a nuclear marker has to satisfy a series of requisites, as its gradual rate of evolution is important. In addition, it is necessary to match the rate of evolutionary change of the marker to the range of species compared. In this case of closely related species such as those of the same genus *Acipenser*, the need for a marker with a high rate of sequence change is fundamental. However, this is not easy in the case of sturgeons. In fact, fish genomes, in general, are characterized by lower mutation rates than in other vertebrate genomes. This phenomenon has been explained in several ways, for example, as a consequence of lower metabolic rates (Martin et al., 1992; Rico et al., 1996; Martin, 1999). However, sturgeons have extremely low rates of evolution even among fishes (De la Herrán et al., 2001a; Krieger and Fuerst, 2002). The reason for this remains unresolved, however, although physiological factors (long generation time, large body size, ectothermy, low metabolic rate; Krieger and Fuerst, 2002) and/or genetic ones (polyploidization events; Ludwig et al., 2001) may be responsible for this slow rate.

Our research group has been working on satellite DNA for more than fifteen years. Several satellite-DNA families of sequences have been isolated in our laboratories. In our experience, these types of sequences have been demonstrated to be successful markers for evolutionary purposes, taxonomic studies, and phylogenetic inference. A satellite-DNA family is composed of short tandem-repeat DNA sequences (about 180–200 bp in length). They are non-coding sequences making up the heterochromatic parts of the chromosomes of eukaryotic genomes (Elder and Turner, 1995; Ugarkovi and Plohl, 2002). That is, they are located mainly at the centromeres and in subtelomeric regions, although other intercalary locations have also been found. A family of tandem repetitive sequences distributes among the chromosomes of one species according to the principle of equilocality (Schweizer and Loidl, 1987). Several satellite-DNA families can be found within the same eukaryotic genome, but each one in a different equilocal region.

Satellite-DNA sequences are highly evolving sequences. Repetitive-DNA families are influenced by several molecular mechanisms of non-reciprocal exchanges (Ohta and Dover, 1984) that can gradually spread a variant sequence throughout a family within a sexual population in a process known in population genetics as molecular drive (Dover, 1986). This could explain the evolutionary pattern of repetitive sequences known as concerted evolution. Concerted evolution leads to high levels of family homogeneity for species-diagnostic mutations. Empirical observations in most species-pair comparisons for different satellite-DNA families indicate that the rate of production of new sequence variants (mutation) is a slower process than their rate of spread, while the general paucity of transition stages indicates that the replacement is relatively fast (Ugarkovi and Plohl, 2002). Thus, when a satellite-DNA family is shared by a group of species, species-specific diagnostic sites can be used as a taxonomic criterion and as markers for the genetic identification of a species. Moreover, the resolving power of these sequences in systematic studies ranges from the identification of conspecific populations (Elder and Turner, 1994) to interfamilial relationships of cetaceans (Arnason et al., 1992; Grétarsdóttir and Arnason, 1992). These sequences are considered a useful tool for identifying phylogenetic relationships (Stepien and Kocher, 1997) and they have been used in recent years for systematic studies (Franck et al., 1994; Garrido-Ramos et al., 1995, 1999; Mestrovic et al., 2000; De la Herrán et al., 2001b).

We have analysed two ancient satellite-DNA families, the *Hind*III and *Pst*I families, found within the genomes of different sturgeon species of the genera *Acipenser*, *Huso* and *Scaphirhynchus* (De la Herrán et al., 2001a; Robles et al., 2004). Therefore, both satellite DNAs are assumed to be more than 100 million years old. This case is exceptional as very few examples of satellite-DNA preservation have been reported (Heikkinen et al., 1995; Mravinac et al., 2002).

The two satellite DNAs of sturgeon are very different in length, sequence, nucleotide composition, and other characteristics. However, they have a high degree of preservation in common and slow rates of sequence change and concerted evolution. In fact, the sequence-change rates of these satellite DNAs are extremely low, between 0.07% and 0.11% per MY. This is an extremely low rate of sequence change compared with the divergence distances found for more recently diversified fish species such as sparids (De la Herrán et al., 2001b). These sequences are evolving in sturgeon at a rate twofold less than the exceedingly slow common cetacean satellite DNA (0.2%; Arnason et al., 1992).

Moreover, sturgeons have undergone slow rates of concerted evolution for satellite-DNA sequences. For example, in appearance, *Pst*I satellite-DNA does not follow concerted evolution, which is reflected in a phylogenetic tree of sequences (Fig. 2.3a). That is, except for the case of the sequences of *Scaphirhynchus albus*, repeat sequences from each species were not grouped by taxonomic affinity, as occurs when concerted evolution is operating (Garrido-Ramos et al., 1999). However, a cladistic association was found among sequences of *A. sturio* and *A. oxyrinchus* on one hand, among sequences of *A. sinensis* and *A. tramonstanus* (Pacific sturgeon species) on the other, and also among sequences of the rest of the species (Atlantic sturgeon species). This resembles the grouping into the four phylobiogeographic clades proposed by Ludwig et al. (2001) (Fig. 2.3b). We have attributed this to the fact that a major factor could be masking the effect of forces leading to concerted evolution and that among the causes to explain the evolution of repetitive sequences in sturgeons, a strong influence could be from the particular evolutionary pathways leading to the current extant species of sturgeons, which must include hybridization and polyploidization events.

Therefore, in appearance, *Pst*I sequences could not be used as a phylogenetic marker for sturgeon species. However, first, we found supporting evidence for the four phylogenetic and biogeographic clades proposed previously by Ludwig et al. (2001). Second, the results provide new evidence for the inclusion of the species of the genus *Huso* within the genus *Acipenser* (Birstein and DeSalle, 1998; De la Herrán et al., 2001a; Ludwig et al., 2001). In fact, the species of *Huso* are more closely related to some species of *Acipenser* than among themselves. Third, Atlantic sturgeons, *A. sturio* and *A. oxyrinchus*, formed an ancient clade that diverged very early, i.e. 90 million years ago, from the other sturgeon species (Birstein and DeSalle, 1998; Ludwig et al., 2001). This is important information, as data gathered by *Pst*I sequences can be used for species differentiation between *A. sturio* and other species, such as *A. naccarii* in taxonomic identification tests (Fig. 2.3a and b). The basal position of the *A. sturio* and *A. oxyrinchus* clade is also supported by the absence of the *Hind*III satellite DNA (Garrido-Ramos et al., 1997; De la Herrán et al., 2001a; Lanfredi et al., 2001; Robles et al., 2004; Fontana et al. 2008). The *Hind*III satellite DNA was present in all sturgeon species analysed except in *A. sturio* and *A. oxyrinchus*. By contrast, the presence of *Hind*III sequences within the genomes of *Scaphirhynchus* supported a close relationship with species of the genus *Acipenser*, as for the *Pst*I tree. Also, as for *Pst*I satellite DNA, *Hind*III sequences of the other species having these sequences are not grouped by taxonomic affinity, although the subdivision between the Atlantic and the Pacific clades was not found for the *Hind*III sequences. However, the sequences of *S. albus* were also grouped together in this case.

However, the most noteworthy aspect of *Hind*III sequences is that the exclusive absence of these repeats within the two species forming the most ancient lineage of sturgeons makes them an excellent diagnostic marker for genetic identification. From a phylogenetic standpoint, the sharing of the same satellite-DNA family can be used for associating related species. If two species share one satellite-DNA family, they must have diverged more recently than other species not having that

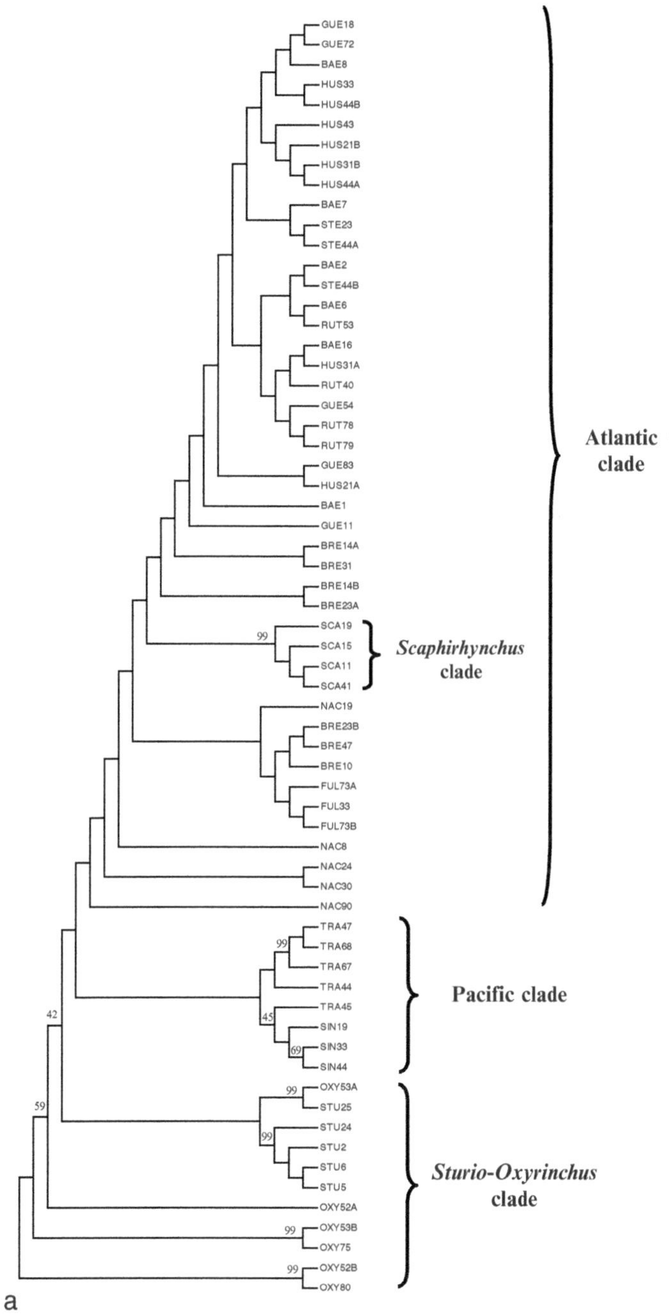

Fig. 2.3 (**a**) Neighbour-joining tree of *Pst*I sequences. In this tree, the names of the sequences correspond to the first three letters of the species names from which they were isolated and the number of the repeat analysed: NAC, *Acipenser naccarii*; BAE, *A. baerii*; GUE, *A. gueldenstaedtii*; STE, *A. stellatus*; RUT, *A. ruthenus*; HUS, *Huso huso*; FUL, *A. fulvescens*; BRE,

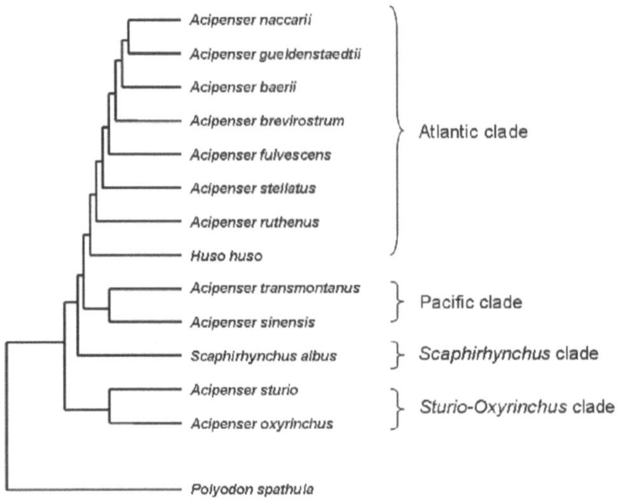

Fig. 2.3 (continued) *A. brevirostrum*; TRA, *A. transmontanus*; SIN, *A. sinensis*; STU, *A. sturio*; OXY, *A. oxyrinchus*; SCA, *Scaphirhynchus albus*. Bootstrap values are shown for main nodes in the tree. (**b**) Phlylogenetic tree based on cytochrome *b* gene sequences according to Robles et al. (2004)

satellite-DNA family. This allows the use of repetitive sequence families as cladistic markers, grouping species having the same satellite-DNA family in a monophyletic lineage apart from the lineage that includes species lacking the satellite-DNA (Arnason and Widegren, 1986; Murata et al., 1993; Van Den Busche et al., 1993; Hartley and Davidson, 1994a,b).

Another repetitive part of nuclear DNA, ribosomal DNA (rDNA), is informative in analysing sturgeons at the molecular level, because it bears unique characteristics, which are to a certain extent ancestral in vertebrates. We examined the structure and the molecular evolution of the 5S ribosomal DNA (rDNA) region in 13 sturgeon species, comparing both the 5S ribosomal RNA (rRNA) genes and the non-transcribed spacer (NTS) sequences between the coding regions (Robles et al., 2005). Using predictive models, we found similar levels of sequence diversity in the coding regions as in the non-coding region but, as expected, fixed inter-specific differences are underrepresented for 5S genes. However, contrary to expectations, we did not find fixed differences between NTS sequences on comparing many pairs of species, especially when those sequences belonged to the same four phylogeographic clades into which the sturgeon is divided, but fixation of mutations and divergence was found between species belonging to different phylogeographic clades. These results resemble those that we found for satellite-DNA sequences (*Hind*III and *Pst*I sequences). Again, our results imply that the evolution of the two parts of the 5S rDNA region cannot be explained exclusively as the outcome of a balance between mutational, homogenizing (i.e. gene conversion as a predominant force in sturgeon) and selective forces. Rather, they suggest, as for satellite-DNA sequences, that other factors (i.e. hybridization) might be superimposed over those forces and thus could to some extent be masking their effects. Nevertheless, NTS sequences can also be used for comparing species of different clades, as when we compare *A. naccarii* with *A. sturio.*

2.3 Use of Nuclear Repetitive-DNA Sequences Begin to Clarify the Distribution of Sturgeon Species in Western Europe

As mentioned above, the presence in recent centuries of *A. naccarii* in the rivers and along the coast of Western Europe is controversial, since the only species considered to inhabit those waters has traditionally been *A. sturio*. However, there are some indications of the presence of the so-called Adriatic sturgeon (considered endemic to that region) in the waters of the Iberian Peninsula (Capello, 1869; Gonçalves, 1942; Garrido-Ramos et al., 1997; Ruiz Rejón et al., 2000) and western Italy (as we have found museums specimens in Livorno and Genoa from rivers of western Italy and catalogued as *A. naccarii*). We were then prompted to test thoroughly for the presence of *A. naccarii* among preserved museum specimens caught in the past in Spanish and Italian rivers.

In the last few years, we have analysed a total of 15 museum sturgeon specimens caught during the last three centuries (Table 2.1; Fig. 2.4) from various rivers in Italy (both from the Tyrrhenian/Ligurian side and from the Adriatic side), France, and Spain (Garrido-Ramos et al., 1997; Ruiz Rejón et al., 2000; De la Herrán et al., 2004; and the new data presented in this chapter for the first time). After a comprehensive and detailed analysis using several methods of DNA extraction and purification (including the most powerful forensic methods) and using several nuclear repetitive markers in comparison with several mitochondrial markers, we have concluded that the species *A. naccarii* inhabits Western Mediterranean rivers such as the Guadalquivir (Garrido-Ramos et al., 1997; Ruiz Rejón et al., 2000; De la Herrán et al., 2004). This species was sympatric with *A. sturio*. Therefore, *A. naccarii* has had a broader distribution area than the Adriatic Sea, extending throughout the Western Mediterranean, and even reaching the waters of the Iberian Peninsula.

Using forensic techniques, we have isolated authentic DNA from several museum sturgeon specimens (Fig. 2.5). Seven of these specimens were caught in the Iberian Peninsula (six from the Guadalquivir and one from the Ebro) all of them catalogued as *A. sturio*. Six specimens were from Italian museums, of which two specimens from the Adriatic coast were catalogued as *A. naccarii*, two specimens from the western coast of Italy were catalogued as *A. sturio* and, notably, two specimens from the western coast of Italy were catalogued as *A. naccarii*. Two specimens of *A. sturio* kept frozen in the CEMAGREF installations (Bordeaux, France) from the Gironde were also analysed.

Table 2.1 List of sturgeon specimens analysed, their current specific status and the results obtained for: (i) presence (+) or absence (-) of the HindIII satellite DNA; (ii) the markers analysed for each specimen and the number (within parentheses) of units sequenced for each nuclear repetitive marker or the number of mitochondrial clones sequenced for each mitochondrial marker.

				Molecular marker and number of sequences analysed						
Specimen	Provenance (year of catch)	Sampling location (preservation)	Classification	*Hind*III (presence/ absence)	*Hind*III	*Pst*I	NTS	Cytochrome *b* (2F-3R)	12S	New status
PSN-34	Po river, Italy	Piscifactoría Sierra, Nevada S.L. (live specimen)	*A. naccarii*	+	11	5	9	1	1	*A. naccarii*
EBD 8173	Guadalquivir river, Alcalá del Río, Sevilla, Spain (1974)	Doñana Biological Station, Sevilla, Spain (ethanol)	*A. sturio*	+	1	4	2	1	2	*A. naccarii*
EBD8401	Guadalquivir river, Coria del Río, Sevilla, Spain (1981)	Doñana Biological Station, Sevilla, Spain (ethanol)	*A. sturio*	+	2	6	6	1	3	*A. naccarii*
EBD8174	Guadalquivir river, Alcalá del Río, Sevilla, Spain (1975)	Doñana Biological Station, Sevilla, Spain (stuffed)	*A. sturio*	+	1	3	5	2	1	*A. naccarii*
	Riío Sevilla, Spain (1975)	Spain Stuffed								
UGP	Guadalquivir river (nineteenth century)	Museum of the Animal Biology Department, Facultad de Ciencias, Universidad de Granada, Spain (stuffed)	*A. sturio*	–	–	6	na*	2	2	*A. sturio*
UGRA1	Guadalquivir river, Sevilla, Spain (nineteenth century)	Museum of the Animal Biology Department, Facultad de Ciencias, Universidad de Granada, Spain (stuffed)	*A. sturio*	–	–	na	na	na	na	*A. sturio*
MNCZ1582	Guadalquivir river, Sevilla, Spain (1936–1944)	Natural Sciences Museum, Madrid, Spain (ethanol)	*A. sturio*	–	–	na	na	na	na	*A. sturio*

GS	Ebro river (eighteenth century)	Gabinete Salvador collection of the Botanical Garden of Barcelona, Spain (stuffed)	*A. sturio*	–	–	7	na*	4	4	*A. oxyrinchus*
CEM-1	Gironde river (1995)	CEMAGREF, Bordeaux, France (frozen)	*A. sturio*	–	–	na	na	na	na	*A. sturio*
CEM-2	Gironde river (1995)	CEMAGREF, Bordeaux France (frozen)	*A. sturio*	–	–	na	na	na	na	*A. sturio*
MZUF5716	Livorno, Italy, Mediterranean region (1908)	Zoological Museum Observatory, Florencia, Italy (stuffed)	*A. sturio*	–	–	na	na	na	na	*A. sturio*
MZUF5714	Salerno, Italy, Mediterranean region (1886)	Zoological Museum Observatory, Florencia, Italy (stuffed)	*A. sturio*	+	na	na	na	na	na	*A. naccarii*
MZUF5704	Livorno, Italy, Mediterranean region (1879)	Zoological Museum Observatory, Florencia, Italy (stuffed)	*A. naccarii*	+	na	na	na	na	na	*A. naccarii*
MSNG40364	Genoa, Italy, Mediterranean region (1861)	Giacomo Doria Civic, Natural History Museum, Italy (stuffed)	*A. naccarii*	+	na	na	na	na	na	*A. naccarii*
MZUF5700	Chioggia, Italy, Adriatic region (1882)	Zoological Museum Observatory, Florencia, Italy (stuffed)	*A. naccarii*	+	na	na	na	na	na	*A. naccarii*
MZUF6472	Chioggia, Italy, Adriatic region (1882)	Zoological Museum Observatory, Florencia, Italy (stuffed)	*A. naccarii*	+	na	na	na	na	na	*A. naccarii*

*na, not analysed.

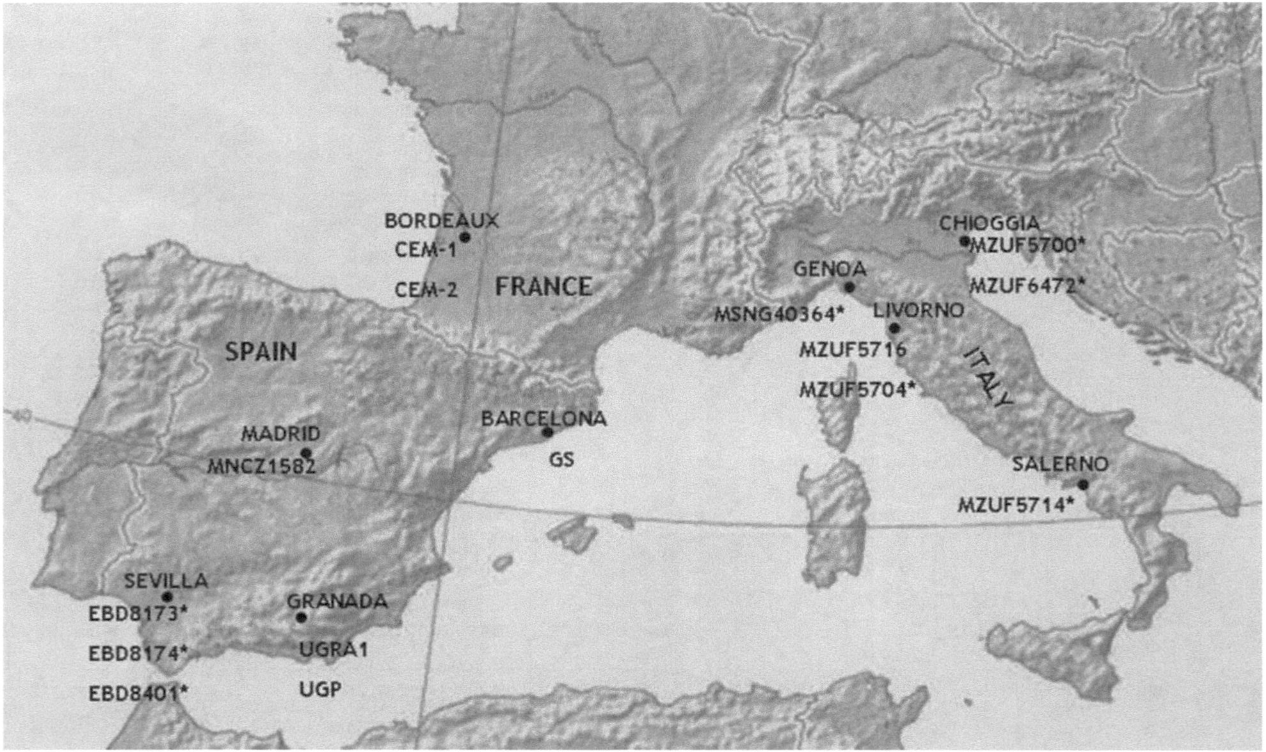

Fig. 2.4 Sampling location of the museum sturgeon specimen analysed. Asterisks indicate specimens having the *Hind*III sequences in their genomes

Fig. 2.5 Some of the museum specimens of sturgeon analysed. From top to bottom: one of the EBD specimens, the UGP specimen and the GS specimen

First, for each individual, we tested the specific status by analysing the presence/absence of a discriminative marker such as the *Hind*III satellite-DNA family. Further, for several selected individuals (according to availability and quality of DNA), we analysed the sequences of three nuclear-DNA markers (the satellite-DNA families *Hind*III and *Pst*I and the NTS spacers of the 5S genes) and two mitochondrial-DNA markers (two fragments, one of the cytochrome *b* and one of the 12S genes).

As mentioned above, sturgeon genomes are characterized by their extremely low rate of sequence evolution and, in addition to repetitive sequences, by their unusually low rate of intra-specific homogenization (and therefore, low rate of concerted evolution). However, we were testing for the specific status of specimens that had been morphologically ascribed to *A. sturio* or to *A. naccarii* or for which there was some controversy concerning assignments to the two species. These two species belong to two separate clades of those comprising the phylogeny of sturgeons. A major differentiation between these two clades is the absence of *Hind*III sequences within the clade to which *A. sturio* belongs and the presence of these types of sequences in the clade to which *A. naccarii* belongs. As observed above, for the remaining nuclear repetitive markers, there are enough levels of family homogeneity for species-diagnostic mutations which clearly differentiate the two species.

In fact, the *Hind*III family of repetitive sequences is present in all the sturgeon species analysed to date except in the case of *A. sturio* and *A. oxyrinchus* (De la Herrán et al., 2001a; Robles et al., 2004) and can therefore be used to discriminate between *A. naccarii* and *A. sturio*. This is a strong cladistic character that does not appear to be affected by differential population distribution (it should not be differentially present or absent in different *A. sturio* populations) because of the absence of *Hind*III sequences from *A. sturio* and *A. oxyrinchus*, two species forming a major clade within the sturgeon phylogeny which first split from the rest of the sturgeon species about 90 million years ago. Furthermore, the diversification between these two species occurred about 15–20 million years ago (Ludwig et al., 2002). In the case of the museum specimens analysed (Table 2.1), we verified by dot-blot hybridization the presence of this satellite-DNA sequence in each one (De la Herrán et al., 2004; new data in this chapter). The hybridization results revealed that the *Hind*III sequences are present within the genomes of three specimens from the Guadalquivir and preserved at the museum of the Biological Station of Doñana (Seville) as *A. sturio*. The same applies to some museum specimens classified as *A. naccarii* and caught in the Adriatic region and also to some specimens of this species caught outside this region, in Livorno and Genoa, as well as to some specimens classified as *A. sturio* from Salerno. On the other hand, we detected no such satellite DNA in other specimens from the Guadalquivir and the Ebro in Spain or from rivers in Western Italy. Of course, we have demonstrated that these DNA samples were in good condition by analysing other regions of the genome (they all showed positive hybridization when probed with ribosomal DNA). These data support the contention that specimens from the Guadalquivir at the Biological Station of Doñana as well as some other specimens classified as *A. sturio* have been misidentified. Furthermore, the specimens caught outside the Adriatic region and classified as *A. naccarii* might be ascribed to this species, according to our results.

Next, we developed a PCR assay for the amplification of *Hind*III, *Pst*I, and NTS repeats from a total of five museum specimens (Table 2.1). Four selected specimens were from the Guadalquivir: three of the specimens are Doñana Station (EBD specimens: EBD-8173, EBD-8174 and EBD-8401) having *Hind*III sequences (as denoted by the dot-blot assay) and one, at the museum of the Animal Biology Department in the University of Granada (specimen UGP), lacks *Hind*III sequences. The fifth animal was at the Botanical Garden of Barcelona from the Gabinete Salvador collection (specimen GS). This specimen was caught during the eighteenth century in Catalonian rivers, possibly in the Ebro, and it too lacks the *Hind*III sequences. Using PCR amplification, we found one *Hind*III monomeric unit from the specimen EBD-8173, one from the EBD-8174 and two from the EBD-8401. Comparing these with the *Hind*III sequences obtained previously from *A. naccarii* (De la Herrán et al., 2001a; EMBL accession numbers: Z499441 and AJ286564 to AJ286573), we found them to be closely similar to the sequences of this latter species. Specifically, mean genetic distances were 0.052 between the *A. naccarii* and the EBD*Hind*III sequences. This value is within the range of intraspecific variation of the *A. naccarii Hind*III sequences (De la Herrán et al., 2001a).

Further analyses were conducted with the *Pst*I sequences (De la Herrán et al., 2004; new data gathered for this chapter). This satellite DNA had been obtained previously from the genome of *A. naccarii*, and the dot-blot hybridization together with the PCR techniques had demonstrated that the *Pst*I satellite DNA sequences are found within the genomes of all sturgeon species belonging to the genera *Acipenser*, *Huso* and *Scaphyrhynchus* analysed to date, including *A. sturio* and *A. oxyrinchus* (Robles et al., 2004). We cloned a total of 13 *Pst*I monomeric units from the three EBD specimens and 13 from the other two specimens. When compared by phylogenetic procedures (Fig. 2.6) with the sequences of *A. naccarii* (EMBL accession numbers AJ550022 to AJ550026) and *A. sturio* (AJ550027 to AJ550031), the sequences isolated from the EBD specimens were clustered with those of *A. naccarii*, but differed completely from the clearly differentiated sequences of *A. sturio* (De la Herrán et al., 2004). The mean genetic distance between the *A. naccarii* and the EBD *Pst*I sequences is 0.140, a value within the range of intraspecific variation of the *A. naccarii Pst*I sequences (Robles et al., 2004). The mean distance between *Pst*I sequences of the EBD specimens and those of *A. sturio* is 0.292, similar to that between *A. naccarii* and *A. sturio* (0.252) (Robles et al., 2004). On the contrary, the *Pst*I sequences isolated from the other two specimens (UGP and GS specimens) clustered with the sequences of *A. sturio* and *A. oxyrinchus*.

A similar pattern was found with the 5S rRNA gene spacer sequences (De la Herrán et al., 2004; Robles et al., 2005). We compared nine spacer sequences of the EBD specimens with those of *A. naccarii* (EMBL accession numbers AJ550032 to AJ550040) and *A. sturio* (AJ550041 to AJ550047). The spacer sequences of the EBD specimens appeared clustered with *A. naccarii*, and the mean distances were 0.047 between the EBDs and *A. naccarii* and 0.170 between the EBDs and *A. sturio* (mean distance between *A. naccarii* and *A. sturio*: 0.130). We have no data for the NTS sequences of the UGP and the GS specimens.

Taken together, all these data support the idea that the specimens EBD-8173, EBD-8174, and EBD-8401 are not *A. sturio*. In fact, three nuclear markers revealed that these specimens may be *A. naccarii*. On the other hand, the other two specimens, one from the Guadalquivir (UGP specimen) and one from the Ebro (GS specimen) should belong to the clade composed by *A. sturio* and *A. oxyrinchus*.

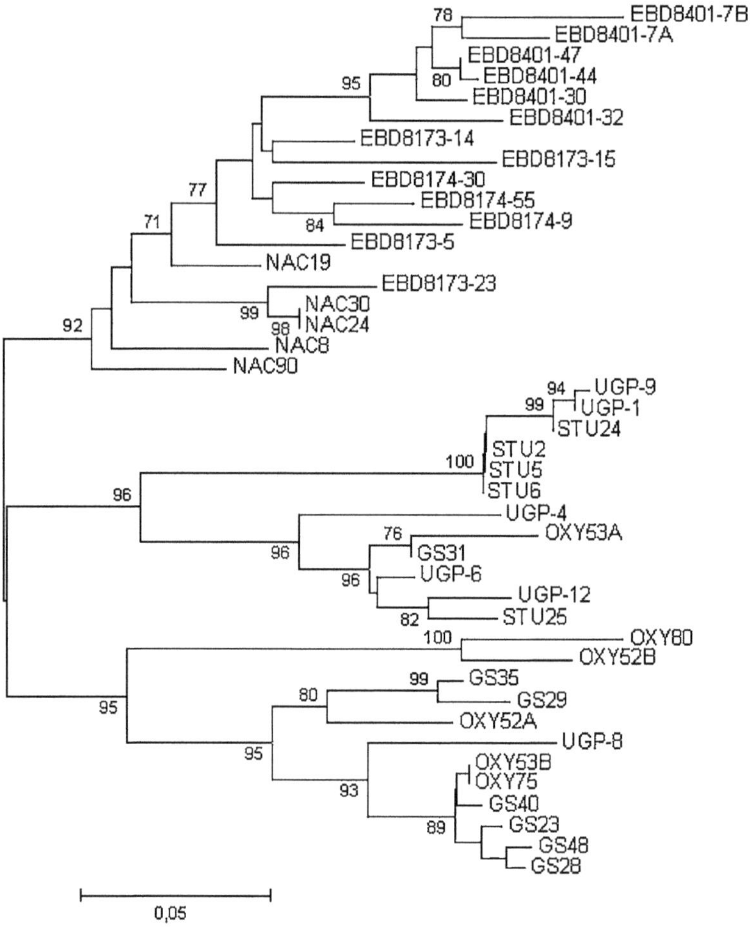

Fig. 2.6 Neighbour-joining tree depicting the close relationships between 13 *Pst*I sequences from the three EBD specimens and *Pst*I sequences of *A. naccarii* (NAC) and between six *Pst*I sequences from the UGP specimen, six sequences of the GS specimen and *Pst*I sequences of *A. sturio* (STU) and of *A. oxyrinchus* (OXY). Bootstrap values are shown for main nodes in the tree

2.4 Mitochondrial-DNA Sequences Confirm the Existence of Three Sturgeon Species (*A. sturio*, *A. naccarii*, and *A. oxyrinchus*) in Western Europe

Our study was completed with the analysis of two mitochondrial DNA regions: one of 212 bp of the cytochrome *b* gene that has 11 diagnostic nucleotide sites for the discrimination between *A. sturio* and *A. naccarii* (Ludwig and Kirschbaum, 1998) and four for the discrimination between *A. sturio* and *A. oxyrinchus*; and another of 139 bp of the 12S gene that differs for three nucleotide sites between *A. sturio* and

A. naccarii (Gasent-Ramírez et al., 2001) and four for the discrimination between *A. sturio* and *A. oxyrinchus*. We obtained both mitochondrial sequences by PCR amplification from the genomes of *A. naccarii*, *A. sturio*, and *A. oxyrinchus*. These sequences coincide with the same sequences found in the GenBank database (Accession numbers: AF283742, AF283730, AF004981, ANY12664, AF308923; AF006154; AJ245838; AF402894; AF125599). Both fragments are diagnostic also for these three species when compared with the rest of the sturgeon species.

Using PCR, we amplified these two mitochondrial DNA fragments from the EBD, the UGP, and the GS specimens (De la Herrán et al., 2004; new data in this chapter). Among the EBD individuals both mitochondrial markers classified two specimens (EBD-8173 and EBD-8401) as *A. naccarii* and the third (EBD-8174) as *A. sturio*. The UGP had the *A. sturio* mitochondrial-DNA sequences, while the GS specimen had the *A. oxyrinchus* mitochondrial-DNA sequences (Fig. 2.7).

Therefore, for two nuclear satellite-DNA families (*Hind*III and *Pst*I) and the spacers of the 5S ribosomal genes, the three EBD specimens from the Guadalquivir had sequences similar to those of *A. naccarii*, but clearly differentiated from the sequences of *A. sturio*. By contrast, both mitochondrial markers (fragments from the cytochrome *b* and the 12S ribosomal RNA genes) classified two specimens (EBD-8173 and EBD-8401) as *A. naccarii* and the third (EBD-8174) as *A. sturio*. Thus, the mitochondrial sequences identified two specimens as *A. naccarii*, while the nuclear data ruled out *A. sturio*, as the sequences resembled those of species within the *A. naccarii* clade. Meanwhile, specimen EBD-8174 was classified as *A. sturio* according to mitochondrial evidence but not according to nuclear sequences, which were also similar to those of *A. naccarii*. The most plausible explanation for this difference is that the EBD-8174 specimen is a hybrid between *A. sturio* and *A. naccarii* or the result of an introgression process. Hybrids have been described for sympatric sturgeon species (Arefjev, 1997) and introgression processes have been described recently in Adriatic sturgeons (Ludwig et al., 2003).

```
Cytb2F-3R

AJ549116-A. naccarii     ATTTCTTGCA ATACACTACA CAGCTGACAT TTCAACAGCC TTCTCCTCTG TTGCCCACAT CTGCCGAGAC GTAAATTACG GATGACTAAT  90
EBD8173-11               .......... .....C.... ....T..... .......... ........T. .T........ .........C .......... ..........
EBD8401-36               .......... .....C.... ....T..... .......... ........T. .T........ .........C .......... ..........
AJ549117-A. sturio       .......... .....T.... ....C..... .......... ........C. .C........ .........T .......... ..........
EBD8174-20               .......... .....T.... ....C..... .......... ........C. .C........ .........T .......... ..........
UGP-31                   .......... .....T.... ....C..... .......... ........C. .C........ .........T .......... ..........
AF308923-A. oxyrinchus   .......... .....T.... ....C..... .......... ........C. .C........ .........T .......... ..........
GS-11                    .......... .....T.... ....C..... .......... ........C. .C........ .........T .......... ..........


AJ549116-A. naccarii     CCGAAATATT CATGCAAACG GGGCCTCCTT CTTCTTTATC TGCTTGTACC TTCACGTAGC ACGAGGTATA TACTATGGTTC   171
EBD8173-11               .......... .......... .......... ......T..C .......... .......... ......T... .....T..T..
EBD8401-36               .......... .......... .......... ......T..C .......... .......... ......T... .....T..T..
AJ549117-A. sturio       .......... .......... .......T.. ......C..T .......... .......... ......C... .....C..C..
EBD8174-20               .......... .......... .......T.. ......C..T .......... ....T..... ......C... .....C..C..
UGP-31                   .......... .......... .......T.. ......C..T .......... .......... ......C... .....C..C..
AF308923-A. oxyrinchus   .......... .......... .......... ......C..T ...C...... .......G.. ...G..C... .....C..C..
GS-11                    .......... .......... .......... ......C..T ...C...... .......G.. ...G..C... .....C..C..
```

Fig. 2.7 Partial sequence alignment (171 bp) of the cytochrome *b* gene of *A. naccarii*, *A. sturio* and *A. oxyrinchus* compared with those sequences isolated from the EBD, UGP and GS specimens. GenBank/EMBL accession numbers for the sequences of *A. naccarii*, *A. sturio* and *A. oxyrinchus* are shown. Shaded nucleotide positions are diagnostic sites for genetic specific identification

From these results, it could be concluded that *A. sturio* and *A. naccarii* were sympatric and could hybridize in the Guadalquivir. Therefore, *A. naccarii* has lived and reproduced until quite recently in that river and we have gained molecular evidence demonstrating that this species has also been captured in other areas outside the Adriatic. Therefore, *A. naccarii* is not endemic to the Adriatic Sea, but rather has had a broader distribution in the Mediterranean.

On the other hand, our results also open up the possibility of the presence of *A. oxyrinchus* in Spanish rivers. Specifically, we found that a museum specimen caught during the eighteenth century in the Ebro was most probably a specimen of this species. This data agree with findings by Ludwig et al. (2002). *A. oxyrinchus* reportedly shifted its Atlantic distribution area going from America to Europe in response to climatic change, even displacing a resident species, *A. sturio*, during the so-called Little Ice Age, a fall in temperature by several degrees of magnitude that occurred in the Northern Hemisphere between the sixteenth and nineteenth centuries, an event that also affected the Iberian Peninsula and during which the Ebro froze seven times (Sala et al., 2001).

2.5 Consequences of These Findings for Taxonomy, Systematics and Genetic Conservation of Sturgeons

At the beginning of this chapter, we examined the geographic distribution of sturgeons around the world. However, today, with these fish disappearing worldwide, the application of molecular techniques is changing these ideas to a certain extent. For example, recently, the presence of *A. oxyrinchus*, a species from the North American side of the Atlantic Ocean, has been reported as having been found in the Baltic Sea during the Middle Ages (Ludwig et al., 2002). Two distinct mitochondrial clades have been recently found in one Po population of *A. naccarii* (the study examined current populations from the River Po in Italy and from the River Buna in Albania), possibly due to secondary ancient introgression of Black Sea *A. gueldenstaedtii* mitochondria in *A. naccarii* individuals (Ludwig et al., 2003). There are also observations in some European regions concerning the introgression of mitochondria in *A. sturio* individuals from rare species. For example, this is the case of *A. gueldenstaedtii* specimens in the Caspian Sea with mitochondria from three different species (*A. gueldenstaedtii*, *A. baerii* and *A. naccarii*) (Birstein et al., 1998, 2005; Jenneckens et al., 2001). Thus, it appears that it is not uncommon for sturgeon species to move and change their distribution areas, which is consistent with the general observation that most sturgeon species inhabit vast areas of continents and catchment basins (Choudhury and Dick, 1998).

In fact, our studies demonstrate the presence of *A. naccarii* and *A. oxyrinchus* in the Iberian Peninsula, two apparently rare species for that region. *A. oxyrinchus* reportedly shifted its Atlantic distribution area, going from America to Europe in response to climatic change, even displacing a resident species, *A. sturio* (Ludwig et al., 2002). This observation could be in agreement with the observations made by our group in the Ebro. While, in this latter case (*A. oxyrinchus*), no evidence of cross-breeding between the two species has been reported, our data reveal that, in the Guadalquivir, *A. sturio* and *A. naccarii* were sympatric and could hybridize.

Our findings imply that the distribution centre of *A. naccarii* would be the Mediterranean Sea and its historic distribution area reached from the Iberian Peninsula to the Adriatic Sea. Also, in such large distribution areas, when a group of organisms approaches extinction, as in the case of the sturgeons under study, remnant populations along the border of the distribution area often survive. This may explain why *A. naccarii* currently survives in greatest relative abundance in the Adriatic–that is, this sea is probably one of the limits of its distribution rather than the centre.

In scientific terms, our studies demonstrate the presence of *A. naccarii* and *A. oxyrinchus* in the Iberian Peninsula and clarifies the distribution of sturgeon species in the Western Mediterranean area, and underscores the need for caution in genetic studies using mitochondrial or nuclear DNA exclusively, as results may be misleading (Garrido-Ramos et al., 1997; Almodóvar et al., 2000; Gasent-Ramírez et al., 2001) for species which, as in the case of sturgeons, are prone to interspecific hybridization.

In the applied sphere, the genetic identification of sturgeon specimens that inhabited the rivers of the Western Mediterranean region has key implications in recovery plans. Sturgeons are almost extinct in all European rivers and most wild populations are not large enough to serve as a source for reintroduction programmes. However, aquaculture can support the recovery plans with reinstated individuals. In fact, individuals from farms have been used for conservation since the middle of the nineteenth century in Russia and EEUU (Billard, 2000). In addition to these two countries, recovery plans using sturgeons from fish farms are currently being implemented in Italy (*A. naccarii*), France (*A. sturio*) (Williot et al., 2001; and the chapter by Williot, this volume), Germany (*A. sturio* and *A. oxyrinchus*; Kirschbaum et al., this volume) and Poland. Our study demonstrates that *A. naccarii* has recently coexisted with *A. sturio*, from the Adriatic Sea to the Iberian Peninsula, a finding that should be taken into account in future sturgeon recovery programmes in Western Europe. In fact, the farming of *A. naccarii* is more advanced than that of *A. sturio* (Williot et al., 2001), and therefore the former could be used as an alternative or complement to the latter for certain recovery programmes in the wild, and of course for commercial production of sturgeons (meat and caviar) in Western Europe.

2.6 Conclusions

In conclusion, a broader vision of the distribution areas of sturgeon species results from the new molecular techniques to test the genetic identification of sturgeon specimens that inhabited the continental and coastal waters of the Northern Hemisphere. These techniques, if applied to museum specimens, could improve our knowledge of the sturgeons available now in those waters and could change our views of the morphological identifications made in the past. Our findings can be summarized as follows:

1. *A. naccarii* is not endemic to the Adriatic region.
2. *A. naccarii* has had a broader distribution area than the Adriatic Sea, extending throughout the Western Mediterranean.
3. *A. naccarii* is autochthonous to the Iberian Peninsula and has been sympatric with *A. sturio,* the other autochthonous Iberian species.
4. The presence of *A. oxyrinchus* is also seen in Spanish rivers (River Ebro). This data suggest that *A. oxyrinchus* shifted its Atlantic distribution area, going from America to Europe in response to climatic change.
5. Sturgeons are almost extinct in all European rivers, and most wild populations are not large enough to serve as a source for recovery programs. The genetic identification of sturgeon species that inhabited Western European rivers has key implications in recovery plans. Aquaculture can supply individuals for reinstatement, as is successfully being undertaken with *A. naccarii,* an autochthonous species of the Iberian Peninsula.

Acknowledgement We are highly indebted to Dr. Alfonso Susanna, Director of the Botanic Institute of Barcelona, for providing material of the Gabinete Salvador sturgeon specimen. This research has been supported by grant number AGL2003-04716 from the Ministerio de Educación y Ciencia.

References

Almodóvar A, Machordom A and Suárez J. (2000) Preliminary results from characterization of the Iberian Peninsula sturgeon based on analysis of the mtDNA cytochrome b. *Boletín Instituto Español de Oceanografía* 16: 17–27.

Arefjev VA. (1997) Sturgeons hybrids: natural reality and practical prospects. *Aquaculture Magazine* 7/8: 52–58.

Arnason U and Widegren B. (1986) Pinniped phylogeny enlightened by molecular hybridization using highly repetitive DNA. *Molecular Biology and Evolution* 3: 356–365.

Arnason U, Grétarsdóttir S and Widegren B. (1992) Mystecete (baleen whale) relationships based upon the sequence of the common cetacean DNA satellite. *Molecular Biology and Evolution* 9: 1018–1028.

Artyukhin EN. (1995) On biogeography and relationships within the genus Acipenser. *Sturgeon Quercus* 3: 6–8.

Billard R. (2000) Esturgeons et caviar. In: Lavoisier (ed.), *Rapport et Contract Life.* B4-3200/94/750, Esturgeons, Paris, 298 pp.

Birstein VJ and DeSalle R. (1998) Molecular phylogeny of Acipenserinae. *Molecular Phylogenetics and Evolution* 9: 141–155.

Birstein VJ and Doukakis P. (2000) Molecular analysis of *Acipenser sturio* L., 1758 and *Acipenser oxyrinchus* Mitchill, 1815: a review. *Boletín Instituto Español de Oceanografía* 16: 61–73.

Birstein VJ, Doukakis P, Sorkin B et al. (1998) Population aggregation analysis of three caviar-producing species of sturgeons and implications for the species identification of black caviar. *Conservation Biology* 12: 766–775.

Birstein VJ, Doukakis P and DeSalle R. (2002) Molecular phylogeny of Acipenseridae: nonmonophyly of Scaphirhynchinae. *Copeia* 2: 287–301.
Birstein VJ, Ruban G, Ludwig A et al. (2005) The enigmatic Caspian Sea Russian sturgeon: How many cryptic forms does it contain? *Systematics and Biodiversity* 3: 203–218.
Capello FB. (1869) Catalogo dos peixes do Portugal que existem no Museo de Lisboa. *Jorn. Sci. Math. Phys. Nat.* Jornal de Sciencias Mathematicas, Physicas e Naturaes da Academia Real das Sciencias de Lisboa. (1ª sér) 2: 131–193.
Choudhury A and Dick TA. (1998) The historical biogeography of sturgeons (Osteichthyes: Acipenseridae): a synthesis of phylogenetics, palaeontology and palaeography. *Journal of Biogeography* 25: 623–640.
De la Herrán R, Fontana F, Lanfredi M et al. (2001a) Slow rates of evolution and sequence homogenization in an ancient satellite DNA family of sturgeons. *Molecular Biology and Evolution* 18: 432–436.
De la Herrán R, Ruiz Rejón C, Ruiz Rejón M et al. (2001b) The molecular phylogeny of the Sparidae (Pisces, Perciformes) based on two satellite DNA families. *Heredity* 87: 691–697.
De la Herrán R, Robles F, Martínez-Espín E et al. (2004) Genetic identification of western Mediterranean sturgeons and its implication for conservation. *Conservation Genetics* 5: 545–551.
Dover G. (1986) Molecular drive in multigene families: how biological novelties arise, spread and are assimilated. *Trends in Genetics* 2: 159–165.
Elder Jr JF and Turner BJ. (1994) Concerted evolution at the population level: pupfish *Hind*III satellite DNA sequences. *Proceedings of the Natural Academy of Sciences USA* 9: 994–998.
Elder Jr JF and Turner BJ. (1995) Concerted evolution of repetitive DNA sequences in eukaryotes. *The Quarterly Review of Biology* 70: 297–320.
Fontana F. (2002) A cytogenetic approach to the study of taxonomy and evolution in sturgeons. *Journal of Applied Ichthyology* 18: 226–233.
Fontana F, Lanfredi M, Kirschbaum F et al. 2008. Comparison of karyotypes of *Acipenser oxyrinchus* and *A. sturio* by chromosome banding and fluorescent *in situ* hybridization. *Genetica* 132: 281–286.
Franck JPC, Kornield I and Wright JM. (1994) The utility of SATA satellite DNA sequences for inferring phylogenetic relationships among the tree major genera of tilapiine fishes. *Molecular Phylogenetics and Evolution* 3: 10–16.
Garrido-Ramos MA, Jamilena M, Lozano R et al. (1995) Phylogenetic relationships of the Sparidae family (Pisces, Perciformes) inferred from satellite-DNA. *Hereditas* 122: 1–6.
Garrido-Ramos MA, Soriguer C, De la Herrán R et al. (1997) Morphometric and genetic analysis as proof of the existence of two sturgeon species in the Guadalquivir river. *Marine Biology* 129: 33–39.
Garrido-Ramos MA, De la Herrán R, Jamilena M et al. (1999) Evolution of centromeric satellite DNA and its use in phylogenetic studies of Sparidae family (pisces, Perciformes). *Molecular Phylogenetics and Evolution* 12: 200–204.
Gasent-Ramírez JM, Godoy JA and Jordano P. (2001) Identificación de esturiones procedentes del Guadaluivir mediante análisis de ADN en especímenes de museo. Medio Ambiente. *Publicaciones de la Consejería de Medio Ambiente de la Junta de Andalucía* 36: 44–49.
Gonçalves BC. (1942) Colcçaõ oceanográfic de D. Carlos I. Peixes *Trav. Stn. Biol. Marit. Lisb.* Travaux de la Station de biologie Maritime de Lisbonne. 46: 1–108.
Grétarsdóttir G and Arnason U. (1992) Evolution of the common cetacean highly repetitive DNA component and the systematic position of *Orcaella brevirostris. Journal of Molecular Evolution* 34: 201–208.
Hartley SE and Davidson WS. (1994a) Distribution of satellite DNA sequences isolated from Arctic char; *Salvelinus alpinus*, in the genus *Salvelinus. Canadian Journal of Fisheries and Aquatic Sciences* 51: 277–283.
Hartley SE and Davidson WS. (1994b) Characterization and distribution of genomic repeat sequences from arctic char (*Salvelinus alpinus*). In: Beamont AR (ed.), *Genetics and Evolution of Aquatic Organisms*. Chapman and Hall, London, pp. 271–279.

Heikkinen E, Launonen V, Muller E et al. (1995) The pvB370 *BamHI* satellite DANN family of the *Drosophila viridis* group and the evolutionary relation to mobile dispersed genetic pDv elements. *Journal of Molecular Evolution* 41: 604–614.

Jenneckens I, Meyer JN, Hörstgen-Schwark G et al. (2001) A fixed allele at microsatellite LS-39 is characteristic fort he black caviar producer *Acipenser stellatus. Journal of Applied Ichthyology* 17: 39–42.

Kirschbaum F, Wuertz S, Williot P et al. (2008) Prerequisites for the restoration of the European Atlantic sturgeon, *Acipenser sturio* and the Baltic sturgeon (*A. oxyrinchus x A. sturio*) in Germany. In: Carmona R et al., editors. Biology, Conservation and Sustainable Development of Sturgeons. Berlin: Springer. p 385–402.

Krieger J and Fuerst PA. (2002) Evidence for a slowed rate of molecular evolution in the Order Acipenseriformes. *Molecular Biology and Evolution* 19: 891–897.

Krieger J, Fuerst PA and Cavender T. (2000) Phylogenetic relationships of North American sturgeons (order Acipenseriformes) based on mitochondrial DANN sequences. *Molecular Phylogenetics and Evolution* 16: 64–72.

Lanfredi M, Congiu L, Garrido-Ramos MA et al. (2001) Chromosomal location and evolution of satellite DNA family in seven sturgeon species. *Chromosome Research* 9: 47–52.

Ludwig A and Kirschbaum F. (1998) Comparison of mitochondrial DNA sequences between the European and the Adriatic sturgeon. *Journal of Fish Biology* 52: 1289–1291.

Ludwig AN, Jenneckens I, Debus L et al. (2000) Genetic analyses of archival specimens of the Atlantic sturgeon *Acipenser sturio* L., 1758. *Boletín Español de Oceanografia* 16: 181–190.

Ludwig A, Belfiore NM, Pitra C et al. (2001) Genome duplication events and functional reduction of ploidy levels in sturgeon (*Acipenser, Huso* and *Scaphirhynchus*). *Genetics* 158: 1203–1215.

Ludwig A, Debus I, Lieckfeldt D et al. (2002) When the American sea sturgeon swam east. *Nature* 419: 447–448.

Ludwig A, Congiu L, Pitra C et al. (2003) Nonconcordant evolutionary history of maternal and paternal lineages in Adriatic sturgeon. *Molecular Ecology* 12: 3253–3264.

Magnin E and Beaulieau G. (1963) Étude morphométrique comparée de l'*Acipenser oxyrinchus* Mitchill du Saint-Laurent et l'*Acipenser sturio* Linné de la Gironde. *Naturaliste Canadian* 90: 5–38.

Martin AP. (1999) Substitution rates of organelle and nuclear genes in sharks: implicating metabolic rte (again). *Molecular Biology and Evolution* 16: 996–1002.

Martin AP, Naylor GJP and Palumbi SR. (1992) Rates of mitochondrial DNA evolution in sharks are slow compared to mammals. *Nature* 357: 153–155.

Martínez-Espín E, Martínez-González LJ, Álvarez JC, Roby RK, Lorente JA. (2008) Forensic strategies for DNA extraction of ancient and degraded sturgeon samples. In: Carmona R et al., editors. Biology, Conservation and Sustainable Development of Sturgeons. Berlin: Springer. p 85–96.

McDowall MR. (1999) Different kinds of diadromy: Different kinds of conservation problems. *ICES Journal of Marine Sciences* 56: 410–413.

Mestrovic M, Mravinac CB, Juan C et al. (2000) Comparative study of satellite sequences and phylogeny of five species from the genus *Polarus* (Insecta, Coleoptera). *Genome* 43: 776–785.

Mravinac B, Plohl M, Meštrović N et al. (2002) Sequence of PRAT satellite DNA 'frozen' in some *Coleopteran* species. *Journal of Molecular Evolution* 54: 774–783.

Murata S, Takasaki N, Saitoh M et al. (1993) Determination of the phylogenetic relationships among Pacific salmonids by using short interspersed elements (SINEs) as temporal landmarks of evolution. *Proceedings of the Natural Academy of Sciences USA* 90: 6995–6999.

Ohta T and Dover G. (1984) The cohesive population genetics of molecular drive. *Genetics* 108: 501–521.

Porres A and Farnós A. (1999) Evolució al segle XX de les poblacions d'esturió (*Acipenser sturio*) al riu Ebre. In: Fernández-Colomé JV, Farnós A (eds.), *Els esturions (el cas del riu Ebre).* Generalitat de Catalunya, Tarragona, pp. 93–112.

Rico C, Rico I and Hewitt G. (1996) 470 million years of conservation of microsatellite loci among fish species. *Proceedings of the Royal Society of London Series B* 263: 549–557.
Rincón PA. (2000) Putative morphometric evidence of the presence of *Acipenser naccarii* Bonaparte, 1836 in Iberian rivers, or why ontogenetic allometry needs adequate treatment. *Boletín Instituto Español de Oceanografía* 16: 217–229.
Robles F, De la Herrán R, Ludwig A et al. (2004) Evolution of ancient satellite DNAs in sturgeon genomes. *Gene* 338: 133–142.
Robles F, De la Herrán R, Ludwig A et al. (2005) Evolution of 5S ribosomal genes in sturgeons. *Genome* 48: 1–11.
Ruiz Rejón M, De la Herrán R, Ruiz Rejón C et al. (2000) Genetic characterization of *Acipenser sturio* L., 1758 in relation to other sturgeon species using satellite DNA. *Boletín Instituto Español de Oceanografía* 16: 231–236.
Sala J, Montón Chiva E, Escrig Barberá J, et al (2001) Nuestro porvenir climático: un escenario de aridez?. Castelló de la Plana, Spain: Colleccio Athenea nº5, Publicacions de la Universitat Jaume I. 224 p.
Schweizer D and Loidl J. (1987) A model for heterochromatin dispersion and the evolution of C-band patterns. In: Hayman DL, Rofe RH, Sharp PJ (eds.), *Chromosomes Today*, vol. 9. Allen & Unwin, London, pp. 61–74.
Stepien CA, Kocher TD. (1997) Molecules and morphology in studies of fish evolution. In: Kocher TD, Stepien SD (eds.), *Molecular Systematics of Fishes*. Academic, San Diego, pp. 1–11.
Svetovidov AN. (1989) Acipenseridae. In: Whitehead PJP, Bauchot ML, Tortonese E (eds.), *Fishes of North-eastern Atlantic and Mediterranean*, 2nd edn. UNESCO, Paris, pp. 200–225.
Ugarković D and Plohl M. (2002) Variation in satellite DNA profiles—causes and effects. *EMBO J* 21: 5955–5959.
Van Den Busche RA, Baker RJ, Wichman HA et al. (1993) Molecular phylogenetics of *Stenodermatini* bat genera: congruence o data from nuclear and mitochondrial DNA. *Molecular Biology and Evolution* 10: 944–959.
Williot P, Rochard E, Rouault T and Kirschbaum F. (2008) *Acipenser sturio* recovery research actions in France. In: Carmona R et al., editors. Biology, Conservation and Sustainable Development of Sturgeons. Berlin: Springer. p 251–68.
Williot P, Sabeau L, Gessner J et al. (2001) Sturgeon farming in Western Europe: recent developments and perspectives. *Aquatic Living Resources* 14: 367–374.

Chapter 3
Morphological and Morphometric Characters in Sturgeon Taxonomy and Phylogeny

Ekaterina D. Vasil'eva

Abstract The variability of traditional morphological characters and craniological measurements, is considered in 19 multi- and low-chromosome sturgeon species, to evaluate the morphological divergence between two lineages with different ploidy levels and the role of morphological characters in taxonomy and phylogeny of Acipenserids. Craniological characteristics are demonstrated to be analogous to traditional morphological ones in their discrepancy with ploidy relations among sturgeon species, but displaying high diagnostic values for some sturgeon species of difficult identification.

Keywords Identification, morphology, phylogeny, sturgeons, taxonomy

3.1 Introduction

Morphological studies on sturgeons have a very long history. They began in the middle of the eighteenth century, when sturgeon species were first described by Linnaeus (1758), and are continuing today by several researchers studying the taxonomic status and morphological variability of different populations. As a result of this long history, a number of sturgeon species, as well as several subspecies and some taxa of superspecial taxonomic ranks, have been described. Many of these taxonomic names are considered synonyms now. The morphological phylogenetic hypothesis about only two groups of sturgeons within the family Acipenseridae currently remains intact. These groups are genera *Huso* and *Acipenser*, which some authors even consider as different Acipenserid subfamilies: Husinae and Acipenserinae (Findeis, 1997; Bemis et al., 1997). The main diagnostic characters of these genera are the shape of the mouth and differences in gill-membrane attachment (Sokolov and Berdichevski, 1989).

E.D. Vasil'eva
Zoological Museum of the Moscow State University, Bolshaya Nikitskaya, 6,
Moscow 125009, Russia

R. Carmona et al. (eds.) *Biology, Conservation and Sustainable Development of Sturgeons*,

Besides basal separation of *Huso huso* (Linnaeus, 1758) and *H. dauricus* (Georgi, 1775) in the genus *Huso*, some attempts have also been made to define the grouping of sturgeon species. For instance, Berg (1911) considered the genus *Acipenser* to be represented by three subgenera. The first of these was the subgenus *Lioniscus* Fitzinger et Heckel, 1836 with the only species *A. nudiventris* Lovetsky, 1828, characterized by continuous lower lip. The second one was the subgenus *Helops* Brandt et Ratzeburg, 1833 with the only species being *A. stellatus* Pallas, 1771, characterized by a greatly elongated sword-shaped snout. Later, Berg (1948) changed the name *Helops* into *Gladostomus* Holly, 1936. And the last subgenus according to Berg (1911, 1948) was *Acipenser* Linnaeus, 1758 sensu stricto, included the other sturgeon species with an interrupted lower lip and a moderately long, conical, blunt or pointed snout. A basically different conception about the general structure of sturgeons from the Black and Caspian Sea was presented by Antoniu-Murgoci (1942), who considered them to be represented by two subgenera, namely the subgenus *Euacipenser* Antoniu-Murgoci with two species (*A. ruthenus* Linnaeus, 1758 and *A. nudiventris*) and the subgenus *Acipenser* with three species (*A. sturio* Linnaeus, 1758, *A. gueldenstaedtii* Brandt, 1833 and *A. stellatus*).

The most splintered Acipenserid system was constructed by Artyukhin (1995). He divided the genus *Acipenser* into seven subgenera. Most of the names of these subgenera were presented earlier by different authors (subgenera *Gladostomus*, *Sturio* Rafinesque, 1810, *Sterleta* Güldenstädt, 1772, *Antaceus* Fitzinger et Heckel, 1836, *Sinosturio* Jaekel, 1929, and *Dinectus* Rafinesque, 1820), but the subgenus *Parasinosturio* with two American North Pacific species *A. medirostris* Ayres, 1854 and *A. transmontanus* Richardson, 1836 was created first by Artyukhin (1995). This scheme was the last attempt to present phylogenetic relations in sturgeons by using mainly morphological characters.

As opposed to morphological data, karyological investigations of Acipenserid fishes revealed them to be represented by two large groups of species with different chromosome numbers, namely species with about 120 chromosomes and species with 240–260 chromosomes. These groups were considered as separate but closely related phylogenetic lineages (Vasil'ev, 1985) which had diverged more than 80 million years ago. According to karyological data the low-chromosome branch was considered to be represented by two subgroups, namely species from the genus *Huso*, and low-chromosome species from the genus *Acipenser*. These subgroups were supposed to have diverged about 10 million years ago. Thus this scheme of phylogenetic relations assumed there was only polyploidization event among Acipenserid sturgeons.

Further studies of DNA content in different sturgeons, as well as investigations of different DNA markers, revealed more complicated phylogenetic relations between sturgeon species (Birstein et al., 1993, 1997, 1999; Birstein and DeSalle, 1998; Comincini et al., 1998; Tagliavini et al., 1999; Krieger et al., 2000; Ludwig et al., 2001; Fain et al., 2001; Robles et al., 2004, 2005). The resulting phylogenetic schemes differ somewhat in relations between particular species but all of them definitely support multiple polyploidization events in Acipenserids and reject the only two phylogenetic branches corresponding to the commonly accepted genera *Huso* and *Acipenser*. Moreover, the shortnose sturgeon *A. brevirostrum* was found to have 362–372 chromosomes and correspond to the hexaploid level (Kim et al., 2005). Thus, the traditional morphological taxonomic hypothesis for Acipenserid fishes accepted the taxa of mixed genetic configurations and ploidy appears to be incongruous with phylogeny in this group.

Standard sets of morphological characters used in taxonomic studies of sturgeons from different areas usually include meristics (such as the number of scutes, rays, and gill rakers), relative size and shape of the snout and mouth, position and morphology of barbels, lip shape, and certain features of the skin and scute morphology, as well as body coloration (Berg, 1948; Vladykov, 1955; Okada, 1959–1960; Vladykov and Greeley, 1963; Svetovidov, 1984; Sokolov, 1989). In several cases, some body proportions are used for species or subspecies identification (Vladykov and Greeley, 1963; Gröger and Debus, 2000; Elvira and Almodóvar, 2000; North et al., 2002) and some cranium structures for phylogenetic constructions (Artyukhin, 2000).

In view of the difficulty of developing a modern morphological tool corresponding to the results of genetic investigations, in the present study the variability of traditional morphological characters, and the newly developed system of craniological measurements (Vasil'eva, 1999, 2004), is contemplated within multi- and low-chromosome sturgeon species to elucidate any morphological divergence between these lineages and to evaluate the role of different morphological characters in the taxonomy and phylogeny of sturgeons.

3.2 Materials and Methods

The material studied included 19 species (Table 3.1), present among 242 specimens from the collections of the Zoological Museum of the Moscow State University, Zoological Institute of Russian Academy of Science (St-Petersburg), and Museé National d'Histoire Naturelle (Paris), as well as fresh sturgeons from the Piscifactoría de Sierra Nevada (Granada, Spain) and the Adygeyskii Sturgeon Hatchery (Krasnodar, Russia). All fishes were subjected to complete meristic (with the exception of stuffed or incomplete specimens) (Holčik et al., 1989) and craniological (Vasil'eva, 1999, 2004) analyses, as well as to the study of traditional morphological characteristics. The craniological analysis included skull measurements specially designed to study museum sturgeons, including stuffed or fluid-preserved specimens, as well as complete or incomplete ones, and this diagnostic approach proved effective for several species, even at the early stages of postlarval ontogeny (Vasil'eva et al., 2001). These measurements are: the distance between the tip of the snout and the angle of infraorbitale accessorium (Ri), the length of the dermocranium from the tip of the snout to the posterior edge of the dermosupraoccipital bone (D), the length of the complex of rostral and medial bones from the anterior edge of the first rostral to the posterior edge of the last medial (Mr), the distance between the anterior edge of the parietal and posterior edge of the dermosupraoccipital bone (Pds), the width of the dermocranium at the level of the end of the angle of infraorbitale accessorium (W), and the distance from the tip of the snout to the base of the barbels (B) (Fig. 3.1). Five craniological

Table 3.1 Materials examined: institutions, number of specimens (in parenthesis); and karyological characters of species

Species	Institution[a] and number of specimens	Karyotype[b]
Huso huso (L.)	K (18); ZMMU (15)	116 ± 4, 118 ± 3
H. dauricus (Georgi)	ZMMU (6)	120
Acipenser stellatus (Pallas)	K (29); ZMMU (10)	118 ± 2, 146 ± 6
A. sturio (L.)	ZIN (7), MNHN (23)	116 ± 4
A. oxyrinchus (Mitchill)	ZIN (3), MNHN (10)	99–112
A. ruthenus (L.)	ZMMU (10)	116 ± 4, 118 ± 2–4
A. nudiventris (Lovetzky)	ZIN (9), MNHN (1)	118 ± 2
A. schrenckii (Brandt)	ZMMU (8), MNHN (1)	238 ± 8
A. sinensis (Gray)	ZIN (2), MNHN (2)	264 ± 4
A. dabryanus (Dumeril)	ZIN (2), MNHN (2)	
A. mikadoi (Hilgendorf)	ZIN (5), MNHN (1)	
A. medirostris (Ayres)	ZIN (1)	249±8
A. baerii (Brandt)	ZMMU (15)	249±5, 246±10
A. brevirostrum (Lesueur)	ZIN (1), MNHN (4))	362–372
A. fulvescens (Rafinesque)	ZIN (2), MNHN (11)	262 ± 6
A. transmontanus (Richardson)	MNHN (1)	271 ± 2.5, 246 ± 10
A. naccarii (Bonaparte)	ZIN (5), MNHN (2), SN (4)	239 ± 7, 246 ± 8
A. gueldenstaedtii (Brandt et Ratzeburg)	ZMMU (4), ZIN (5), MNHN (10)	250 ± 8
A. persicus (Borodin)	ZMMU (8); MNHN (3)	258 ± 2

[a]Institutions: ZMMU, Zoological Museum of the Moscow State University; ZIN, Zoological Institute of Russian Academy of Science (St-Petersburg); MNHN, Museé National D'Histoire Naturelle (Paris); SN, the Piscifactoria de Sierra Nevada (Granada, Spain); K, the Adygeyskii Sturgeon Hatchery (Krasnodar, Russia).

[b]Data from Fontana and Colombo (1974); Fontana et al. (1974, 1997, 2001, 2004); Vasil'ev (1985); Ráb (1986); Pirogovskii et al. (1989); Sokolov and Vasil'ev (1989a,b); Vlasenko et al. (1989a); Yu et al. (1989); Birstein et al. (1997); Vaneenennaam et al. (1998); Chicca et al. (2002); Kim et al. (2005).

indices (%) were calculated on the basis of these measurements: Ri/D (Ri: D), Mr/D (Mr: D), Pds/Ri (Pds: Ri), W/Ri (W: Ri), and B/Ri (B: Ri).

For univariate and multivariate analyses, the mean values of the characteristics and their standard deviations and errors were calculated for each species sample. A cluster analysis employed for comparison of the species was applied to the matrix of indices mean values calculated previously for each of the species. The Euclidian distance was used as an estimate of differences, and UPGMA algorithm was used for graphic representation of the distance matrix (program Statistica for Windows 5.5).

3.3 Results and Discussion

The results of our investigations revealed most of the studied characteristics (both traditional and craniological ones) to be mosaic distributed within Acipenserid lineages with different chromosome numbers. Some characters were peculiar to only

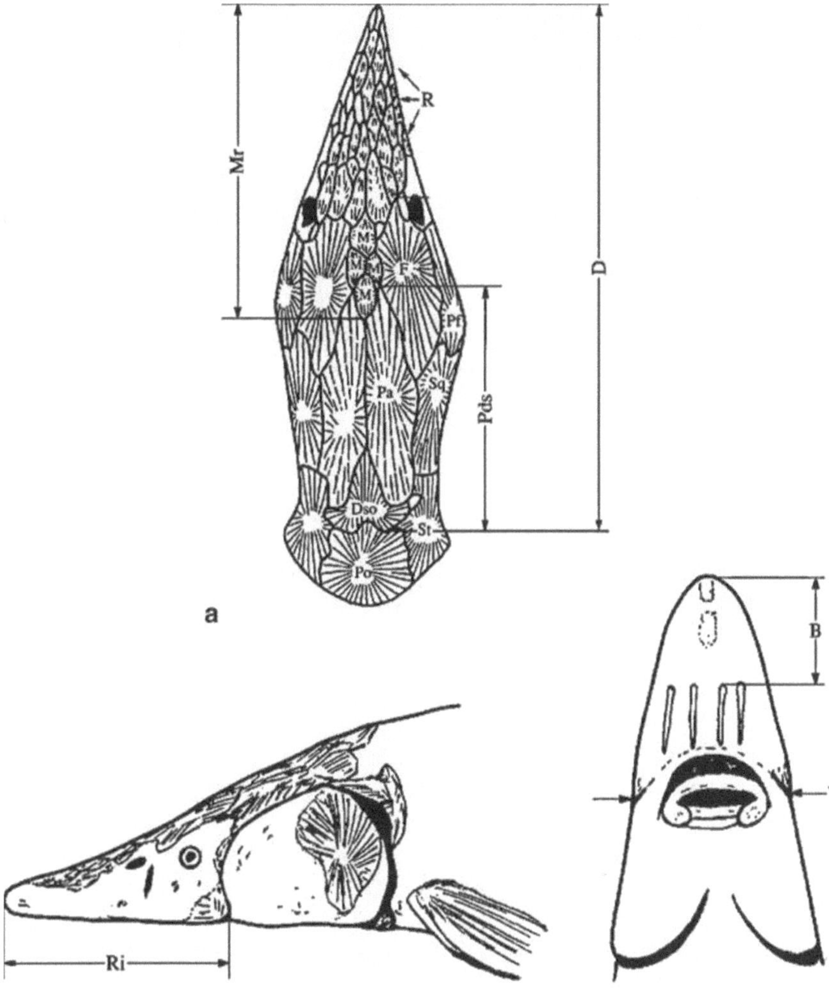

Fig. 3.1 The scheme of measurements of skull characters used in sturgeon study: a, dorsal view; b, lateral view; c, ventral view. D, Dermocranium length from the tip of rostrum to posterior edge of dermosupraoccipital bone (mm); Mr, length of the complex of rostral (R) and medial (M) bones (mm); Pds, distance from anterior end of parietal bone to posterior edge of dermosupraoccipital bone (mm); Ri, distance from the tip of rostrum to posterior corner of infraorbitale accessorium (mm); B, distance from the tip of snout to the base of barbells (mm); W, dermocranium width at the level of corners of infraorbitale accessorium (mm)

one or several species but not for any lineage as a whole. Simultaneously, the craniological characteristics appeared to be quite analogous to traditional morphological ones in their discrepancy with ploidy relations among sturgeon species: the similarity relations indicated by craniological characteristics also revealed some subclusters to be represented by both low- and multi-chromosome species (Figs. 3.2 and 3.3).

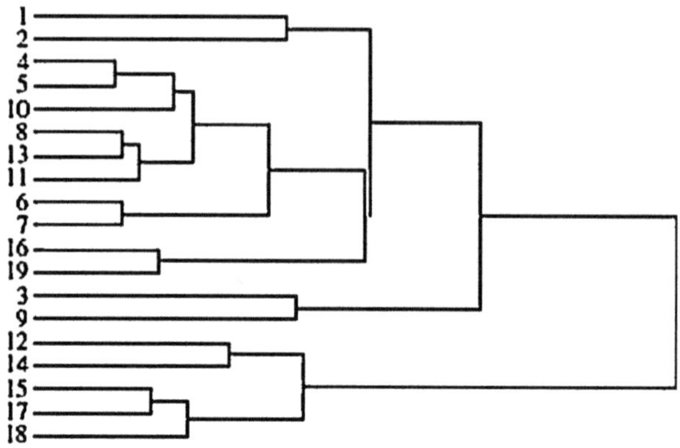

Fig. 3.2 Similarity among 19 sturgeon species with different ploidy by craniological indices as revealed by cluster analysis. Species with 2n about 120: (1) *Huso huso*, (2) *H. dauricus*, (3) *Acipenser stellatus*, (4) *A. sturio*, (5) *A. oxyrinchus*, (6) *A. ruthenus*, (7) *A. nudiventris*; species with 4n about 240: (8) *A. schrenckii*, (9) *A. sinensis*, (12) *A. medirostris*, (13) *A. baerii*, (15) *A. fulvescens*, (16) *A. transmontanus*, (17) *A. naccarii*, (18) *A. gueldenstaedtii*, (19) *A. persicus*; species with 6n about 360: (14) *A. brevirostrum*; species with uncertain ploidy: (10) *A. dabryanus*, (11) *A. mikadoi*

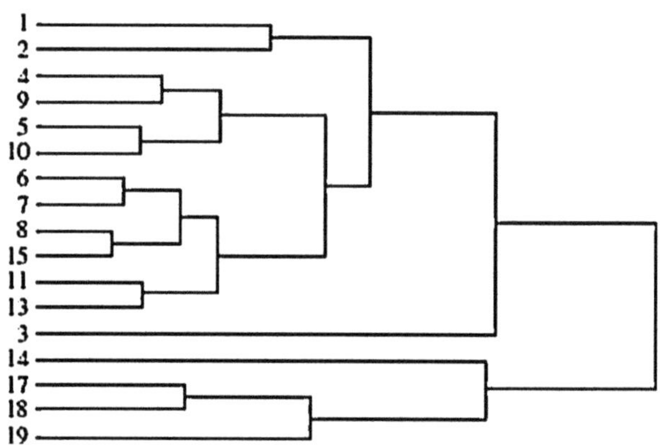

Fig. 3.3 Similarity among 17 sturgeon species represented by specimens with similar body lengths by craniological indices as revealed by cluster analysis. The species designations are the same as in Fig. 3.2

These results allowed two possible interpretations. First, multi- and low-chromosome lineages of sturgeons are homogeneous, but the characters studied (both traditional and newly formulated craniological ones) are useless for species separation. Second, multi- and low-chromosome lineages are really heterogeneous and include several groups of related species. The latter conclusion seems to be the only probable explanation according to zoogeographic, genetic, and ecological data and to the results of DNA investigations. This conclusion agrees with the working hypothesis elaborated by Vasil'ev (1999) to explain the polyploid evolution in this group of fishes. This hypothesis assumes the origin of multi-chromosome sturgeon species through hybridization between different pairs of low-chromosome sturgeons. In that case, even multi-chromosome species should be morphologically more similar to their parental low-chromosome species than to other multi-chromosome sturgeons.

In spite of the absence of direct correlation between morphological divergence and ploidy level of species, craniological characteristics have some advantage over traditional morphological ones. First, in craniological cladograms *H. huso* and *H. dauricus* do not form a separate cluster but are included in the same cluster with most of the other sturgeon species (Figs. 3.2 and 3.3). This result conforms to both karyological and molecular phylogenetic constructions in Acipenserid fishes.

Second, craniological indices demonstrate their high diagnostic values for some sturgeon species that are difficult to identify. For example, most experts are well aware of the problem of correct identification of Persian, *A. persicus* Borodin, 1897, and Russian, *A. gueldenstaedtii* Brandt et Ratzeburg, 1833, sturgeons. The main diagnostic characters of these species are the shape of the snout and the peculiar coloration (Vlasenko et al., 1989a,b), but these characteristics are subject to ontogenetic and environmental variability. As a result, some specimens or even some populations may be misidentified. Contrary to traditional diagnostic features, craniological characteristics allow distinguishing even young specimens of Russian and Persian sturgeons, reared under artificial conditions and devoid of typical coloration (Vasil'eva et al., 2001). The multivariate analysis of 12-month-old Russian sturgeons and six-month-old and 12-month-old Persian sturgeons, as well as their androgenetic nucleocytoplasmic hybrids and true hybrids of the same ages, by principal-component analysis based on craniological indices, demonstrates good separation of the two species mentioned above (Fig. 3.4). It was found that 12-month-old Russian and Persian sturgeons differ significantly in most of these characteristics. For example, the Pds/Ri index varied in four Russian sturgeons from 79.1% to 97.8% (mean value 87.5%), while in five Persian sturgeons the variation was from 68.9% to 78.3% (mean value 74.4%). Thus, this characteristic allows the separation of two species of the same size at a young age.

Moreover, according to craniological data *A. gueldenstaedtii* seems to be more closely related to *A. naccarii*, than to *A. persicus* (Fig. 3.3). This result contradicts the traditional morphological relations of these sturgeons but corresponds to the karyological and mtDNA data (Vasil'ev, 1985; Tagliavini et al., 1999; Birstein et al., 2000; Ludwig et al., 2001).

Another noteworthy result from craniological data involves the close relationships between the group including *A. gueldenstaedtii*, *A. naccarii*, and *A. persicus* and the American species *A. brevirostrum* Lesueur, 1818 (Fig. 3.3), despite their tetraploid–hexaploid relations. The same relations between *A. brevirostrum* and *A. gueldenstaedtii*—*A. persicus* group were defined by cytochrome *b* gene analyses (Birstein et al., 1997; Birstein and DeSalle, 1998; Ludwig et al., 2001; Robles et al., 2004) and they contradict the phylogenetic constructions prepared by Artyukhin (1995) on the basis of morphological characters.

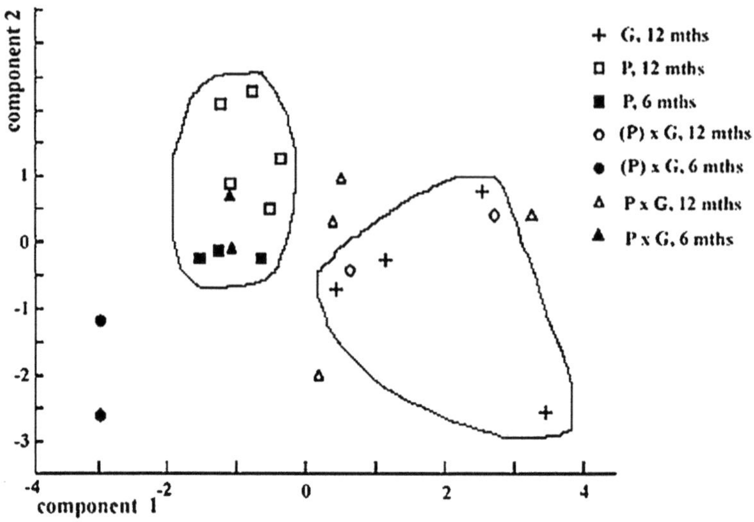

Fig. 3.4 Distribution of specimens of Persian (P) and Russian (G) sturgeons, their androgenetic, (P) × G, and true P × G hybrids of different ages in the space of the first two axes of multidimensional scaling (from Vasil'eva et al., 2001)

Despite certain advantages of the craniological characteristics for the recognition of sturgeon species, genetic analysis seems to be the only appropriate tool for the establishment of phylogenetic relations between them. At the same time, recent genetic methods need further development, primarily in the field of interpretation in relation to real evolutionary trends. Also, the recent state of knowledge needs precise genetic data taken from correctly identified sturgeon species. But correct identification of several morphologically similar sturgeons needs further development of morphological tools and close collaboration between geneticists and morphologists. Otherwise, genetic data may result in erroneous conclusions, such as 16-ploidy level of *A. mikadoi* (Birstein, 1993) or the existence of two sibling species within the Russian sturgeon, *A. gueldenstaedtii* (Birstein et al., 2000). In addition, the problem of the correct identification of specimens using the most up-to-date genetic analyses under recent conditions of occurrence and even the predominance of several exotic and introduced species among natural sturgeon populations (Hernando et al., 1999; Arndt et al., 2000; Keszka and Heese, 2003) and quite common interspecific hybridization in nature, as well as between cultural progenies (Podushka, 1999; Ilyasov, 2000) entering natural waters, exists.

3.4 Conclusions

1. Craniological characteristics appear to be quite analogous to traditional morphological ones in their discrepancy with ploidy relations among sturgeon species.

2. Multi- and low-chromosome sturgeon lineages are heterogeneous; both include several groups of related species.
3. Craniological indices demonstrate their high diagnostic value for some sturgeon species that are difficult to identify.
4. Genetic analysis seems to be the only appropriate tool for the establishment of phylogenetic relations among sturgeon species, but genetic studies need correct identification of materials and thus need further development of morphological diagnostic tools and close cooperation between geneticists and morphologists.

Acknowledgements I am extremely grateful to the Organizing Committee of the 1st International workshop on the biology, conservation and sustainable development of sturgeons in southern Europe (Bioresturgeons 2005), and its sponsors, for making it possible for me to participate in this meeting. This study was partially supported by the Russian Foundation for Scientific Schools (grant no. 2210.2008.4) and for Basic Research (project no. 07-04-00219-a).

References

Antoniu-Murgoci A. 1942. Contributions a l'étude des Acipensérides de Roumanie. *Ann Sci Univ Jassy* 28:289–385.

Arndt GM, Gessner J, Anders E, Spratte S, Filipiak J, Debus L, Scora K. 2000. Predominance of exotic and introduced species among sturgeons captured from the Baltic and North Seas and their watersheds, 1981–1999. Dokl [Symposium on Conservation of the Atlantic sturgeon *Acipenser sturio* L., 1758 in Europe, Madrid, 6–11 September 1999] *Bol Inst Esp Oceanogr* 16(1–4):29–36.

Artyukhin EN. 1995. On biogeography and relationships within the genus *Acipenser. Sturgeon Quart* 3(2):6–8.

Artyukhin EN. 2000. *Acipenser* taxonomy and geographical distribution of sturgeons. The International Conference: Sturgeons on the Threshold of the XXIst Century, Astrakhan, 11–15 September 2000. Book of Abstracts, Astrakhan, pp. 18–20 (in Russian).

Bemis WE, Findeis EK, Grande L. 1997. An overview of Acipenseriformes. In: Birstein VJ, Waldman JR and Bemis WE (eds.), *Sturgeon Biodiversity and Conservation.* Kluwer Academic Publishers, Dordrecht, pp. 25–71.

Berg LS. 1911. *Ryby (Marsipobranchii i Pisces). Fauna Rossii i sopredel'nykh stran.* 1. Sankt-Petersburg, 337 pp (in Russian).

Berg LS. 1948. *Freshwater Fishes of the USSR and Neighbouring Countries*, Part 1. Academy of Sciences USSR, Moscow-Leningrad, 466 pp (in Russian).

Birstein VJ. 1993. Sturgeons and paddlefishes: threatened fishes in need of conservation. *Conserv Biol* **7**:773–787.

Birstein VJ, DeSalle R. 1998. Molecular phylogeny of Acipenserinae. *Mol Phylogenet Evol* 9:141–155.

Birstein VJ, Poletaev AI, Goncharov BF. 1993. The DNA content in Eurasian sturgeon species determined by flow cytometry. *Cytometry* 14:377–383.

Birstein VJ, Hanner R, DeSalle R. 1997. Phylogeny of the Acipenseriformes: cytogenetic and molecular approaches. *Environ Biol Fish* 48:127–155.

Birstein VJ, Doukakis P, DeSalle R. 1999. Molecular phylogeny of Acipenseridae and black caviar species identification. *J Appl Ichthyol* 15:12–16.

Birstein VJ, Doukakis P, DeSalle R. 2000. Polyphyly of mtDNA lineages in the Russian sturgeon, *Acipenser gueldenstaedtii*: forensic and evolutionary implications. *Conserv Genet* 1:81–88.

Chicca M, Suciu R, Ene C, Lanfredi M, Congiu L, Leis M, Tagliavini J, Rossi R, Fontana F. 2002. Karyotype characterization of the stellate sturgeon, *Acipenser stellatus* by chromosome banding and fluorescent in situ hybridization. *J Appl Ichthyol* 18:298.

Comincini S, Lanfredi M, Rossi R, Fontana F. 1998. Use of rapid markers to determine the genetic relationships among sturgeons (Acipenseridae, Pisces). *Fish Sci* 64(1):35–38.

Elvira B, Almodóvar A. 2000. Morphology and taxonomy of the Atlantic sturgeon, *Acipenser sturio* from Spain. *Folia Zool* 49:221–230.

Fain SR, LeMay JP, Shafer JA, Hoesch RM, Hamlin BC, Straughan DJ. 2001. DNA sequence identification of sturgeon caviars traveling in World Trade. Fourth International Symposium on Sturgeon, Oshkosh, Wisconsin, 8–13 July 2001. Extended Abstracts, LE1.

Findeis EK. 1997. Osteology and phylogenetic interrelationships of sturgeons (Acipenseridae). In: Birstein VJ, Waldman JR and Bemis WE (eds.), *Sturgeon Biodiversity and Conservation*. Kluwer Academic Publishers, Dordrecht, pp. 73–126.

Fontana F, Colombo G. 1974. The chromosomes of Italian sturgeons. *Experientia* 30:739–742.

Fontana F, Janković D, Živkovoić S. 1974. Somatic chromosomes of *Acipenser ruthenus* L. *Arh Biol Nauka* 27:77–81.

Fontana F, Rossi R, Lanfredi M, Arlati G, Bronzi P. 1997. Cytogenetic characterization of cell lines from three sturgeon species. *Caryologia* 50:91–95.

Fontana F, Tagliavini J, Congiu L. 2001. Sturgeon genetics and cytogenetics: recent advancements and perspectives. *Genetica* 111:359–373.

Fontana F, Bruch RM, Binkowki FP, Lanfredi M, Chicca M, Beltrami N, Congiu L. 2004. Karyotype characterization of the lake sturgeon, *Acipenser fulvescens* (Rafinesque 1817) by chromosome banding and fluorescent in situ hybridization. *Genome* 47:742–746.

Gröger J, Debus L. 2000. Morphometric comparison of *Acipenser sturio* L. populations based on mixed estimation and morphometric measurements. *Arch Fish Mar Res* 48(2):175–193.

Hernando JA, Vasil'eva ED, Arlati G, Vasil'ev VP, Santiago JA, Belysheva-Polyakova L, Domezain A, Soriguer MC. 1999. New evidence for a wider historical area of two species of European sturgeons: *Acipenser naccarii* and *Huso huso* (Acipenseridae). *J Ichthyol* 39(9):803–806.

Holčik J, Bănărescu P, Evans D. 1989. General introduction to fishes. In: Holčik J (ed.), *The Freshwater Fishes of Europe*, vol. 1 (Pt. 2). AULA-Verlag, Wiesbaden, pp. 18–147.

Ilyasov YuI. 2000. Hybridization problems in sturgeon culture. The International Conference: Sturgeons on the Threshold of the XXIst Century, Astrakhan, 11–15 September 2000. Book of Abstracts, Astrakhan, p. 303 (in Russian).

Kim DS, Nam YK, Noh JK, Park CH, Chapman FA. 2005. Karyotype of North American shortnose sturgeon *Acipenser brevirostrum* with the highest chromosome number in the Acipenseriformes. *Ichthyol Res* 52:94–97.

Krieger J, Fuerst P, Cavender TM. 2000. Phylogenetic relationships of the North American sturgeons (order Acipenseriformes) based on mitochondrial DNA sequences. *Mol Phylogenet Evol* 16(1):64–72.

Keszka S, Heese T. 2003. Occurrence of exotic Russian sturgeons, *Acipenser gueldenstaedtii* Brandt et Ratzeburg, 1832 (Actinopterygii: Acipenseridae) in the Baltic Sea. *Acta Ichthyol Pisc* 33(2):173–177.

Linnaeus C. 1758. Systema Naturae. Ed. X (Systema naturae per regna tria naturae, secundum classes, ordines, genera, species, cum characteribus, differentiis, synonymis, locis. Tomus I. Editio decima, reformata.). Salvius, Holmiae. 824 p.

Ludwig A, Belfiore NM, Pitra C, Svirsky V, Jenneckens I. 2001. Genome duplication events and functional reduction of ploidy levels in sturgeon (*Acipenser*, *Huso* and *Scaphirhynchus*). *Genetics* 158:1203–1215.

North JA, Farr RA, Vescei P. 2002. A comparison of meristic and morphometric characters of green sturgeon *Acipenser medirostris*. 4. International Symposium on Sturgeon, Oshkosh, Wisconsin, 8–13 July 2001. *J Appl Ichthyol* 18(4–6):234–239.

Okada Y. 1959–1960. *Studies on the Freshwater Fishes of Japan*. Prefectural University of Mie, Tsu, 860 pp.

Pirogovski MI, Sokolov LI, Vasil'ev VP. 1989. *Huso huso* (Linnaeus, 1758). In: Holčik J (ed.), *The Fr eshwater Fishes of Europe*, vol. 1 (Pt. 2). AULA-Verlag, Wiesbaden, pp. 156–200.

Podushka CB. 1999. [The danger of non-controlled hybridization in sturgeon farming]. *Tez. Dokl. 1 nauchno-prakt. Konf. "Problemy sovremennogo tovarnogo osetrovodstva", 14–15 March, 1999*. Astrakhan:140 (in Russian).

Ráb P. 1986. A note on the karyotype of the starlet, *Acipenser ruthenus* (Pisces, Acipenseridae). *Folia Zool* 35:73–78.

Robles F, de la Herrán R, Ludwig A, Ruiz Rejón C, Ruiz Rejón M, Garrido-Ramos MA. 2004. Evolution of ancient satellite DNAs in sturgeon genomes. *Gene* 338:133–142.

Robles F, de la Herrán R, Ludwig A, Ruiz Rejón C, Ruiz Rejón M, Garrido-Ramos MA. 2005. Genomic organization and evolution of the 5S ribosomal DNA in the ancient fish sturgeon. *Genome* 48:18–28.

Sokolov LI. 1989. *Acipenser* Linnaeus, 1758. In: Holčik J (ed.), *The Freshwater Fishes of Europe*, vol. 1 (Pt. 2). AULA-Verlag, Wiesbaden, pp. 201–205.

Sokolov LI, Berdichevskii LS. 1989. Acipenseridae Bonaparte, 1831. In: Holčik J (ed.), *The Freshwater Fishes of Europe*, vol. 1 (Pt. 2). AULA-Verlag, Wiesbaden, pp. 150–153.

Sokolov LI, Vasil'ev VP. 1989a. *Acipenser nudiventris* Lovetsky, 1828. In: Holčik J (ed.), *The Freshwater Fishes of Europe*, vol. 1 (Pt. 2). AULA-Verlag, Wiesbaden, pp. 206–226.

Sokolov L.I., Vasil'ev V.P. 1989b. *Acipenser baeri* Brandt, 1869. In: Holčik J (ed.), *The Freshwater Fishes of Europe*, vol. 1 (Pt. 2). AULA-Verlag, Wiesbaden, pp. 263–284.

Svetovidov AN. 1984. Acipenseridae. In: Whitehead PJP, Bauchot M-L, Hureau J-C, Nielsen J and Tortonese E (eds.), *Fishes of the North-Eastern Atlantic and the Mediterranean*, vol. 1. UNESCO, Paris, pp. 220–225.

Tagliavini J, Conterio F, Gandolfi G, Fontana F. 1999. Mitochondrial DNA sequences of six sturgeon species and phylogenetic relationships within Acipenseridae. *J Appl Ichthyol* 15:17–22.

Vaneenennaam AL, Murray JD, Medrano JF. 1998. Mitotic analysis of the North-American White sturgeon, *Acipenser transmontanus* Richardson (Pisces, Acipenseridae), a fish with a very high chromosome number. *Genome* 41:266–271.

Vasil'ev VP. 1985. *Evolution Karyology of Fishes*. Nauka, Moscow, 300 pp (in Russian).

Vasil'ev VP. 1999. Polyploidization by reticular speciation in Acipenseriform evolution: a working hypothesis. *J Appl Ichthyol* 15:29–31.

Vasil'eva ED. 1999. Some morphological characteristics of Acipenserid fishes: considerations of their variability and utility in taxonomy. *J Appl Ichthyol* 15:32–34.

Vasil'eva ED. 2004. Morphological data corroborating the assumption of independent origins within octoploid sturgeon species. *J Ichthyol* 44 (Suppl. 1):63–72.

Vasil'eva ED, Grunina AS, Recoubratsky AV. 2001. The pattern of manifestation of some morphological characters in androgenetic nucleo-cytoplasmatic hybrids of the Persian *Acipenser persicus* and Russian *A. gueldenstaedtii* sturgeons in the postlarval period. *J Ichthyol* 41(6):454–460.

Vladykov VD. 1955. A comparison of Atlantic sea sturgeon with a new subspecies from the Gulf of Mexico (*Acipenser oxyrhynchus desotoi*). *J Fish Res Board, Canada* 12(5):754–761.

Vladykov VD, Greeley JR. 1963. Order Acipenseroidei. In: Olson YH (ed.), *Fishes of the Western North Atlantic*, Number 1, Part 3. Memoir Sears Foundation for Marine Research, New Haven, pp. 24–60.

Vlasenko AD, Pavlov AV, Sokolov LI, Vasil'ev VP. 1989a. *Acipenser gueldenstaedti* Brandt, 1833. In: Holčik J (ed.), *The Freshwater Fishes of Europe*, vol. 1 (Pt. 2). AULA-Verlag, Wiesbaden, pp. 294–344.

Vlasenko AD, Pavlov AV, Vasil'ev VP. 1989b. *Acipenser persicus* Borodin, 1897. In: Holčik J (ed.), *The Freshwater Fishes of Europe*, vol. 1 (Pt. 2). AULA-Verlag, Wiesbaden, pp. 345–366.

Yu X, Zhou T, Li Yu, Li K, Zhou M. 1989. *Chromosomes of Chinese Fresh-water Fishes*. Science Press, Beijing, China, 179 pp (in Chinese).

Chapter 4
Molecular Markers and the Study of Phylogeny and Genetic Diversity in North American Sturgeons and Paddlefish

Jeannette Krieger and Paul A. Fuerst

Abstract For a number of reasons, including their threatened or endangered status and importance in caviar production, much effort has been and is being expended worldwide on the study of the genetic variation of sturgeons and paddlefish. Presented here is a review of the genetic studies that have been conducted on the ten North American acipenseriform taxa and the types of molecular markers that have been used in these studies. The results that have been obtained from this research are invaluable for guiding conservation efforts by increasing our understanding of the relationships among species within the group, identifying intraspecific population structure and shedding light on acipenseriform life history traits.

Keywords Acipenseriformes, North America, sturgeon, paddlefish, nuclear DNA, mtDNA, molecular markers, phylogeny, population genetics

4.1 Introduction

Currently, most members of the order Acipenseriformes worldwide are considered threatened, endangered or, in some cases, even extinct. As a result, these fish have become the focus of research on their general biology, aimed at survival and recovery, including the study of their conservation and genetics. Within the area of conservation genetics, molecular markers play an important role in helping us understand sturgeons and paddlefish at many different levels. An appreciation of evolutionary relationships obtained from species divergence of genes among acipenseriform taxa can guide decisions about the focus of conservation efforts, while information about intraspecific genetic variability can identify populations that

J. Krieger, P.A. Fuerst
Departments of Evolution, Ecology and Organismal Biology and Molecular Genetics,
The Ohio State University, Columbus, Ohio 43210, USA

R. Carmona et al. (eds.) *Biology, Conservation and Sustainable Development of Sturgeons*,

should be managed as distinct stocks or provide information about populations that may be suitable genetic sources for restocking depauperate areas. The aim of this paper is to review the types of genetic studies that have been conducted on the ten acipenseriform taxa found in North America, describing the kinds of molecular markers that have been used in these studies and the results that have been obtained by using genetic approaches to study populations.

4.2 Rates of Molecular Evolution in the Acipenseriformes

Following genetic and karyotypic studies, it was proposed that sturgeons and paddlefish have a reduced rate of molecular evolution (Birstein and Vasiliev, 1987; Brown et al., 1996; see Krieger and Fuerst, 2002a for a review). Krieger and Fuerst (2002a) tested this hypothesis using relative-rate tests on sequences of 21 proteins and two ribosomal RNA (rRNA) genes to examine the rates of acipenseriform molecular evolution relative to that of the teleosts. Sequences from both the nuclear and mitochondrial genomes were included. Using two different relative rate tests (the two-cluster relative-rate test of Takezaki et al., 1995 and the nonparametric relative-rate test of Tajima, 1993), sturgeon/paddlefish sequences were compared with teleost sequences relative to sequences from an outgroup (either Chondrichthyes, Dipnoi, Petromyzontiformes or Rodentia). Seventy of 81 comparisons (86%) between individual taxonomic groups showed the acipenseriform sequence evolving more slowly than the teleost sequence. The rate of change in acipenseriforms may be almost 50% slower than in teleosts (Krieger and Fuerst, 2002a). Characteristics of sturgeons and paddlefish that have been proposed previously to account for slow rates of molecular evolution in other taxa include long generation time (Kohne, 1970; Li et al., 1987; Ohta, 1993; Mooers and Harvey, 1994; Li et al., 1996) and low metabolic rate (Martin and Palumbi, 1993; Martin, 1999). The relatively slow rate of molecular evolution in acipenseriforms has implications for efforts to differentiate between species, identify stocks within species or determine evolutionary relationships among species.

4.3 Molecular Phylogeny of North American Acipenseriformes

4.3.1 Nuclear 18S rRNA Gene

The classification of the group has, historically, been based on morphological characteristics. When molecular tools were applied in early DNA based efforts to clarify the phylogeny of sturgeons, the nuclear 18S rRNA gene was one of the first genes investigated. Birstein et al. (1997) determined partial 18S rRNA gene sequences

(230 base pairs or bp) in eight species of acipenseriforms, including four North American species. They found that these sequences provided little resolution of phylogenetic relationships because of a low degree of sequence divergence among species. Later, Krieger and Fuerst (2002b) examined the phylogenetic utility of the complete 18S rRNA gene in North American sturgeons and paddlefish. After sequencing the entire gene in the ten North American species, intraindividual variation of the 18S rRNA gene was discovered in all North American sturgeons, but not in paddlefish or in other non-acipenseriform species examined (gar, bowfin, bichirs, redfish and lungfish) (Krieger and Fuerst, 2002b). Sturgeons seem to have the normal structure of the rRNA transcription unit found in other eukaryotes. Three rRNA genes are found in a single transcription unit that is tandemly repeated in the genomes of eukaryotes, so that many copies of each gene are present in an individual. Despite the presence of multiple copies, intraindividual variation of the sequence of the 18S rRNA gene is an unusual phenomenon; a process called concerted evolution homogenizes repetitive genes both within an individual and a species (Arnheim et al., 1980). Sturgeons represent the first example of 18S rDNA intraindividual variation in a vertebrate, and Krieger and Fuerst (2002b, 2004) have proposed that the presence of extra rDNA arrays in the sturgeon genome due to a polyploid ancestry allowed the formation of multiple sequence variants in sturgeons.

The intraindividual variation of a few sturgeon species has been characterized. In early attempts, the 5 half of the gene from *Acipenser oxyrinchus oxyrinchus* and *A. transmontanus* and the entire gene from *A. brevirostrum* were studied with limited success. Subsequent efforts have focused on *A. fulvescens*. Thirty DNA clones containing full-length genes isolated from *A. fulvescens* were screened by sequencing and found to contain at least 17 different sequence variants (Krieger and Fuerst, 2002b, 2004). Both individual nucleotide substitutions and insertion/deletion (indel) events were observed in the isolated sequence variants.

In phylogenetic analyses, the sturgeon sequences form a monophyletic clade compared to paddlefish or other fishes. However, these sequences were not phylogenetically useful within sturgeons, since the intraindividual variation present in sturgeons affects the utility of this gene to distinguish species or identify species relationships. The sequence variants isolated from a particular sturgeon species do not always cluster together. Instead, they intermix with variants from other sturgeon species (see Fig. 4 in Krieger and Fuerst, 2002b). At higher taxonomic levels the acipenseriform sequence variants do retain accurate phylogenetic information. The lake sturgeon sequence variant that is most similar to the *Polyodon spathula* sequence and the sequence variant most divergent from the paddlefish sequence were both used to investigate the stability of the position of the order Acipenseriformes with the Actinopterygii. When one of these two sequence variants was used with the *P. spathula* and *Scaphirhynchus platorynchus* sequences to represent the order Acipenseriformes in phylogenetic analyses, the placement of the order was consistent in relation to other fish groups, regardless of which of the two lake sturgeon sequence variants was included. The inclusion of single lake sturgeon variants alone also produced the same placement for the order, showing Acipenseriformes clustered with the Polypteriformes (Krieger and Fuerst, 2002b).

The main question related to the phenomenon of intraindividual variation is why there are so many 18S rDNA sequence variants in sturgeons when most other species examined possess only one. One possibility is that some of the sequence variants are non-functional pseudogenes, while one or a few are functional sequences. Sequence comparison, phylogenetic analyses, and relative-rate analyses were conducted with the 17 18S rRNA gene sequences isolated from *A. fulvescens* to search for evidence that some variants may be pseudogenes. In addition, RNA analysis was used to examine which sequence variants are expressed in the fish. The results from these analyses provided some evidence supporting the pseudogene hypothesis, but were not conclusive in proving any of the variants to actually be pseudogenes. The question remains open.

Phylogenetic analysis of the lake sturgeon sequence variants and the single *P. spathula* sequence indicates the presence of two major clusters of sequences. Two lake sturgeon sequence variants were grouped with the paddlefish sequence, and were named paddlefish-like alleles, while 15 sequence variants clustered together, but separate from the paddlefish sequence, and were designated non-paddlefish-like alleles (see Fig. 1 in Krieger and Fuerst, 2004). The split of the sequence variants into two groups, with one group more closely related to the functional single paddlefish sequence, suggested that paddlefish-like alleles may be functional while non-paddlefish-like alleles may be pseudogenes.

Nevertheless, sequence comparison did not provide any obvious evidence that would clearly indicate pseudogene status for any of the lake sturgeon sequence variants (Krieger and Fuerst, 2004). None of the observed sequence variants have extensive regions of non-homology or indel events when compared to the paddlefish. The number of differences among the various lake sturgeon sequence variants ranges from 6 to 72. Variable sites are distributed relatively evenly throughout the gene, with proportionally fewer variable sites located in regions of the molecule that are essential for the formation of proper secondary structure and thus functionality. Most sites showing indel events are located in regions of the molecule that form hairpin and internal loops, regions that are considered less essential for secondary structure formation and, consequently, ribosome function. In addition, 63% of the variable sites found in the lake sturgeon variants are considered either minimally conserved or non-conserved sites in a generalized eukaryotic 18S rRNA secondary structure model map of site conservation (Cannone et al., 2002). Each of these observations suggests that overall nucleotide variation among lake sturgeon 18S rRNA gene variants has occurred in the presence of natural selection, and that the variants carry changes that have little or minimal effects on a functional molecule. However, it should be noted that the remaining 37% of variable sites occur at positions that are considered to be universally conserved, highly conserved, or conserved in the secondary structure site conservation model. Further examination of nucleotide sites in regions thought to be important for the normal function of the ribosomal small subunit in the ribosome during translation (A-site, P-site, E-site and helix 27) showed almost no changes in the sequence variants and compared to eukaryotic sequences that are consistent with a standard structural diagram of eukaryotic phylogenetic conservation, indicating sequence conservation in functional sites in the lake sturgeon variant alleles. Exceptions were observed at one position in the ribosomal A-site sequence and two positions in the P-site sequence, all three of which are considered universally conserved in eukaryotes. These latter variations were observed in the three lake sturgeon sequence variants that show the largest number of changes at universally conserved sites, making them the most probable candidates to be non-functional pseudogenes.

The results of the relative-rates test also gave some support to the pseudogene hypothesis, as did reverse transcription polymerase chain reaction (RT-PCR) analysis (Krieger and Fuerst, 2004). A relative-rate analysis was carried out to determine if the non-paddlefish-like alleles are under reduced selective constraints as compared to the paddlefish-like alleles, as would be expected if the non-paddlefish-like alleles are non-functional pseudogenes. The non-paddlefish like alleles were found to be evolving more quickly than the paddlefish-like alleles, and so appear to be under reduced selection pressure as expected of pseudogenes. Reverse transcription of lake sturgeon 18S rRNA followed by DNA sequencing of the RT-PCR product produced a single readable sequence, indicating that only one variant is expressed in the RNA of lake sturgeons in quantities large enough to be detected during sequencing. This result suggested that despite the large number of 18S rDNA alleles found in the lake sturgeon genome, there may be only one functional, expressed variant and that the other alleles may be unexpressed pseudogenes. The detected variant was identical in sequence to one of the clones isolated from the lake sturgeon genome, a paddlefish-like allele that differs by only one base from the paddlefish sequence. Although concerted evolution has failed to homogenize the numerous rDNA sequence variants found in sturgeon, expression of these variants appears to be under selective pressure.

In a more recent investigation (Krieger et al., 2006), intraindividual variation of the 18S rRNA gene has been found in all 14 Eurasian sturgeon species studied, but no variation was found in the Chinese paddlefish. The observation that intraindividual variation is present in sturgeons but absent in paddlefishes suggests that intraindividual variation arose within the Acipenseriformes after the divergence of the sturgeon and paddlefish families. Phylogenetic analysis of two polymorphic but readable segments of 18S rRNA gene sequence from 24 species of Acipenseriformes again provided no resolution of species relationships. However, patterns of 18S rDNA indel events observed within sturgeon species did correlate with proposed evolutionary relationships of certain species. For example, relevant to the North American species, *A. sturio* and *A. oxyrinchus* cluster together in molecular phylogenetic analyses (Birstein and DeSalle, 1998; Fontana et al., 2001; Ludwig et al., 2001; Birstein et al., 2002; Krieger et al., 2008). The presence of a unique indel event common to only *A. sturio*, *A. o. oxyrinchus* and *A. o. desotoi* supports the close relationship of these species. In addition, within the two readable polymorphic regions of the 18S rRNA gene that were analyzed for phylogenetic utility, there were three unique polymorphic sites shared by the three species. They also uniquely lacked variation at one other site where variation is found in all other species of *Acipenser* and *Huso*.

4.3.2 Mitochondrial DNA (mtDNA)

Although the search for nuclear markers remains an important focus for the study of sturgeon evolution, difficulties with identifying useful nuclear DNA sequences have been encountered. This appears to be due to the high ploidy level of sturgeons and the apparent slowed rates of nucleotide evolution. The lack of useful nuclear gene information means that most phylogenetic studies rely on mitochondrial DNA (mtDNA). The advantages of using mtDNA for phylogenetic studies in sturgeons and paddlefish are: (1) mtDNA is maternally inherited, essentially as a haploid molecule, and avoids potential problems caused by high ploidy levels in the nuclear genome; (2) animal mtDNA evolves more rapidly than nuclear DNA, and (3) complete mtDNA sequences are available from many fish species, including seven acipenseriform species (*P. spathula*, *Psephurus gladius*, *Scaphirhynchus albus*, *Huso huso*, *A. transmontanus*, *A. stellatus* and *A. dabryanus*), which facilitates the design of PCR and DNA sequencing primers. Many studies of acipenseriform phylogeny based on mitochondrial DNA sequences have been carried out on various subsets of species (for example: Brown et al., 1996; Birstein and DeSalle, 1998; Tagliavini et al., 1999; Krieger et al., 2000; Zhang et al., 2000; Ludwig et al., 2001; Birstein et al., 2002). However, the focus of this paper is on North American species, so the discussion will be limited to two studies that examined solely North American species. The sequence of the mitochondrial DNA control region and whole mtDNA restriction fragment length polymorphisms (RFLPs) were examined for their phylogenetic utility by Brown et al. (1996) in four species of sturgeon. They found that *A. transmontanus* and *A. medirostris* were the most closely related of the group examined, with *A. fulvescens* branching next, followed by *A. oxyrinchus*. These early results were consistent with a subsequent larger phylogenetic study by Krieger et al. (2000) that included all North American acipenseriform species and was based on four complete combined mitochondrial gene sequences (12S rRNA, cytochrome *c* oxidase subunit II, $tRNA_{Asp}$ and $tRNA_{Phe}$ genes). In the latter study, analyses using neighbor-joining, maximum parsimony, and maximum likelihood all produced similar topologies. Sister-species relationships were identified between *A. fulvescens* and *A. brevirostrum*, two eastern North American species, and between *A. transmontanus* and *A. medirostris*, two western North American species. The two *A. oxyrinchus* subspecies (*A. o. oxyrinchus* and *A. o. desotoi*), also identified as sister taxa, were very similar, separated by only three nucleotide differences in the four genes and were taxonomically located in a position basal to all other *Acipenser* species. However, on the basis of this study, it could not be determined which group held the most basal position within the North American Acipenseridae: the three species of *Scaphirhynchus* or the pair of *A. oxyrinchus* subspecies. In addition, nucleotide sequences for these four genes in all three *Scaphirhynchus* species were identical, providing no evidence for the separation of species with the genus. This tree topology agreed with that produced by a study in a larger number of sturgeon species of whole mitochondrial cytochrome *b* gene sequences (Ludwig et al., 2001), but differed slightly from the results of two studies of partial mitochondrial genes (also in larger numbers of sturgeon species) conducted by Birstein and DeSalle (1998) and

Birstein et al. (2002). In contrast to the finding that *A. fulvescens* and *A. brevirostrum* are sister species (Krieger et al., 2000), the results of Birstein and DeSalle (1998) and Birstein et al. (2002) suggested a distant relationship between lake and shortnose sturgeon. Recently, we have completed the most extensive study of sturgeon phylogeny, using data from eight mitochondrial genes (Krieger et al., 2008). In this latest study, *A. fulvescens* and *A. brevirostrum* are again identified as closely related, and the basal clade remains unresolved (Krieger et al., 2008).

4.4 Species Distinction in North America

In the United States, the amount of protection provided to a species under the Endangered Species Act is partially dependent upon its taxonomic status. Almost all North American acipenseriform species are either threatened or endangered, and there are two groups of North American sturgeons that include very closely related species or subspecies, one being *A. oxyrinchus*, with two subspecies, and the other being *Scaphirhynchus*, with three species. There has been controversy over the separate status of these taxa, prompting more focused genetic analysis of these fish.

4.4.1 A. oxyrinchus Subspecies

Gulf sturgeon (*A. o. desotoi*) and Atlantic sturgeon (*A. o. oxyrinchus*) have been considered separate subspecies, on the basis of differing habitat ranges and life history characteristics. The Atlantic sturgeon is found along the east coast of North America from the St. Lawrence river in Quebec to the St. Johns river in Florida, while the Gulf sturgeon is restricted to the Gulf of Mexico and its tributaries (Smith, 1985). Atlantic sturgeon mostly inhabit coastal marine waters, then return to natal rivers to spawn (Dovel and Berggren, 1983), while Gulf sturgeon live in their natal rivers for much of the year and only make brief wintertime excursions into the Gulf of Mexico (Wooley and Crateau, 1985). As yet, only one separating morphological character has been identified, i.e., relative spleen length (Wooley, 1985).

In the United States, the Gulf sturgeon is considered threatened, while the Atlantic sturgeon is not. Two genetic studies addressing the question of differentiation between Atlantic and Gulf sturgeons have been conducted. Ong et al. (1996) examined 203 bp of mitochondrial control region sequence, and King et al. (2001) studied seven microsatellite loci in the two taxa. Three fixed differences were detected between Gulf and Atlantic sturgeon mitochondrial control region sequences (Ong et al., 1996), while considerable divergence between the nuclear genomes of these two taxa was demonstrated with microsatellite DNA allele frequencies and diversity (King et al., 2001). The average genetic distance between the subspecies was determined to be two to three times that seen between

A. o. oxyrinchus populations (King et al., 2001). Therefore data from both mitochondrial and nuclear markers agree with the limited morphological evidence and with geographic distribution to support the subspecific status of the Atlantic and Gulf sturgeon and their separate management.

4.4.2 *Scaphirhynchus* Species

Shovelnose sturgeons (*S. platorynchus*) have been found throughout the Mississippi river basin and the Rio Grande river and are sympatric with the pallid sturgeon (*S. albus*) in the Missouri and Lower Mississippi rivers (to which the pallid sturgeon are restricted) (Lee, 1980a,b; Carlson et al., 1985), while the allopatric Alabama sturgeons (*S. suttkusi*) are found only in the Mobile river drainage (Burke and Ramsey, 1995). Morphological evidence supporting the presence of three separate species has been provided (Bailey and Cross, 1954; Williams and Clemmer, 1991; Mayden and Kuhadja, 1996), but their taxonomic distinction has been called into question, especially that of the sympatric shovelnose and pallid sturgeon.

Both the pallid and Alabama sturgeons are listed as endangered, and numerous studies have attempted to demonstrate genetic differentiation among the three *Scaphirhynchus* species. An allozyme study in shovelnose and pallid sturgeons by Phelps and Allendorf (1983) found the two species to be electrophoretically indistinguishable. Based on the banding patterns produced by PCR amplification with primers for three nuclear genes, Genetic Analyses, Inc. (1994) found greater differentiation between the Alabama sturgeon and the other two species than between the shovelnose and pallid sturgeon, and it was concluded that the latter two may not be distinct species. Campton et al. (2000) were able to find evidence distinguishing all three species by examining mitochondrial control region haplotypes and their frequencies, and also identified a unique substitution and haplotype that differentiated the Alabama sturgeon from the other taxa. Simons et al. (2001) conducted phylogenetic analyses of the mtDNA control region and cytochrome *b* gene sequences in all three species. Only samples of the Alabama sturgeon were recovered as a monophyletic group, and they found only low levels of sequence divergence among the species, which they interpreted to have been caused by a combination of a slow rate of molecular evolution and hybridization between shovelnose and pallid sturgeons. Based on data from six polymorphic microsatellite loci, Tranah et al. (2001) provided evidence that sympatric populations of pallid and shovelnose sturgeons at three different locations, chosen to represent the extreme of the species range, were genetically distinct from each other but that hybridization might be occurring. Recently, Tranah et al. (2004) provided additional genetic evidence from mitochondrial control region and microsatellite data that pallid and shovelnose sturgeons in the lower Mississippi river are reproductively distinct populations that are experiencing hybridization. Based on current knowledge, it appears that although Alabama and shovelnose sturgeons were initially considered conspecific (Chermock, 1955), the Alabama sturgeon is the most genetically distinct taxon in the group, while distinguishing between shovelnose and pallid sturgeon remains problematic (at least in some areas of their range) possibly due to hybridization. Future analysis of additional markers (most likely microsatellites) may be able to provide more evidence about the levels of differentiation between pallid and shovelnose sturgeons.

4.5 Population Genetic Studies Within North American Acipenseriform Species

Population studies on North American acipenseriform species examined stock structure, levels of genetic diversity within populations, evolutionary relationships among geographic populations, gene flow and homing fidelity. Most studies have utilized one or more of the four types of markers: allozymes, whole mtDNA RFLPs, mtDNA control region sequence polymorphism, and DNA microsatellites. At present, each of the North American species has been examined in at least one population study, in spite of the difficulty of collecting specimens, although sample sizes in some studies were small. Mitochondrial DNA RFLPs and control region sequence have been the most widely used markers to date. Recently, though, much effort has been devoted to developing disomic microsatellite markers for use in North American sturgeon and paddlefish species (May et al., 1997; McQuown et al., 2000; King et al., 2001; Pyatskowit et al., 2001; Heist et al., 2002; Henderson-Arzapalo and King, 2002; McQuown et al., 2002; Welsh et al., 2003; Welsh and McClain, 2004; Welsh and May, 2006). This is because of the great potential information that microsatellite markers can provide for population comparisons. Microsatellite development has been hampered in sturgeons because of the polyploid nature of the sturgeon genome. Many potential microsatellite markers have reduced utility because they appear as more than disomic, thus complicating the interpretation of genetic inheritance and genetic variation. Nevertheless, through much effort, a battery of new disomic microsatellite markers have been developed for use in sturgeons.

As a result, microsatellite markers are being increasingly used in sturgeon population studies; papers studying all the North American species have been published, reporting results obtained utilizing these nuclear markers. Varying levels of diversity and stock structuring have been detected with molecular markers in the different species. Each species and population has unique life histories and influences and must therefore be studied and considered individually when making conservation plans. However, some generalizations about each species are beginning to emerge.

Some aspects of the population genetics of sturgeons have been reviewed by Wirgin et al. (1997) and by Robinson and Ferguson (2004) in a recent comprehensive review of population genetic studies conducted up to 2002 on North American sturgeon and paddlefish. Some earlier results are summarized below, together with an update on results that have been reported since 2002.

4.5.1 Sturgeons of Eastern North America

Molecular studies have been extensively pursued to investigate the three species of *Acipenser* that are found in eastern North America. The shortnose sturgeon (*A. brevirostrum*) and the two subspecies of the Atlantic sturgeon (*A. o. oxyrinchus* and *A. o. desotoi*) are anadromous and found in the Atlantic Ocean. The lake sturgeon (*A. fulvescens*) occurs in freshwater habitats. All three have been studied, using both mitochondrial DNA and microsatellite analysis to determine the degree of population differentiation and structure, and the degree of homing fidelity for reproductive individuals.

4.5.1.1 *A. brevirostrum*

The shortnose sturgeon (*A. brevirostrum*) is currently found in 19 estuary systems along the east coast of North America, from the St. John river in New Brunswick to the St. Johns river in Florida (Scott and Crossman, 1973; Taubert, 1980). It is a semi-anadromous fish, in that it spends more time in the rivers than other anadromous species, such as the sympatric Atlantic sturgeon. It was once fished commercially, but is now considered endangered. Each of the 19 populations is considered a distinct management unit by the US National Marine Fisheries Service, the entity that is responsible for the protection and recovery of this species (U.S. National Marine Fisheries Service, 1996, 1998). However, prior to 1995 no morphological or genetic studies had been carried out to support the designation of the 19 populations as separate units. To remedy this deficiency, a series of population genetic studies using mtDNA control region sequences have been conducted on the shortnose sturgeon (Walsh et al., 2001; Grunwald et al., 2002; Quattro et al., 2002; Waldman et al., 2002; Collins et al., 2003; Wirgin et al., 2005). These investigations studied various numbers of shortnose sturgeon populations and individuals, but came to similar conclusions. Shortnose sturgeons show moderate to high haplotypic diversity, and large numbers of mitochondrial haplotypes are observed within populations, ranging from 15 to 30. There was no evidence of genetic bottlenecking, and larger than expected effective population sizes (Quattro et al., 2002) were observed for this endangered species. Different haplotype frequencies between populations lead to the conclusion that the shortnose sturgeon has a strong stock structure along the Atlantic coast. In some cases, different haplotypes (private haplotypes) were found in different populations, so that populations from most rivers were genetically distinct. For example, Grunwald et al. (2002) found significant genetic differences among all populations sampled from 11 rivers and estuaries on the east coast of North America, and five regional groupings of populations were identified. Low gene flow and high homing fidelity (at least for females) were indicated by limited haplotype sharing among populations and a large number of private haplotypes.

All the available evidence in the case of the shortnose sturgeon suggests the presence of distinct river-specific populations and supports the conservative strategy of separate management of most shortnose sturgeon populations (Walsh et al., 2001; Grunwald et al., 2002; Quattro et al., 2002; Waldman et al., 2002; Collins et al., 2003; Wirgin et al., 2005). Waldman et al. (2002) estimated the rate of migration in *A. brevirostrum*. Levels of migration were insufficient to break down population differences between drainage systems, and were found to be intermediate between levels seen in Gulf sturgeon (lowest) and Atlantic sturgeon (highest among the trio of taxa being compared).

4.5.1.2 *A. oxyrinchus*

The two subspecies of *A. oxyrinchus* provide an interesting comparison, with respect to management issues. The Atlantic sturgeon (*A. o. oxyrinchus*) is not considered endangered (even though its populations have been reduced tremendously compared to historical levels), while the Gulf sturgeon (*A. o. desotoi*) has been listed as threatened. Both subspecies have been studied using molecular methods, primarily by mitochondrial DNA analysis.

Bowen and Avise (1990) were the first to apply molecular techniques, using mtDNA RFLP analysis, to show that Atlantic and Gulf sturgeons showed significant genetic differences from one another and should be considered genetically distinct populations. They also noted that, compared to other fishes, *A. oxyrinchus* populations showed only low levels of genetic variability. Based on the levels of genetic variation, they estimated that the effective population size of the Gulf sturgeon was of the order of 50 individuals, a very small number and a level that could potentially lead to further rapid loss of genetic variation without careful stock management.

Miracle and Campton (1995) confirmed the low levels of variability in Gulf sturgeons, using mtDNA RFLPs and mt-control region sequences, suggesting that the subspecies has undergone a genetic bottleneck. Recently, Dugo et al. (2004) used DNA microsatellites to examine stock substructuring in the Gulf sturgeon. They found significant evidence of multiple genetic stocks, even within a single river drainage system, and suggested that there may be a more widespread east-west population structure between the rivers draining into the Gulf of Mexico. Waldman et al. (2002), using data from mt-control region sequences, estimated that Gulf sturgeon populations from eight river systems exchanged genes at rates substantially below the rates seen in Atlantic sturgeons.

Population structure in the Atlantic sturgeon has been investigated in several studies (Waldman et al., 1996a,b, 2002; Wirgin et al., 2000, 2002). The earliest three studies used mitochondrial haplotype analysis, while the last two incorporated microsatellite DNA analysis to study populations. Atlantic sturgeons showed substantial haplotype diversity in most populations, although no variation was detected in the northern populations of the St. Lawrence river and St. John river in Canada. Each of these studies came to the conclusion that Atlantic sturgeon populations were highly structured by the river drainage system, with substantial differentiation from north to south. Levels of gene flow (migration) between drainage systems were estimated to be greater, however, than those measured in equivalent populations of shortnose sturgeon from the same river systems (Waldman et al., 2002). Levels of migration were substantially greater in Atlantic sturgeons than in Gulf sturgeons.

4.5.1.3 *A. fulvescens*

The lake sturgeon (*A. fulvescens*) is one of the few species of the genus *Acipenser* that lives almost exclusively in freshwater, migrating between lakes and rivers. It has an extensive distribution, and can be found in the Great Lakes, Hudson/James Bay and Mississippi watersheds in North America. Although once abundant throughout its range, it is currently considered threatened and populations have been extirpated or severely reduced in some areas. Knowledge of the genetic diversity and stock structure of the species is invaluable to inform management decisions, especially in terms of restocking efforts in the rivers where there are still remnants of the original population.

Three early studies examined the population genetics of lake sturgeons using whole mtDNA RFLPs (Guénette et al., 1993; Ferguson et al., 1993; Ferguson and Duckworth, 1997). In contrast to the studies on the shortnose sturgeon, these studies found low genetic diversity, as only two or three haplotypes were detected, as well as little genetic divergence. There was also evidence of high gene flow, which is consistent with the life history of the species, as it migrates long distances and with the lack of population structure detected between or within drainages (Ferguson et al., 1993). The results of these studies did not support the separate management of lake sturgeon populations, although the marker system used (mtDNA RFLPs) did not possess the resolution to identify management units (Ferguson and Duckworth, 1997).

The desire for lake sturgeon conservation in the Great Lakes has prompted new study to reexamine lake sturgeon population genetics using more variable markers, especially microsatellite loci. McQuown et al. (2003) studied population differences with seven microsatellite loci. They found high levels of genetic diversity, in contrast to the earlier mtDNA analyses, and observed a general division of the population into at least three large differentiated groups, providing the first genetic evidence to support local management of lake sturgeon. However, some of the loci used by McQuown et al. (2003) were not ideal, since they showed polyploid (tetrasomic) inheritance. In a study funded by the Great Lakes Fishery Trust, Welsh and McClain (2004) standardized the use and scoring of 13 microsatellite loci for the study of the Great Lakes sturgeon. Standardization of markers was conducted to establish a common set of loci for use, and their proper analysis in this species, so that data collected during different studies by different laboratories can be combined or compared, which is a significant step toward organized cooperative conservation efforts (Welsh and May, 2006).

The 13 standard microsatellite markers were then used to conduct a population genetic analysis of lake sturgeon adults collected from 19 spawning sites throughout the Great Lakes basin (Welsh and McClain, 2004). They found a range of genetic diversity among populations in terms of the average number of alleles observed. The two rivers examined in the Hudson Bay drainage had the lowest diversity (3.61 and 3.85 average alleles) while the sample from the St. Lawrence river showed the highest diversity (5.38 average alleles). Relatively high levels of heterozygosity were maintained in all populations, again despite anthropogenic pressures. Populations from some rivers were found to be genetically indistinguishable (the Bad and White rivers, and the Detroit, St. Clair and Lower Niagara rivers), but the Bad and White rivers were also found to be genetically unique from the rest of the Great Lakes. Most spawning populations were found to be distinct from other populations, indicating that there is substantial genetic structuring and possibly spawning site fidelity in lake sturgeons. Phylogenetic analyses showed a well-supported split between the Hudson Bay and Great Lakes populations, as well as moderately supported separations within the Lake Superior basin.

The data indicate the presence of within-lake basin differences, but also possible genetic exchange among the Great Lakes sturgeons, which illustrates the importance of genetic assessment of basins and coordination among basins when considering management plans for this species (Welsh and McClain, 2004). The genetic diversity and structuring observed in this study are in contrast to the results obtained from previous studies of lake sturgeon genetic diversity based on mtDNA and illustrate the potential usefulness of hypervariable microsatellite markers.

Very recently, another study combining mitochondrial DNA analysis with microsatellites to analyze population differentiation in lake sturgeons in the upper Great Lakes basin has appeared (DeHaan et al., 2006). Eight of the standardized microsatellite loci (Welsh and May, 2006) were used to study 11 populations. DeHaan et al. (2006) report that genetic diversity in lake sturgeons, as measured by microsatellites, is high, but probably below that reported in Atlantic sturgeons. Both microsatellite and mtDNA support the existence of at least three population clusters which have a geographic component (Lake Superior, western Lake Michigan, and northern Lake Michigan – Lake Huron, the Hudson Bay, and St. Lawrence populations were not included in this analysis).

Many parts of the range of lake sturgeons have not yet been adequately surveyed, so much remains uncertain about the wide patterns of population divergence in the species. The use of a standardized set of microsatellite loci, and more sensitive mtDNA analysis, based on sequencing, rather than on RFLP techniques, should yield substantial information that can be used for species and population management.

4.5.2 Sturgeon of Western North America

Two North American species, the white sturgeon (*A. transmontanus*) and green sturgeon (*A. medirostris*), exist in river drainages connecting to the Pacific Ocean.

Compared to the eastern species, few population studies have been conducted yet on these taxa.

4.5.2.1 *A. transmontanus*

Brown et al. (1992) used mtDNA RFLP variation to characterize regional differences between white sturgeon populations in the Columbia and Fraser river drainages. While all but one of the ten RFLP genotypes were shared between populations, the frequency of different classes differed between the two drainages. In particular, the Columbia river population showed lowered levels of variability. This observation was interpreted to be associated with a possible population bottleneck due to anthropogenic exploitation, which would explain why the Columbia river population showed a lower level of variation even though sturgeon from the Columbia river are hypothesized to be the source of recolonization of the Fraser river drainage following the last glaciation. Brown et al. (1993) also examined sequence variation in the mitochondrial control region in the same two populations. Again, the Columbia river population showed lower levels of intrapopulational diversity. The populations shared most haplotypes and showed only minimal divergence. Smith et al. (2002) applied both mitochondrial sequence analysis and microsatellite analysis to study diversity more intensively in the Fraser river drainage. Their results indicate that there is significant differentiation within the Fraser river drainage; at least four subpopulations appear to exist. Clearly, much more work needs to be done on other river systems to better understand the population interrelationships of white sturgeons, and fully characterize the variation in this species.

4.5.2.2 *A. medirostris*

The green sturgeon is the least studied of the North American taxa. Brown et al. (1996) showed that green sturgeon populations contain variations for mtDNA RFLP length, with the suggestion that intraspecific levels of diversity were slightly lower than those seen in three other species (*A. transmontanus*, *A. fulvescens* and *A. brevirostrum*). Israel et al. (2004) used six microsatellite loci to study four population samples of *A. medirostris,* from the Columbia river in the north to San Pablo Bay in the south. They also applied mtDNA analysis to differentiate green sturgeon individuals from sympatric white sturgeons. Significant population heterogeneity was observed among green sturgeons, but the significant differences did not show clear geographic patterning, indicating that additional sampling will be necessary to fully understand the population-structural dynamics of the green sturgeon. It appears likely that additional breeding populations exist along the Pacific coast that may be interacting with the populations sampled in this study. It is imperative to develop additional disomic microsatellite markers for use on green sturgeons, to prevent problems in interpretation arising from the use of tetrasomic markers.

4.5.3 *Genetic Studies on Other North American Acipenseriforms*

Molecular analyses of other taxa have not been performed to the levels that have been applied for the three eastern species of *Acipenser*. However, studies on each of the species have provided some insight into population variation and structure.

4.5.3.1 *P. spathula*

Paddlefish (*P. spathula*) have been studied to a lesser degree than sturgeons. The earliest genetic analysis of population variability was done by Carlson et al. (1982), using allozymes. Very low levels of genetic variability were observed both within and between five populations extending from Montana to Alabama. Epifanio et al. (1996) extended the allozyme studies, added mtDNA analysis, and expanded the populations studied. They observed higher levels of heterozygosity than first reported, and a small degree of population differentiation as measured by allozymes, suggesting there are two major population clusters, fish from the Mississippi river–Pearl river and fish from the Alabama river. Mitochondrial analysis, surprisingly, showed less evidence of differentiation, with most haplotypes being shared between the regions. Szalanski et al. (2000) applied mtDNA RFLP analysis on paddlefish in the Missouri river, finding substantial variability and also lack of changes in haplotype frequencies over several years.

While microsatellite markers have been developed that can be used for paddlefish (Heist et al., 2002), we are unaware of any reported population studies using microsatellites.

4.5.3.2 *Scaphirhynchus* species

The pallid sturgeon (*S. albus*) and Alabama sturgeon (*S. suttkusi*) are considered endangered in the United States, while the shovelnose sturgeon (*S. platorynchus*) is relatively common throughout its habitat in the Missouri and lower Mississippi rivers. The first genetic analysis of *Scaphirhynchus* was reported by Phelps and Allendorf (1983), who compared *S. albus* and *S. platorynchus* using allozymes, and found the two species have low levels of genetic variation, and were electrophoretically indistinguishable. Mitochondrial DNA analysis further illustrated the problems of distinguishing these taxa, as discussed above. It has been used to examine population differentiation within *S. albus* and *S. platorynchus*. Levels of genetic variation in the mitochondrial D-loop were found to be lower than that in Atlantic or shortnose sturgeons (Campton et al., 1995). Differences between southern and northern populations of each species existed, but were at levels equivalent to the differences between the two *Scaphirhynchus* species (Tranah et al., 2001). The addition of microsatellite analysis (Tranah et al., 2004) showed greater ability to distinguish species, and gave increased weight to the possibility of hybridization in southern populations of the Mississippi river basin. They also suggest that the population differentiation of the northern and southern populations observed using mitochondrial markers can be measured using microsatellite loci, especially in the pallid sturgeon.

4.6 Conservation Management Implications of Genetic Studies

Almost all the molecular population studies suggest some degree, often marked, of population differentiation between groups. This often occurs at the level of difference between river drainages, or between lakes in the Great Lakes basin. Whatever the ultimate cause, population differentiation has implications for potential restocking of areas in which reduced populations remain. If restocking is performed using a generic hatchery stock that is genetically different from the resident population, there is the possibility of outbreeding depression caused by hybridization between genetically differentiated groups. The continued development of new molecular techniques to more accurately measure population similarity should allow for better management decisions in the future.

4.7 Conclusion

Traditionally, allozymes, mtDNA RFLPs, mtDNA sequences and nuclear microsatellite loci have been applied to analyze the phylogeny and population genetics of species in the Acipenseriformes. The majority of studies in sturgeons and paddlefish have used mtDNA methods, partly because of difficulties in analyzing the nuclear DNA of acipenseriforms related to polyploidy. However, the challenge for the future will be the isolation of new nuclear markers in the Acipenseriformes. Some microsatellites exhibit extremely high levels of allelic variation, which is especially useful in species like sturgeons that show low overall levels of genetic variation, are recently derived, or possess geographically proximate populations for which genetic differentiation may be limited and whose significance may be difficult to quantify. As a result, much effort has gone into isolating and developing microsatellite primers with disomic inheritance patterns in sturgeon. Although often useful across species, not all markers are useful for all species and there is still a need to develop sufficient numbers of markers for those species not yet adequately covered. Microsatellite data, especially when the loci are disomic, have provided evidence for genetic differentiation and diversity in some species where mtDNA diversity was previously found to be low (as in lake sturgeons). A few researchers have been able to use both mitochondrial and nuclear DNA in population studies to increase the amount of information obtained (for example, Smith et al., 2002; Wirgin et al., 2002). New nuclear markers, including both microsatellite loci and unique protein coding loci, will also be useful for further resolution of acipenseriform evolutionary relationships, since mtDNA has probably reached its limit of usefulness for this task and additional gene sequences from the mitochondrial genome may be unable to provide more information or clarity to unanswered questions.

In summary, studies using molecular markers are getting close to a definitive phylogeny of the sturgeons. The genomes of acipenseriforms seem to be evolving at a rate slower than that in teleosts. Information about the population substructure within species is increasing. Use of the most sensitive molecular markers shows that all species of North American sturgeons and the North American paddlefish exhibit significant evidence of population substructure that must be accounted for in management plans. Much work remains to be done to genetically define the limits of populations as management units and to understand the effects of natal watershed fidelity, anthropogenic modification of the environment, and the consequences of population augmentation by stocking on the future of these ancient but elegant fishes.

Acknowledgments We thank Arne Ludwig, Kathrin Hett and Vadim Birstein for their collaboration on the Eurasian 18S rRNA gene work.

References

Arnheim N, Krystal M, Schmickel R, Wilson G, Ryder O, Zimmer E. 1980. Molecular evidence for genetic exchanges among ribosomal genes on nonhomologous chromosomes in man and apes. *Proc Natl Acad Sci USA* 77:7323–7327.

Bailey RM, Cross FB. 1954. River sturgeons of the American genus *Scaphirhynchus:* characters, distribution and synonymy. *Mich Acad Sci, Arts Lett* 39:169–208.

Birstein VJ, DeSalle R. 1998. Molecular phylogeny of Acipenserinae. *Mol Phylogenet Evol* 9:141–155.

Birstein VJ, Vasiliev VP. 1987. Tetraploid–octoploid relationships and karyological evolution in the order Acipenseriformes (Pisces): karyotypes, nucleoli, and nucleolus-organizer regions in four acipenserid species. *Genetica* 72:3–12.

Birstein VJ, Hanner R, DeSalle R. 1997. Phylogeny of the Acipenseriformes: cytogenetic and molecular approaches. *Environ Biol Fish* 48:127–155.

Birstein VJ, Doukakis P, DeSalle R. 2002. Molecular phylogeny of Acipenseridae: nonmonophyly of Scaphirhynchidae. *Copeia* 2:287–301.

Bowen BW, Avise JC. 1990. Genetic-structure of Atlantic and Gulf of Mexico populations of sea bass, menhaden, and sturgeon—influence of zoogeographic factors and life-history patterns. *Mar Biol* 107:371–381.

Brown JR, Beckenbach AT, Smith MJ. 1992. Influence of Pleistocene glaciations and human intervention upon mitochondrial-DNA diversity in white sturgeon (*Acipenser transmontanus*) populations. *Can J Fish Aquat Sci* 49:358–367.

Brown JR, Beckenbach AT, Smith MJ. 1993. Intraspecific DNA sequence variation of the mitochondrial control region of white sturgeon (*Acipenser transmontanus*). *Mol Biol Evol* 10:326–341.

Brown JR, Beckenbach K, Beckenbach AT, Smith MJ. 1996. Length variation, heteroplasmy and sequence divergence in the mitochondrial DNA of four species of sturgeon (*Acipenser*). *Genetics* 142:525–535.

Burke JS, Ramsey JS. 1995. Present and recent historic habitat of the Alabama sturgeon, *Scaphirhynchus suttkusi* Williams and Clemmer, in the Mobile Basin. *Bull Alab State Mus Nat Hist* 17:17–24.

Campton DE, Garcia AI, Bowen BW, Chapman FA. 1995. Genetic evaluation of pallid, shovelnose, and Alabama sturgeon (*Scaphirhynchus albus*, *S. platorynchus*, and *S. suttkusi*) based on control region (D-loop) sequences of mitochondrial DNA. Final Report to the U.S. Fish and Wildlife Service, Bismarck, North Dakota, 35 pp.

Campton D, Bass AL, Chapman FA, Bowen BW. 2000. Genetic distinction of pallid, shovelnose and Alabama sturgeon: emerging species and the US Endangered Species Act. *Conserv Genet* 1:17–32.

Cannone JJ, Subramanian S, Schnare MN, Collett JR, D'Souza LM, Du Y, Feng B, Lin N, Madabusi LV, Müller KM, Pande N, Shang Z, Yu N, Gutell RR. 2002. The comparative RNA Web (CRW) site: an online database of comparative sequence and structure information for ribosomal, intron and other RNAs. *BMC Bioinformat* 3:2 (correction: *BMC Bioinformat* 3:15.)

Carlson DM, Kettler MK, Fisher SE, Whitt GS. 1982. Low genetic variability in paddlefish populations. *Copeia* 3:721–723.

Carlson DM, Pflieger WL, Trial L, Haverland PS. 1985. Distribution, biology and hybridization of *Scaphirhynchus albus* and *S. platorynchus* in the Missouri and Mississippi rivers. *Environ Biol Fish* 14(1):51–59.

Chermock RL. 1955. First record of the shovelnose sturgeon, *Scaphirhynchus platorhynchus*, from Tombigbee River Alabama. *Copeia* 1955:154.

Collins MR, Cooke D, Post B, Crane J, Bulak J, Smith TIJ, Greig TW, Quattro JM. 2003. Shortnose sturgeon in the Santee-Cooper reservoir system, South Carolina. *Trans Am Fish Soc* 132:1244–1250.

DeHaan PW, Libants SV, Elliot RF, Scribner KT. 2006. Genetic population structure of remnant lake sturgeon populations in the upper Great Lakes basin. *Trans Am Fish Soc* 135:1478–1492.

Dovel WL, Berggren TJ. 1983. Atlantic sturgeon of the Hudson estuary, New York. *NY Fish Game J* 30:140–172.

Dugo MA, Kreiser BR, Ross ST, Slack WT, Heise RJ, Bowen BR. 2004. Conservation and management implications of fine-scale genetic structure of Gulf sturgeon in the Pascagoula River, Mississippi. *J Appl Ichthyol* 20:243–251.

Epifanio JM, Koppelman JB, Nedbal MA, Philipp DP. 1996. Geographic variation of paddlefish allozymes and mitochondrial DNA. *Trans Am Fish Soc* 125:546–561.

Ferguson MM, Duckworth GA. 1997. The status and distribution of lake sturgeon, *Acipenser fulvescens*, in the Canadian provinces of Manitoba, Ontario and Quebec: a genetic perspective. *Environ Biol Fish* 48:299–309.

Ferguson MM, Bernatchez L, Gatt M, Konkle BR, Lee S, Malott ML, McKinley RS. 1993. Distribution of mitochondrial DNA variation in lake sturgeon (*Acipenser fulvescens*) from the Moose River basin, Ontario, Canada. *J Fish Biol* 43(Suppl. A):91–101.

Fontana F, Tagliavini J, Congiu L. 2001. Sturgeon genetics and cytogenetics: recent advances and perspectives. *Genetica* 111:359–373.

Genetic Analyses, Inc. 1994. Genetic studies of *Scaphirhynchus spp*. Unpublished report for the U.S. Army Corps of Engineers, Omaha District; U.S. Fish and Wildlife Service, Bismarck, North Dakota; U.S. Army Corps of Engineers, Mobile District.

Grunwald C, Stabile J, Waldman JR, Gross R, Wirgin I. 2002. Population genetics of shortnose sturgeon *Acipenser brevirostrum* based on mitochondrial DNA control region sequences. *Mol Ecol* 11:1885–1898.

Guénette S, Fortin R, Rassart E. 1993. Mitochondrial DNA variation in lake sturgeon (*Acipenser fulvescens*) from the St. Lawrence River and James Bay drainage basins in Quebec, Canada. *Can J Fish Aquat Sci* 50:659–664.

Heist EJ, Nicholson EH, Sipiroski JT, Keeney DB. 2002. Microsatellite markers for the American paddlefish (*Polyodon spathula*). *Conserv Genet* 3:205–207.

Henderson-Arzapalo A, King TL. 2002. Novel microsatellite markers for Atlantic sturgeon (*Acipenser oxyrinchus*) population delineation and broodstock management. *Mol Ecol Notes* 2:437–439.

Israel JA, Cordes JF, Blumberg MA, May B. 2004. Geographic patterns of genetic differentiation among collections of green sturgeon. *N Am J Fish Manag* 24:922–931.
King TL, Lubinski BA, Spidle AP. 2001. Microsatellite DNA variation in Atlantic sturgeon (*Acipenser oxyrinchus oxyrinchus*) and cross-species amplification in the Acipenseridae. *Conserv Genet* 2:103–109.
Kohne DE. 1970. Evolution of higher-organism DNA. *Quart Rev Biophys* 33:327–375.
Krieger J, Fuerst PA. 2002a. Evidence for a slowed rate of molecular evolution in the order Acipenseriformes. *Mol Biol Evol* 19(6):891–897.
Krieger J, Fuerst PA. 2002b. Evidence of multiple alleles of the nuclear 18S ribosomal RNA gene in sturgeon (Family: Acipenseridae). *J Appl Ichthyol* 18:290–297.
Krieger J, Fuerst PA. 2004. Diversity of nuclear 18S rRNA gene sequences within individuals in Lake Sturgeon (*Acipenser fulvescens*). *J Appl Ichthyol* 20:433–439.
Krieger J, Fuerst PA, Cavender TM. 2000. Phylogenetic relationships of the North American sturgeons (Order Acipenseriformes) based on mitochondrial DNA sequences. *Mol Phylogenet Evol* 16(1):64–72.
Krieger J, Hett AK, Fuerst PA, Birstein VJ, Ludwig A. 2006. Unusual intraindividual variation of the nuclear 18S rRNA gene is widespread within the Acipenseridae. *J Hered* 97(3):218–225.
Krieger J, Hett AK, Fuerst PA, Artyukhin E, Ludwig A. 2008. The molecular phylogeny of the order Acipenseriformes revisited. *J. Appl Ichthyol* 24 (Suppl. 1):36–45.
Lee DS. 1980a. *Scaphirhynchus albus* (Forbes & Richardson) pallid sturgeon. In: DS Lee, CR Gilbert, CH Hocutt, RE Jenkins, DE McAllister, JR Stauffer (eds.), *Atlas of North American Freshwater Fishes*. North Carolina State Museum of Natural History, Raleigh, NC, p. 43.
Lee DS. 1980b. *Scaphirhynchus platorhynchus* (Rafinesque) shovelnose sturgeon. In: DS Lee, CR Gilbert, CH Hocutt, RE Jenkins, DE McAllister, JR Stauffer (eds.), *Atlas of North American Freshwater Fishes*. North Carolina State Museum of Natural History, Raleigh, NC, p. 44.
Li W-H, Tanimura M, Sharp PM. 1987. An evaluation of the molecular clock hypothesis using mammalian DNA sequences. *J Mol Evol* 25:330–342.
Li W-H, Ellesworth DL, Krushkal J, Chang BH-J, Hewett-Emmett D. 1996. Rates of nucleotide substitution in primates and rodents and the generation-time effect hypothesis. *Mol Phylogenet Evol* 5:182–187.
Ludwig A, Belfiore NM, Pitra C, Svirsky V, Jenneckens I. 2001. Genome duplication events and functional reduction in ploidy levels in sturgeon (*Acipenser*, *Huso* and *Scaphirhynchus*). *Genetics* 158:1203–1215.
Martin AP. 1999. Substitution rates of organelle and nuclear genes in sharks: implicating metabolic rate (again). *Mol Biol Evol* 16:996–1002.
Martin AP, Palumbi SR. 1993. Body size, metabolic rate, generation time, and the molecular clock. *Proc Natl Acad Sci USA* 90:4087–4091.
May B, Krueger CC, Kincaid HL. 1997. Genetic variation at microsatellite loci in sturgeon: primer sequence homology in *Acipenser* and *Scaphirhynchus*. *Can J Fish Aquat Sci* 54:1542–1547.
Mayden RL, Kuhadja BR. 1996. Systematics, taxonomy and conservation status of the endangered Alabama sturgeon, *Scaphirhynchus suttkusi* Williams and Clemmer (*Actinopterygii, Acipenseridae*). *Copeia* 2:241–273.
McQuown E, Sloss BL, Sheehan RJ, Rodzen J, Tranah G, May B. 2000. Microsatellite analysis of genetic variation in sturgeon: new sturgeon primer sequences for *Scapirhynchus* and *Acipenser. Trans Am Fish Soc* 139:1380–1388.
McQuown E, Gall GAE, May B. 2002. Characterization and inheritance of six microsatellite loci in lake sturgeon. *Trans Am Fish Soc* 11:299–307.
McQuown E, Krueger CC, Kincaid HL, Gall GAE, May B. 2003. Genetic comparison of lake sturgeon populations: differentiation based on allelic frequencies at seven microsatellite loci. *J Great Lakes Res* 29:3–13.
Miracle AL, Campton DE. 1995. Tandem repeat sequence variation and length heteroplasmy in the mitochondrial DNA D-loop of the threatened Gulf of Mexico sturgeon, *Acipenser oxyrhynchus desotoi*. *J Hered* 86:22–27.

Mooers AØ, Harvey PH. 1994. Metabolic rate, generation time and the rate of molecular evolution in birds. *Mol Phylogenet Evol* 3:344–350.
Ohta T. 1993. An examination of the generation time effect on molecular evolution. *Proc Natl Acad Sci USA* 90:10676–10680.
Ong T-L, Stabile J, Wirgin I, Waldman JR. 1996. Genetic divergence between *Acipenser oxyrinchus oxyrinchus* and *A. o. desotoi* as assessed by mitochondrial DNA sequencing analysis. *Copeia* 2:464–469.
Phelps SR, Allendorf FW. 1983. Genetic identity of pallid and shovelnose sturgeon (*Scaphirhynchus albus* and *S. platorynchus*). *Copeia* 3:696–700.
Pyatskowit JD, Krueger CC, Kincaid HL, May B. 2001. Inheritance of microsatellite loci in the polyploid lake sturgeon (*Acipenser fulvescens*). *Genome* 44:185–191.
Quattro JM, Greig TW, Coykendall DK, Bowen BW, Baldwin JD. 2002. Genetic issues in aquatic species management: the shortnose sturgeon (*Acipenser brevirostrum*) in the southeastern United States. *Conserv Genet* 3:155–166.
Robinson MR, Ferguson MM. 2004. Genetics of North American Acipenseriformes. In: GTO LeBreton, FWH Beamish, RS McKinley (eds.), *Sturgeons and Paddlefish of North America*. Kluwer Academic Publishers, Dordrecht, the Netherlands, pp. 217–230.
Scott WB, Crossman EJ. 1973. *Freshwater Fishes of Canada*. Fisheries Research Board of Canada Bulletin, Ottawa, 184, 966 pp.
Simons AM, Wood RM, Heath LS, Kuhajda BR, Mayden RL. 2001. Phylogenetics of *Scaphirhynchus* based on mitochondrial DNA sequences. *Trans Am Fish Soc* 130:359–366.
Smith CT, Nelson RJ, Pollard S, Rubidge E, McKay SJ, Rodzen J, May B, Koop B. 2002. Population genetic analysis of white sturgeon (*Acipenser transmontanus*) in the Fraser River. *J Appl Ichthyol* 18(4–6):307–312.
Smith TIJ. 1985. The fishery, biology, and management of Atlantic sturgeon, *Acipenser oxyrhynchus*, in North America. *Environ Biol Fish* 14:61–72.
Szalanski AL, Bischof R, Mestl G. 2000. Population genetic structure of Nebraska paddlefish based on mitochondrial DNA variation. *Trans Am Fish Soc* 129:1060–1065.
Tagliavini J, Conterio F, Gandolfi G, Fontana F. 1999. Mitochondrial DNA sequences of six sturgeon species and phylogenetic relationships within Acipenseridae. *J Appl Ichthyol* 15:17–22.
Tajima F. 1993. Simple methods for testing molecular clock hypothesis. *Genetics* 135:599–607.
Takezaki N, Razhetsky A, Nei M. 1995. Phylogenetic test of the molecular clock and linearized trees. *Mol Biol Evol* 12:823–833.
Taubert BD. 1980. Reproduction of shortnose sturgeon (*Acipenser brevirostrum*) in Holyoke Pool, Connecticut River, Massachusetts. *Copeia* 1980:114–117.
Tranah GJ, Kincaid HL, Krueger CC, Campton DE, May B. 2001. Reproductive isolation in sympatric populations of pallid and shovelnose sturgeon. *Trans Am Fish Soc* 21:367–373.
Tranah G, Campton DE, May B. 2004. Genetic evidence for hybridization of pallid and shovelnose sturgeon. *J Hered* 95(6):474–480.
U.S. National Marine Fisheries Service. 1996. *Status Review of Shortnose Sturgeon in the Androscoggin and Kennebec Rivers*. Northeast Regional Office, National Marine Fisheries Service, Gloucester, MA.
U.S. National Marine Fisheries Service. 1998. *Final Recovery Plan for the Shortnose Sturgeon (Acipenser brevirostrum)*. National Marine Fisheries Service, Silver Springs, MD.
Waldman JR, Hart JT, Wirgin II. 1996a. Stock composition of the New York Bight Atlantic sturgeon fishery based on analysis of mitochondrial DNA. *Trans Am Fish Soc* 125:364–371.
Waldman JR, Nolan K, Hart J, Wirgin II. 1996b. Genetic differentiation of three key anadromous fish populations of the Hudson River. *Estuaries* 19:759–768.
Waldman JR, Grunwald C, Stabile J, Wirgin I. 2002. Impacts of life history and biogeography on the genetic stock structure of Atlantic sturgeon *Acipenser oxyrinchus oxyrinchus*, Gulf sturgeon *A. oxyrinchus desotoi*, and *A. brevirostrum*. *J Appl Ichthyol* 18:509–518.
Walsh MG, Bain MB, Squiers Jr T, Waldman JR, Wirgin I. 2001. Morphological and genetic variation among shortnose sturgeon *Acipenser brevirostrum* from adjacent and distant rivers. *Estuaries* 24(1):41–48.

Welsh A, May B. 2006. Development and standardization of disomic microsatellite markers for lake sturgeon genetic studies. *J Appl Ichthyol* 22:337–344.

Welsh AB, McClain JR. 2004. Development of a management plan for lake sturgeon within the Great Lakes based on population genetics structure. Final project report, Great Lakes Fishery Trust Project Number 2001.75, September 29, 2004.

Welsh AB, Blumberg M, May B. 2003. Identification of microsatellite loci in lake sturgeon, *Acipenser fulvescens*, and their variability in green sturgeon, *A. medirostris*. *Mol Ecol Notes* 3:47–55.

Williams JD, Clemmer GH. 1991. *Scaphirhynchus suttkusi*, a new sturgeon (Pisces: Acipenseridae) from the Mobile Basin of Alabama and Mississippi. *Bull Alab State Mus Nat Hist* 10:17–31.

Wirgin II, Stabile JE, Waldman JR. 1997. Molecular analysis in the conservation of sturgeons and paddlefish. *Environ Biol Fish* 48:385–398.

Wirgin I, Waldman J, Rosko J, Gross R, Collins MR, Rogers SG, Stabile J. 2000. Genetic structure of Atlantic sturgeon populations based on mitochondrial DNA control region sequences. *Trans Am Fish Soc* 129:476–486.

Wirgin I, Waldman J, Stabile J, Lubinski B, King T. 2002. Comparison of mitochondrial DNA control region sequence and microsatellite DNA analyses in estimating population structure and gene flow rates in Atlantic sturgeon *Acipenser oxyrinchus*. *J Appl Ichthyol* 18:313–319.

Wirgin I, Grunwald C, Carlson E, Stabile J, Peterson DL, Waldman J. 2005. Range-wide population structure of shortnose sturgeon *Acipenser brevirostrum* based on sequence analysis of the mitochondrial DNA control region. *Estuaries* 28(3):406–421.

Wooley CM. 1985. Evaluation of morphometric characters used in taxonomic separation of Gulf of Mexico sturgeon, *Acipenser oxyrhynchus desotoi*. In: FP Binkowski, SI Doroshov (eds.), *North American Sturgeons: Biology and Aquaculture Potential*. Dr W. Junk Publishers, Dordrecht, pp. 97–103.

Wooley CM, Crateau E. 1985. Movement, microhabitat, exploitation, and management of Gulf of Mexico sturgeon, Apalachicola River, Florida. *N Am J Fish Manag* 5:590–605.

Zhang S, Zhang Y, Zheng X, Chen Y, Deng H, Wang D, Wei Q, Zhang Y, Nie L, Wu Q. 2000. Molecular phylogenetic systematics of twelve species of Acipenseriformes based on mtDNA *ND4L-ND4* gene sequence analysis. *Sci China (Series C)* 43(2):129–137.

Chapter 5
Forensic Strategies Used for DNA Extraction of Ancient and Degraded Museum Sturgeon Specimens

E. Martinez-Espin, L.J. Martinez-Gonzalez, J.C. Alvarez, R.K. Roby, and J.A. Lorente

Abstract The procedure for the successful extraction of nucleic acids depends on the type of sample being extracted and the purpose of the extraction. Protocols for the extraction of high copy number DNA samples vary significantly from those of degraded DNA samples. This study involved the development of a DNA extraction protocol that includes a cleaning procedure designed to remove external contaminants (e.g. biological, environmental, or chemical). This protocol, used to test ancient human bones, was used for testing ancient and degraded DNA samples from sturgeons.

We analysed three specimens captured in the Guadalquivir River from the mid-1970s to the early 1980s (preserved at the Biological Station of Doñana in Seville [EBD], Spain), and also included one from the Ebro river (stored at the Botanical Institute of Barcelona), and one from the collections at the Department of Zoology at the Science Faculty of the University of Granada (UGP).

The sturgeon samples were processed with forensic scientific techniques. In forensic sciences, the external components of a sample must be cleaned to ensure the results are from the specimen being tested rather than the contaminants from handling the specimens or external exposure. The samples were cleaned to remove potential contaminants and subjected to an organic extraction procedure.

A cleaning procedure was selected to obtain the individual sturgeon DNA and the samples were sufficiently cleaned and the contaminants removed from the sample. This extraction ensures reliable amplification and cloning.

Keywords Ancient DNA, forensic strategies, DNA extraction, sturgeon samples, contamination, museum specimens, inhibitors

---

E. Martinez-Espin, L.J. Martinez-Gonzalez, J.C. Alvarez, R.K. Roby, and J.A.Lorente
Laboratory of Genetic Identification, Dept. de Medicina Legal. Facultad de Medicina. Universidad de Granada, E-18012 Granada, Spain

R. Carmona et al. (eds.) *Biology, Conservation and Sustainable Development of Sturgeons*,

5.1 Introduction

With the advent of polymerase chain reaction (PCR) technology over 20 years ago, significant advances have been made in molecular biology techniques for studying ancient DNA and forensic DNA, as cited in scientific literature. The ancient DNA and forensic DNA communities share some scientific approaches and methodologies (Jobling et al., 2004). With the introduction of highly sensitive PCR based profiling techniques, forensic scientists are able to obtain DNA profiles from a wide range of samples that were previously undetectable, such as single hairs, cigarette butts, saliva on a postage stamp, fragments of bone, touch DNA, etc. In both forensic analysis and ancient DNA analysis, the template may be present in very small quantities, may be degraded, or may be chemically altered. These conditions can cause challenges (Capelli et al., 2003) for molecular techniques. Issues of contamination are constantly evaluated through the use of proper controls when working with low copy or degraded DNA. Scientists working with challenging DNA samples utilize DNA databases of laboratory personnel, reagent controls, positive and negative controls, and, often, vial controls. Additional measures taken to control contamination include the design of the laboratory and preparing reagents. The laboratory facilities are designed to separate the rooms for DNA extraction and for PCR amplification and analysis. Additionally, some laboratories have control of air flow with negative pressure and positive pressure, as well as vestibules. With the increased sensitivity in molecular biological test methods, low copy number (LCN), can yield full microsatellite profiles. Studies by van Oorschot and Jones (1997) on touch surface or LCN testing and studies by Findlay et al. (1997) with increasing PCR cycle number for single cells can often yield full microsatellite profiles with both methods. Frequently, partial profiles are produced from LCN testing and single cell analysis. Samples with <100 pg total DNA can produce profiles where allele 'drop-out' is observed. Allele drop-out means that an expected peak from a sample fails to amplify and is not visualized on an electropherogram because of the very low levels of input DNA. Guidelines for such work include duplicate detection of every allele before reporting the profile (Gill et al., 2000). The use of such profiles in criminal casework requires effective monitoring.

One of the values of museum or ancient samples is that they may provide additional biological information about endemic species or demonstrate evolutionary divergence and phylogenetic relationships (Donoghue et al., 1989). Because they are older, ancient samples or fossil DNA sequences should be less divergent than extinct sequences and should, therefore, have value for relating more species. When compared with extinct DNA, ancient DNA sequences may also provide an insight into the pattern of molecular evolutionary changes through time.

Ancient DNA sequence data can be used for estimating the rate and pattern of molecular change through time (Moran et al., 1993; Ochman and Wilson, 1987). To study this pattern, it might be possible to compare typical pairwise distance derived from nucleotide sequence data measured between extinct members and between the older members.

In particular, ancient DNA is very important for studying extinct threatened species as it is a way of understanding the sturgeons at the molecular level. Sturgeons are threatened with extinction throughout the world (McDowall, 1999). Since the wild populations of sturgeons are currently at critically low levels, recovery efforts focus on sturgeons raised on fish farms. If sturgeons from fish farms are to be released into the wild, it is important to have accurate identification of the sturgeon species inhabiting the river. In Western European rivers, the only recognized native species, *Acipenser sturio*, is currently restricted to the Gironde River and is difficult to rear in captivity. However, in the Iberian Peninsula, although the only species traditionally recognized was *A. sturio*, *A. naccarii*, a species previously considered endemic to the Adriatic region, was reported by Capello (1869) and Gonçalves (1942). Unfortunately, morphological identification is difficult as the two species differ only in the position of the barbels and the size and the shape of their snouts (Svetovidov, 1989). Furthermore, only a few museum specimens are available for reference.

In this study, we analysed three museum specimens from the Biological Station of Doñana in Seville (EBD) of sturgeons caught in the Spanish Guadalquivir River from the mid-1970s to the early 1980s (see illustrations, Figs. 5.1, 5.2 and 5.3). By analysing one nuclear marker, it was recently reported that two of these specimens could be considered *A. naccarii* (Garrido-Ramos et al., 1997). And very recently, de la Herrán et al. (2004) have found that using two mitochondrial DNA markers, one of those specimens could be assigned to *A. sturio*. Therefore, we have now undertaken a comprehensive genetic identification of these specimens to clarify the taxonomic status of the sturgeons inhabiting the Guadalquivir River. For this, we have used forensic techniques (Anderson et al., 1999) to obtain DNA from these specimens and also studied other museum specimens from European rivers from the Western

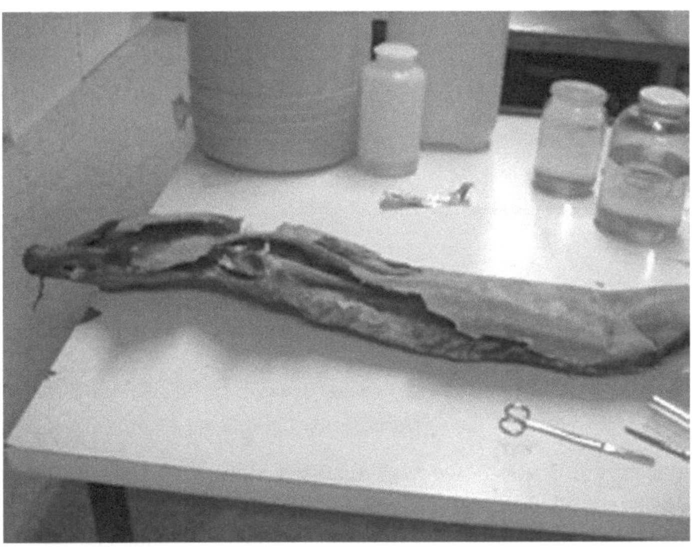

Fig. 5.1 Sturgeons caught in the Spanish Guadalquivir River EBD-8173

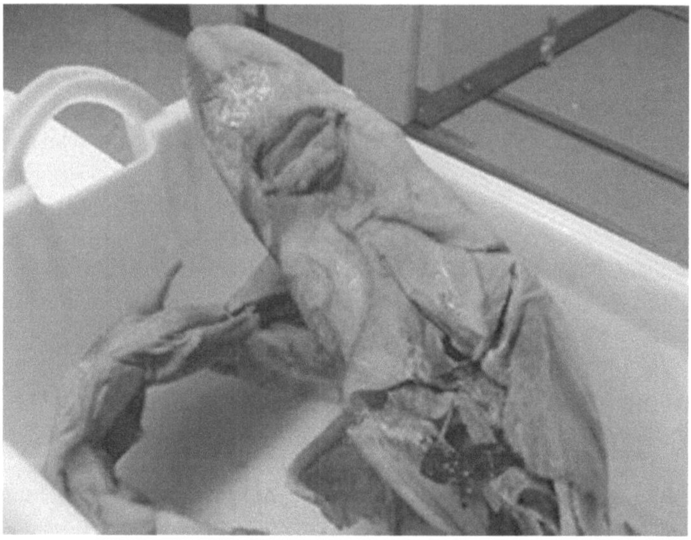

Fig. 5.2 Specimen sturgeon EBD-8174

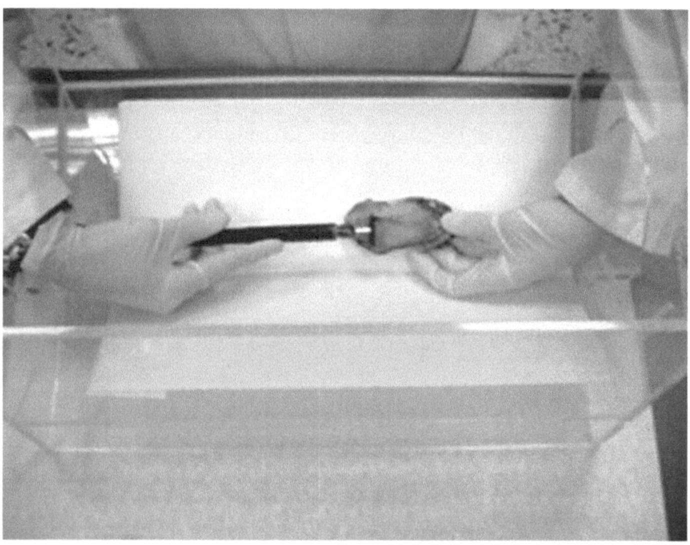

Fig. 5.3 Specimen sturgeon EBD-8401

Mediterranean. Furthermore, we have analysed more sturgeons: the first one from the Ebro river, stored in the Botanical Museum of Barcelona, and one from the collections held at the Zoology Department, University of Granada (UGP).

This research was conducted to demonstrate that DNA could be successfully extracted from museum specimens of sturgeons, which may be considered both ancient and degraded DNA samples.

As mentioned above, this study involved the development of a DNA extraction protocol that includes a cleaning procedure designed to remove external contaminants (e.g. biological, environmental, chemical, etc.). This protocol, used to analyse ancient human bones, was now used to analyse ancient and degraded sturgeon DNA samples.

5.2 Forensic Issues

A forensic DNA laboratory often has to deal with DNA samples that are less than ideal. The recovered biological sample may be limited in quantity, degraded, and/or exposed to environmental insults. Thus, the techniques employed are critical for the efficient recovery of DNA from degraded and limited samples.

The rate of DNA degradation varies with exposure to light, humidity, and temperature. Bacterial and fungal contamination followed by the growth of these microorganisms may rapidly lead to physical, chemical, and biochemical degradation of high molecular weight genomic DNA of the specimen of interest (Bender et al., 2004).

5.2.1 Low Levels and Degraded DNA

Numerous advances have been made in the last 15 years in terms of sample processing, speed, and sensitivity in molecular biology techniques, including extraction techniques. Instead of requiring large samples of biological tissue with well-preserved DNA, tiny amounts of sample, as little as a single cell in some cases, can yield a useful DNA profile.

Different extraction procedures are used for different forensic specimens. For example, there are specific procedures for the extraction of DNA from blood, tissues, bone, teeth, or mixtures of semen and other cells.

Environmental exposure can degrade DNA molecules by randomly breaking them into smaller pieces. DNA molecules can be degraded by water and enzymes such as nucleases that fragment the DNA. Both agents are ubiquitous in nature. It can be very difficult, if not impossible, to extract genetic information from severely degraded DNA samples.

In living cells, the integrity of DNA molecules is continually maintained by enzymatic repair processes (Lindahl, 1993). After the death of an organism, cell compartments that normally sequester catabolic enzymes break down. As a consequence, the DNA is rapidly degraded by enzymes such as lysosomal nucleases. In addition, the DNA molecule faces onslaught by bacteria, fungi, and insects that feed on and degrade macromolecules. Under rare circumstances, such as when a tissue becomes rapidly desiccated after death or the DNA becomes adsorbed to a mineral matrix, it may escape enzymatic and microbial degradation. On such occasions, slower but still relentless chemical processes start affecting the DNA. Many of these processes are similar or identical to those that affect the DNA in the living cell. However, after death, they are not counterbalanced by cellular repair processes and, thus, damage accumulates progressively until the DNA loses its integrity and decomposes with an irreversible loss of nucleotide sequence information (Pääbo et al., 2004).

Exposure to water is probably the single most destructive force acting on the DNA molecule. Water has been shown to initiate strand breakage by attacking the base-sugar bonds. Where the base is lost, the chain is weakened, and eventually cleaved, as reflected in studies by Eglinton and Logan (1991) and Lindahl (1993). Given these facts, a crucial step in the preservation of DNA is relatively rapid dehydration of tissues (Cano, 1996).

Another consideration in the long-term preservation of DNA is the pH of the environment. Acid environments may increase the rate of degradation of DNA molecules as H^+ ions can attack the OH^- groups of the sugar and the nitrogenous bases, contributing to strand breakage. Bone cellular death promotes an alkaline environment, which can favour preservation of DNA (Adegoke et al., 1991; Varricchio, 1993).

Exposure to ultraviolet light also causes extensive damage and degradation of DNA. Limiting the exposure of the sample to ultraviolet light, such as a rapid burial of an organism, is important to minimize the consequences of UV damage to DNA.

An ethidium-bromide stained agarose yield gel may be run to evaluate the quality and quantity of the DNA sample. Degraded DNA appears as a smear of DNA as compared to a high molecular weight band of DNA. There are qPCR techniques that can also measure the degradation of a sample.

The chances of successful amplifications of ancient DNA samples increase as the size of the designed amplicon is decreased. As a general rule, it is recommended that the selected primer pair amplify a region of the desired gene to measure ≤200 bp. As the ancient DNA becomes damaged and degraded, the resulting fragment length becomes smaller. Thus, amplification of small DNA segments will be more successful than that of larger segments (Handt et al., 1994).

5.2.2 Contamination Issues

The sensitivity of PCR and its ability to amplify low quantities of DNA can be a problem if proper care is not taken. Validated laboratory protocols must be followed so that contamination from higher concentrations of DNA, such as the DNA from the analyst, can be avoided.

Contamination implies that accidental transfer of DNA can occur. There are three potential sources of contamination when performing PCR: sample contamination with genomic DNA from the environment, contamination between samples during preparation, and contamination of a sample with amplified DNA from a previous PCR reaction (Lygo et al., 1994). The first source of contamination depends largely on sample collection and the care taken by the collection team. ‘Substrate controls’ are initiated to monitor contamination caused by environmental contributions (Gill, 1997). Contamination caused by transfer of DNA between samples during preparation and contamination caused by amplified DNA products in the laboratory can be controlled and even eliminated by using appropriate laboratory procedures and designated work areas.

Separation of pre-PCR and post-PCR laboratory space is essential (Carracedo et al., 2000). PCR product contaminants from post-PCR reactions/amplicons can easily be carried on the clothes or person of a scientist and be detected in reagents and on equipment. PCR product contaminants from these sources can then contaminate pre-PCR areas, potentially causing the generation of false results. Physical separation of processing samples with low and high quantities of DNA is also important. These practices include the use of dedicated lab areas, lab coats, disposable gear (such as gloves, caps and/or sleeves) and positive displacement or aerosol-resistant pipette tips. Only one case and only one item from that case should be extracted and amplified prior to known reference samples or the evidence and the references can be processed in a physically separate room in the laboratory. Pre- and post-amplification areas should be physically separated, work surface areas should be thoroughly cleaned before and after use, and workspace under dedicated hoods should be employed when possible. In addition, exposing all appropriate materials and reagents to UV light should be carried out.

Because of the sensitivity of detection with ancient DNA analysis, low levels of exogenous DNA contamination can be observed; reliable results can be obtained with low levels of contamination. Contamination must be monitored. Reagent blanks and negative controls (samples containing all reagents, except template DNA) are one way in which contamination from the extraction process through the analysis is monitored. A PCR negative control monitors the presence of exogenous DNA from amplification through analysis. Both these controls should be processed in a similar manner to that of an evidence sample.

Laboratory contamination is monitored with negative controls, as described above, that test for contamination of PCR reagents and supplies. Basically, a negative control is processed in parallel with the forensic case evidence, or in this case, ancient DNA samples. If any detectable PCR products are observed in the negative control, then the sources of contamination should be sought out and eliminated before proceeding further (Butler, 2005).

5.2.3 PCR Inhibition

Another challenge in amplifying DNA samples is the presence of inhibitors in the extracted DNA sample. Inhibitors can be from many sources such as melanin, soil, and proteins. Inhibitors can interfere with the cell lysis necessary for the DNA extraction process (Vaneedchoutte and Van Eldere, 1997). Inhibitors hinder polymerase activity, thus preventing enzymatic amplification of the target DNA (Wilson, 1997). Inhibitors may also cause false-negative results. False-positive results may be caused by cross-contamination (Rys and Persing, 1993) due to the highly sensitive nature of the PCR process.

Samples containing PCR inhibitors often produce partial profile results that look similar to a degraded DNA sample. PCR inhibitors may be removed or their effects reduced by the addition of bovine serum albumin (BSA), by diluting the extracted DNA which thus dilutes the quantity of inhibitors present, or by performing a separation step to the amplification process to separate the extracted DNA from the inhibiting compound. This step can be accomplished by performing a more pure extraction procedure (e.g. phenol/chloroform/isoamyl alcohol extractions) or purification of DNA using Centricon-100 and Microcon-100 filters (Comey et al., 1994) as they have low-melt agarose gel plugs (Moreira,1998).

5.3 Sturgeon Ancient DNA Extraction

5.3.1 Forensic Processing of DNA Sturgeon Samples

We analysed the three specimens captured in the Guadalquivir River from the mid-1970s to early 1980s and preserved at the Biological Station of Doñana in Seville. The specimens were labelled as EBD-8173, EBD-8174 and EBD-8401. We have also analysed one specimen from the Ebro River (it was labelled GS) and two from the collections at the Department of Zoology at the University of Granada (UGP and UGG). We wished to determine whether the six specimens were *A. sturio* or a different species. Each specimen was analysed as described above. Only one specimen was cleaned and processed at a time. And a negative control was included with the analysis of each specimen.

The samples labelled EBD-8173 and EBD-8401 were stored in ethanol. The ethanol evaporated and the specimen was completely desiccated. The other samples were stored in a dry condition with no preservatives. We took different samples from each specimen: scales, skin and tissues (four different tissue samples from each specimen) and initiated the same process for all the samples.

Next, we cleaned the tissue samples prior to the DNA extraction procedure in a polymethacrylate (PMMA) box (see Fig. 5.4). For cleaning, we used a miniature Dremel drill to eliminate the exposed outer parts of the samples, which were in contact with the environment or with ethanol, and cleaned to remove potential contaminants (see Fig. 5.5). Then, the completely dried tissue samples were pulverized in liquid nitrogen using a Freezer Mill. After pulverization, the powdered sample was transferred to a sterile 15-mL conical polypropylene tube.

5.3.2 DNA Extraction

We performed an organic extraction procedure, in which 300 μL of extraction buffer (10 mM Tris, pH 8; 100 mM NaCl; 10 mM EDTA, pH 8; 2% sodium dodecyl sulfate (SDS)), 25 μL 10 mg/mL proteinase K, and 12 μL 1 M dithiothreitol (DTT) were added to 0.25 g pulverized sturgeon tissue into a 1.5 mL Eppendorf tube.

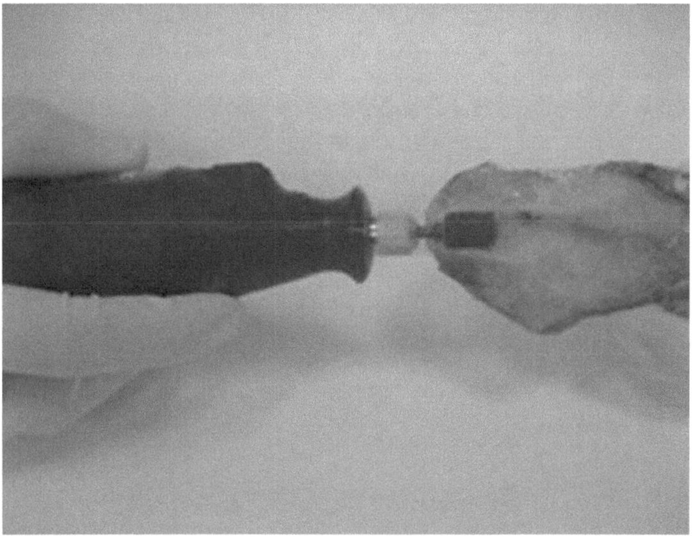

Fig. 5.4 Treatment for the samples in a polymethacrylate (PMMA) box; the samples were cleaned in the boxy by sanding with an aluminum oxide bit.

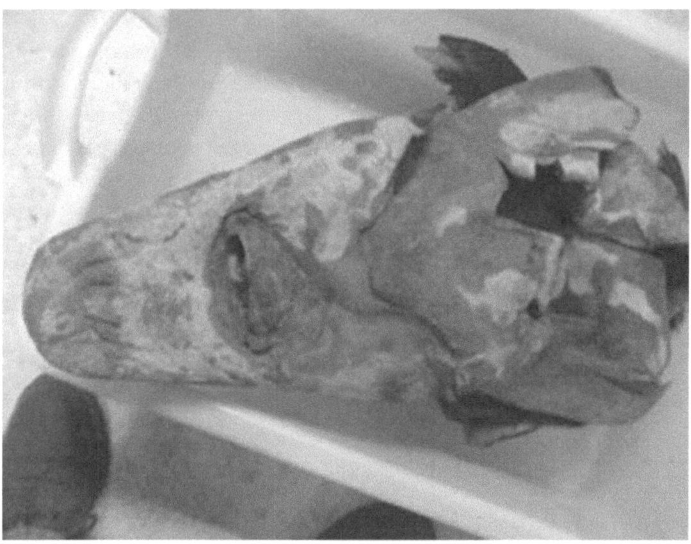

Fig. 5.5 A miniature Dremel drill was used to eliminate the exposed outer tissue by high speed sanding.

After an overnight incubation at 56°C with gentle stirring, samples were centrifuged at 15,700×*g*. The supernatant was cleaned by adding phenol/chloroform/isoamyl alcohol (25:24:1), vortexed at a low speed until the mixture attained a milky emulsion. We then spun the tube in a microcentrifuge at 15,700×*g* for 3 min. Finally, we used a series of phenol/chloroform washes successively until it cleared (less than three times).

Purification of the extracts was done using sterile water washes in Microcon YM-100 Millipore centrifugal filter units. We transferred the aqueous phase obtained in the preceding step and spun it in a microcentrifuge at 500×*g* for 25 min. We carefully removed the concentrator unit from the assembly and discarded the fluid. After returning the concentrator to the top of the filtrate cup, we added 200 μL sterile double-distilled water, spun the assembly at 500×*g* for 25 min. After two washes, the concentrate was transferred to a new 1.5 mL Microcon tube and diluted to a final volume of 50 μL of sterile water. Finally, we inverted the concentrator onto the new tube and spun the assembly in a microcentrifuge at 500×*g* for 5 min. Later we discarded the concentrator, and obtained the purified DNA in the tube.

5.4 Conclusions

The cleaning procedures used and the organic extraction process described in this paper were successful and helped us to produce sequence information on each of the extracts. Individual sturgeon DNA was extracted from the four samples taken from each of the six sturgeon specimens (a total of 24 extracts). The organic extraction of DNA described in this paper from ancient and degraded sturgeon museum specimens yielded a sufficient quantity of authentic DNA to allow successful PCR-based DNA testing to be performed. DNA sequence information was obtained from all the specimens tested. Neither biological contaminants from other DNA species nor chemical contaminants inhibited the successful sequencing from each of the extracts. All extracts from the same specimen produced identical sequence information in the overlapping regions. This extraction procedure allowed successful amplification and cloning by the staff of the Genetics Department at the University of Granada (de la Herrán et al., 2004; Robles et al., 2004, 2005).

The technologies available for molecular forensic DNA testing are continuing to evolve and provide more sensitive tools for molecular DNA testing.

References

Adegoke JA, Ighavini BO, Onuigbo RO. 1991. Characteristic features of the sonicated DNA of *Agama agama* L. (Reptilia, Agamidae) on hydroxyapatite columns, using mouse DNA as a reference. *Genetica* 83: 171–180.

Anderson TD, Ross JP, Roby RK, Lee DA, Holland MM. 1999. A validation study for the extraction and analysis of DNA from human nail material and its application to forensic casework. *J. Forensic Sci.* 44(5): 1053–1056.

Bender K, Farfán MJ, Schneider PM. 2004. Preparing of degrades human DNA under controlled conditions. *Forensic Sci. Int.* 139: 135–140.

Butler JM. 2005. *Forensic DNA Typing: Biology, Technology, and Genetics of STR Markers*, 2nd edn. Elsevier Academic, New York, pp. 145–180.

Cano RJ. 1996. Analysing ancient DNA. *Endeavour* 20(4): 162–167.

Capelli C, Tschentscher F, Pascali VL. 2003. 'Ancient' protocols for the crime scene? Similarities and differences between forensic genetics and ancient DNA analysis. *Forensic Sci. Int.* 131: 59–64.
Capello FB. 1869. Catalogo dos peixes do Portugal que existem no Museu de Lisboa. *J. Sci. Math. Phys. Nat.* 2: 131–193.
Carracedo A, Bär W, Lincoln P, Mayr W, Morling N, Olaisen B, Schneider P, Budowle B, Brinkmann B, Gill P, Holland M, Tully G, Wilson M. 2000. DNA commission of the international society for forensic genetics: guidelines for mitochondrial DNA typing. *Forensic Sci. Int.* 110(2): 79–85; *Int. J. Legal Med.* 113(4): 193–196; *Vox Sang.* 79(2): 121–125.
Comey CT, Koons BW, Presley KW, Smerick JV, Sobieralski CA, Stanley DM and Baechtel FS. 1994. DNA extraction strategies for amplified fragment length polymorphism analysis. *J. Forensic Sci.* 39: 1254–1269.
de la Herrán R, Robles F, Martínez-Espín E, Lorente JA, Ruiz Rejón C, Garrido-Ramos MA, Ruiz Rejón M. 2004. Genetic identification of western Mediterranean sturgeons and its implication for conservation. *Conserv. Genet.* 5: 545–551.
Donoghue MJ, Doyle JA, Gauthier J, Kluge AG, Rowe T. 1989. The importance of fossils in phylogeny reconstruction. *Annu. Rev. Ecol. Syst.* 20: 431–460.
Eglinton G, Logan GA. 1991. Molecular preservation. *Phil. Trans. R. Soc. London Ser.* 333: 315–328
Findlay I, Frazier R, Taylor A, Urquhart A. 1997. Single cell DNA fingerprinting for forensic applications. *Nature* 389: 555–556.
Garrido-Ramos MA, Soriguer MC, de la Herrán R, Jamilena M, Ruiz Rejón C, Domezain A, Hernando JA, Ruiz Rejón M. 1997. Morphometric and genetic analysis as proof of the existence of two sturgeon species in the Guadalquivir River. *Mar. Biol.* 129: 33–39.
Gill P. 1997. The utility of 'substrate controls' in relation to 'contamination'. *Forensic Sci. Int.* 85: 105–111.
Gill P, Whitaker J, Flaxman C, Brown N, Buckleton J. 2000. An investigation of the rigor of interpretation rules for STRs derived from less than 100pg of DNA. *Forensic Sci. Int.* 112: 17–40.
Gonçalves BC. 1942. Colecção oceanográfica de D. Carlos I. Peixes. *Trav. Stn. Biol. Marit. Lisb.* 46: 1–108.
Handt O, Höss M, Krings M, Pääbo S. 1994. Ancient DNA: methodological challenges. *Experientia* 50: 524–529.
Jobling MA, Hurles ME, Tyler-Smith C. 2004. *Human Evolutionary Genetics Origins, Peoples and Disease*. Garland Publishing, New York, pp. 473–478.
Lindahl T. 1993. Instability and decay of the primary structure of DNA. *Nature* 362: 709–715.
Lygo JE, Jhonson PE, Holdaway DJ, Woodroffe S, Whitaker JP, Clayton TM, Kimpton CP, Gill P. 1994. The validation of short tandem repeat (STR) loci for use in forensic casework. *Int. J. Legal Med.* 107(2): 77–89.
McDowall RM. 1999. Different kinds of diadromy: different kinds of conservation problems. *ICES J. Mar. Sci.* 56(4): 410–412.
Moran N, Munson MA, Baumann P, Ishikawa H. 1993. A molecular clock in endosymbiotic bacteria is calibrated using the insect hosts. *Proc. R. Soc. London Ser. B.* 253: 167–171.
Moreira D. 1998. Efficient removal of PCR inhibitors using agarose embedded DNA preparations. *Nucleic Acids Res.* 26: 3309–3310.
Ochman H, Wilson AW. 1987. Evolution in bacteria: evidence for a universal substitution rate in cellular genomes. *J. Mol. Evol.* 26: 74–86.
Pääbo S, Poinar H, Serre D, Jaenicke-Després V, Hebler J, Rohland N, Kuch M, Krause J, Vigilant L, Hofreiter M. 2004. Genetic analyses from ancient DNA. *Annu. Rev. Genet.* 38: 645–679.
Robles F, de la Herrán R, Ludwig A, Ruiz Rejon C, Ruiz Rejon M, Garrido-Ramos MA. 2004. Evolution of ancient satellite DNAs in sturgeon genomes. *Gene* 338(1): 133–142.
Robles F, de la Herrán R, Ludwig A, Ruiz Rejon C, Ruiz Rejon M, Garrido-Ramos MA. 2005. Genomic organization and evolution of the 5s ribosomal DNA in the ancient fish sturgeon. *Genome* 48(1): 18–28.

Rys PN, Persing DH. 1993. Preventing false positives: quantitative evaluation of three protocols for inactivation of polymerase chain reaction amplification products. *J. Clin. Microbiol.* 31: 2356–2360.

Svetovidov AN. 1989. Acipenseridae. In: *Fishes of North-eastern Atlantic and Mediterranean*, Whitehead PJP, Bauchot ML, Tortonese E (eds.), 2nd edn. UNESCO, Paris, pp. 220–225.

Vaneechoutte M, Van Eldere J. 1997. The possibilities and limitations of nucleic acid amplification technology in diagnostic microbiology. *J. Med. Microbiol.* 46: 188–194.

van Oorschot RAH, Jones MJ. 1997. DNA fingerprints from fingerprints. *Nature* 387: 767.

Varricchio D. 1993. Bone microstructure of the Upper Cretaceous dinosaur *Troodon formosus*. *J. Vert. Paleontol.* 113: 99–104.

Wilson IG. 1997. Inhibition and facilitation of nucleic acid amplification. *Appl. Environ. Microbiol.* 63(10): 3741–3751.

Chapter 6
Mechanisms of Polyploid Evolution in Fish: Polyploidy in Sturgeons

Victor P. Vasil'ev

Abstract Possible ways of polyploid speciation in fish are analysed. According to this analysis, the autopolyploid origin of bisexual species seems to be practically improbable, whereas their allopolyploid origin is quite probable, as it has been confirmed by data on reticular speciation in vertebrates and experimental crossings in fish. The most probable hybrid origin of polyploid sturgeon species is confirmed by various data. The species with 120-chromosomes belong to the ancient tetraploids, but they have passed through significant diploidization that has resulted in their practically functional diploid state. Therefore two scales of ploidy levels should be distinguished in Acipenseriformes: the 'evolutionary scale', which presumes diploid–tetraploid–octoploid-12-ploid relationships among these species, and the 'recent scale', which presumes diploid–tetraploid–hexaploid relationships. At least three independent polyploidization events have taken place in sturgeon evolution, but in fact there seem to have been many more.

Keywords Polyploid speciation, fish, Acipenseriformes

6.1 Introduction

Polyploidy is widely distributed among plants, where it represents one of the speciation pathways and plays an important part in evolution. Almost half the angiosperm plants and 95% of filicals are of polyploid origin, and most of them are allopolyploids (Grant, 1977). It is generally accepted that polyploidy signifies three or more ploidy levels. At the same time, the origin of sexual reproduction about one milliard years ago (Smith, 1978) resulted in the first diploid zygotes. This process should be considered the first stage of polyploidization among mega evolution events.

V.P. Vasil'ev
Institute of Problems of Ecology and Evolution RAS, Leninsky Prospect, 33,
117071 Moscow, Russia

R. Carmona et al. (eds.) *Biology, Conservation and Sustainable Development of Sturgeons*,

Formerly, polyploidy was considered of no importance in vertebrate evolution and also quite impossible among these animals. Muller (1925) considered a developed mechanism of sex determination a substantial barrier for the origin of polyploid species, and he was perhaps the first who advanced arguments for the impossibility of polyploidy among bisexual animals. Later, White (1946) analysed data that demonstrated possible participation of polyploidy in the evolution of bisexual animals and concluded they were unconvincing. For example, he examined Svärdson's (1945) concept of polyploid evolution in salmonids based on the multiple series from ten chromsomes in these fishes, and noted the coincidence of these data. In all fairness, it should be mentioned that in this case White (1946) was correct: recent evidence of tetraploid origin in salmonids shows a very distant connection to Swardson' hypothesis. Finally, Mayr (1963) presented the same conclusion about the insignificance of polyploidy in speciation among bisexual animals.

At the same time, both White (1946) and Mayr (1963) assumed polyploidy in parthenogenetic species, on the basis of numerous data. It should be mentioned that distinct proof of the close relation between unisexual reproduction (parthenogenesis and gynogenesis), hybridization, and polyploidy of vertebrates was disseminated only about 1965–1970. Of late, the hypothesis on the impossibility of polyploid evolution in bisexual animals is only of historical interest. The development of cytogenetic methods and their wide utilization in taxonomic, phylogenetic, and evolution investigations of vertebrates from the beginning of the 1960s has resulted in the discovery of polyploidy not only in parthenogenetic forms, but also in bisexual species (fish, amphibians) in which origin of polyploidy is of main significance.

The mechanisms of polyploidization (including successive events resulting in ploidy increase) are well known in unisexual vertebrates (Vasil'ev, 1985; Vrijenhoek, 1989), but remain unclear in bisexual vertebrates. This is caused by the ancient origin of polyploid bisexual species (usually tens or even hundreds of millions of years ago) and the subsequent evolutionary changes in polyploid genomes, as well as the uncertainty of the specific structure of animal groups subjected to polyploidization events. Thus, the greatest step in understanding the origin of bisexual polyploids is the presentation of a hypothesis concerning the type of polyploidy (allo- or autopolyploidization) based on any data on karyotype structure, meiosis, or genetic markers (Vasil'ev, 1985).

This paper is dedicated to the analysis of possible mechanisms of the origin of polyploid vertebrates. In addition to polyploidization modes, polyploid speciation is examined, with special attention to the polyploid evolution of acipenseriform fish.

6.2 Materials and Methods

This study represents the development of hypotheses on polyploid evolution in fish, previously elaborated on the basis of karyological investigations in several taxonomic groups, including Acipenseriformes, and data from the literature (see Vasil'ev, 1985). The present analysis is performed taking into consideration the recent available data on the genomes of polyploid fish forms obtained by different genetic methods. The review of the analysis is presented below.

6.3 Results and Discussion

About 200, of the nearly 25,000 karyologically studied fish species, have a polyploid origin. In amphibians, polyploid bisexual species occur more rarely than in fish, but it appears that both these groups had the same number of primary polyploid species (Vasil'ev et al., 1991). These primary polyploid species yielded 'bunches' of new polyploid species of the same ploidy level through divergent evolution. In amphibians this process practically did not occur. To date, polyploid origin has been proved for several groups of fishes (Table 6.1). Of these, only two groups, namely salmonids and catostomids, are represented exclusively by polyploids, while other families include species with different ploidy levels. Besides those presented in Table 6.1, polyploid species are assumed to be found in siluriforms, especially in callichthyids (Vasil'ev, 1985; Vasil'ev et al., 1991). This table demonstrates the occurrence of polyploidy in different fish groups and at the same time proves the absence of truly marine fishes in this group, as they are all freshwater fishes or anadromous forms that reproduce in fresh water. It should be mentioned that hybridization is more infrequent among marine fish than among freshwater fish. Besides, freshwater differs from marine environments in the probability of creating small isolates, for example during seasonal water-level changes. Such freshwater isolates represent strong factors for interspecific hybridization and further fixation of evolutionary events.

Most polyploid species are tetraploids. Hexaploidy was discovered in three cyprinid genera, namely *Barbus*, *Varicorhinus* and *Schizothorax*, as well as in one sturgeon species (Tables 6.1 and 6.2). This ploidy level deserves special attention,

Table 6.1 Fish taxa from polyploid origin[a]

Family, genus	Karyotype[a]	Ploidy level	Family, genus	Karyotype[a]	Ploidy level
Protopteridae	68	4*n*	*Myxocyprinus*	100	4*n*
Acipenseridae[b]	~120	2*n**	*Schizothorax*	98	4*n*
	~250	4*n**		148	6*n*
	362–370	6*n**	*Barbus*	100	4*n*
Salmonidae	52–92	4*n*		150	6*n*
Catostomidae	96–100	4*n*	*Varicorhinus*	150	6*n*
Cyprinidae			Cobitidae		
Cyprinus	100	4*n*	*Botia*	98–100	4*n*
Carassius	100	4*n*	*Misgurnus*	100	4*n*
Tor	100	4*n*	*Cobitis*	86–98	4*n*

[a]Data from Vasil'ev (1985); Zan et al. (1986); Yu et al. (1989); Vasil'ev et al. (1991); Golubtsov and Krysanov (1993).

[b]Publications used for karyological data in Acipenseriformes are presented in Table 6.2.

*Ploidy levels in acipenseriform fishes are discussed in the text.

Table 6.2 Chromosome numbers and DNA content in different acipenseriform species

Species	Chromosome number[a]	DNA content, pg[b]	Ploidy level: Evolutionary scale	Ploidy level: Recent scale
Polyodon spathula	120	3.17–4.89	4	2
Psephurus gladius	Unknown	4.11	4	2
Scaphirhynchus platorynchus	~112	4.73	4	2
Pseudoscaphirhynchus kaufmanni	Unknown	3.47	4	2
Acipenser sturio	~121	3.60	4	2
A. nudiventris	~118	3.88–4.04	4	2
A. ruthenus				
Volga River	~118	3.74	4	2
Danube River	~146			
A. stellatus	~118	3.74	4	2
A. oxyrinchus	Unknown	4.55	4	2
Huso huso	~118	3.17; 3.6	4	2
H. dauricus	Unknown	3.78	4	2
A. gueldenstaedtii	~250	7.87	8	4
A. persicus	~258	Unknown	8	4
A. baerii	~249	8.30	8	4
A. naccarii	~246	5.7–6.3	8	4
A. brevirostrum	362–370	13.08	12	6
A. transmontanus	~271	9.57	8	4
A. sinensis	~264	9.07	8	4
A. dabryanus	Unknown	8.26	8	4
A. fulvescens	~262	8.90	8	4
A. mikadoi	~250[c]	14.33	8	4
A. medirostris	~249	8.82	8	4
A. schrenckii	~238	6.07	8	4

[a]Data from Fontana and Colombo (1974); Vasil'ev et al. (1980); Vasil'ev (1985); Yu et al. (1989); Fontana (1994); Fontana et al. (1996, 1997, 1998, 2004); Van Eenennaam et al. (1998b); Chicca et al. (2002); Kim et al. (2005).

[b]Data from Blacklidge and Bidwell (1993); Birstein et al. (1993, 1997, review); Vasil'ev (1999, review); Zhang et al. (1999); Fontana et al. (2001, review); Yin et al. (2004).

[c]This data was obtained by Arefjev (personal communication) as a result of his study of karyotype in hybrid *A. baerii* × *A. mikadoi.*

because hexaploidy obviously cannot have resulted from simple autopolyploid duplication of chromosome number, as this process results in tetraploids, octoploids and 16-ploids.

In salmonids polyploidy is dated from 25 to 100 million years ago (Allendorf and Thorgaard, 1984; Behnke, 1992). The most exact time of tetraploidy in salmonids appears to be that suggested by Osinov and Lebedev (2004), who used paleontological and molecular data. According to these authors, sister groups of salmoniform and esociform fishes diverged about 85–95 million years ago. Further division of salmoniforms into three subfamilies (Salmoninae, Coregoninae,

Thymallinae) occurred about 38–46 million years ago, significantly later than the origin of their tetraploid ancestor. Thus, these authors believe that the tetraploid salmonid ancestor appeared immediately after the separation of salmoniform and esociform taxa. Some authors assume salmoniforms to be of autopolyploid origin (Ohno, 1970; Allendorf and Thorgaard, 1984), but Allendorf and Thorgaard (1984) do not rule out the possibility of segmental polyploidy. They consider the formation of multivalents in meiosis, as well as the presence of non-diverged pairs of duplicated loci and tetrasomic heritability of some of them in salmonids, as the main evidence of autopolyploidy in this group (Allendorf and Thorgaard, 1984). But other authors assume salmonids to be segmental tetraploids, based on the data on meiosis and the results of the analyses of several loci (Vasil'ev, 1977; Johnson et al., 1987; Altukhov et al., 1997; Osinov and Lebedev, 2004).

Catostomids is another family in which all species possibly originated from the same tetraploid ancestor. Uyeno and Smith (1972) date the origin of their polyploidy about 50 million years ago based on paleontological data. The formation of only bivalents in the meiosis of catostomids presumes their allopolyploid origin (Ferris and Whitt, 1980).

Cobitidae is the only fish family with evident polyploid events at least in three phyla. This family includes two subfamilies, Cobitinae and Botiinae, both including species of polyploid origin. All the studied Botiinae species are of tetraploid origin (see Vasil'ev et al., 1991). In the subfamily Cobitinae, bisexual species of polyploid origin are found in *Misgurnus* from the Far East and Europe and in *Cobitis* from Japan (Raicu and Taisescu, 1972; Ueno and Ojima, 1976; Ojima and Takai, 1979). (Tetraploid forms from genus Cobitis are not described as taxonomic species in spite of their phylogenetic independence.) Both the genera include bisexual diploid species closely related to bisexual tetraploids that suggest independent tetraploidization events in these genera. The allotetraploid origin of the Japanese *Cobitis* species appears to be most probable, from the available karyological and morphological (taxonomic relations) data (Vasil'ev, 1984, 1985). Strong evidence of the hybrid origin of Japanese tetraploid *Cobitis* species has been obtained by mtDNA analysis (Saitoh et al., 2000). This is the only case with strong evidence of the type of polyploidization.

As a whole, however, available data on the evolution of polyploid species do not present distinct proof about possible mechanisms of tetraploidization in fish, either allo- or autopolyploidy and a definite way of occurrence. In the case of allopolyploidy, only segmental allopolyploidy (the summarizing of genomes with partially homeologous chromosomes) might have occurred.

Groups of morphologically similar, obviously homeologous chromosomes are very often revealed among quite numerous groups of fish and amphibians even by traditional methods. For example, the presence of those chromosomes allows the distinction of different cyprinid subfamilies by karyological analysis (Vasil'ev, 1985). All loach species from subfamily Cobitidae have two characteristic submetacentric chromosomes that are easily identified (the real number of homeologous chromosomes is evidently somewhat more). A summary of those karyotypes would result in the origin of segmental allopolyploids. At the same time, the combination of more divergent karyotypes is not likely to result in the origin of a stable polyploid form, as it is impossibile to integrate the function of those genomes. In fact, the absence of morphologically homeologous chromosomes in karyotypes of any fish groups presumes their rather large genetic divergence according to taxonomic data. In this sense, it should be mentioned that the formation of only bivalents in meiosis of several polyploid fish and amphibian species did not result only from ancient allopolyploidy, but also from strongly advanced diploidization and (or) by factors regulating meiosis.

The following analysis describes another way to resolve the problem of allo- or autopolyploidy in vertebrates, taking into consideration the type of sex determination and possible modes for the isolation of the resulting polyploids. The last point seems extremely important, as polyploidization does not explain the origin of the isolating mechanisms themselves.

Autopolyploid bisexual vertebrate species may be formed from three main mechanisms of autopolyploidization.

1. A diploid egg may be fertilized by diploid sperm (Fig. 6.1). Such an event has an infinitesimal possibility. For example, spontaneous egg diploidization (usually by the junction of pronuclei in haploid egg as a result of suppressing the second meiotic division) occurs regularly in several fish species, but with low frequencies: it reaches 3% in common carp, but according to our data it is close to 0% in acipenserids. There are no data on spontaneous diploidization in sperm, but peculiarities of cytological processes of sperm development allow the consideration of its frequency to be close to 0. Moreover, this way of tetraploidization will result in unisexual progeny: all-male with male heterogamety or all-female with female heterogamety.
2. A normal diploid zygote may be subjected to further endoreduplication of chromosomes during the first division (Fig. 6.1). For the production of both male and female progeny, the simultaneous origin of at least two tetraploid specimens of different sexes is necessary. Autotetraploid specimens spontaneously originated in this way are found in different mammal species (see Dyban and Baranov, 1978): in mice, rats, rabbits, pigs, large horned animals, and even in humans, these being the species most often subjected to cytogenetic studies. All these tetraploids perish at early stages of development. Spontaneously originated autotetraploid specimens were not found in fish or amphibians, which is probably due more to the absence of species often subjected to cytogenetic studies than to the actual situation in these groups. Spontaneous autotetraploids seem to occur in fish and amphibians with similar frequencies as in mammals. Moreover, experimental results demonstrate that autotetraploids are viable in many amphibians and most fishes. At present, autotetraploids are found in ten Caudate and several Anuran species (Kawamura, 1984). However, F_2 was not obtained in all cases, since F_1 progenies were completely sterile or they were partially fertile but unisexual. At the same time, the crossing of autotetraploid and diploid specimens resulted in triploids in several experiments.

Of late, investigations in artificial polyploidy in fish are being widely developed, as a result of their high economic importance. The autotetraploids were obtained from

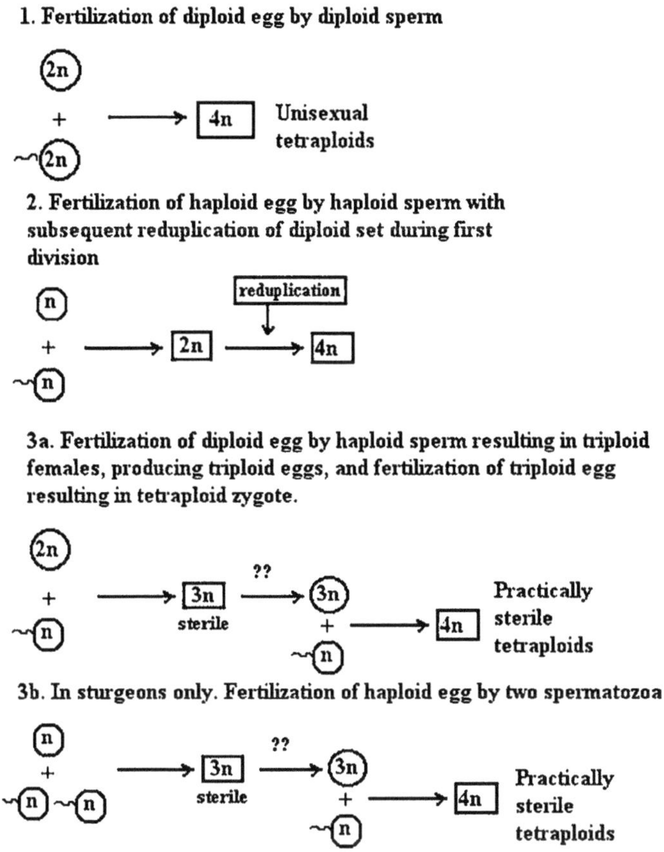

Fig. 6.1 Possible mechanisms of the origin of autotetraploid specimens. Gametes are in ovals, specimens in rectangles

several studies, but only in one case were they reared to sexual maturity (Chourrout et al., 1986). The survival of autotetraploid rainbow trouts, obtained by high hydrostatic pressure during the first division, was significantly lower than the same of diploids. And only males survived up to maturity. F_2 autotetraploid progeny was obtained by the fertilization of eggs from diploid females with diploid sperm from autotetraploid males and subsequent thermal shock only.

3. The third way of tetraploidization (Fig. 6.1) includes two stages. The first stage results in the origin of autotriploid specimens producing triploid eggs. The fertilization of these eggs at the second stage results in the origin of tetraploid specimens. Spontaneously originated autotriploids are periodically found during karyological studies in different fish species (see Vasil'ev, 1985). Possibly, they resulted from the fertilization of diploidized egg (see the first way of tetraploidization). To date, well-known technologies to obtain autotriploid progenies (usually by using thermal shock after insemination to depress the second division) have been developed and applied in aquaculture. The sterility of these progenies

increases their meat productivity. However, due to sterility, the origin of autotetraploids by this method seems problematic.

Thus it can be seen that the origin of autopolyploids by the first and the third methods is practically impossible. Only the second method deserves attention, but this method has two important limitations: (1) at least two tetraploid specimens with different sex should originate at the same time; (2) these tetraploid specimens should be isolated from diploids by ecological barriers (for example, by the presence of small isolates). Moreover, the probability of meiotic disturbances with significant and negative influences on fertility is quite high. In any case, this probability is much stronger than in allotetraploids.

According to available data, the origin of polyploid species by hybridization seems the most realistic. There are two main ways of allotetraploidization: direct and indirect, related with clonal heritability. Three main mechanisms are basic for the direct method of allotetraploidization (Fig. 6.2).

1. The fertilization of a diploid egg from one species with a diploid sperm from another species (Fig. 6.2). This, as in the case of autopolyploidy (Fig. 6.1), will result in unisexual progeny, but the probability of this process is infinitesimal.
2. The hybridization between diploid species with subsequent production of unreduced gametes and their further junction (Fig. 6.2). The production of unreduced gametes is the main characteristic of clonal vertebrates that have hybrid origin and are known among fish, amphibians and reptiles. Actually, this characteristic ensures clonal heritability. At present, about 80 clonal vertebrate species are known (see Vrijenhoek et al., 1989). This method of producing unreduced gametes is based on endoreduplication of chromosomes before the first meiotic division and is typical for many parthenogenetic animals (see Vasetskyi, 1977). Another method to produce unreduced gametes is made possible by the absence of the first meiotic division. Among fish, it is found in triploid all-female crucian carp, *Carassius auratus* (Cherfas, 1966). It should be mentioned that the production of unreduced gametes by interspecific hybrids is quite usual and typical not only for known clonal forms. At present, artificial interspecific hybrids producing unreduced eggs are known among different fish groups: salmonids, cyprinids, cyprinodontids, orysiatids, and centrarchids (Ojima et al., 1975; Cherfas et al., 1981; Gomelskyi et al., 1985; Dawley et al., 1985; Dawley, 1987, 1992; Sakaizumi et al., 1993; Kurita et al., 1995; Dannewitz and Jannson, 1996). However, the origin of tetraploid species by this method (Fig. 6.2) seems improbable, since only one sex is fertile in interspecific hybridization (Haldane, 1922; Nurelli, 1998).
3. The origin of tetraploid specimens by interspecific hybridization and duplication of chromosome number during the first division (Fig. 6.2), but this seems improbable for the same reasons as the origin of autotetraploids in the same way (Fig. 6.1).

The hypothesis of indirect polyploidization should be considered as the most suitable among the various hypotheses on the origin of bisexual polyploid vertebrate species. It presumes two similar ways. According to the first, bisexual tetraploid

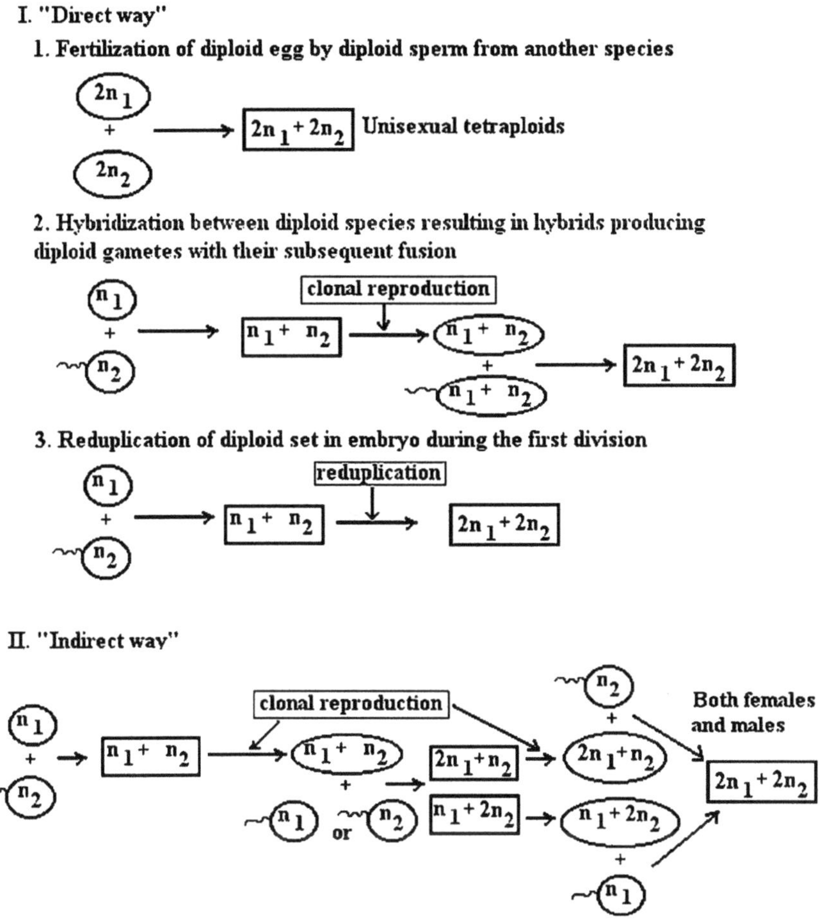

Fig. 6.2 Possible mechanisms of the origin of allotetraploid specimens. Gametes are in ovals, specimens in rectangles

vertebrates may result from hybridization between diploid bisexual species and triploid gynogenetic (parthenogenetic) forms of hybrid origin (Fig. 6.2). At present, allotetraploid forms that originated in this way are known among different vertebrates. At first they were found in the Volga river basin in the diploid–polyploid complex of fish from the genus *Cobitis* (Vasil'ev and Vasil'eva, 1982; Vasil'ev et al., 1989). This complex includes two diploid bisexual species (*C. taenia* and *C. melanoleuca*), one triploid all-female and two tetraploid forms (one is represented by both females and males). The triploid form has been proven to be of hybrid origin, and the tetraploid forms from hybridization between triploids and bisexual diploid species, *C. taenia* and *C. melanoleuca*. Sometime later, the complex included two diploid species (*C. rossomeridionalis* and *C. melanoleuca*) and one tetraploid form of hybrid origin was found in the Don river basin (Vasil'ev et al., 2005). Tetraploid specimens were also found in other unisexual–bisexual complexes of *Cobitis*. Tetraploid forms were also found in unisexual–bisexual complexes of crucian carp, *C. a. langsdorfii*, in Japan (see Vasil'ev, 1985) and in complexes of amphibians from the genus *Ambystoma* (Lowcock et al., 1987). The processes of natural origin of clonal triploid and tetraploid forms, as well as their stages (the origin of diploid clonal hybrids leading to triploid clonal lineages by backcrossing and then the obtaining of tetraploid specimens by crossings among triploids) were experimentally reproduced in hybrids between crucian carp and common carp (Cherfas and Emelyanova, 1986). Instead of the natural sperm-blocking observed in natural gynogenesis, the authors used UV inactivated sperm. Thus, the origin of tetraploid specimens by this method (Fig. 6.2) seems quite probable. One of the important problems in this method is the restoring of bisexuality after obtaining the tetraploid level. This, as well as the ecological isolation of the bisexual tetraploid, seems fully unrealistic, based on the data on the tetraploid *Cobitis* form representing both females and males (Vasil'ev et al., 1989).

Another way to produce tetraploid bisexual species is similar to the previous one, but it lacks the stage of clonal lineages (Fig. 6.2). Natural gynogenesis presumes the coincidence of two apparently quite independent events: (1) the production of unreduced eggs usually based on premeiotic endoreduplication of chromosomes and (2) blocking of fertilization of the egg (females of natural clonal species use sperm from related bisexual species for their own reproduction, but this sperm only stimulates egg development). The coincidence of these events is very rare; therefore the number of clonal fish species is not large, but the production of unreduced gametes by hybrids seems quite usual from the data on artificial hybridization (see above). If several males and females from different species appear in any isolate, the process of tetraploidization will start automatically: (a) hybridization, (b) the production of diploid eggs by diploid hybrids, (c) further hybridization between diploid hybrids and initial males resulting in triploids, and (d) other stages leading to tetraploids (Fig. 6.2). This process may be called 'automatic polyploidization'. In contrast to the case with natural clonal species, this process does not need sperm blocking to boost the success of the origin of tetraploids. Subsequently the success of the formation of bisexual tetraploids depends on the cytogenetics of development and the type of sex determination in parental diploid species and tetraploids. The latter may be a significant barrier for the origin of two sexes at the tetraploid level.

In vertebrates, sex is often determined by a Y-chromosome in the case of male heterogamety and a W-chromosome in female heterogamety. In male heterogamety, tetraploids originating from hybridization between triploids (XXX) and bisexual diploids (XY) will include males (XXXY) as well as females (XXXX). In female heterogamety, tetraploids will always be female, because their triploid mothers will always contribute at least one W-chromosome; male tetraploids (ZZZZ) are impossible. In fact, tetraploid bisexual species of fishes (salmonids, common carp, and crucian carp) have male heterogamety (Yamamoto and Kajishima, 1968; Golovinskaya, 1969; Thorgaard, 1977, 1978; Johnston et al., 1979; Hunter et al., 1982). In fact, male heterogamety in genus *Cobitis* (Saitoh et al., 1984; Vasil'eva and Vasil'ev, 1998) enables one of the tetraploid forms to represent both females and males. In strong female heterogamety, the origin of bisexual tetraploid by the above-mentioned method is impossible. However, in fish, the balance between sex chromosomes and autosomes often determines the sex. In such cases, it is very difficult to predict the probability of tetraploid bisexuality on the basis of the schemes presented.

The possible ways in which bisexual tetraploid species might have originated demonstrate the practical impossibility of autopolyploidy, since it needs several very improbable coinciding events, including the formation of physical barriers between parental diploids and tetraploids; in some cases, the significant barrier is the unisexuality of the resulting tetraploids. On the contrary, the origin of bisexual tetraploids in an indirect way through repeated hybridizations (Fig. 6.2) seems the most probable, more so given that most of the stages of this process occur in nature and are reproduced by experiments. The hypothesis on 'automatic polyploidization' (Fig. 6.2) seems the most plausible.

Three discrete groups with different chromosome numbers represent Acipenseriform species. The first of these includes species with about 112–146 chromosomes, the second group includes species with about 250–270 chromosomes, and the third one includes species with 360–370 chromosomes (Table 6.2). About half of these chromosomes are represented by very small chromosomes classified as microchromosomes. Their high numbers cause problems for their exact calculation. In addition, some of microchromosomes of Acipenseriformes are probably accessory chromosomes (B-chromosomes); the number of such chromosomes may be subjected to polymorphism (Vasil'ev, 1985). The analysis of synaptonemal complexes in the white sturgeon, *Acipenser transmontanus*, demonstrated the probable presence of one to seven accessory chromosomes in this species (Van Eenennaam et al., 1998a). It should be added that the number of B-chromosomes could vary on intra- and inter-population levels in animals and plants (Orlov and Bulatova, 1983). Nevertheless, despite the problems of exact chromosome calculation in Acipenseriformes, the difference among the three chromosome groups is obvious. Therefore, even the first data on the presence of two Acipenserid groups, differing by twice the number of chromosomes, allowed the hypothesis on diploid-polyploid relations among these fishes (Nikoljukin, 1972) to be presented.

The large amounts of recently accumulated diverse data confirm polyploid evolution in acipenseriform. The problems of ploidity levels, as well as the ways of polyploidization, are unresolved questions related to polyploid evolution in this group. The first problem is difficult to solve because of the ancient date of the first polyploidization event that occurred at the level of the ancestors of all recent Acipenseriformes; this means that species with chromosome numbers about 120 passed through the stage of tetraploidization. Several data testify to this. First, the karyotype with 120 chromosomes is relatively large, not typical to diploid fish species; their DNA content is also about twice as large as in most fish species (see Vasil'ev, 1985). The other evidence is the general structure of acipenseriform karyotypes: the authors studying the karyotype of *Polyodon spathula* succeeded in a rather satisfactory division into groups with four chromosomes (Dingerkus and Howell, 1976). The same procedure, with some assumptions, is also applicable for *Acipenser sturio*, *A. ruthenus*, and *Huso huso* (Vasil'ev, 1985). Arefjev (1983) obtained better evidence. He conducted the karyogram analysis of the first 24 chromosomes from *A. nudiventris* (the analysis of higher chromosome numbers is more difficult; the measurement of small chromosomes has low accuracy) and revealed ten pairs organized into groups with four very similar chromosomes. He could not find homeologists only for two pairs of chromosomes. This was possibly caused by the origin of one of these pairs through centric fusion of large acrocentric chromosomes. That is, four such chromosomes present in karyotypes of *P. spathula* and *Scaphirhynchus platorynchus* (Ohno et al., 1969; Dingerkus, Howell, 1976), karyotypes of *A. ruthenus* and *H. huso* have only two such chromosomes, while they are absent in karyotypes from *A. nudiventris* and *A. sturio*. The accidental nature of such satisfactory resolution of karyotypes into groups with four morphologically similar chromosomes is practically improbable according to recent knowledge of karyograms from numerous fish species. The best results of karyotype resolution into four-chromosome groups may be obtained for *P. spathula* and *S. platorynchus* (Vasil'ev, 1985). This is not remarkable, as these species appear to be the most ancient among recent Acipenseriformes.

Several data on allozyme variability and molecular genetic markers also confirm the tetraploid origin of the 120-chromosomal species. For example, *Acipenser stellatus* has 31% duplicated allozyme loci (Ryabova and Kutergina, 1990). The investigations on the *Sox*-gene family (these genes belong to the ancient gene group) in the genome of the European Atlantic sturgeon (*A. sturio*) showed that three (*Sox*4, *Sox*17, and *Sox*21) out of eight studied genes exist in duplicated state (Hett and Ludwig, 2005). At the same time, several studies revealed many 120-chromosome sturgeon species reaching functional diploidization in many genes. The analysis of 35 allozyme loci in *P. spathula* revealed only two loci expressing as duplicated loci. This number comes to 5.7% only (Carlson et al., 1982) and practically corresponds to the diploid level. Duplicated loci were discovered in genes of protein hormones and proteins from several 120-chromosome species: *P. spathula*, *Scaphirhynchus albus*, *A. ruthenus*, *Huso dauricus* (Alabyev et al., 2000; Andoh et al., 2000; Kim et al., 2000; Trabucchi et al., 2002).

The investigations of six microsatellite loci in 20 acipenseriform species (Ludwig et al., 2001) indicated that all the studied 120-chromosome species (*S. platorynchus*, *H. huso*, *A. sturio*, *A. stellatus*, *A. ruthenus*), as well as the species supposed to have the same chromosome numbers (*S. albus*, *H. dauricus*, *A. oxyrinchus*) could be classified as functional diploid species according to the loci studied. Most of the species, with about 250 chromosome numbers (*A. baerii*, *A. gueldenstaedtii*, *A. naccarii*, *A. persicus*, *A. transmontanus*), were classified as functional tetraploids; *A. mikadoi* with about 250 chromosomes had loci with tetra- and octosomic heritability and was supposed to belong to the octoploid ploidity level (Ludwig et al., 2001). *A. brevirostrum* (chromosome number is about 362–370) was demonstrated to belong to tetra- or octoploid species, since it had loci with disomic, tetrasomic, and octosomic heritability. Thus, all acipenseriform species passed through significant functional diploidization;

moreover, the 120-chromosome species are more advanced in this sense than species with 250-chromosomes and one species with 362–370 chromosomes.

Three ploidy levels in Acipenseriformes are distinguished, according to recent available data. The first group includes species with about 120 chromosomes, which have an ancient tetraploid origin. These species reach the functionally diploid level in many, possibly most, of the genetic markers and in several cytogenetic characters. That is why some authors consider these species to be of diploid origin (Fontana et al., 2001, 2003, 2004). Nevertheless, the general karyotype structure and the presence of duplicated loci in their genomes allow the contention of the tetraploid origin of these species (Vasil'ev, 1985; Birstein, 2005). The second polyploidization level (the origin of the species from octoploid with about 250 chromosomes) was probably passed when the 120-chromosome species went through the long process of functional diploidization. Otherwise, tetrasomic heritability in the most studied loci (and even disomic heritability in several cases) of the 250- and 360–370-chromosome species is hardly acceptable. It should also be added that the analysis of synaptonemal complexes in *A. transmontanus* did not reveal multivalents. This result supposes the process of chromosomal (structural) diploidization to be near completion (Van Eenennaam et al., 1998a). However, the absence of multivalents in meiosis may be related not only to the completion of diploidization, but also to the hybrid origin of polyploids. The third polyploidization event resulted in the origin of *A. brevirostrum* with 362–370 chromosomes. Thus, acipenseriform fish passed through three stages of polyploidization in their evolution: from the diploid level to the tetraploid, and from the tetraploid level to the octoploid, and from the octoploid level to 12-ploid.

However, another scale of ploidy levels also can be discussed. Since the 120-chromosome species practically reach the level of functional diploidization, and the 250-chromosome species are nearly the level of functionally tetraploid species (incidentally, *A. brevirostrum* with 362–370 chromosomes is in the same state), the scale with diploid–tetraploid–hexaploid relations in Acipenseriformes may be accepted, on the basis of the recent state of genomes.

The first scale of diploid–polyploid relations in acipenseriform fish may be called the 'evolutionary scale' of ploidy levels. Attached to this scale, the above-mentioned relations between Acipenseriformes are diploid–tetraploid–octoploid–12-ploid with diploid species recently having disappeared. The second 'recent scale' presumes this relation to be diploid–tetraploid–hexaploid.

The hypothesis on polyploidy resulted from reticular evolution with the corresponding hybrid origin of polyploid species presented earlier (Vasil'ev, 1999). This hypothesis was based on the following facts: polyploidization events resulted from reticular speciation in nature, significant variability of DNA-content in polyploid sturgeon species, and independent origin of at least some of the 250-chromosome species. At present, data on the slow rates of evolution and sequence homogenization in an ancient satellite DNA family (*Hind*III), as well as on the evolution of 5S rDNA gene, have been documented (de la Herrán et al., 2001; Robles et al., 2005). The authors consider that these data may be interpreted as the result of hybridization.

Unexpected new data on karyotype of *A. brevirostrum* (Kim et al., 2005), which has hexaploid origin according to 'recent scale', obviously testify to a hybrid origin of this species. In fact, in the case of autopolyploidy, hexaploidy may arise in two ways: (1) by the fertilization of a diploidized egg from tetraploid species (~250 chromosomes) by sperm from the same species (~125 chromosomes); (2) by fertilization of a normal egg (~125 chromosomes) by two normal spermatozoa (~125 + 125 chromosomes) or by one diploidized spermatozoon (~250 chromosomes). However, both these ways are practically improbable for two reasons. First, the probability of spontaneous egg diploidization in sturgeons is close to 0 (Van Eenennaam et al., 1996; our unpublished data). Second, autohexaploid specimens that originated in this way will be sterile, like the artificially obtained autotriploid fishes. Thus, *A. brevirostrum* originated from hybridization between tetraploid (according to 'recent scale') species with about 250 chromosomes and diploid species with about 120 chromosomes. The most probable scheme of this process is demonstrated in Fig. 6.2 ('indirect way'); its intermediate stage may or may not be related to clonal lineages ('automatic polyploidization').

The theory of the hybrid origin of polyploid Acipenseriform species allows a new view of the results of mtDNA-marker studies. In fact, if tetraploid sturgeon species have originated by hybridization, their phylogenetic relations obtained by mtDNA-markers should reflect relations between females of hybridized diploid species. For these relations to be defined, mtDNA-markers are used in studies of natural clonal vertebrate species originated by hybridization. The analyses of mtDNA-markers (cytochrome *b*, 12S, and 16S mtrRNA) demonstrated hexaploid *A. brevirostrum* to be included in the same clade with the Atlantic group of multichromosome species: *A. gueldenstaedtii*, *A. persicus*, *A. naccarii*, and *A. baerii* (Birstein and DeSalle, 1998; Ludwig et al., 2001). Therefore, it should be assumed that *A. brevirostrum* originated by hybridization between a female (or several females) from these tetraploid species or closely related extinct species. Since *A. brevirostrum* and the four tetraploid species mentioned above belong to the same Atlantic group and are distributed in water bodies of the Old and New Worlds, hexaploidy of *A. brevirostrum* obviously arose before continental division, about 50 million years ago (Monin, 1977). It is also possible to consider that females of *H. huso* and/or *A. ruthenus*, or any ancient forms closely related to them, participated in hybrid tetraploidization in the Ponto-Caspian (including *A. baerii*) sturgeon group. Similarly, in the Pacific sturgeon group including tetraploids and *H. dauricus* (Ludwig et al., 2001), females of *H. dauricus* or a *H. dauricus*-like form must have participated in the hybridization. This is why more extensive studies including both mtDNA and nuclear genes are necessary for a better understanding of the phylogenetic relationships among sturgeons.

According to karyological and molecular-genetic (Birstein and DeSalle, 1998) data, at least three polyploidization events by 'recent scales' occurred in Acipenseriformes: (1) the origin of the Atlantic tetraploid group; (2) the origin of hexaploid *A. brevirostrum*; and (3) the origin of the Pacific tetraploid group. Birstein (2005) considers seven polyploidization events in phylogenetic sturgeon lineages: two polyploidizations are presumed for *A. mikadoi* lineage, three among the Atlantic group (the first, common for *A. persicus*, *A. naccarii*, *A. baerii*, *A. gueldenstaedtii*, and *A. brevirostrum*; the second, in *A. brevirostrum* lineage; and the third in *A. fulvescens*), and two in the Pacific lineage (one for the *A. transmontanus–A. schrenkii–A. medirostris* group, and another in *A. sinensis*). Therefore, at least two species, *A. brevirostrum* and *A. sinensis*, are considered to be autopolyploids. However, if tetraploids have a hybrid origin with intermediate triploid forms (Fig. 6.2, 'indirect way'), which seems most probable, these triploid forms, as well as the resulting tetraploids, should have genomes formed from different haploid sets of closely related ancestral diploid species. Thus, they arose by the junction of phylogenetic lineages, but did not appear in different phylogenetic lineages. In this case, it should be considered that there were multiple tetraploidization events in this fish group and not just three. This situation is observed in known clonal triploid and tetraploid vertebrate forms (see Vrijenhoek, 1989). For example, at least five stable or arising de novo tetraploid forms are found in diploid–polyploid (clonal-bisexual) complexes of fishes from the genus *Cobitis* (Cobitidae), and all these forms differ in the combinations of haploid genomes from four diploid bisexual species.

Hybrid origin by reticular speciation may have one more unexpected result. Different combinations of nuclear and mitochondrial genomes can result in tetraploid forms having similar nuclear genomes, but different mt genomes (Fig. 6.3). In this sense, the results of mtDNA (D-loop) and nuclear DNA (microsatellite and amplified fragment length polymorphism) analyses in Adriatic sturgeon, *A. naccarii* (Ludwig et al., 2003) are of great interest. According to mtDNA analysis, one of the three studied haplogroups of *A. naccarii* is very closely related to one lineage of its sister species *A. gueldenstaedtii*, and their relations do not correlate to geographical distribution. The results of nuclear-marker analysis, however, demonstrate their relations to be strongly correlated to geographical distribution. The authors consider these results to be caused either by a postglacial introgression or to an ancestral polymorphism but not by a hybrid population. Another noteworthy example was presented by studies in Russian sturgeon, *A. gueldenstaedtii*, from the Caspian Sea. The sequence of three mtDNA sites (cytochrome *b*, NADH5, control region) demonstrated the obvious polyphyletic nature of these mitochondrial DNA markers in this species (Birstein et al., 2000). One of its mtDNA lineages was phylogenetically related to *A. naccarii* and *A. persicus*, while another lineage clustered together with *A. baerii*. In a recent paper with the outstanding title 'The enigmatic Caspian Sea Russian sturgeon: how many cryptic forms does it contain?' the authors (Birstein et al., 2005) note a unique state of phylogenetic clade *A. gueldenstaedtii–A. baerii–A. naccarii*, discerned by mtDNA analysis (cytochrome *b*, D-loop) and propose several hypotheses to explain the existence of three mtDNA lineages (two main clades with one of them represented by two subclades) in Caspian Russian sturgeon. The main hypothesis presumes the independent origin of three genetic forms of Caspian Russian sturgeon from three related 'pure' ancestral species and their further dispersion (selected in relation to discovered haplotypes) in water systems connected with the Caspian Sea in the past (Birstein et al., 2005). However, a hybrid origin of tetraploid forms by reticular speciation seems to be able to explain this 'enigmatic' situation.

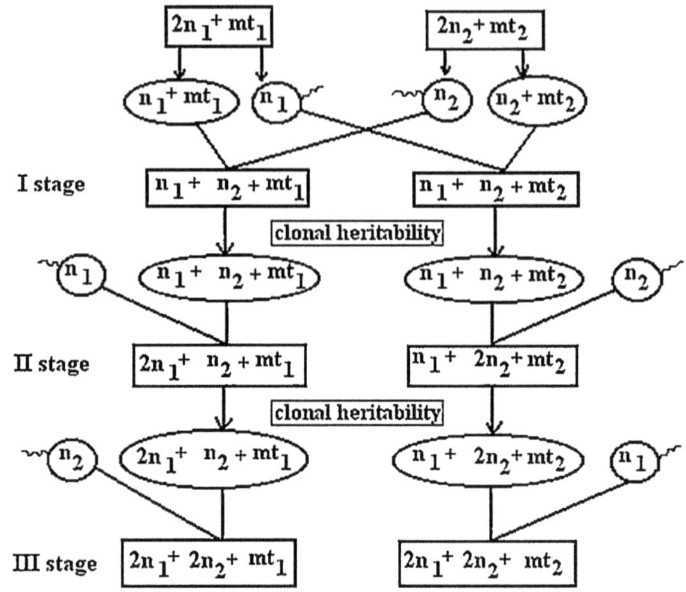

Fig. 6.3 Possible ways to form nuclear and mt genomes by reticular speciation process (other ways including three parental diploid bisexual species are also probable). Gametes are in ovals, specimens in rectangles

The combination of nuclear and mitochondrial genomes arising from the complete cycle of reticular speciation (from diploid bisexual species through hybridization, clonal heritability and triploidy to tetraploid bisexual species) may result in the origin of a new tetraploid species with two different mtDNA genomes (Fig. 6.3). If more than two related diploid bisexual species participate in reticular speciation (this situation is often observed in known clonal bisexual fish complexes), this process should result in several tetraploid species with different combinations of haploid nuclear and mitochondrial genomes. On the other hand, the observed polyphyly of mtDNA lineages in *A. gueldenstaedtii* and *A. naccarii* may testify to independent origins of tetraploid species belonging to the Atlantic sturgeon clade by multiple tetraploidization events.

6.4 Conclusions

1. The analysis of possible paths of the origin of bisexual tetraploid species demonstrates the practical impossibility of autopolyploidy, as this mode requires the coincidence of several improbable events, including an appearance of physical barriers between parental diploids and arising tetraploids. In some cases the mode of sex determination represents a significant barrier for autotetraploid speciation.

2. The origin of bisexual tetraploids by an 'indirect' method with repeated hybridization seems more probable, since most of the stages of this process take place in nature and were reproduced in experiments. Two modes of the 'indirect' method are possible. The first of them includes the origin of intermediate diploid and triploid clonal lineages and needs a sperm-block system (natural gynogenesis). The second mode is realized without clonal lineages, by clonal heritability, with the production of unreduced diploid eggs in the first hybrid generation (the hypothesis of 'automatic polyploidization').
3. The analysis of ploidy levels based on karyotype structure in 120-chromosome sturgeon species, as well as molecular-genetic data, admits the contention of the ancient tetraploid origins of these species. However, these species passed through significant functional diploidization and reached a practically diploid state. In this context, two scales of ploidy levels should be introduced. (1) The 'evolutionary scale' presumes a diploid–polyploid relationship among sturgeons to be diploid–tetraploid–octoploid–12-ploid with recently disappeared parental diploid species. (2) The 'recent scale' presumes a diploid–tetraploid–hexaploid relationship among sturgeons.
4. The data on interspecific variability of DNA content, as well as revealed phylogenetic relations (see Vasil'ev, 1999) and the results of studies on nuclear and mt markers testify to the hybrid origin of polyploid (according to 'recent scale') species.
5. The 'recent scale' presumes at least three independent polyploidization events in Acipenseriformes: (1) the origin of the Atlantic group of tetraploid species (*A. gueldenstaedtii*, *A. baerii*, *A. naccarii*); (2) the origin of hexaploid species *A. brevirostrum* from the Atlantic species group; and (3) the origin of the Pacific group of tetraploid species (*A. transmontanus*, *A. medirostris*, *A. sinensis*, *A. schrenckii* and, possibly, *A. fulvescens* and *A. mikadoi*). However, according to mtDNA structure, at least the tetraploid species included in the Atlantic group should have independent origins by hybridization between several 120-chromosome species; thus multiple tetraploidization should have taken place in this group.

Acknowledgements I am extremely grateful to the Organizing Committee of the 1st International workshop on the biology, conservation, and sustainable development of sturgeons in southern Europe (Bioresturgeons 2005) and its sponsoring entities for making it possible for me to participate in this meeting. This study was partially supported by the Russian Foundation for Basic Research (project no. 06-04-48416), as well as by RAS Programs 'Biodiversity and genofond dynamics', 'Biodiversity', and 'Biological resources of Russia'. I am very thankful to Dr. Ekaterina Vasil'eva who translated the text of this paper and prepared the illustrations.

References

Alabyev BY, Gusenkov SV, Najakshin AM, Mechetina LV, Taranin AV. 2000. CD3-epsilon homologues in the chondrostean fish. *Acipenser ruthenus. Immunogenetics* 51:1012–1020.

Allendorf FW, Thorgaard GH. 1984. Polyploidy and the evolution of salmonid fishes. In: Ryman N, Utter F (eds.), *The Evolutionary Genetics of Fishes*. University of Washington Press, Seattle, pp. 333–344.

Altukhov YuP, Salmenkova EA, Omelchenko VT. 1997. *Population Genetics of Salmonid Fish*. Nauka, Moscow [in Russian, English summary].
Andoh T, Nagasawa H, Matsubara T. 2000. Multiple forms of glucagon and insulin in the kaluga sturgeon, *Huso dauricus*. *Peptides* 21:1785–1792.
Arefjev VA. 1983. Polykaryogrammic analysis of the ship *Acipenser nudiventris* Lovetsky (Acipenseridae, Chondrostei). *Voprosy Ikhtiol* 23:209–218 [in Russian].
Behnke RJ. 1992. Native trout of western North America. *Am Fish Soc Monogr* 6:275.
Birstein VJ. 2005. Phylogeny and evolution of Acipenseriformes: new molecular and genetic data create new puzzles. In: Gall JaM, Kolchisky AI (eds.) *Evolutionary Biology: History and Theory*. SPb IH RAS "Nestor-Istoriya", S-Petersburg 3, pp 231-269.
Birstein VJ, DeSalle R. 1998. Molecular phylogeny of Acipenserinae. *Mol Phylogenet Evol* 9(1):141–155.
Birstein VJ, Poletaev AI, Goncharov BF. 1993. The DNA content in Eurasian sturgeon species determined by flow cytometry. *Cytometry* 14:377–383.
Birstein VJ, Hanner R, DeSalle R. 1997. Phylogeny of the Acipenseriformes: cytogenetic and molecular approaches. *Environ Biol Fish* 48:127–155.
Birstein VJ, Doukakis P, DeSalle R. 2000. Polyphyly of mtDNA lineages in the Russian sturgeon, *Acipenser gueldenstaedtii*: forensic and evolutionary implications. *Conserv Genet* 1:81–88.
Birstein VJ, Ruban GI, Ludwig A, Doukakis P, DeSalle R. 2005. The enigmatic Caspian Sea Russian sturgeon: how many cryptic forms does it contain? *Syst Biodiver* 3(2):203–218.
Blacklidge KH, Bidwell CA. 1993. Three ploidy levels indicated by genome quantification in Acipenseriformes of North America. *J Hered* 84:427–430.
Carlson DM, Kettler MK, Fisher SE, Whitt GS. 1982. Low genetic variability in paddlefish populations. *Copeia* 3:721–725.
Cherfas NB. 1966. Analyses of meiosis in unisexual and bisexual forms of the crucian carp. *Trudy Vniiprkh* 14:63–82 [in Russian].
Cherfas NB, Emelyanova OV. 1986. The role of the distant hybridization in the origin of all-female complexes in fish (the results of investigations in natural populations and crossing experiments). In: Strunnikov VA (ed.), *Biology of Development and Managements in Heritability*. Nauka, Moscow, pp. 82–104 [in Russian].
Cherfas NB, Gomelskyi BI, Emelyanova OV. 1981. Triploidy in back-cross hybrids between crucian carp and common carp. *Genetica* 17(6):1136–1139 [in Russian].
Chicca M, Suciu R, Ene C, Lanfredi M, Congiu L, Leis M, Tagliavini J, Rossi R, Fontana F. 2002. Karyotype characterization of the stellate sturgeon, *Acipenser stellatus* by chromosome banding and fluorescent in situ hybridization. *J Appl Ichthyol* 18:298.
Chourrout D, Chevassus B, Krieg F et al. 1986. Production of second generation triploid and tetraploid rainbow trout by mating tetraploid males and diploid females—potential of tetraploid fish. *Theor Appl Genet* 72(2):193–206.
Dannewitz J, Jannson H. 1996. Triploid progeny from a female Atlantic salmon ?D7; brown trout hybrid backcrossed to a male brown trout. *J Fish Biol* 48(1):144–146.
Dawley RM. 1987. Hybridization and polyploidy in a community of three sunfish species (Pisces: Centrarchidae). *Copeia* (2):326–335.
Dawley RM. 1992. Clonal hybrids of the common laboratory fish *Fundulus heteroclitus*. *Proc Natl Acad Sci USA* 89:2485–2488.
Dawley RM, Graham JH, Schultz RJ. 1985. Triploid progeny of pumpkinseed × green sunfish hybrids. *J Hered* 76:251–257.
Dingerkus G, Howell WM. 1976. Karyotypic analysis and evidence of tetraploidy in the North American paddlefish, *Polyodon spathula*. *Science* 194:842–844.
Dyban AP, Baranov VS. 1978. *Cytogenetics of Development in Mammals*. Nauka, Moscow, 216 pp [in Russian].
Ferris SD, Whitt GS. 1980. Genetic variability in species with extensive gene duplication: the tetraploid catostomid fishes. *Am Nat* 115(5):650–666.
Fontana F. 1994. Chromosomal nucleolar organizer regions in four sturgeon species as markers of karyotype evolution in Acipenseriformes (Pisces). *Genome* 37:888–892.
Fontana F, Colombo G. 1974. The chromosomes of Italian sturgeons. *Experientia* 30:739–742.

Fontana F, Lanfredi M, Rossi R, Bronzi P, Arlati G. 1996. Karyotypic characterization of *Acipenser gueldenstaedti* with C-, $AgNO_3$, and fluorescence banding techniques. *Ital J Zool* 63:113–118.
Fontana F, Rossi R, Lanfredi M, Arlati G, Bronzi P. 1997. Cytogenetic characterization of cell lines from three sturgeon species. *Caryologia* 50:91–95.
Fontana F, Tagliavini J, Congiu L, Lanfredi M, Chicca M, Laurente C, Rossi R. 1998. Karyotypic characterization of the great sturgeon, *Huso huso*, by multiple staining techniques and fluorescent in situ hybridization. *Mar Biol* 132:495–501.
Fontana F, Tagliavini J, Congiu L. 2001. Sturgeon genetics and cytogenetics: recent advancements and perspectives. *Genetica* 111:359–373.
Fontana F, Lanfredi M, Congiu L, Leis M, Chicca M, Rossi R. 2003. Chromosomal mapping of 18S-28S and 5 rRNA genes by two-colour fluorescent in situ hybridization in six sturgeon species. *Genome* 46:473–477.
Fontana F, Bruch RM, Binkowski FP, Lanfredi M, Chicca M, Beltrami N, Congiu L. 2004. Karyotype characterization of the lake sturgeon, *Acipenser fulvescens* (Rafinesque 1817) by chromosome banding and fluorescent in situ hybridization. *Genome* 47(4):742–746.
Golovinskaya KA. 1969. Artificial gynogenesis in carp. In: *Genetics, Selection and Hybridization in Fishes*. Nauka, Moscow, pp. 79–84 [in Russian].
Golubtsov AS, Krysanov EY. 1993. Karyological study of some cyprinid species from Ethiopia. The ploidy differences between large and small *Barbus* of Africa. *J Fish Biol* 42:445–455.
Gomelskyi BI, Cherfas NB, Emelyanova OV. 1985. On capacity of crucian carp × common carp hybrids to produce diploid sperm. *Dokl AN USSR* 282(5):1255–1258 [in Russian].
Grant V. 1977. *Organismic Evolution*. W.H. Freeman, San Francisco.
Haldane JBS. 1922. Sex ratio and unisexual sterility in hybrid animals. *J Genet* 12:101–109.
De la Herrán R, Fontana F, Lanfredi M, Congiu L, Leis M, Rossi R, Rejón MR, Garrido-Ramos MA. 2001. Slow rates of evolution and sequence homogenization in an ancient satellite DNA family of sturgeons. *Mol Biol Evol* 18(1):432–436.
Hett AK, Ludwig A. 2005. SRY-related (*Sox*) genes in the genome of European Atlantic sturgeon (*Acipenser sturio*). *Genome* 48:181–186.
Hunter GA, Donaldson EM, Goetz FW, Edgell PR. 1982. Production of all-female and sterile groups of coho salmon (*Oncorhynchus kisutch*) and experimental evidence for male heterogamety. *Trans Am Fish Soc* 111(3):367–372.
Johnston R, Simpson TH, Walker AF. 1979. Sex reversal in salmonid culture. 3. The production and performance of all-female populations of brook trout. *Aquaculture* 18(3):241–252.
Johnson KR, Wright JE, May B. 1987. Linkage relationships reflecting ancestral tetraploidy in salmonid fish. *Genetics* 116:579–591.
Kawamura T. 1984. Review polyploidy in amphibians. *Zoll Sci* 1(1):1–15.
Kim JB, Gadsboll V, Whittaker J, Barton BA, Conlon JM. 2000. Gastroenteropancreatic hormones (insulin, glucagon, somatostatin, and multiple forms of PYY) from the pallid sturgeon, *Scaphirhynchus albus* (Acipenseriformes). *Gen Comp Endocrinol* 120:535–563.
Kim DS, Nam YK, Noh JK, Park CH, Chapman FA. 2005. Karyotype of North American shortnose sturgeon *Acipenser brevirostrum* with the highest chromosome number in the Acipenseriformes. *Ichthyol Res* 52:94–97.
Kurita J, Oshiro T, Takashima F, Sakaizumi M. 1995. Cytogenetic studies on diploid and triploid oogenesis in interspecific hybrid fish between *Oryzias latipes* and *O. curvinotus*. *J Exp Zool* 273:234–241.
Lowcock LA, Licht LE, Bogart JP. 1987. Nomenclature in hybrid complexes of *Ambystoma* (Urodela: Ambystomatidae): No case for the erection of hybrid 'species'. *Syst Zool* 36:328–336.
Ludwig A, Belfiore NM, Pitra C, Svirsky V, Jenneckens I. 2001. Genome duplication events and functional reduction of ploidy levels in sturgeon (*Acipenser*, *Huso* and *Scaphirhynchus*). *Genetics* 158:1203–1215.
Ludwig A, Congiu L, Pitra C, Fickel J, Gessner J, Fontana F, Patarnello T, Zane L. 2003. Nonconcordant evolutionary history of maternal and paternal lineages in Adriatic sturgeon. *Mol Ecol* 12:3253–3264.
Mayr E. 1963. *Animal Species and Evolution*. Harvard University Press, Cambridge, MA.

Monin AC. 1977. *The History of the Earth*. Nauka, Leningrad [in Russian].
Muller HJ. 1925. Why polyploidy is rarer in animals than in plants. *Am Nat* 59:346–353.
Nikoljukin NI. 1972. *Distant Hybridization in Fish*. Pischevaya promyshlennost, Moscow [in Russian].
Nurelli M. 1998. The causes of Haldane's rule. *Science* 282(5390):889–890.
Ohno S. 1970. *Evolution by Gene Duplication*. Springer-Verlag, Berlin-Heidelberg-New York.
Ohno S, Muramoto J, Stenius C, Christian L, Kittrel WA, Atkin NB. 1969. Microchromosomes in holocephalian, chondrostean and holostean fishes. *Chromosoma* 26:35–40.
Ojima Y, Takai A. 1979. The occurrence of polyploid in the Japanese common loach *Misgurnus anguillicaudatus*. *Proc Jpn Acad Ser B* 55(10):487–491.
Ojima Y, Hayashi M, Ueno K. 1975. Triploidy appeared in the back-cross offspring from funa-carp crossing. *Proc Jpn Acad* 51(8):702–706.
Orlov VN, Bulatova NSh. 1983. *Comparative Cytogenetics and Karyosystematics in Mammals*. Nauka, Moscow [in Russian].
Osinov AG, Lebedev VS. 2004. Salmonid fish (Salmonidae, Salmoniformes): position in the suborder Protacanthopterygii, basic evolutionary history, molecular dating. *Voprosy Ikhtiol* 44(6):738–765 [in Russian].
Raicu P, Taisescu E. 1972. *Misgurnus fossilis*, a tetraploid fish species. *J Hered* 73(a):92–94.
Robles F, de la Herrán R, Ludwig A, Rejón GR, Rejón MR, Garrido-Ramos MA. 2005. Genomic organization and evolution of the 5S ribosomal DNA in the ancient fish sturgeon. *Genome* 48:18–28.
Ryabova GD, Kutergina IG. 1990. Analysis of allozyme variability in the stellate sturgeon, *Acipenser stellatus* (Pallas) from the northern Caspian Sea. *Genetica* 26:902–911 [in Russian, with English summary].
Saitoh K, Takai A, Ojima Y. 1984. Chromosomal study on the three local races of the striated spined loach (*Cobitis taenia striata*). *Proc Jap Acad* 60(Ser B):187–190.
Saitoh K, Kobayashi T, Ueshima R, Numachi KI. 2000. Analyses of mitochondrial and satellite DNKs on spined loaches of the genus *Cobitis* from Japan have revealed relationships among population of three diploid-tetraploid complexes. *Folia Zool* 49(Suppl.):3–7.
Sakaizumi M, Shimizu Y, Matsuzaki T, Hamaguchi S. 1993. Unreduced diploid eggs produced by interspecific hybrids between *Oryzias latipes* and *O. curvinotus*. *J Exp Zool* 266:312–318.
Smith JM. 1978. *The Evolution of Sex*. Cambridge University Press, Cambridge.
Svärdson G. 1945. Chromosome studies on Salmonidae. *Rep Inst Freshwater Res Drottingholm* 23:1–151.
Thorgaard GH. 1977. Heteromorphic sex chromosomes in male rainbow trout. *Science* 196:900–902.
Thorgaard GH. 1978. Sex chromosomes in the sockeye salmon: A Y-autosome fusion. *Can J Genet Cytol* 20:349–354.
Trabucchi M, Tostivint H, Lihrmann I, Sollars C, Vallarino M, Dolres RM, Vaudry H. 2002. Polygenic expression of somatostatin in the sturgeon *Acipenser transmontanus*: molecular cloning and distribution of the mRNAs encoding two somatostatin precursors. *J Comp Neurol* 443:332–345.
Ueno K, Ojima Y. 1976. Diploid-tetraploid complexes in the genus *Cobitis* (Cobitidae, Cyprinidae). *Proc Jpn Acad* 52(8):446–449.
Uyeno T, Smith GB. 1972. Tetraploid origin of the karyotype of catostomid fishes. *Science* 175(4022):644–646.
Van Eenennaam AL, Van Eenennaam JP, Medrano JF, Doroshov SI. 1996. Rapid verification of meiotic gynogenesis and polyploidy in white sturgeon (*Acipenser transmontanus* Richardson). *Aquaculture* 147:177–189.
Van Eenennaam AL, Murray JD, Medrano JF. 1998a. Synaptonemal complex—analysis in spermatocytes of white sturgeon, *Acipenser transmontanus* Richardson (Pisces, Acipenseridae), a fish with a very high chromosome number. *Genome* 41(1):51–61.
Van Eenennaam AL, Murray JD, Medrano JF. 1998b. Mitotic analysis of the North-American White sturgeon, *Acipenser transmontanus* Richardson (Pisces, Acipenseridae), a fish with a very high chromosome number. *Genome* 41:266–271.

Vasetskyi SG. 1977. IV. Meiotic divisions. In: Detlaf TA (ed.), *Problems of Biology of Development. Recent Problems of Oogenesis*. Nauka, Moscow, pp. 145–172 [in Russian].

Vasil'ev VP. 1977. On polyploidy in fish and some problems of karyotype evolution in salmonids. *Zhurn Obtschei Biolog* 38(3):380–392 [in Russian].

Vasil'ev VP. 1984. Several aspects of chromosome differentiation in fish. In: *Biological Bases for Fisheries: Genetics and Selection*. Nauka, Leningrad, pp. 166–180 [in Russian].

Vasil'ev VP. 1985. *Evolutionary Karyology of Fishes*. Nauka, Moscow, 300 pp [in Russian].

Vasil'ev VP. 1999. Polyploidization by reticular speciation in Acipenseriform evolution: a working hypothesis. *J Appl Ichthyol* 15:29–31.

Vasil'ev VP, Vasil'eva ED. 1982. A new diploid–polyploid complex in fishes. *Dokl AN USSR* 266:250–252 [in Russian].

Vasil'ev VP, Sokolov LI, Serebryakova EV. 1980. Karyotypes of the Siberian sturgeon, *Acipenser baeri*, of the Lena River and some aspects of karyotype evolution in Acipenseriformes. *Voprosy Ikhtiol* 20:814–822 [in Russian].

Vasil'ev VP, Vasil'eva ED, Osinov AG. 1989. Evolution of a diploid-triploid-tetraploid complex in fishes of the genus *Cobitis* (Pisces, Cobitidae). In: Dawley RM, Bogart JP, (eds.) Evolution and Ecology of Unisexual Vertebrates. Bulletin 466. New York State Museum. Albany. N.Y. p 153–169.

Vasil'ev VP, Osinov AG, Vasil'eva ED. 1991. On the problem of reticular speciation in vertebrates: diploid-triploid-tetraploid complex in genus *Cobitis* (Cobitidae). V. The origin of even-polyploid species. *Voprosy Ikhtiol* 31(2):202–215 [in Russian].

Vasil'ev VP, Lebedeva EB, Vasil'eva ED, Levenkova ES, Ryskov AP. 2005. A unique diploid–tetraploid unisexual–bisexual complex in fish (Pisces, Cobitidae). *Dokl RAS* 401(4):559–561 [in Russian].

Vasil'eva ED, Vasil'ev VP. 1998. Sibling-species in genus *Cobitis* (Cobitidae). 1. *Cobitis rossomeridionalis* sp. nova. *Voprosy Ikhtiol* 38(5):604–614 [in Russian].

Vrijenhoek RC. 1989. Genetic and ecological constraints on the origins and establishment of unisexual vertebrates. In: Dawley RM, Bogart JP (eds.), *Evolution and Ecology of Unisexual Vertebrates*, Bulletin 466. New York State Museum, Albany, NY, pp. 24–31.

Vrijenhoek RC, Dawley RM, Cole CJ, Bogart JP. 1989. A list of the known unisexual vertebrates. In: Dawley RM, Bogart JP (eds.), *Evolution and Ecology of Unisexual Vertebrates*, Bulletin 466. New York State Museum, Albany, NY, pp. 19–23.

White MJD. 1946. The evidence against polyploidy in sexually reproducing animals. *Am Nat* 80:610–618.

Yamamoto TO, Kajishima T. 1968. Sex hormone induction of sex reversal in the goldfish and evidence for male heterogamety. *J Exp Zool* 168:215–222.

Yin H-B, Sun Z-W, Sun D-J. 2004. Comparative study of DNA-content in five cultivated sturgeon species and kaluga sturgeon. *J Shanghai Fish Univ* 13(2):111–114 [in Chinese].

Yu X, Zhou T, Li Yu, Li K, Zhou M. 1989. *Chromosomes of Chinese Fresh-water Fishes*. Science Press, Beijing, China, 179 pp [in Chinese].

Zan RG, Song Z, Liu WG. 1986. Studies on karyotype and nuclear DNA contents of some cyprinoid fishes, with notes on fish polyploids in China. In: Uyeno T, Arai R, Taniuchi T, Matsuura K (eds.), *Indo-Pacific Fish Biology. Proceedings of the Second International Conference on Indo-Pacific Fishes*. The Ichthyological Society of Japan, Tokyo, pp. 877–885.

Zhang S-M, Yan Y, Deng H, Wang D-Q, Wei Q-W, Wu O-J. 1999. Genome size ploidy characters of several species of sturgeons and paddlefishes with comment on cellular evolution of Acipenseriformes. *Acta Zool Sin* 45:200–206.

Part II
Biology and Aquaculture

Chapter 7
Histological, Histochemical and Ultrastructural Changes in the Digestive Tract of Sturgeon *Acipenser naccarii* During Early Ontogeny

M.V. Ostos-Garrido, J.I. Llorente, S. Camacho, M. García-Gallego, A. Sanz, A. Domezain, and R. Carmona

Abstract Besides species diversification, current priorities in aquiculture include the development of suitable inert diets to enable the artificial feeding of fish from the onset of exogenous feeding. Achieving this aim demands knowledge of the structural and functional characteristics of the digestive tract throughout ontogeny. This chapter provides a histological, histochemical and ultrastructural analysis of the digestive tract of *Acipenser naccarii* sturgeons during different ontogenic development stages (lecithotrophic, lecithoexotrophic and exotrophic periods). These data are essential to our understanding of the digestive capacity of this species during the first stages of life. The knowledge gained will be useful for the investigation and design of appropriate diets to optimize the farming of this species, an activity of major ecological and economic importance.

Keywords *Acipenser naccarii*, digestive tract, histochemistry, histology, ontogeny, ultrastructure

7.1 Introduction

Structural and functional knowledge of the digestive features of a fish species during ontogenic development is an essential prerequisite for the design of suitable inert diets, a prime objective in aquiculture.

Although the general development patterns of Acipenserids have been described (Dettlaff et al., 1993), the existence of inter-species temporal differences means that

M.V. Ostos-Garrido, J.I. Llorente, S. Camacho and R. Carmona
Department of Cell Biology, University of Granada, 18071 Granada, Spain

M. García-Gallego, A. Sanz
Department of Animal Biology, University of Granada, 18071 Granada, Spain

A. Domezain
I + D Department, Sierra Nevada Fish Farm Ltd., Riofrío, 18313 Granada, Spain

the specific chronology of embryonic development must be established in each species.

Acipenser naccarii is native to the Iberian Peninsula, and its presence is of major ecological and economic importance. The optimization of a breeding and fattening protocol for the fry of this species would allow the generation of stock for possible reintroductions into the natural environment. To design this protocol, knowledge of the histological, histochemical, and ultrastructural characteristics of the various gastrointestinal tract regions and an analysis of the modifications that appear during the development of this system is required. The aim is to obtain data that can elucidate the structures involved in larvae digestion and establish the nutritional needs and feeding habits of *A. naccarii* fry in order to improve farming techniques.

7.2 Material and Methods

7.2.1 Experimental Animals

Starting with wild *A. naccarii* parentals, Sierra Nevada Fish Farm (Riofrío, Granada) obtained a generation of sturgeons (F1) raised and reproduced in captivity, producing a second generation (F2) that was used in the present study.

Fertilized eggs were incubated at 15±1°C in McDonald jars with closed circuit and thermoregulators; hatching occurred after 8 days. The animals were then transferred to polyester pools where they were kept at the same incubation temperature (15±1°C) with an artificial photoperiod of 12 h light:12 h dark. Once the reserves of the yolk sac were exhausted, around day 7 post-hatching (PH), the animals were fed with live prey. Inert food was introduced 1 month after fecundation.

7.2.2 Histological, Histochemical and Ultrastructural Studies

Samples from two consecutive years (2002 and 2003) were studied to compare results between two different reproductive cohorts. For 15 days after hatching, samples were taken every 8 h for examination under optical microscopy (OM), transmission electron microscopy (TEM), and field emission scanning electron microscopy (FESEM). From day 15 to day 30 PH, sampling was performed on alternate days. Each sample comprised 11 specimens: 5 for OM, 3 for TEM and 3 for FESEM.

Whole animals were anaesthetized with MS-222, fixed in Bouin's fluid, embedded in paraffin and dorsoventrally sectioned for histological and histochemical study; 5 μm-thick sections were obtained with a Leica RM 2135 microtome and stained with hematoxylin–eosin (H-E), for examination of the general structure of the tissue. Periodic Acid-Schiff (PAS), Alcian Blue (AB) pH 2.5 and pH 1, and combined PAS-AB pH 2.5 techniques were used to analyse mucosubstances.

For the TEM ultrastructural study, portions from the different gastric and intestinal compartment regions were fixed in a 1:1 mixture of 1.5% glutaraldehyde and 1% paraformaldehyde in 0.05 M cacodylate buffer at 7.4 pH for 2 h at 4°C. Samples were then post-fixed in a mixture of 2% osmium tetroxide and 1% ferrocyanide in 0.05 M cacodylate buffer (pH 7.4) for 2 h at 4°C and block staining was performed with 2% uranyl acetate in aqueous solution. After ultrathin sections were obtained, they were placed on copper grids and observed under a Zeiss-902 transmission electron microscope at the Scientific Instrumentation Centre of the University of Granada.

For the FESEM study, fish were fixed with 4% glutaraldehyde, dehydrated with acetone, rinsed in mail acetate, critical point dried with CO_2, coated with carbon, and visualized using a LEO field-emission GEMINI-1530 microscope at the SIC-UGR. In some cases, samples had to be lightly coated with gold to increase the conductivity.

7.3 Results and Discussion

Different phases of embryonic digestive tract development can be histologically characterized according to the feeding pattern of the embryo (Balon, 1975, 1999). Thus, the following phases can be differentiated: a lecithotrophic or endogenous feeding period, which includes embryo and free-embryo phases, when the animals feed exclusively on yolk reserves; a lecithoexotrophic or mixed-feeding period, when animals consume both yolk and food of exogenous origin; and an exotrophic period, when animals feed solely on exogenous food.

7.3.1 Lecithotrophic Period

A. naccarii hatch around day 8 post-fecundation. At hatching, embryos have a partially differentiated digestive system represented by a large ventral yolk sac and spiral valve anlage (Fig. 7.1a). This initial degree of development has also been observed in other sturgeon (Dettlaff et al., 1993; Gisbert and Doroshov, 2003) and fish species.

In just-hatched embryos, the oropharyngeal cavity is completely full of yolk material, and the digestive tract is not connected with the exterior. Hence, embryos feed exclusively on endogenous food reserves, producing an accumulation of yolk material in supranuclear vacuoles of yolk sac endodermal cells (Fig. 7.1b) and spiral valve enterocytes (Fig. 7.1c). This phenomenon results from the holoblastic development of Acipenserid eggs and the participation of yolk-rich endodermal cells in the formation of the alimentary canal (Dettlaff et al., 1993).

The typical bag-shaped yolk sac, which occupies most of the body cavity, is formed by a large central mass of intra-embryonic yolk surrounded by endodermal

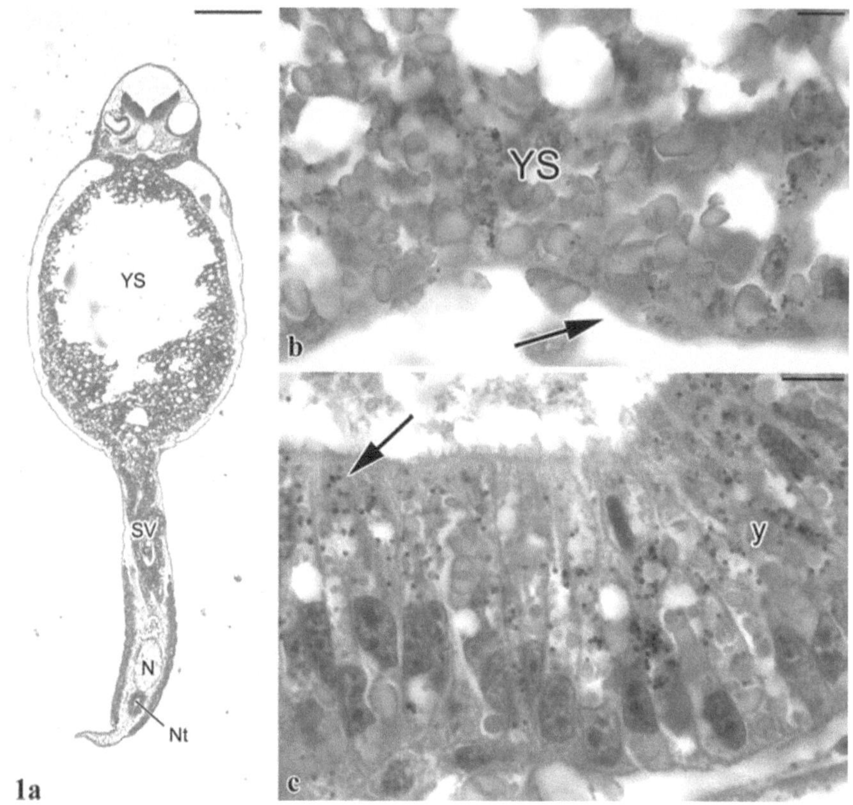

Fig. 7.1 Micrographs showing different histological aspects of just-hatched *Acipenser naccarii* embryo under optical microscopy (H-E). (a) Dorsoventral section of just-hatched *Acipenser naccarii* embryo. (b) Detail of yolk sac endodermal epithelium (arrow). (c) Detail of spiral valve epithelium showing large concentration of granules of both yolk (y) and melanin (arrows). YS, yolk sac; SV, spiral valve; Nt, neural tube; N, notochord. Scale bars: a, 0.5 mm; b, c, 10 μm

epithelium undergoing differentiation (Fig. 7.1b). The spiral valve anlage appears in the posterior part of the body cavity and is formed by a cylindrical epithelium in differentiation, with cells characterized by a basophilic cytoplasm, nuclei in basal position and several apical vacuoles containing numerous yolk and melanin granules (product of yolk degradation) (Fig. 7.1c).

During the free-embryo phase, embryos undergo numerous anatomic, structural and functional changes in their digestive system, preparing them for the capture and assimilation of exogenous food.

Opening of the mouth and anus is observed during this phase, together with zonal differentiation of the gastrointestinal tract. Mouth opening occurs around day 3 PH (Fig. 7.2a). Towards the end of the lecithotrophic period, ciliated cells and mucous cells can be seen in the oropharyngeal cavity alongside typical epithelial cells with microridges at their apical pole (Fig. 7.2b). Ciliated cells remain distributed

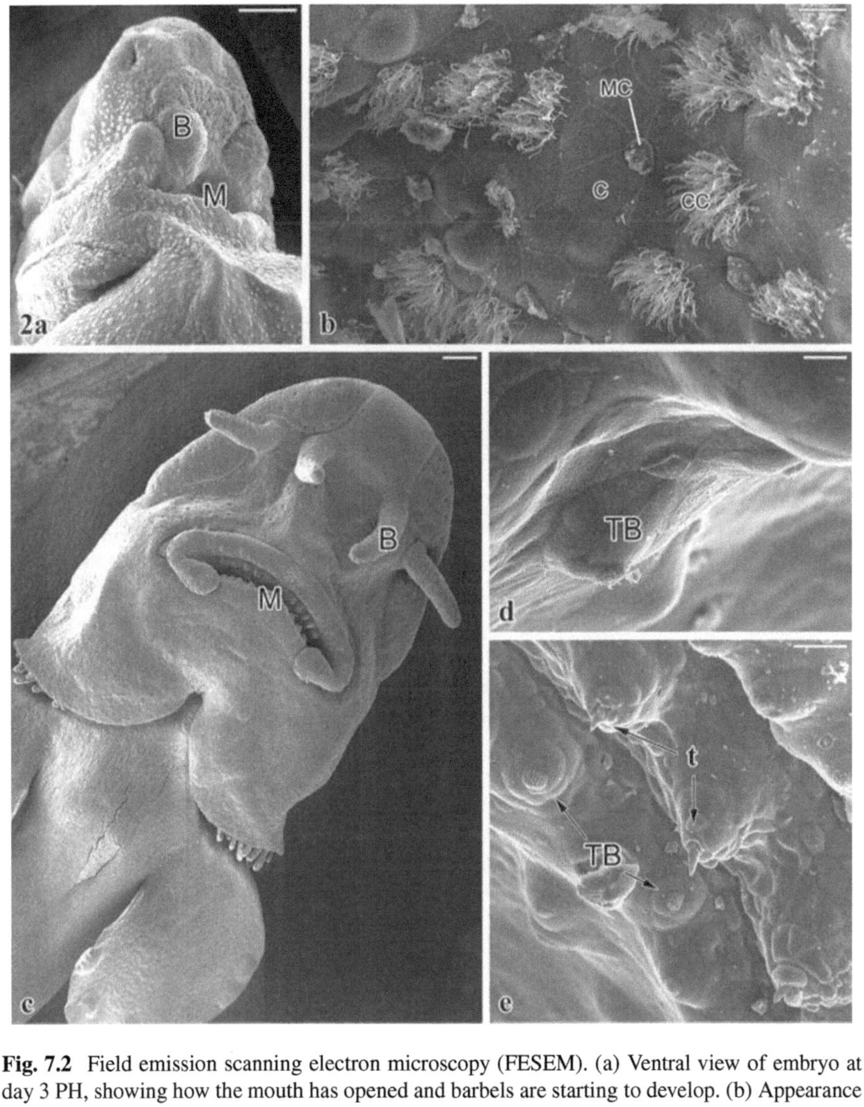

Fig. 7.2 Field emission scanning electron microscopy (FESEM). (a) Ventral view of embryo at day 3 PH, showing how the mouth has opened and barbels are starting to develop. (b) Appearance of oropharyngeal cavity epithelium of free embryo at day 5 PH, showing presence of cells with microridges at apical pole (C), ciliated cells (CC) and mucous cells (MC). (c) Ventral view of embryo at day 7 PH. (d) Detail of taste bud (TB) in oropharyngeal cavity. (e) Detail of teeth (t) that have traversed the oropharyngeal cavity epithelium. M, mouth; B, barbells. Scale bars: a, c, 200 μm; b, d, e, 10 μm

throughout the oropharyngeal cavity until the start of exogenous feeding. Various authors (Iwai and Rosenthal, 1981; Morrison, 1993) have related the presence of this cell type to the facilitation of food passage towards the stomach.

Mucous cells begin to appear around day 5 PH, with basal nucleus and a cytoplasm filled with mucous granules of different sizes and densities, but it is on day 6

PH that they are positive to the combined PAS-AB pH 2.5 technique, demonstrating production by these cells of neutral and acid mucosubstances, e.g. sialomucins.

Taste buds appear on days 3–4 PH, localized on lips and barbels (Fig. 7.2c). They are small and not very numerous at this time, increasing in number and size as the animal grows. Around day 6 PH, these sensory structures are very abundant in the mouth epithelium, localized on the upper and lower lip, oropharyngeal cavity, and branchial chamber (Fig. 7.2d). They already contain taste cells and are functional. Since these chemoreceptor structures are involved in food localization, it can be deduced the fish will respond positively to exogenous feeding from this time on.

One of the main characteristics of sturgeons is the presence of dental structures during the initial development stages and their absence at adult stages. This phenomenon has also been described in other species, e.g. *Hoplosternum littorale*. According to Brannon et al. (1984), the presence of teeth at the start of development (in *Acipenser transmontanus*) is related to the type of diet (benthic invertebrates), allowing fragmentation of food to ease its passage to the gastric portion. The differentiation of dental germs begins with a thickening and later evagination of part of the buccal epithelium towards day 3 PH. Around day 9 PH, teeth have already traversed the buccopharyngeal epithelium and present the characteristic structure described for teleosts (Fig. 7.2e), which is retained until the adult stage.

Development of the oesophagus starts with its posterior part and proceeds towards the buccopharynx. It remains undifferentiated during the first post-hatching days. Towards day 6–7 PH, communication between oesophageal tract and stomach can be observed (Fig. 7.3). Shortly before the start of exogenous feeding, two portions can

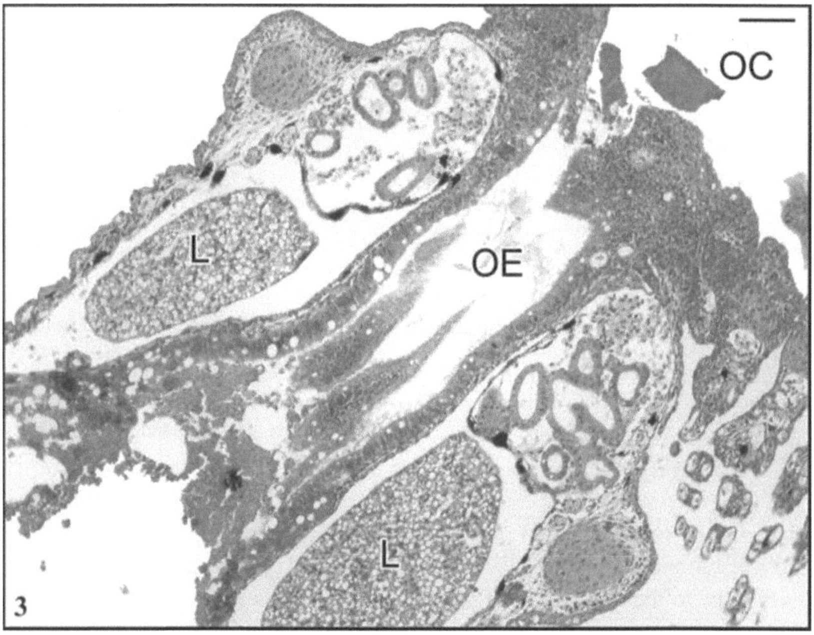

Fig. 7.3 Dorsoventral section of anterior body of free embryo at day 7 PH, showing oropharyngeal cavity (OC), oesophagus (OE) and liver (L). H-E. Scale bar: 100 μm

be differentiated by their cellular characteristics: an anterior portion, showing an epithelium of cylindrical cells with microvilli and numerous secretory mucous cells responsible for producing neutral and acid mucosubstances; and a posterior portion with fewer mucous cells but a high concentration of ciliated cells, which have a food transport function. Gisbert and Doroshov (2003) proposed a lubrication function for these mucosubstances. In a study of pleuronectids, Murray et al. (1996) reported that mucosubstances produced by oesophageal mucous cells may play a major role in the pregastric digestion of food, besides their lubrication effect.

Towards day 3 PH, the yolk sac undergoes a series of modifications leading to acquisition of the caecal stomach structure. The first event observed is the emergence of a furrow in the dorsal-posterior region of the yolk sac (Fig. 7.4a) that progresses anteroventrally and eventually divides the alimentary canal into two well-defined portions: gastric anlage and intestinal anlage. The yolk sac then gradually adopts a more tubular shape, and a proliferation of smooth muscle fibres at the junction with the anterior intestine is observed around day 4 PH. The stomach is not yet divided into its two constituent portions, but it is independent from the intestinal portion. Between days 5 and 6 PH, the front surface of the gastric tract folds until it acquires a U-shape, with the portion corresponding to the glandular stomach forming the descending branch, and the pyloric region rising almost to the level of the oesophagus opening to constitute the ascending branch. The curve stretches to form the closed portion, creating the caecal stomach (Fig. 7.4b). At this stage, the pyloric portion can be readily identified by the presence of a wall of highly developed smooth muscle (Fig. 7.4b). This portion of the gastric compartment,

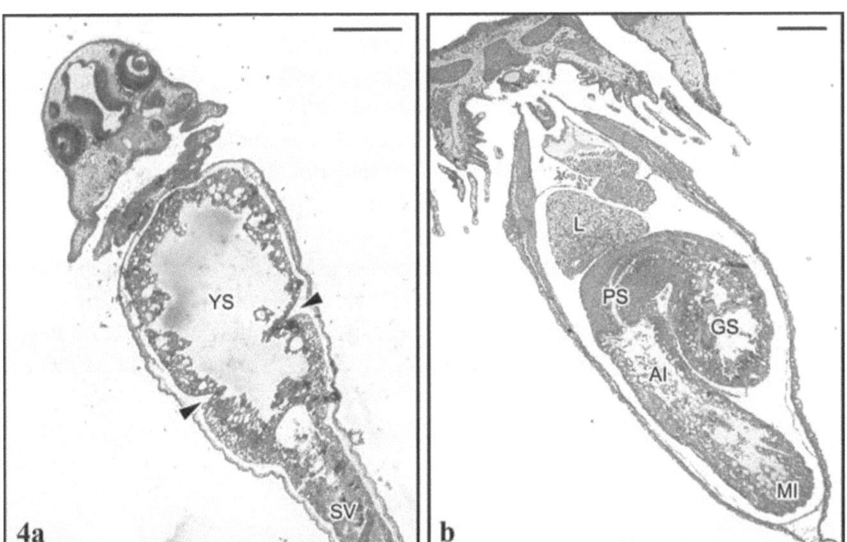

Fig. 7.4 **(a)** Dorsoventral section of free embryo at day 3 PH showing furrow in yolk sac (arrowheads). H-E. **(b)** Dorsoventral section of free embryo at day 7 PH. PAS-AA pH 2.5. YS, yolk sac; SV, spiral valve; GS, glandular stomach; PS, pyloric stomach; AI, anterior intestine; MI, medial intestine; L, liver. Scale bars: a, 500 μm; b, 250 μm

as in other fish species, is involved in the mechanical digestion of food (Buddington and Christofferson, 1985). The phenomenon of gastric compartment torsion has also been observed in other sturgeon species. This was detected by Gisbert et al. (1998) around day 5 PH in *Acipenser baeri* and by Gisbert and Doroshov (2003) around day 6 PH in *Acipenser medirostris*.

At this stage of development, the epithelium of the two stomach regions is columnar and mucosecretory and characterized by PAS positivity of the apical portion of the cytoplasm. Transmission electron microscopy (TEM) studies reveal that the epithelium is formed by two types of cylindrical cells with basal nucleus, those with apical microvilli and those with cilia. Abundant large lipid drops can be observed throughout the gastric epithelium. A series of cellular aggregates, which will give rise to gastric glands, begin to appear in the chorion of the glandular stomach between days 6 and 7 PH. The cytoplasm of glandular cells, designated oxyntic-peptic cells, are PAS-negative, whereas the coating epithelial cells are highly PAS-positive. The latter already present abundant microvilli and numerous ribosomes, lengthened mitochondria and vesicles that are restricted to the apical portion and related to the PAS positivity observed under optical microscopy (OM). Neutral glycoconjugates are commonly present in the gastric epithelium of fish and have been described in species with different feeding habits. According to Grau et al. (1992), their presence may be related to the absorption of easily digestible substances, e.g. disaccharides and short-chain fatty acids. Gisbert and Doroshov (2003) proposed that they may have a protective function against chemical and physical damage.

Lipid drops are relatively abundant in the basal cytoplasm of oxyntic-peptic cells region. We highlight the onset of a tubular-vesicular network in the apical cytoplasm along with numerous supranuclear mitochondria and a well-developed juxtanuclear Golgi apparatus. The absence of secretion granules may indicate that the stomach is not completely functional at this stage. According to Gisbert and Williot (2002), temporal variations in gastric gland emergence among sturgeon species are caused by differences in egg development temperature and/or the amount of yolk.

Intestine differentiation, which starts around day 3 PH with a furrow in the dorsal-posterior region of the yolk sac (see above), progresses in a posterior-anterior direction and the anterior or pyloric intestine is the last to be differentiated. Around day 5 PH, there is a generalized increase in intestine length, and the pyloric and middle intestine can be clearly distinguished by the orientation of the mucosa folds (Fig. 7.4b). The first goblet cells appear in the intestine around day 6 PH and are observed in the pyloric intestine from day 7 PH, specializing in neutral and mostly acid mucosubstance production, as indicated by the colour shown with the PAS-AB pH 2.5 combined technique. This composition in mucopolysaccharides has also been observed in other members of the Acipenseriforme group, e.g. Siberian sturgeon (Gisbert et al., 1998, 1999) and green sturgeon (Gisbert and Doroshov, 2003).

At this time (day 6 PH), an evagination of the anterior intestine appears and starts to be organized into blind sacs to form the pyloric caeca.

Ultrastructural characteristics of intestinal enterocytes and pyloric caeca at the end of the free-embryo period indicate a relationship with the lipid absorption

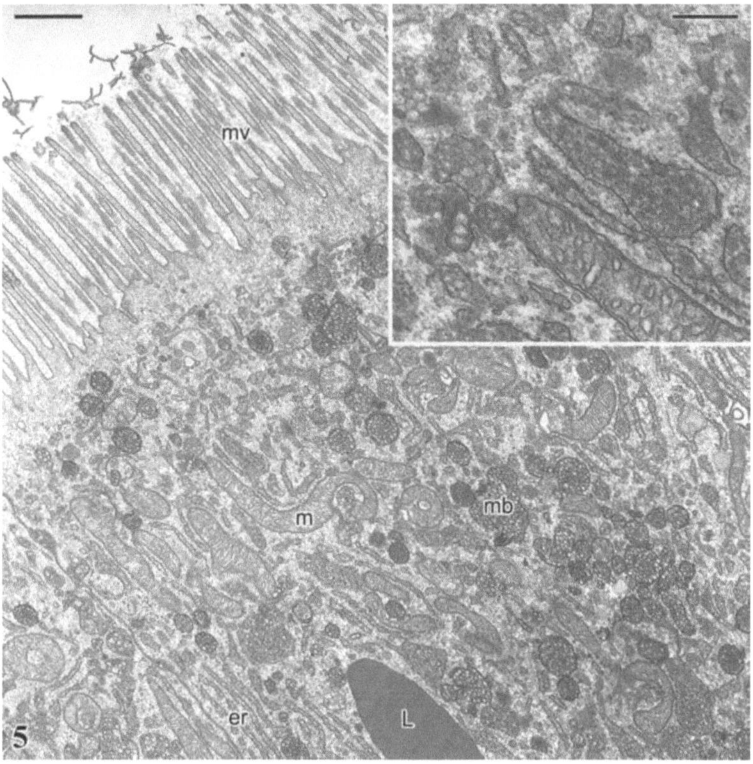

Fig. 7.5 Ultrastructure of enterocyte of free embryo at day 7 PH. TEM. mv, microvilli; m, mitochondria; mb, multivesicular bodies; er, endoplasmic reticulum; L, lipids. Scale bar: 1 μm. Inset: detail of previous image, showing presence of lipid particles within dilated cavities of endoplasmic reticulum. TEM. Scale bar: 0.3 μm

processes (Fig. 7.5). Long apical microvilli show basal invaginations involved in endocytosis processes. The apical cytoplasm presents numerous mitochondria and bodies of multivesicular appearance formed by small lipid particles. Cavities of the endoplasmic reticulum are highly dilated and contain abundant small lipid particles (Fig. 7.5, inset) that move towards and become incorporated in Golgi complex sacules. Once modified, they become detached as large vesicles that migrate towards the lateral plasma membrane, where they discharge their content via exocytosis into intercellular spaces. These small particles can be considered very low density lipoproteins (VLDL). Some of the absorbed lipids accumulate as small drops that coalesce to form large drops distributed throughout the cytoplasm, functioning as energy reserves. These lipid deposits are considered temporary stores of re-esterified fatty acids (triglycerides) that are accumulated when the rate of fatty acid absorption exceeds that of export by enterocytes (Bergot, 1981; Sheridan, 1988). Besides enterocytes and goblet cells, cells bearing cilia and microvilli on their apical surface can also be seen in the intestinal epithelium.

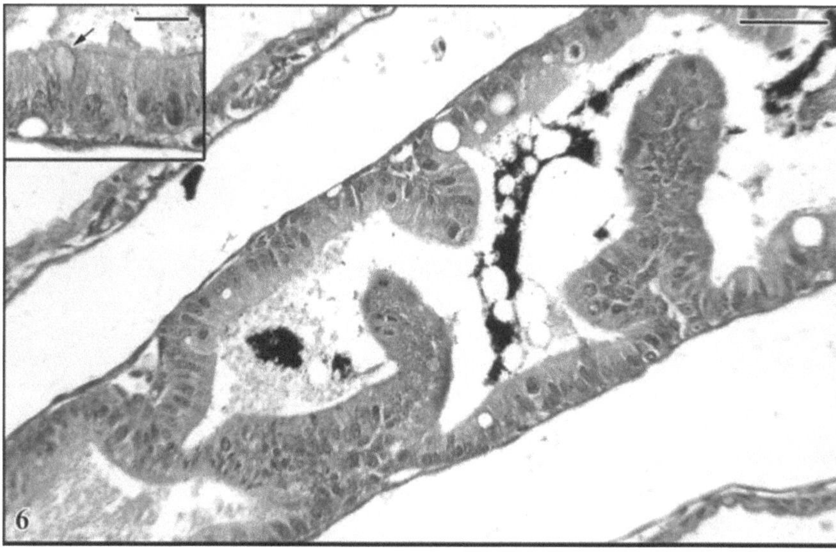

Fig. 7.6 Panoramic image of spiral valve in free embryo at day 3 PH. AA pH 2.5. Scale bar: 50 μm. Inset: detail of globet cell (arrow). Scale bar: 20 μm

Spiral valve differentiation, an anatomic adaptation to enlarge the absorption surface area, takes place very early in the development of the animal, i.e. around day 2–3 PH. The anatomical structure of this intestinal segment is very similar to that found in juvenile specimens (Fig. 7.6). At the beginning of the free-embryo period, this structure is lined by a simple columnar epithelium with some interposed ciliated cells. The first goblet cells appear around day 3 PH and are small and not very numerous in the anterior region, increasing in number towards the posterior region and strongly alcianophilic at pH 2.5 and pH 1, indicating the presence of acid mucosubstances of sialomucin and sulphomucin type. On day 7 PH, these cells are already producing neutral mucosubstances, as shown by their reaction to the PAS-AB pH 2.5 combined technique. This variation in mucosubstance production can be attributed to changes in embryo nutritional needs or feeding habits (Domeneghini et al., 1998). Tibbetts (1997) suggested that while sulphomucin secretion assists in the capture of small food particles, sialomucins and neutral glycoproteins may have a local protection function against the attack of digestive enzymes.

As a sub-product of yolk reserve utilization, an accumulation of melanin is produced in the final portion of the spiral valve. The continuous deposit of melanin granules forms a mass that ultimately occupies the entire lumen of the posterior portion of the spiral valve, i.e. 'the melanin plug' which will be expelled immediately after the first exogenous feeding (Fig. 7.7).

Ultrastructural studies reveal the presence of columnar enterocytes (Fig. 7.8) with long apical microvilli and numerous interdigitations in lateral surfaces. In the

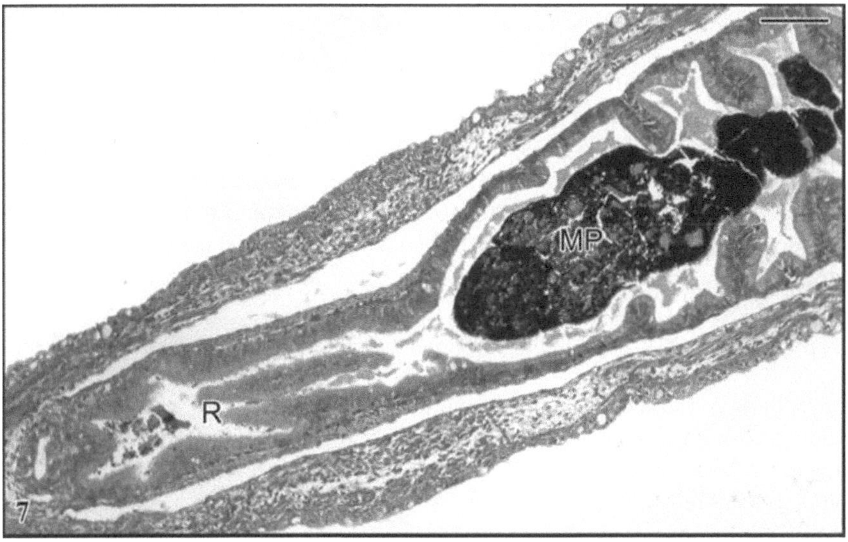

Fig. 7.7 Final portion of spiral valve showing melanin plug (MP). H-E. R, rectum. Scale bar: 100 μm

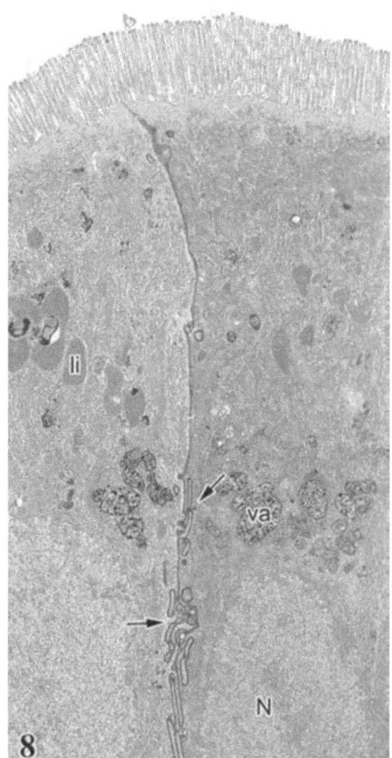

Fig. 7.8 Detail of spiral valve epithelium in free embryo at day 7 PH, showing enterocytes with basal nucleus (N) and lateral surfaces with abundant interdigitations (arrows). TEM. li, lysosomes; va, vacuoles with lipoprotein content. Scale bar: 2 μm

cytoplasm, we highlight the presence of numerous lysosomes, mitochondrias and vacuoles loaded with lipoprotein particles in supranuclear position and related to the Golgi apparatus. Interpolated among these cells can be seen other cells with cilia and microvilli at their apical pole. The enterocytes in this region show a high development of organelles involved in lipid metabolism. In their cytoplasm, alongside lipid drops of varied shape and distribution (apical and basal cytoplasm), there are resynthesized lipoprotein particles at the endoplasmic reticulum level. After their modification and packaging in the Golgi complex, these particles are unloaded towards lateral and basal surfaces, passing from there to the blood via sinusoidal capillaries.

Between days 4–5 PH, the final portion of the spiral valve differentiates, stretching and narrowing to form the rectal anlage. At this time point, the lining epithelium is formed by ciliated columnar cells arranged in a single layer and interposed with some goblet cells. During the following days, the rectum continues to increase in length and degree of epithelial differentiation, with a rising number of goblet and ciliated cells.

7.3.2 Lecithoexotrophic Period

The Lecithoexotrophic Period commences when free embryos consume exogenous food for the first time, around day 8 PH in *A. naccarii*. Animals show a type of mixed feeding during this period, ingesting exogenous food (Artemias) while also feeding on the remaining yolk reserves. This phase is considered critical for development, since it has a major influence on the survival and growth of free embryos.

When *A. naccarii* embryos begin exogenous feeding, no important changes are observed in the histological and histochemical characteristics of the gastrointestinal tract. There is a generalized increase in the number of mucosubstance-producing cells in all digestive compartments, especially in the most distal portion of the spiral valve. In this structure, the melanin plug is expelled around day 9 PH, and continuity is established among the different parts of the gastrointestinal tract. At this time, remnants of exogenous food can be identified in both gastric and intestinal portions (including the spiral valve).

In the oropharyngeal cavity of day 9 PH embryos, the teeth have a conical shape and have pierced the buccopharyngeal epithelium, presenting the characteristic structure described for teleosts. The stomach is anatomically and histologically very similar to juvenile and adult specimens, with features that indicate a high degree of gastric specialization. The glandular stomach (with well-developed gastric glands) would be involved in the beginning of the chemical digestion of food by synthesis of HCl and pepsinogen, whereas the pyloric portion would facilitate food fragmentation by contraction of its highly developed muscular walls.

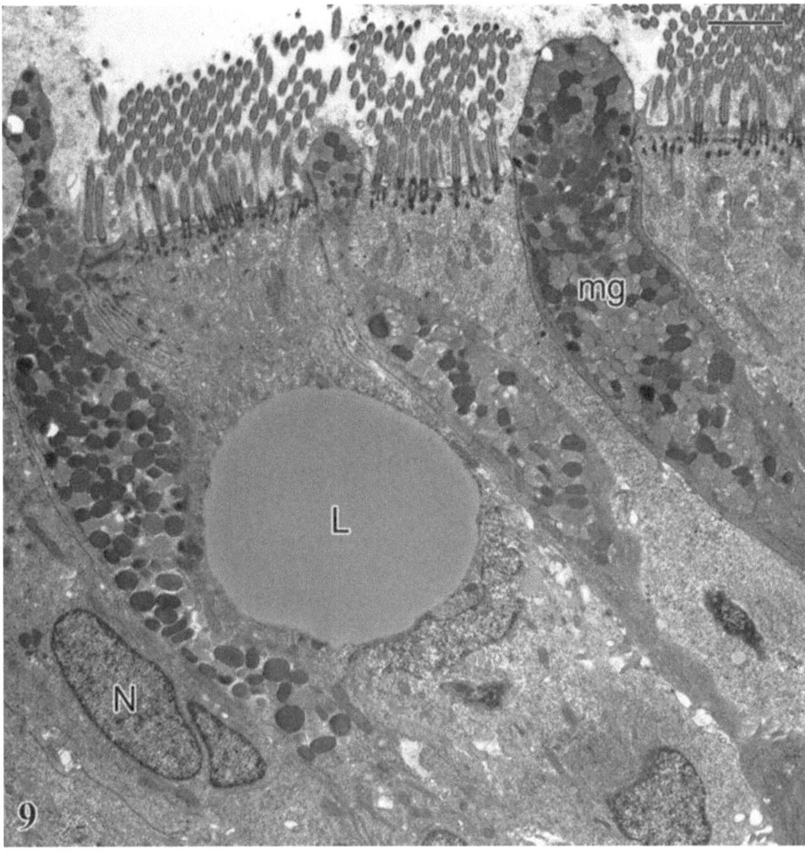

Fig. 7.9 Partial view of oesophageal epithelium of *Acipenser naccarii* juvenile. TEM. N, nucleus; mg, mucous granules; L, lipid droplet. Scale bar: 2 μm

An increase in the length and degree of folding of the mucosa in the intestine affects its different portions, including the pyloric caeca, thereby enlarging the absorptive surface area.

7.3.3 *Exotrophic Period*

During this period, animals acquire an anatomically similar digestive system to that of adults and inert diets are introduced.

The oropharyngeal cavity shows an increase in the number, thickness and length of dental structures.

In the oesophagus, the cylindrical epithelium shows a large predominance of ciliated cells and numerous mucous cells with the same histochemical characteristics as described in the lecithoexotrophic period. These mucous cells (Fig. 7.9) are

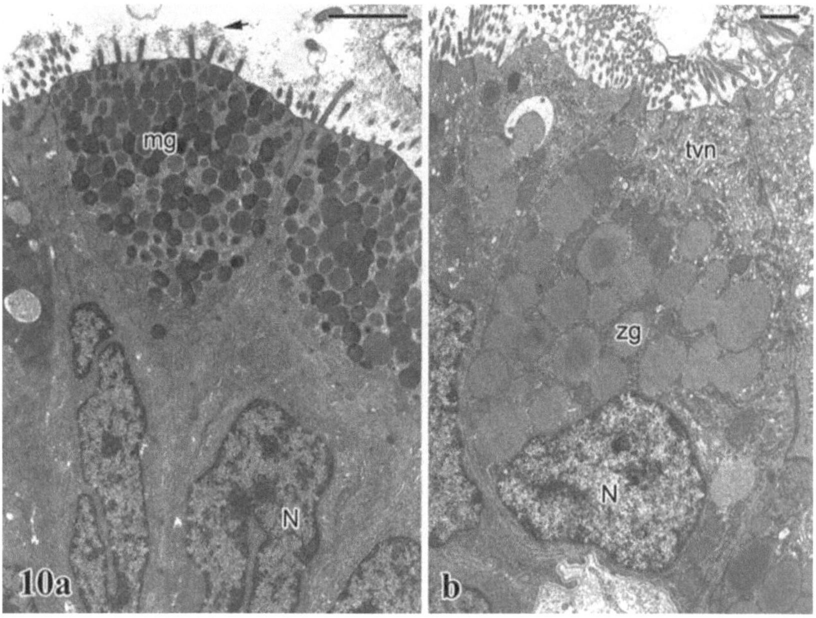

Fig. 7.10 Glandular stomach of Acipenser naccarii juvenile under transmission electron microscopy. (a) Detail of gastric epithelium, showing ultrastructural characteristics of mucosecretory cells with apical microvilli and prominent glycocalix (arrow). (b) Oxyntic-peptic cell of juvenile at 36 days PH, showing a highly developed tubular-vesicular network (tvn) and abundant zymogen granules (zg). N, nucleus; mg, mucous granules. Scale bars: a, 2 μm; b, 1 μm

columnar, with abundant granulations of varied electronic density that unload their secretion products into the lumen by an apocríne mechanism.

Mucosecretory cells of the gastric epithelium present short microvilli on their apical surface with a prominent glycocalix (Fig. 7.10a). In the cytoplasm, the most noteworthy finding is the presence of abundant electrodense granulations, responsible for the PAS-positivity observed in these cells under OM. Oxyntic-peptic cells show a highly developed tubular-vesicular network that is involved in the production of HCl and abundant zymogen granules, indicating the total functionality of the organ during this exogenous feeding period (Fig. 7.10b).

Enterocytes of the spiral valve present a nucleus in basal position and small supranuclear vacuoles, which considerably increase towards the posterior region of the spiral valve (Figs. 7.11a and 7.11b), where a large rise is also observed in the number of goblet cells, which preserve the same histochemical characteristics as at earlier ages.

Ultrastructural studies demonstrate a significant predominance of ciliated cells over enterocytes at this level. The latter show ultrastructural characteristics (deep invaginations among apical microvilli, numerous vesicles and vacuoles of various shapes in the apical cytoplasm, presence of lysosomes, etc.) related to the capture and subsequent degradation of dietary proteins (Fig. 7.11c).

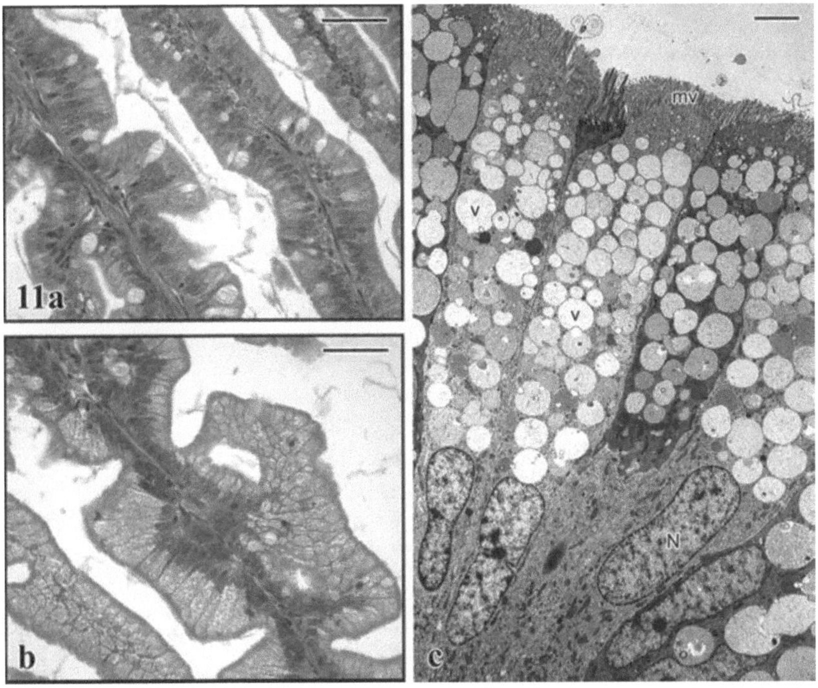

Fig. 7.11 Micrographs showing different histological aspects of spiral valve epithelium of Acipenser naccarii juvenile at 36 days PH. (a) Detail of anterior portion of spiral valve epithelium observed by OM. H-E. (b) Detail of posterior portion of spiral valve epithelium observed by OM. H-E. (c) Panoramic image of spiral valve epithelium of Acipenser naccarii juvenile, showing ultrastructural characteristics of enterocytes, related to the capture and subsequent degradation of dietary proteins. TEM. N, nucleus; mv, microvilli; v, vacuoles. Scale bars: a, b, 50 μm; c, 5 μm

7.4 Conclusions

Results obtained from this anatomical and histological analysis lead to the conclusion that the digestive system in *A. naccarii* develops in an asynchronous manner and proceeds from the distal to the proximal portion, with the spiral valve as the first structure to develop and the glandular stomach as the last. Findings also demonstrate that the yolk lipid component is digested in both anterior intestine and spiral valve during the lecithotrophic or endogenous feeding period but only in the intestine at later development stages. The spiral valve, an adaptation of the intestinal tract that enlarges the absorptive surface area, is specialized in protein capture via pinocytosis during final development phases.

Histochemical studies show that goblet cells in the spiral valve epithelium initially produce acid mucosubstances. However, in anticipation of the complete morphofunctional development of the organ, which occurs when the melanin plug is expelled, they also secrete neutral mucosubstances with important protection and lubrication functions. Mucous cells of the gastric epithelium specialize in neutral mucosubstance production at the end of the endogenous feeding period, coinciding with the emergence of the tubular-vesicular network in oxyntic-peptic cells. In the latter, zymogen granules are only observed from day 19 PH, indicating that acid digestion of food is not assured in *A. naccarii* individuals at the commencement of exogenous feeding.

One month after fecundation, the histological, histochemical and ultrastructural characteristics of the gastrointestinal tract in *A. naccarii* indicate the acquisition of a functionally developed digestive system characteristic of juvenile specimens.

Acknowledgements This study was supported by the MYCT AGL2000-0382-p4-05 project.

References

Balon EK (1975) Terminology in fish development. *J Fish Res Board Can* 32:1663–1670.

Balon EK (1999) Alternative ways to become a juvenile or a definitive phenotype (and on some persisting linguistic offenses). *Env Biol Fishes* 56:17–38.

Bergot F (1981) Fat absorption. In: Fontaine M. (ed.), *Nutrition des Poissons, Actes du Colloque.* CNERNA, Paris, pp. 123–129.

Brannon EL, Melby CL, Brewer SD (1984) Columbia River white sturgeon (Acipenser transmontanus) enhancement. Final report to the Bonneville Power Administration. Portland. Oregon.

Buddington R, Christofferson J (1985) Digestive and feeding characteristics of the chondrosteans. *Env Biol Fishes* 14:31–41.

Dettlaff T, Ginsburg A, Schamalhausen O (1993) *Sturgeon Fishes. Developmental Biology and Aquaculture.* Springer-Verlag, Berlin, 300 pp.

Domeneghini C, Pannelli A, Straini R, Beggetti A (1998) Gut glycoconjugates in *Sparus aurata*, L. (Piosces, Teleostei). A comparative histochemical study in larval and adult ages. *Histopathology* 13:359–372.

Gisbert E, Doroshov S (2003) Histology of the developing digestive system and the effect of food deprivation in larval green sturgeon (*Acipenser medirostris*). *Aquat Living Resour* 16:77–89.

Gisbert E, Williot P (2002) Advances in larval rearing of Siberian sturgeon. *J Fish Biol* 6:1071–1092.

Gisbert E, Rodríguez A, Castelló-Orvay F, Williot P (1998) A histological study of the development of the digestive tract of Siberian sturgeon (*Acipenser baerii*) during early ontogeny. *Aquaculture* 167(3–4):195–209.

Gisbert E, Sarasquete M, Williot P, Castelló-Orvay F (1999) Histochemistry of the development of the digestive system of Siberian sturgeon during early ontogeny. *J Fish Biol* 55:596–616.

Grau A, Crespo S, Sarasquete M, González de Canales M (1992) The digestive system of the amberjack *Seriola dumerili*, Risso: a light and scanning microscopic study. *J Fish Biol* 41:287–303.

Iwai T, Rosenthal J (1981) Ciliary movements in guts of early clupeoid and salangid larvae. *Mar Ecol Prog Ser* 4:365–367.

Morrison CM (1993) Histology of the Atlantic cod, *Gadus morhua:* An atlas. Part four. Eleutheroembryo and larva. *Can Spec Pub Fish Aquat Sci* 119:1–504.

Murray H, Wright GM, Goff GP (1996) A comparative histological and histochemical study of the post-gastric alimentary canal from three species of pleuronectid, the Atlantic halibut, the yellowtail flounder and winter flounder. *J Fish Biol* 48:187–206.

Sheridan MA (1988) Lipid dynamics in fish: aspects of absorption, deposition and mobilization. *Comp Biochem Physiol* 90B:629–690.

Tibbetts IR (1997) The distribution and function of mucous cells and their secretions in the alimentary tract of *Arrhamphus sclerolepis krefftii. J Fish Biol* 50:809–820.

Chapter 8
The Developmental Anatomy of the Heart of the Sturgeon *Acipenser naccarii*

J.M. Icardo, A. Guerrero, A.C. Durán, E. Colvee, A. Domezain, and V. Sans-Coma

Abstract We review the anatomic development of the sturgeon's (*Acipenser naccarii*) heart. Attention has been focussed on the main developmental events that take place during the embryonic and early post-hatching periods. The study examines identification of the early heart tube, cardiac loop formation, and the transformation of the tubular heart into a multi-chambered organ in a temporal sequence. Also included are the development of the heart valves and that of the epicardium. Many of these processes have been followed into adulthood to illustrate the maturation of the different structures with age. On the whole, sturgeon heart formation appears to share many developmental mechanisms with other vertebrates. This indicates the conservation of the mechanisms along the phyletic scale. The development of the *A. naccarii* heart appears to be a very slow process in relation both to other sturgeon species and to other fish classes. This should allow detailed investigation of specific morphologic events. Many of the developmental changes experienced by the heart could well prove useful in establishing the chronology of both embryonic and juvenile specimens.

List of abbreviations used in the figure legends A: atrium; Ao: ventral aortal; A–V: atrioventricular canal; C: conus; dpf: days post-fertilization; dph: days post-hatching; E: endocardium; HBF: head body fold; L: liver; M: myocardium; OFT: outflow tract; OP: olfactory placode; olfactory pit; P: prosencephalon; PC: developing pericardial cavity; PR: pituitary rudiment; R: rhombencephalon; SEM: scanning electron microscope; SV: sinus venosus; T: trunk; TL: tail; TS:, transitional segment; V: ventricle; v: developing valves; I, II, and III, branchial arches

J.M. Icardo, E. Colvee
Department of Anatomy and Cell Biology, University of Cantabria, 39011 Santander, Spain

A. Guerrero, A.C. Durán and V. Sans-Coma
Department of Animal Biology, University of Málaga, 29071 Málaga, Spain

A. Domezain
Department R&D, Piscifactoría de "Sierra Nevada", 18313 Riofrío, Granada, Spain

R. Carmona et al. (eds.) *Biology, Conservation and Sustainable Development of Sturgeons*,

Keywords Sturgeon, *Acipenser naccarii*, heart development, cardiac loop

8.1 Introduction

Sturgeons belong to one of the most ancient groups of Osteichthyes, having thrived in both freshwater and seawater from the low Jurassic (Bemis et al., 1997). The long survival of this fish is explained by a number of specific factors, such as the wide range of spawning temperatures or ecological plasticity (see Dettlaff et al., 1993), and, concomitantly, the cardiovascular system of sturgeons evolved to match the specific physiologic requirements. As a general rule, the heart (and the entire cardiovascular system) is designed to fit the characteristics of any species or group of species. On the other hand, the heart is the first organ to develop, and early heart contraction is necessary to supply oxygen and nutrients to the rapidly growing embryo (Burggren et al., 1997).

A thorough knowledge of the anatomy and the embryology of the sturgeon heart is of great interest from several perspectives. Structural analyses of both embryos and adults would not only enhance our biological knowledge but could also be of comparative significance. For instance, the sturgeons have preserved primitive features that relate them to the Chondrichthyans, and the egg structure is very similar to that of the paraphyletic group of amphibians (see Dettlaff et al., 1993). The classic description of the sturgeon heart includes four chambers arranged in series: sinus venosus, atrium, ventricle and outflow tract (OFT). The last is formed by a long conus arteriosus, which contains the conus valves, and is followed distally by the so-termed intrapericardial segment of the ventral aorta (Santer, 1985; Farrell and Jones, 1992). However, detailed structural descriptions of the sturgeon heart are scarce, and existing reports cover only certain differentiative (Khloponin, 1979) and structural (Mykeblust and Kryvi, 1979) aspects of ventricular myocytes, the vascularization pattern (Romenskii, 1978), and the existence of lymphoid-like tissue in the subepicardium (Scatizzi, 1933; Fange, 1986). Thus, a comprehensive view of the anatomy and structure of the sturgeon heart is still lacking. The paucity of fundamental data is even more marked in the developing embryo. Hardly anything is known of the development of the sturgeon heart except for the existence of an S-shaped tube that could be used for embryo staging (Dettlaff et al., 1993), and timing data on the onset of the heartbeat. Thus, it may come as no surprise that, even in the adult heart, many morphologic traits have yet to be fully characterized.

In recent years, we have conducted a systematic study of the anatomy and embryology of the heart of the autochthonous sturgeon *Acipenser naccarii*. Many of the results have been published already (Icardo et al., 2002a,b; Guerrero et al., 2004; Icardo et al., 2004), and several others await further elaboration. We provide here a summary of our morphologic findings, hoping to stimulate research in this area. The functional analysis of the heart (and of the entire cardiovascular system), and the responses of the heart to different experimental settings, can be found elsewhere (for instance, see Maxime et al., 1995, 1998).

8.2 Materials and Methods

This study has been carried out on embryos and alevins of the autochthonous sturgeon *A. naccarii* (Bonaparte 1836) obtained from Sierra Nevada Fishery at Riofrío, Granada, Spain. Embryos and alevins were maintained at 16°C during the first 10 months of development. Thereafter, the mean water temperature was 15°C. A time period of 8 days spans egg fertilization and the time of hatching. Embryos were collected at days 4–8 post-fertilization (dpf). Most alevins were collected between days 1 and 28 post-hatching (dph). The first day post-hatching coincided with the time of mass hatching (Dettlaff et al., 1993). Data obtained from alevins older than 28 dph, and from adult specimens, have also been included. This has allowed us to follow several morphogenetic events into adulthood.

It should be underscored that the precise staging of *A. naccarii* is yet to be done (for staging of other sturgeon species, see Dettlaff et al., 1993). Such a work is beyond the scope of this review. Due to the lack of detailed information, we have based our observations on developmental days. These observations have been performed on specimens collected at different years, showing a close correlation between the developmental features and the developmental time in all the cases. To help in future studies of fine stage characterization, heart and body development have been presented together in previous reports (Icardo et al., 2004; also, see Fig. 8.1). The illustrations that accompany this work have been obtained by using standard light- and electron-microscope techniques. Full description of the techniques used can be found elsewhere (Icardo et al., 2002a,b; Guerrero et al., 2004; Icardo et al., 2004).

8.3 The Egg and the Heart: The Embryonic Period

The developmental period studied here starts at 4 dpf. On this day, the egg is spherical (Fig. 8.1a). The dorsal side of the embryo shows the developing neural tube. The head appears spread on the surface of the embryo, being separated from the embryonic body by the head-body fold. The pituitary anlage, the eye rudiments and the olfactory placode appear on the head surface (Fig. 8.1a and b). The first pair of branchial arches are seen in the neck region. The lateral plates of the mesoderm are seen to migrate under the ectoderm, being fused rostral to the head. Since the lateral plates carry the bilateral heart primordium in all vertebrates, the ectoderm was opened in this area. This exposed the primitive heart tube, which is contained within the developing pericardial cavity (Fig. 8.1b and c). The heart appears attached to the embryonic body through the first pair of aortic arches cranially, and through the vitellin veins caudally. As occurs in all vertebrates (Icardo, 1997), the early heart shows a tubular structure. It is formed by the myocardium on the outside and endocardium on the inside. A layer of acellular cardiac jelly appears sandwiched between the two tissue layers (Icardo et al., 2004). All the heart cells (indeed, all embryonic cells) show numerous yolk particles and lipid droplets in their cytoplasm. These inclusions will remain in the cells until the third or fourth day after hatching.

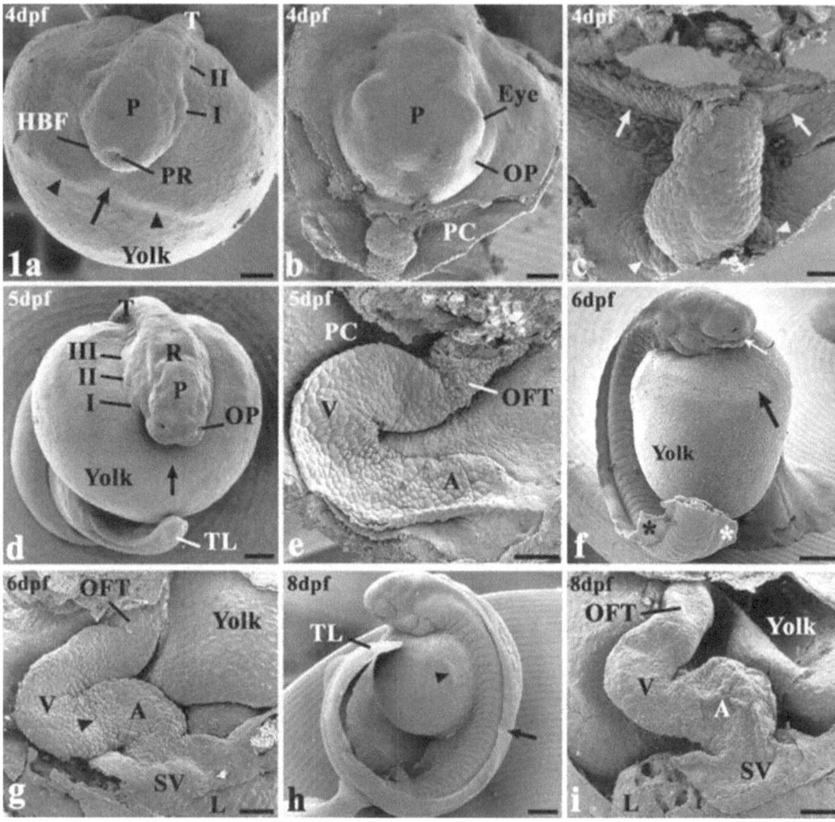

Fig. 8.1 SEM composite illustrating the development of the sturgeon embryo. (a) Arrowheads indicate the migrating lateral plates of the mesoderm. Arrow indicates heart position. (b) Opening of the primitive pericardial cavity exposes the primitive, straight heart tube. (c) In this micrograph, the heart tube is bent forward, remaining attached to the body through the first pair of aortic arches (arrows) cranially, and through the two developing vitellin veins (arrowheads) caudally. (d) The mouth is delineated, and the trunk and somites are very prominent. Arrow indicates heart location under the ectoderm. (e) The heart adopts a C-shape. (f) Part of the tail has been cut off. The anlage of the dorsal and ventral fins (asterisks) are much more prominent than in the previous stage. Black arrow indicates the flattened area overlying the pericardial cavity. White arrow indicates the superior and inferior lips. (g) The outflow tract and atrium appear in contact. Arrowhead indicates the atrioventricular segment. (h) The yolk sac is greatly reduced in size. Arrow indicates the dorsal fin. Arrowhead indicates the anlage of the lateral fin. (i) At this stage the heart loses the C-shape, straightens and unloops (see text for details). Magnification bars: a, 200 μm; b, 100 μm; c, 50 μm; d, 200 μm; e, 100 μm; f, 300 μm; g, 100 μm; h, 300 μm; i, 100 μm (reproduced from Icardo et al., 2004, with modifications)

At 5 dpf the prominent neural tube appears flanked by the somites (Fig. 8.1d). In the head, the prosencephalon and the rhombencephalon have developed as local expansions of the anterior neural tube, the eyes are less prominent because of the invagination of the lens placode, and the olfactory pits appear in an anterior and lateral position. The tail has elongated and its tip is separated from the body by the formation of a body fold. Opening of the pericardial cavity as described above exposes the heart, which at this stage has lengthened and is bent to the right to display a C-shape (Fig. 8.1e). Like all vertebrate hearts, the developing sturgeon heart undergoes a process of bending and rightward rotation known as cardiac looping (see Patten, 1922). Concomitant with looping, the first regional divisions appear. These are, from cranial to caudal, the outflow tract, the ventricle and the atrium. The onset of heart beat could not be set very precisely. However, the heart is vigorously beating at this stage. At 6 dpf the head is more prominent (Fig. 8.1f). The superior and inferior lips delineate the mouth commissure. The anlages of the dorsal and ventral fins are more apparent than in the previous stage, and the trunk musculature shows a clear segmentation. Reduction of the yolk, together with expansion of the pericardial cavity, separates the heart from the ectoderm and allows for easier dissection. In fact, the entire heart area is flattened. The heart has adopted an S-shape in such a way that half of the tube is apposed to the other half (Fig. 8.1g). The cranial end is formed by the outflow tract. The ventricle has relatively reduced its length, and is separated from the atrium by a discrete atrioventricular sulcus. A new segment, the sinus venosus, appears now, interposed between the atrium and the hepatic anlage. Between 7 and 8 dpf there are no drastic external changes. The embryonic body enlarges progressively (Fig. 8.1h) and the yolk sac becomes reduced to a small sphere in front of the developing abdominal cavity. The anlage of the lateral fin appears as a discrete ectodermal ridge protruding from the yolk sac. The most striking feature at these stages is the unfolding of the heart. Just before hatching, the heart straightens and unloops (Fig. 8.1i).

The above observations represent the average obtained from the study of 6–8 specimens per developmental day. It should be stressed that the morphology of a single embryo does not always correlate with the developmental period. This variability seems to be common to all vertebrates, has been reported for other sturgeon species (Dettlaff et al., 1993), and is more patent when the embryo is younger. In fact, significant developmental differences between single specimens have been detected at least until the tenth day post-hatching (for instance, compare Fig.8.1b and c). In this section, the main external features of the embryo have been related to the developmental modifications experienced by the heart. Our work may help to establish the detailed chronology of the *A. naccarii* embryo.

Another feature that should be stressed at this point is the very slow rate of *A. naccarii* development. This is remarkable when comparisons are made with other fish species. For instance, the formation of the cardiac loop, observed here at 5 dpf, occurs 36 h post-fertilization in the teleost *Danio rerio* (Alexander and Stainier, 1999). In relation to other sturgeon species, *A. naccarii* requires approximately double the time to undergo basic developmental events. In *A. gueldenstaedti colchicus*, the straight heart tube and the cardiac loop form 53 and 60 h post-fertilization, respectively (Dettlaff et al., 1993) (versus 4 and 5 dpf in *A. naccarii*). The slow rate of *A. naccarii* development offers a great advantage. It allows step-by-step analysis of the occurrence of many developmental mechanisms, which would be much more difficult at higher developmental rates.

8.4 The Post-Hatching Heart: From Alevins to Adulthood

8.4.1 External Modifications

A striking feature observed in this period is the unlooping of the heart (Icardo et al., 2004). At day 1 post-hatching (dph), the heart takes the shape of an elongated tube lying on the endoderm that covers the yolk sac. More strikingly, the cardiac tube undergoes a secondary process of looping that carries the heart into a C-shape again. At 2 dph, the ventricle is located to the right of the midline, the atrium is growing cranially, and the sinus venosus is in direct contact with the developing liver, draining the liver sinusoids (Fig. 8.2a). The two vitellin veins empty into the lateral ends of the sinus venosus. Subsequently, the heart experiences a counter clockwise movement (Fig. 8.2b). As a result (1) the ventricle is displaced to the left to occupy a caudal position; (2) the atrium 'ascends' to locate first to the right of the outflow tract and, then, dorsal to it; (3) the atrioventricular area becomes a distinct segment, and (4) the sinus venosus comes to occupy a dorsal position with respect to the ventricle. This process of chamber morphogenesis occurs between 3 and 9 dph. Thus, at 9–10 dph, the external anatomy of the heart (Fig. 8.2c) is comparable to that of the adult (Fig. 8.2d).

The process of chamber morphogenesis and spatial rearrangement that occurs in the sturgeon during the post-hatching period is similar to that observed in other vertebrates (for instance, see Icardo and Manasek, 1991; Icardo, 1996). However, the formation of a primary and a secondary cardiac loop has not been detected in any other species. The unlooping of the early heart occurs around the time of hatching. It may simply result from the heart being passively deformed by the straightening of the embryonic body and the increase in length of the pericardial cavity. However, the presence of the primary loop is intriguing. It consists of bending and rightward rotation, which coincides with the first step of looping in avians and mammals (see Männer, 2004). The absence of any associated morphogenetic process indicates that it is a process intrinsic to the heart. Curiously, the properties that determine this appear to have been conserved phylogenetically: when the early heart tube of the avian embryo is excised and cultured, it undergoes a process of bending and rightward rotation (Nakamura and Manasek, 1978). Thus, the sturgeon heart appears to constitute an excellent model to study the determinants of the early looping. The secondary loop observed in the sturgeon represents the entire loop process (Männer, 2004) of avian and mammals. It is very plausible that the completion of looping is only possible in association with other morphogenetic processes. In fact, the secondary loop occurs in the sturgeon in association with the formation of the alimentary canal, i.e. the formation of the foregut in birds and mammals.

Another important process occurring during the early post-hatching period is the transformation of the outflow tract (OFT) into a two-segment chamber. Initially, the entire OFT is covered by an external myocardial layer positive for myosin, as demonstrated by the MF-20 antibody (Guerrero et al., 2004). However, the distal portion of the OFT loses the skeletal muscle identity and becomes

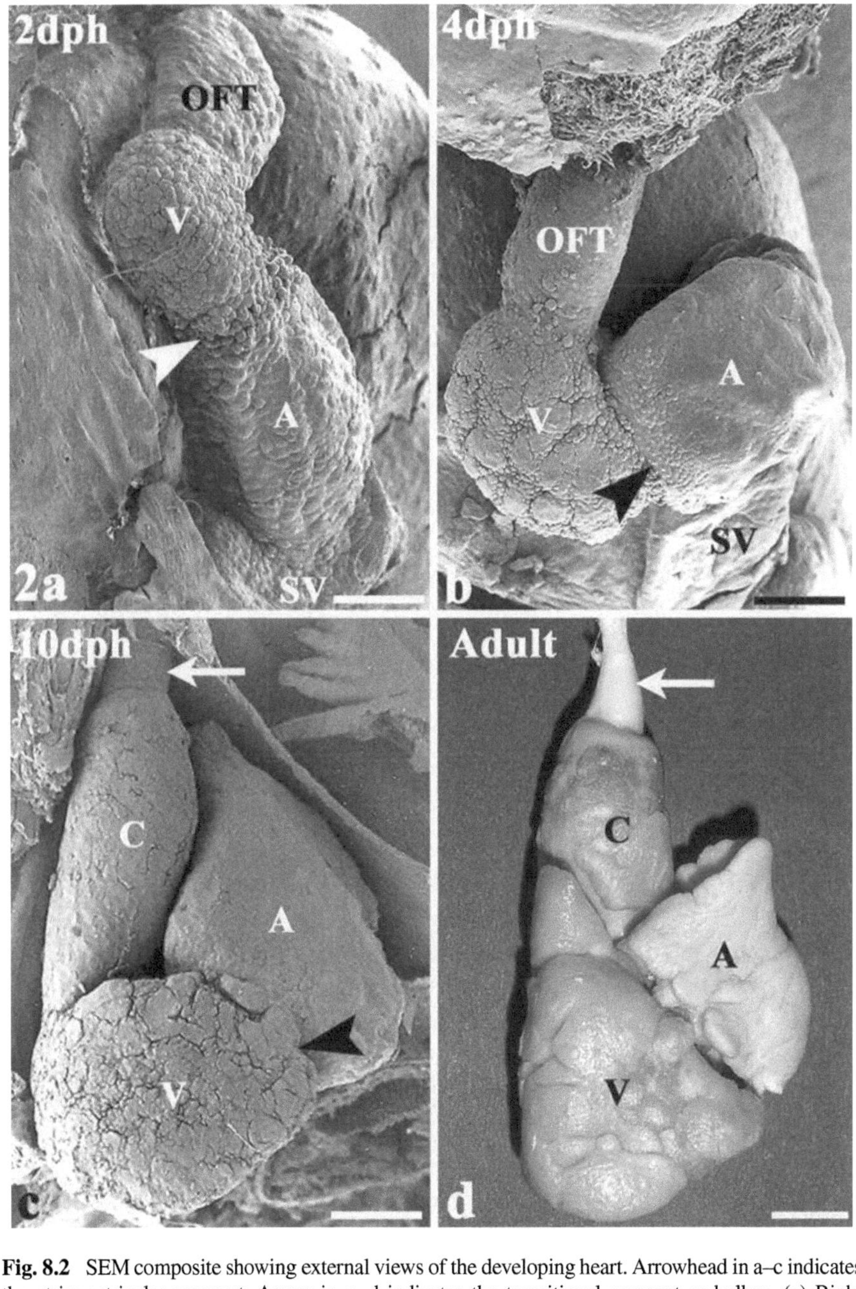

Fig. 8.2 SEM composite showing external views of the developing heart. Arrowhead in a–c indicates the atrioventricular segment. Arrow in c, d indicates the transitional segment or bulbus. (a) Right lateral view. The heart has unlooped. Only the ventricular segment is deviated to the right. (b) Frontal view. The heart is experiencing a counter clockwise movement: the ventricle is locating caudally; the atrium is 'ascending' to locate to the left of the OFT; the sinus venosus is being displaced dorsally. (c) Left lateral view. The epicardium has covered conus and ventricle. Note the cylinder appearance of the transitional segment. (d) Left lateral view. The surface of conus and ventricle shows a cobblestone appearance due to the development of the subepicardial tissue. The transitional segment is cone-shaped. Magnification bars: a–c, 100 μm; d, 1 cm (a–c, reproduced from Guerrero et al., 2004)

positive for β-actin antibody, which identifies smooth muscle cells (Fig. 8.3a and b). This shift in antibody reactivity occurs between 4 and 5 dph and coincides with distinct morphologic and histological changes (Guerrero et al., 2004). Externally, the distal OFT becomes narrower than the rest of the OFT and shows a more regular calibre (Fig. 8.2c). Internally, it progressively adopts an arterial-like organization (Fig. 8.3c and d). In the adult heart, this segment shows organized layers of smooth muscle cells, collagen and elastin (Icardo et al., 2002b). Since the organization of these layers is different on each side of the pericardial attachment, the distal OFT portion has recently been termed the transitional segment (Icardo et al., 2002b), suggesting a different origin from the rest of the ventral aorta, or bulbus arteriosus (Guerrero et al., 2004), by homology with the teleost bulbus. The early changes observed in the distal OFT have been suggested to represent a phenotypic transformation, from a myocardial to a smooth muscle-like phenotype (Guerrero et al., 2004). However, recent comparative studies have cast some doubts on this

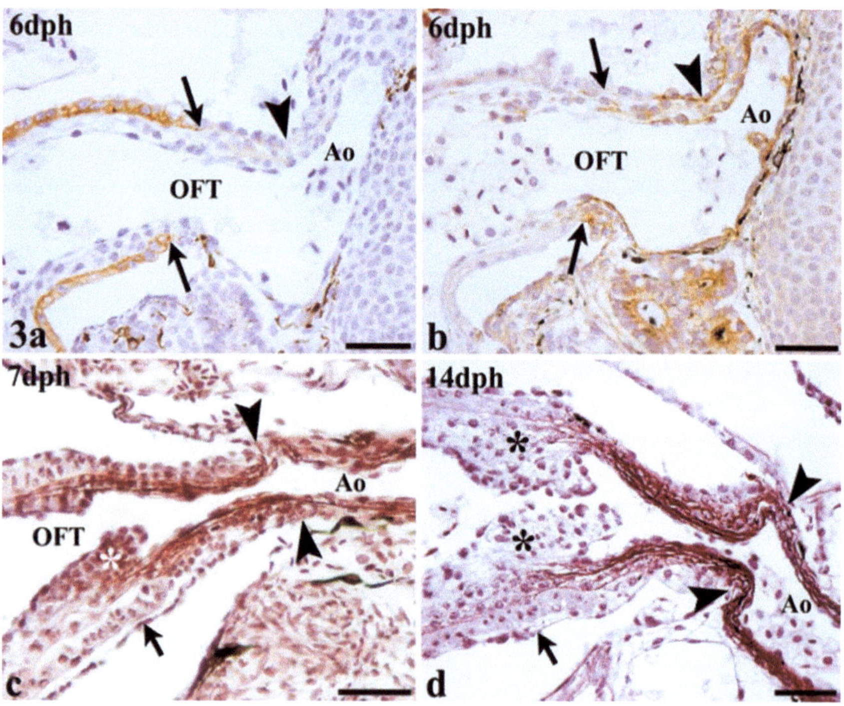

Fig. 8.3 Composite showing the transformation of the distal portion of the OFT. Large arrows in (a) and (b) mark the separation between the two OFT portions. Small arrows in (c) and (d) indicate the epicardium. Arrowheads indicate the boundary between the OFT and the ventral aorta. Asterisks in (c) and (d) indicate the endocardial cushions. (a) MF20 immunostaining. The distal portion of the OFT is negative for MF20; the proximal portion is positive. (b) β-actin immunostaining. The distal portion of the OFT is positive for this antibody; the proximal portion is negative. Note the complementary staining pattern. (c) Orcein staining for elastin. Elastic fibres are present in the aortic wall and in the cardiac OFT down to the base of the endocardial cushions. (d) Orcein staining. Note the better organization of the elastic material. Magnification bars: a, b, 100 μm; c, d, 50 μm

interpretation. It has been reported that the teleost bulbus, the supposed homologue of the transitional segment, never expresses myosin markers (Grimes et al., 2006). Thus, a phenotypic transformation could not occur during embryogenesis. If these findings can be applied to the sturgeon heart, what we may be seeing is merely the growth of a new segment that is incorporated at the arterial end of the developing heart through the pharyngeal mesenchyme. The new segment would originate from a secondary heart field (see Abu-Issa et al., 2004), and through the incorporation of cells of neural crest origin. While the existence of a secondary heart field has not been shown in any fish species as yet, neural crest derived cells reach the arterial pole of the heart in other vertebrates (Kirby et al., 1983), including the teleost group (Li et al., 2003).

8.4.2 Internal Modifications

We deal here with the modifications experienced by the inner surface of the heart. The development of the ventricular trabeculations, the transformation of the outflow tract into a two-segment chamber, and the development of the cardiac valves, are reviewed in this section.

Through the early post-hatching period the heart maintains the tubular structure seen during the embryonic period. However, the tubular organization is soon to undergo dramatic changes. In the ventricular chamber, endocardial evaginations progress through the cardiac jelly and reach the myocardium. Then, disruption of the myocardium by the endocardial invasion resuts in the primitive trabeculae (Fig. 8.4a). This process starts at 3–4 dph and progresses from the ventricular apex up. Trabeculation in the sturgeon resembles the trabeculation process observed in other vertebrates (Icardo and Fernández-Terán, 1987). Once the first trabeculations have appeared, the ventricular wall is formed by an external, one-cell-thick myocardial layer (Fig. 8.4a and b), and a system of trabeculae that radiate from the cono-ventricular region (Fig. 8.4b). At 45 dph, the external myocardial layer thickens and is formed by a two-cell-thick layer. For a still undetermined period of time, the ventricular wall maintains this architectural organization. It is only much later that the external myocardium compacts (Fig. 8.4c) and the ventricular wall adopts the typical adult configuration: a thick, compact layer on the outside and a complex trabecular system on the inside (Fig. 8.4d). The development of a compact layer in the sturgeon ventricle is a late feature in comparison with other vertebrates (see Icardo and Manasek, 1991). However, it constitutes a very important event from the physiologic viewpoint. The existence of a compact layer accounts for an increase in the mechanical performance of the heart. For instance, adult eels (*Anguilla anguilla*) show higher cardiac output and better responses to increased afterload than juvenile eels. The improved cardiac performance relates to the increase in the compacta thickness that occurs in the eels with age (Cerra et al., 2004). Most probably, this is also the case in sturgeons. However, studies of cardiac function have only been carried out in adult hearts (Crocker et al., 2000), and comparative studies are still lacking. The development of the ventricular compacta

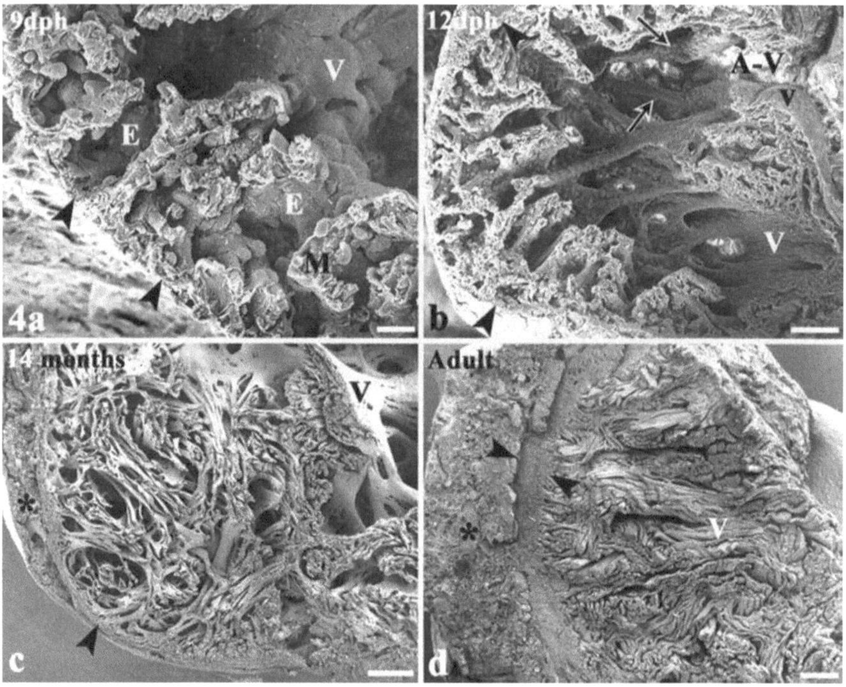

Fig. 8.4 SEM composite depicting the development of the ventricular chamber. Arrowheads indicate the external myocardial layer in all micrographs. Asterisks in (c) and (d) indicate the subepicardial space. (a) The early trabeculae are formed by a core of myocardium surrounded by endocardium. (b) The trabeculae keep a general radial orientation. In the atrioventricular canal, the developing valves show a system of chordae tendineae (black and white arrowheads). The chordae appear as linear protrusions arising from the ventricular wall. (c) The external myocardial layer shows an increase in thickness. However, most of the ventricular wall is formed by a complex system of trabeculae. (d) Adult. The compacta is well developed. The ventricular trabeculations are thicker and appear more compacted than in the previous figure. Magnification bars: a, 10μm; b, 50μm; c, 20μm; d, 300μm

occurs parallel to the acquisition of a collagenous skeleton in the myocardium (Icardo et al., 1996). The collagen surrounding the myocytes is involved in the maintenance of the passive properties of the myocardium and in myocardial mechanics (Weber, 1989). The parallel development of the two systems, the ventricular compacta and the collagen scaffold, should account for an improved cardiac performance in the sturgeon with age.

The OFT of the sturgeon heart is another segment undergoing profound remodelling with age. The transformation of the OFT into a proximal, myocardial portion or conus, and a distal, arterial portion or transitional segment (or bulbus) has been commented on above. The next step is the development of endocardial cushions, the forerunners of the heart valves. The endocardial cushions appear as accumulations of cardiac jelly at specific heart locations. In the conus, endocardial cushions appear at 3–4 dph. They are arranged into proximal and distal rows. Each row is formed by 4–5 cushions. The cushions are initially acellular, but they are soon invaded by cells derived from the endocardium. Histological examination of the developing cushions shows the presence of endocardial cells projecting cytoplasmatic processes and migrating into the underlying cardiac jelly. Endocardial cushion activation, cell migration into the matrix, and transformation of the epithelial cells into mesenchyme, is a sequence common to all the vertebrates (Markwald et al., 1977; Kinsella and Fitzharris, 1980; Icardo, 1989). As a result of this process, the acellular cushions become transformed into mesenchymatous structures (Fig. 8.5a). Cushion formation is followed by a process of excavation that takes place at the distal end of each cushion (Fig. 8.5a). Excavation of the distal cushions starts at 8 dph, these cushions being transformed into the distal valve row of the adult heart.

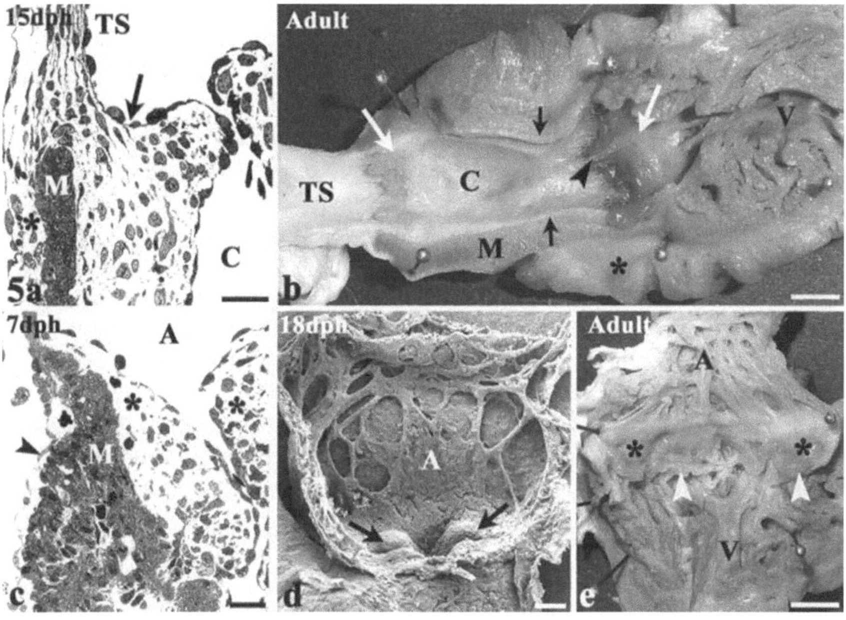

Fig. 8.5 Composite showing several aspects of the developmental anatomy of the sturgeon heart. (a) Semithin longitudinal section. The distal cushions are being excavated (arrow). Note the cranial border of the myocardial cuff. The transitional segment is formed by cells oriented longitudinally. The subepicardial tissue (asterisk) is a loose mesenchyme. (b) The conus and ventricle have been opened longitudinally. White arrows indicate the proximal and distal valve rows. Note the absence of valves in between. Small black arrows indicate the subendocardial elastic tissue. Arrowhead indicates valve chordae. The subepicardium (asterisk) is mostly occupied by adipose tissue. (c) Semi-thin longitudinal section through the atrioventricular canal. The cushions (asterisks) are mesenchymatous structures growing toward each other. Arrowhead indicates the epicardium. (d) SEM micrograph. The lateral wall of the atrium has been eliminated. Note the presence of numerous pectinate muscles. Arrows indicate the presence of four cushions in the atrioventricular canal. (e) Longitudinal section of the heart at the level of the atrioventricular region. The atrioventricular valve (asterisks) is exposed. Arrowheads indicate the free border of leaflets. Many pectinate muscles end in the atrial margin of the valve. Magnification bars: a, c, d, 10 μm; b, e, 0.5 cm

Excavation of the proximal cushions starts at 10 dph. Strikingly, these cushions are excavated at the distal end and in the middle portion. Thus, they constitute the anlage of the proximal and the middle valve rows (Icardo et al., 2004). The mechanisms of valve excavation are not yet understood. It is also unclear whether there is a real process of excavation in the sturgeon or whether it is the result of the distal growth of the valve tissue. In addition, the excavation process is quite uniform in the distal row. In the adult, the distal valves have a regular semilunar appearance (Fig. 8.5b). However, excavation is a highly irregular process at the level of the proximal cushions. It may start at a different level on each cushion, or be more accentuated in some cushions than in others, or even fail. This results in the wide variety of valve morphologies observed in the proximal and middle valve rows of the adult heart (Icardo et al., 2002a). It must be stressed that cushions never develop in the middle portion of the conus. This results in the presence of a long conus segment devoid of valve structures (Fig. 8.5b). This feature has also been observed in other sturgeon species, as in *A. sturio* and *A. huso* (Stöhr, 1876), suggesting that it is a conserved feature within the Acipenseridae (Icardo et al., 2002a). The conus valves retain their mesenchymatous appearance throughout life, being formed in the adult by a loose connective tissue that contains fibroblasts, collagen and some elastic material (Icardo et al., 2002a). In the adult, the conus valves also show a variable number of supporting chordae (Icardo et al., 2002a). These chordae extend from the inner side of the valve to the conus wall (Fig. 8.5b), being formed by a dense collagen core covered by the endocardium. The chordae system associated with the conus valves of the sturgeon is similar to that reported for the conus valves of elasmobranches (Hamlett et al., 1996). The presence of this chordae system appears to be a common trait of phylogenetically basal fish groups. The arterial valves of birds and mammals lack any chordae.

Cardiac valves also develop at the atrioventricular orifice. As in the conus, the atrioventricular valves develop from a set of endocardial cushions. In the sturgeon, two large cushions (ventral and dorsal) and two smaller (lateral) cushions appear at 4 dph. Then, the cushions are invaded by endocardially derived cells, grow toward each other (Fig. 8.5c), and appear fully developed at 10 dph. Curiously, the location and size of the cushions mimic the developmental pattern of birds and mammals (see for review Icardo, 1984; Icardo and Manasek, 1991). The difference here is that the opposite cushions do not fuse (Fig. 8.5d), and the canal is never divided. Progressively, the cushions become thinner, undergo a process of excavation on the ventricular face, and appear to be transformed directly into the atrioventricular leaflets. This process has been completed by 18–20 dph in most cases. The system of chordae tendineae develops at 22–24 dph. Preliminary observations in the sturgeon indicate that the mode of formation of the chordae tendineae is similar to that observed in several elasmobranches (Hamlett et al., 1996). The early chordae appear as ridges that originate from the ventricular wall underlying the developing leaflets (Fig. 8.4b). These ridges become independent from the ventricular wall through the development of confluent perforations. The developing chordae establish the connection between the valve leaflets and the ventricular myocardium. Papillary muscles are absent or appear reduced to discrete muscle protrusions. In the adult heart, the number and shape of the leaflets varies (Fig. 8.5e). This variation could be due to partial fusion of the developing cushions.

Very little is known about the development of the venous pole of the heart. The atrium is always a thin-walled chamber. At 7–8 dph, the development of the pectinate muscles starts. These muscles appear first as discrete ridges arising from the myocardial wall. Soon, they protrude into the atrial lumen (Fig. 8.5d) and form a complicated system of trabeculae that run toward the atrioventricular aperture (Fig. 8.5e). The sinus venosus is also a thin-walled chamber. It receives the ducts of Cuvier at its lateral ends, and communicates with the developing liver through a wide opening. Thus, the liver sinusoids empty directly into the sinus venosus (Fig. 8.1i). This anatomic relationship is maintained, at least through the 28-dph stage. Whether this relationship is maintained in the adult heart is unknown as yet.

The sinoatrial opening is also guarded by a valve system. The composition of the adult sinoatrial valve in the sturgeon is unknown as yet. In elasmobranchs, the valve leaflets have been reported to contain collagen, fibroblasts (Hamlett et al., 1996), and muscle (Gallego et al., 1997). In the dogfish (*Scyliorhinus canicula*), the sinoatrial valves develop as lateral infoldings of the myocardial wall (Gallego et al., 1997). The formation of these infoldings is asymmetric, and the sinoatrial opening is displaced to the right. Preliminary observations indicate that the sinoatrial valve of the sturgeon develops in a similar way, with the infoldings located in dorsal and ventral positions.

8.5 Extra-Cardiac Origin of Newly Recruited Cells

Up to this point we have been dealing mostly with the origin and developmental fate of the myocardium and endocardium. However, cardiogenesis is a very complex process. In the course of development, new cell types are incorporated into the developing heart through well-defined patterns of cell migration. This incorporation takes place in a coordinated temporal-spatial sequence, establishing new cellular systems that become integrated in the developing heart (Icardo and Manasek, 1991). We exclude from this consideration the conduction system cells. To the best of our knowledge, the development of this system has not been studied in the sturgeon. However, the conduction system cells differentiate from the pre-existing myocardium in all the vertebrates.

The formation of the epicardium (the visceral pericardium) constitutes the best known example of the addition of new cell types to the growing heart. In all vertebrates, the epicardium originates from an extracardiac source, the proepicardium, located in the vicinity of the sinus venosus (see Männer et al., 2001). In the sturgeon, we have located this organ at the junction between the sinus venosus and the hepatic anlage. It is formed by numerous villous protrusions that grow across the pericardial cavity and at 4–5 dph reach the dorsal surface of the heart. Preliminary observations indicate that the transfer of cells from the villi to the heart surface occurs by direct contact between the villi and the myocardial surface, and through the liberation of numerous free-floating vesicles that are secondarily attached to the outer myocardium. This dual mode of formation is common to all vertebrates, though the prevalence of one or other of the two mechanisms appears to be quite specific (Männer et al., 2001). The sturgeon conus and ventricle are entirely covered by the epicardium during the second week of post-hatching development (Fig. 8.2c) (we have no data on the time of the epicardial covering for the atrium and sinus venosus). A sequence that is also common to all vertebrates is the development of a subepicardial space (Fig. 8.5a and c). This space subsequently becomes populated by mesenchyme cells and by vessels. Small vessels can be identified in the sturgeon subepicardium as early as 9–10 dph. The vessels invade the conus myocardium at 22–23 dph. The vascularization of the ventricular myocardium occurs much later. This should occur at the time of the formation of the compacta, though the exact developmental period remains unknown (see above). In the sturgeon, as well as in the elasmobranchs (Santer, 1985), small vessels also reach the trabecular laycr.

With the formation of the epicardium, the sinus venosus becomes attached to the dorsal side of the ventricle by the development of the so-termed sinu-ventricular ligament. This ligament contains a large vein that connects the coronary vascular tree with the sinus venosus. It may be a way of draining the venous blood from the heart into the main circulatory stream. However, the fate of the sinu-ventricular ligament in the sturgeon is unknown. This ligament is a transient structure in amphibian and chick embryos, persisting through life in crocodiles (revised in Männer et al., 2001). It has yet to be verified whether the sinu-ventricular ligament is a transient structure in the sturgeon or whether it remains through adult life.

Another interesting feature of the subepicardium is the presence, in the conus and in the ventricle, of nodular structures that contain lympho-hemopoietic, thymus-like tissue (Icardo et al., 2002b). It is precisely the presence of these nodes, and that of the surrounding adipose tissue, that gives to the heart surface a cobblestone appearance (Fig. 8.2d). The sturgeon subepicardium appears to be a complex organ involved in the production of blood cells of the white line, and in the establishment of the cell-mediated immune responses. We do not know the exact time at which the precursor lymphoid cells colonize the subepicardium. A small number of developing nodes can be observed at 90 dph. Full development is observed months after hatching (Fig. 8.4c). Although the sturgeons have a cervical thymus (Fange, 1986), the development of the subepicardial lympho-hemopoietic tissue may well help in mediating resistance to infections and the effectiveness of the different vaccines. As occurs with the thymus of many fish and of other vertebrates, this tissue atrophies when the animal reaches sexual maturation (Icardo et al., 2002b). In the adult, the subepicardial space is mostly occupied by adipose tissue (Figs. 8.4d and 8.5b).

Cells of neural crest origin reach the developing heart in birds and mammals, where they contribute to the morphogenesis of the outflow tract (Kirby et al., 1983; Jiang et al., 2000). In teleosts, cardiac neural crest cells invade all the chambers of the heart contributing to both myocardial and smooth muscle cells

(Li et al., 2003; Sato and Yost, 2003). On comparative grounds, it can be hypothesized that neural crest cells also invade the developing sturgeon heart. The longitudinal orientation of cells observable in the distal segment of the early outflow tract, as well as the presence of melanocytes in this region, suggests the incorporation of new cells through the pharyngeal mesenchyme. The possible existence of a secondary heart field contributing to the distal outflow tract has been discussed above. Finally, the sturgeon subepicardium is also invaded by unmyelinated (autonomic) nerve fibres. These are most probably involved in the control of the vagal tone of the heart. If cardiac innervation in the sturgeon is similar to that of the elasmobranches, they should lack adrenergic innervation (Burggren et al., 1997).

8.6 Conclusions

The emergence of form and function in the heart is a synchronized process which occurs very early in development. It is necessary to supply the tissues of the rapidly growing embryo. This chapter summarizes several of the most salient points of the developmental anatomy of the sturgeon heart. Our knowledge is still incomplete, especially if we compare it with the tantalizing amount of data that refer to the development of the heart in birds and mammals. On the other hand, we are just beginning to grasp the specifics of sturgeon heart development. Many of the developmental events described here—the formation of the cardiac loop, the development of the heart valves, the formation of the epicardium and the development of the coronary vascular bed—are similar to those occurring in the heart of other vertebrates. Thus, many of the developmental mechanisms appear to be similar, if not identical. This indicates the conservation of these mechanisms along the phyletic scale. The development of the *A. naccarii* heart appears to be an extremely slow process in relation to other sturgeon species, and to other fish classes as well. This should allow detailed investigation of specific morphologic events. Many of the developmental changes experienced by the heart could well prove useful in establishing the chronology of both embryonic and juvenile specimens. Future studies, including experimental work and comparative analyses of gene expression, will hopefully contribute to our understanding of the developmental anatomy of the sturgeon heart.

Acknowledgement Supported by grants CGL2004-06306-C02-01/BOS and AGL2000-0382-P4-05 from the Ministerio de Educación y Ciencia.

References

Abu-Issa R, Waldo K, Kirby ML. 2004. Heart fields: one, two or more? *Dev Biol* 272:281–285.

Alexander J, Stainier DYR. 1999. Mutations affecting cardiac development in zebrafish. In: Harvey RP, Rosenthal N (eds.), *Heart Development*. Academic, San Diego, pp. 91–110.

Bemis WE, Findis EK, Grande L.1997. An overview of Acipenseriformes. *Environ Biol Fish* 48:25–71.
Burggren WW, Farrell A, Lillywhite H. 1997. Vertebrate cardiovascular systems. In: Dantzler WH (ed.), *Handbook of Physiology, Sect. 13, Comparative Physiology*, vol. 1. Oxford University Press, New York, pp. 215–308.
Cerra MC, Imbrogno S, Amelio D, Garofalo F, Colvee E, Tota B, Icardo JM. 2004. Cardiac morphodynamic remodelling in the growing eel (*Anguilla anguilla* L). *J Exp Biol* 207:2867–2875.
Crocker CE, Farrell AP, Gamperl AK, Cech Jr JJ. 2000. Cardiorespiratory responses of white sturgeon to environmental hypercapnia. *Am J Physiol Regul Integr Comp Physiol* 279:R617–R628.
Dettlaff TA, Ginsburg AS, Schmalhausen OI. 1993. *Sturgeon Fishes. Developmental Biology and Aquaculture*. Springer-Verlag, Berlin, 300 pp.
Fange R. 1986. Lymphoid organs in sturgeons (Acipenseridae). *Vet Immunol Immunopathol* 12:153–161.
Farrell AP, Jones DR. 1992. The heart. In: Hoar WS, Randall DJ, Farrell AP (eds.), *Fish Physiology, Vol. XII, The Cardiovascular System*, Part A. Academic, San Diego, pp. 1–87.
Gallego A, Durán AC, de Andrés AV, Navarro P, Muñoz-Chápuli R. 1997. Anatomy and development of the sinoatrial valves in the dogfish (*Scyliorhinus canicula*). *Anat Rec* 248:224–232.
Grimes AC, Stadt HA, Sheperd IT, Kirby ML. 2006. Solving an enigma: arterial pole development in the zebrafish heart. *Develop Biol* 290:265–276.
Guerrero A, Icardo JM, Durán AC, Gallego A, Domezain A, Colvee E, Sans-Coma V. 2004. Differentiation of the cardiac outflow tract components in alevins of the sturgeon *Acipenser naccarii* (Osteichthyes, Acipenseriformes). Implications for heart evolution. *J Morphol* 260:172–183.
Hamlett WC, Schwartz FJ, Schmeinda R, Cuevas E. 1996. Anatomy, histology, and development of the cardiac valvular system in elasmobranches. *J Exp Zool* 275:83–94.
Icardo JM. 1984. The growing heart: an anatomical perspective. In: Zak R (ed.), *The Growth of the Heart in Health and Disease*. Raven Press, New York, pp. 41–80.
Icardo JM. 1989. Changes in endocardial cell morphology during development of the endocardial cushions. *Anat Embryol* 179:443–448.
Icardo JM. 1996. Developmental biology of the vertebrate heart. *J Exp Zool* 275:144–161.
Icardo JM. 1997. Morphogenesis of vertebrate hearts. In: Burggren WW, Keller B (eds.), *Development of Cardiovascular Systems. Molecules to Organisms*. Cambridge University Press, New York, pp. 114–126.
Icardo JM, Fernández-Terán MA. 1987. Morphologic study of ventricular trabeculation in the embryonic chick heart. *Acta Anat* 130:264–274.
Icardo JM, Manasek FJ. 1991. Cardiogenesis: developmental mechanisms and embryology. In: Fozzard HA, Haber E, Jennings RB, Katz AM, Morgan HE (eds.), *The Heart and Cardiovascular Systems. Scientific Foundations*. Raven Press, New York, pp. 1563–1586.
Icardo JM, Colvee E, Tota B. 1996. Morphological organization of the sturgeon (*Acipenser naccarii*) heart with special reference to the collagenous architecture. In: VII International Symposium on Fish Physiology, Oslo, Norway, p. 93.
Icardo JM, Colvee E, Cerra MC, Tota B. 2002a. Structure of the conus arteriosus of the sturgeon (*Acipenser naccarii*) heart. I. The conus valves and the subendocardium. *Anat Rec* 267: 17–27.
Icardo JM, Colvee E, Cerra MC, Tota B. 2002b. The structure of the conus arteriosus of the sturgeon (*Acipenser naccarii*) heart. II. The myocardium, the subepicardium and the conus-aorta transition. *Anat Rec* 268:388–398.
Icardo JM, Guerrero A, Durán AC, Domezain A, Colvee E, Sans-Coma V. 2004. The development of the sturgeon heart. *Anat Embryol* 208:439–449.
Jiang X, Rotwitch DH, Soriano P, McMahon AP, Sucov HM. 2000. Fate of the mammalian neural crest. *Development* 127:1607–1616.
Khloponin PA. 1979. Morphologic aspects of cardiac myocyte differentiation in Black-sea-Sea of Azov sturgeon. *Arkh Anat Gistol Embriol* 77:44–51 (in Russian).

Kinsella MG, Fitzharris TP. 1980. Origin of cushion tissue in the developing chick heart: cinematographic recordings of in situ formation. *Science* 207:1359–1360.

Kirby ML, Gale TF, Stewart DE. 1983. Neural crest cells contribute to normal aortopulmonary septation, *Science* 220:1059–1061.

Li YX, Zdanowicz M, Young L, Kumiski D, Leatherbury L, Kirby ML. 2003. Cardiac neural crest in zebrafish embryos contributes to myocardial cell lineage and early heart function. *Dev Dyn* 226:540–550.

Männer J. 2004. On rotation, torsion, lateralization, and handedness of the embryonic heart loop: new insights from a simulation model for the heart loop of chick embryos. *Anat Rec* 278A:481–492.

Männer J, Pérez-Pomares JM, Macías D, Muñoz-Chapuli R. 2001. The origin, formation and developmental significance of the epicardium: a review. *Cell Tissue Organ* 169:89–103.

Markwald RR, Fitzharris TP, Manasek FJ. 1977. Structural development of endocardial cushions. *Am J Anat* 148:85–120.

Maxime V, Nonnotte G, Peyraud C, Williot P, Truchot JP. 1995. Circulatory and respiratory effects of an hypoxic stress in the Siberian sturgeon. *Resp Physiol* 100:203–212.

Maxime V, Nonnotte G, Williot P. 1998. Adaptations respiratoires et circulatoires de l'sturgeon sibérien à une hypoxie environnementale. *Bull Fr Pêche Piscic* 351:377–391.

Mykeblust R, Kryvi H. 1979. Ultrastructure of the heart of the sturgeon *Acipenser stellatus* (Chondrostei). *Cell Tissue Res* 202:431–43.

Nakamura A, Manasek FJ. 1978. Experimental studies of the shape and structure of isolated cardiac jelly. *J Embryol Exp Morphol* 43:167–183.

Patten BM. 1922. The formation of the cardiac loop in the chick. *Am J Anat* 30:373–397.

Romenskii O. 1978. Blood supply of the compact and spongy myocardium of fish, amphibians and reptiles. *Arkh Anat Gistol Embriol* 75:91–95 (in Russian).

Santer RM. 1985. Morphology and innervation of the fish heart. *Adv Anat Embryol* 89:1–102.

Sato M, Yost HJ. 2003. Cardiac neural crest contributes to cardiomyogenesis in zebrafish. *Dev Biol* 257:127–139.

Scatizzi I. 1933. L'organo linfomieloide dello sturione. *Arch Zool Ital* 18:1–26.

Stöhr P. 1876. Ueber den Klappenapparat im Conus arteriosus der Selachier und Ganoiden. *Morph Jb* 2:197–228.

Weber KT. 1989. Cardiac interstitium in health and disease: the fibrillar collagen network. *J Am Coll Cardiol* 13:1637–1652.

Chapter 9
Observations on the Brain Development of the Sturgeon *Acipenser naccarii*

A. Gómez, E. Durán, F.M. Ocaña, F. Jiménez-Moya, C. Broglio, A. Domezain, C. Salas, and F. Rodríguez

Abstract The remarkable range of evolutionary diversification of ray-finned fishes, reflected in the number of species and environmental adaptations, is also evident in their brain structure. Among the ray-finned fishes, the Chondrostei, an early branch of the actinopterygian line, are usually considered the extant taxa most closely related to the primitive actinopterygians living in the Paleozoic Era, and sharing with them numerous characteristics. Unfortunately, current knowledge of the ontogenesis of the central nervous system and behaviour of the chondrosteans is rudimentary, despite their evolutionary importance. Thus, the study of the morphologic and functional organization of the chondrostean lineage brain is an essential step in identifying the primitive and derived traits. The present work summarizes recent data on the gross morphology and cytoarchitecture of the brain of the sturgeon *Acipenser naccarii*, during ontogenesis. In addition, the main changes in the development of the five main brain subdivisions have been compared to the onset of different types of behaviour that provide a rough index of sensory and motor maturation.

Keywords Brain, development, *Acipenser naccarii*, sturgeon, sensory and neural maturation, cytoarchitecture, neuroanatomy

9.1 Introduction

Actinopterygian fishes present remarkable evolutionary diversification, evident also in their brain characteristics (Northcutt and Davis, 1983; Meek and Nieuwenhuys, 1998; Nieuwenhuys, 1998; Butler and Hodos, 2005; Ito et al., 2007). Among the actinopterygian fishes, chondrosteans are considered the closest of the extant taxa to

---

A. Gómez, E. Durán, F.M. Ocaña, F. Jiménez-Moya, C. Broglio, C. Salas and F. Rodríguez
Laboratorio de Psicobiología. Universidad de Sevilla, Sevilla, Spain

A. Domezain
Departamento de Investigación y Desarrollo, Piscifactoría de Sierra Nevada, S.L. Camino de la Piscifactoría 2, Riofrío, Granada, Spain

R. Carmona et al. (eds.) *Biology, Conservation and Sustainable Development of Sturgeons*,

the primitive actinopterygians living in the Paleozoic era, and sharing with them many morphological and functional traits (Lauder and Liem, 1983; Bemis et al., 1997). As in all vertebrates, the differentiation of specific brain regions in the actinopterygian fishes is determined by genetic and adaptive factors.

Given the strategic position that sturgeons occupy in the actinopterygian radiation, a detailed description of the normal development of the sturgeon brain and behaviour will provide essential information not only to increase our knowledge on the neurobiology of this species but also to understand the evolution of the vertebrate brain in general. Moreover, as chondrosteans comprise a number of endangered species, additional data on their normal developmental pattern is essential to develop adequate strategies for their conservation and for the achievement of optimal standards of culture in hatcheries and farms.

Unfortunately, current knowledge concerning the brain organization and development in the Chondrostei, and in particular the Acipenseridae, is notably meagre, as it is based almost entirely on anatomical studies carried out during the last years of the nineteenth and the beginning of the twentieth centuries. Thus, von Kupffer (1893, 1906) analysed the development of the central nervous system in the *Acipenser sturio*, and Goronowitsch (1888) and Johnston (1898a,b, 1901), studied the brain of *Acipenser ruthenus* and *Acipenser rubicundus*, respectively. The studies on the ontogenesis of the brain in the twentieth century were restricted to those carried out by Bergquist (1932) on the development of the diencephalon of *A. ruthenus*, and those by Nieuwenhuys (1962, 1963, 1964) on the development of the telencephalon in several species. Johnston (1898b, 1911) provided the initial data on the anatomical organization of the telencephalon, which remained the only one until the studies by Nieuwenhuys (1964) and Northcutt and Braford (1980). Moreover, the studies probing brain organization in chondrosteans amounted to only a few in the twentieth century (see Nieuwenhuys, 1998). It is noteworthy that there is an increasing number of recent experimental studies which are providing essential data on the general brain organization (Rupp and Northcutt, 1998; Huesa et al., 2006), connectivity (Repérant et al., 1982; Albert et al., 1999; Ito et al., 1999; Yamamoto et al., 1999; Huesa et al., 2000, 2003, 2006) and neurochemical patterns (Kotrschal et al., 1985; Oka et al., 1989; González et al., 1992; Leprêtre et al., 1993; Chiba and Honma, 1994; Adrio et al., 2002, 2005; Trabucchi et al., 2002; Piñuela and Northcutt, 2007) in several species of sturgeons, although the new data on the brain development in these species is still virtually nonexistent.

In this work, we summarize recent data compiled in a systematic study of brain development in *Acipenser naccarii*, a sturgeon species autochthonous of Southern Europe (Garrido-Ramos et al., 1997, 2008; De la Herrán et al., 2004), carried out with specimens obtained from the Sierra Nevada fish farm of Granada, Spain. We have analysed the development of the five main brain subdivisions (Fig. 9.1) in specimens ranging from pre-hatching periods to 1 year of age, by means of several morphological and morphofunctional techniques (Vázquez et al., 2002a,b), which have added detailed information on the neurogenesis and the neuronal and glial cytoarchitecture during brain development. We have also pointed out the relationships

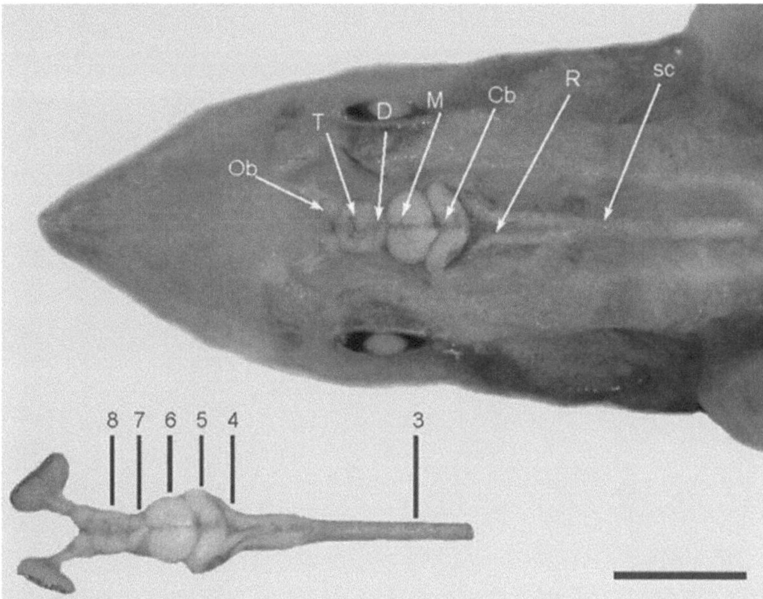

Fig. 9.1 Dorsal view of the head of a 4-month-old *Acipenser naccarii* with its brain exposed in situ. The insert at the bottom shows a dorsal view of the brain, indicating the approximate positions of the transverse sections illustrated in Figs. 9.3, 9.4, 9.5, 9.6, 9.7 and 9.8. Magnification bar =5 mm Abbreviations: *Cb*, cerebellum; *D*, diencephalon; *M*, mesencephalon; *Ob*, olfactory bulb; *R*, rhombencephalon; *sc*, spinal cord; *T*, telencephalon

between the onset of the main changes in brain development with the onset of different types of behaviour that provide a rough index of sensory and motor maturation.

9.2 Development of the Neural Tube and Early Brain Morphology

As in all vertebrates, the neural tube of the *A. naccarii* has a flattened, circular shape and a horizontal longitudinally oriented narrow cavity in the early embryonic stage, but the overall brain organization, consisting of prosencephalon, mesencephalon, and rhombencephalon, is already defined in 2-day sturgeon embryos, with their boundaries signalled by the plica encephali ventralis in the midbrain region (Nieuwenhuys, 1998). At hatching time, the boundaries between the telencephalon and the diencephalon are not clearly defined, but the posterior tuberculum and the primordia of the epiphysis are already evident in the dorsal diencephalon (Fig. 9.2). Although the ventral diencephalon and in particular the ventral hypothalamus develop relatively late, the adenohypophysis is already

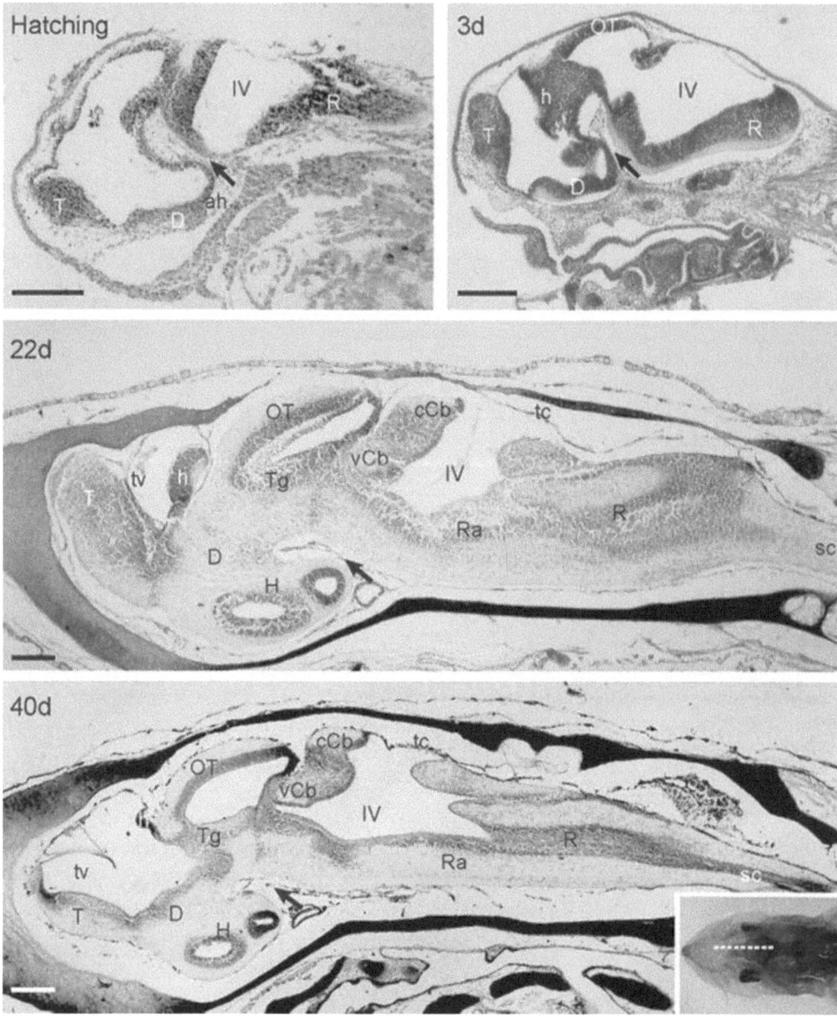

Fig. 9.2 Photomicrographs of Nissl-stained mid-sagittal sections of the entire brain of *Acipenser naccarii* at different developmental ages (the post-hatching day is indicated in each photograph). The arrow marks the position of the plica encephalis ventralis. The insert in the bottom right shows a dorsal view of the head of a larva, indicating the section level. Magnification bars =250 µm. Abbreviations: *ah*, adenohypophysis; *cCb*, corpus cerebelli; *D*, diencephalon; *h*, habenula; *H*, hypothalamus; *R*, rhombencephalon; *OT*, optic tectum; *Ra*, raphe nucleus; *sc*, spinal cord; *T*, telencephalon; *Tg*, tegmentum; *tc*, tela choroidea; *tv*, telencephalic ventricle; *vCb*, valvula cerebelli; *IV*, fourth ventricle

evident on the floor of the diencephalon at hatching time (Grandi and Chicca, 2004, see Fig. 9.2).

In 3-day-old larvae the increase in the depth of the plica encephali produces a transitory but very pronounced curvature in the ventral rhombencephalon (Fig. 9.2).

Whereas in sturgeons the neuropore of the rostral neural tube remains open for a long time, and the telencephalon continues growing and everting over the fish's lifetime (Nieuwenhuys, 1964), the eversion process of the telencephalic hemispheres begins very early in the development, being clearly evident in 10-day-old larvae of *A. naccarii* (see below). In 22-day larvae, notable changes have taken place, as the five main brain subdivisions and the spinal cord already present the adult morphologic pattern (Figs. 9.1 and 9.2). In the mesencephalon, which maintains in great part the appearance of the neural tube in chondrosteans (Nieuwenhuys, 1998), the ventral zone develops earlier (day 22) than the tectum (day 40, Fig. 9.2). In 22-day larvae the dorsal rhombencephalon has diverged, leading to the transformation of the roofplate in a wide tela ependymalis (von Kupffer, 1906; Nieuwenhuys, 1998), and the cerebellum, originated in a double, bilateral anlage, has fused medially (Fig. 9.2). In the *A. naccarii* brain, most of the neurons do not migrate far from the ventricular surface, and the normal adult histological pattern is already evident in 40-day-old specimens (Fig. 9.2). In fact, in the adult brain, only in the cerebellum and the dorsal telencephalon have the neurons spread over the entire width of the wall. Nevertheless, other groups of migrated cells appear in the very large inferior oliva and the nucleus tori lateralis (Vázquez et al., 2002a, see below).

9.3 Development of the Spinal Cord

In 3-day-old embryos the primordium of the spinal cord shows the three main zones, the floor and the dorsal plates, with large cubic cells (Nieuwenhuys, 1998), and the lateral plates, slightly thicker, with elongated elements and a pseudostratified organization (Fig. 9.3). Soon, the ventral and dorsal plates become thicker, gradually diminishing the extent of the neural canal. During the first days of post-hatching development, large dorsal cells appear but these will have completely disappeared in 9-day-old larvae. It has been reported that in *A. sturio*, these large dorsal cells, conspicuous in 2-day-old larvae, are already absent in 28-day-old larvae (von Kupffer, 1906; Nieuwenhuys, 1998). The large cells of the lateral plates and the enlargement of the layer of fibres situated in the periphery can be clearly observed at hatching time in *A. naccarii* (Fig. 9.3).

In chondrosteans, the adult pattern of organization in the grey matter of the spinal cord includes three zones, the dorsal and ventral horns and the intermediate zone. The entire ventrolateral zone of the ventral horn is considered the spinal motor column, as it contains large neurons with distinct morphologies (Fig. 9.3), at least some of which are motoneurons. The other grey zones are composed mostly of small cells, the large ones being completely absent in the dorsal horn (Fig. 9.3). It is important to note that in chondrosteans, as well as in the actinopterygian fish in general, no direct telencephalon-spinal cord connectivity has been observed (Nieuwenhuys, 1998; Butler and Hodos, 2005). Among the descending fibre systems in the ventral funicular the thick fibres of the paired Mauthner cells are particularly conspicuous and clearly evident in 14-day old larvae of *A. naccarii*.

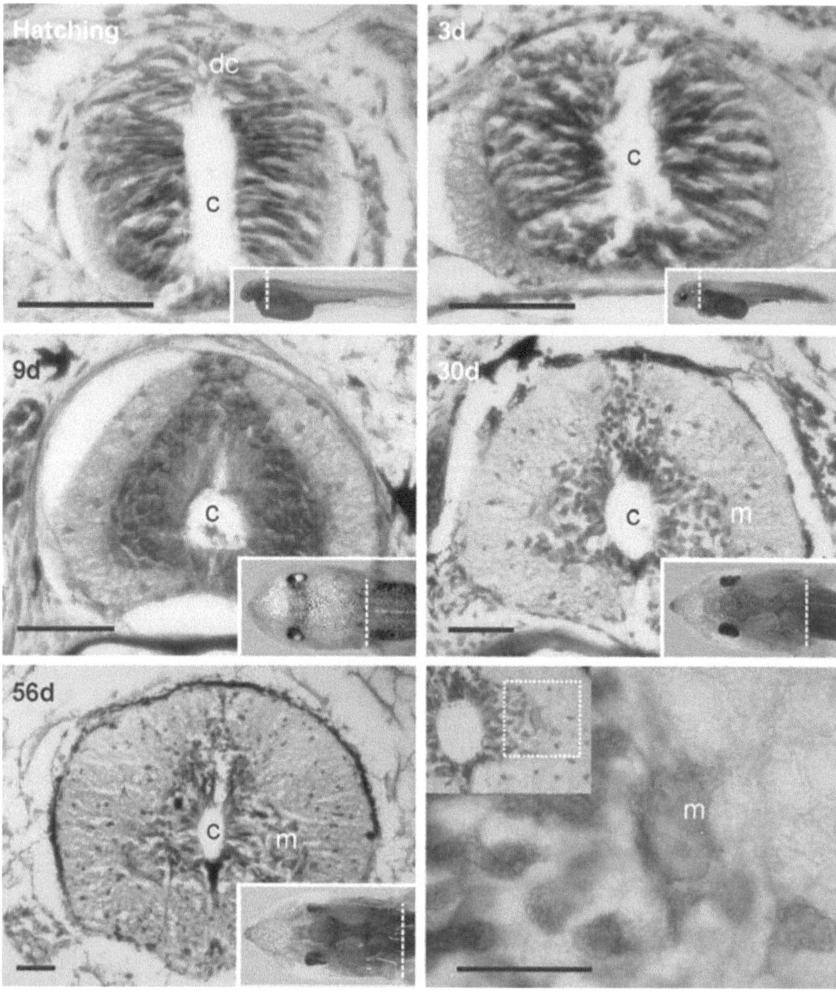

Fig. 9.3 Photomicrographs of Nissl-stained transverse sections through the spinal cord of *Acipenser naccarii* at different developmental stages (the PH day is indicated in each photograph). The level of the section is indicated on the lateral or dorsal view of the larva at the bottom of each photograph. All sections were taken approximately at the level of the pectoral fin. The figure at the bottom right shows at higher magnification the ventrolateral zone in 30-day-old larva. Magnification bars =50 μm. Abbreviations: *c*, canal medullar central; *dc*, dorsal cells; *m*; motoneurons

In 30-day-old larvae, the spinal cord already presents the adult pattern of organization, showing thick and well-myelinized fibres. Also at this time, the pectoral fins appear fully developed with their rays completely formed (Hernando et al., 2005a), and the caudal fin is already separated from the dorsal and anal ones, although its development concludes about day 60, when the rays are clearly visible in the whole fin (Hernando et al., 2005b). This stage in neural and fin development coincides with a marked improvement in swimming ability, and also with the onset of the locomotion patterns and the spatial distribution of exploratory activity typical of juvenile *A. naccarii*. These advances not only allow larvae to disperse actively over a wide rearing area, but also probably improve their foraging capacities and chances of avoiding predators.

9.4 Development of the Hindbrain

The hindbrain reaches a notable size in sturgeons (see Fig. 9.1). Associated with the formation of the plica encephalis ventralis in the midbrain, a remarkable curving appears on the floor of the rhombencephalon that disappears later in development (Nieuwenhuys, 1998; Vázquez et al., 2002a; Fig. 9.2). Dorsally, the walls of the rhombencephalon separate progressively, leading to the transformation of the dorsal plate in a wide tela ependymalis by the 40-day stage. In 4-day-old larvae the ventricular surface of the floor plate already shows a protuberance containing the medial longitudinal fascicle (Fig. 9.4). In the juvenile, the embryonic mantle becomes a narrow and dense zone of periventricular grey.

Three distinct bilateral indentations (the sulcus limitans of His, the sulcus intermedius ventralis, and the sulcus intermedius dorsalis), which run along most of the length of the hindbrain of chondrosteans, mark the border between four longitudinal cell zones or columns. Thus, the grey matter in the rhombencephalon is organized in four longitudinal zones that appear as intraventricular protrusions (Fig. 9.4). The sulcus limitans of His, which is evident in 10-day-old larvae, enables identification of the dorsolateral (alar) plate, and the ventromedial (basal) plate. The sulci intermedius dorsalis and the intermedius ventralis subdivide the alar and the basal plates into two subzones each: the dorsalis and the intermediodorsalis areas, and the intermedioventralis and ventralis areas, respectively. Each of these is specifically related to one of the following categories: somatosensory, viscerosensory, visceromotor, or somatomotor. All of the primary afferents, both the somatosensory and the viscerosensory, are situated in the alar plate.

Within the somatosensory centres, those related to the octavus and the lateral line nuclei constitute the octavolateral area, which can be subdivided in a dorsal, an intermediate, and a ventral zone (Larsell, 1967; McCormick, 1982). The dorsal and the intermediate zones receive first-order inputs from the lateral line nerves, while the ventral zone represents the goal area of the octavus nerve. The dorsal area comprises the dorsal nuclei of the octavolateral area, on the upper surface of the alar plate, forming a distinct protuberance that in sturgeons extends from the auricle of the cerebellum to the level of the entrance of the posterior lateral line nerve (Nieuwenhuys, 1998, see Figs. 9.2 and 9.4). In chondrosteans, the sensory receptors of the lateral line system include the neuromasts (mechanoreceptors) and also organs that closely resemble the ampullae of Lorenzini (Teeter et al., 1980), the electroreceptors of the cartilaginous fishes. In sturgeons the development of the ampullary organ takes considerably longer than neuromast development. Nevertheless, it has been observed that when the *A. naccarii* begins to feed actively,

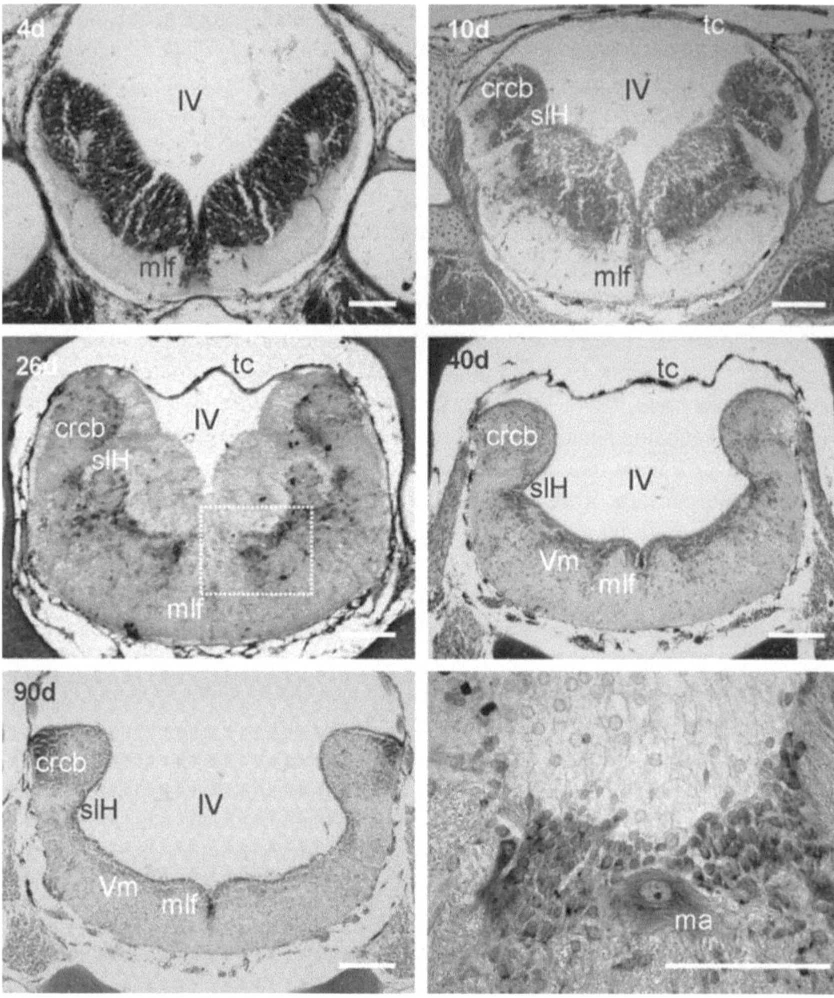

Fig. 9.4 Photomicrographs of Nissl-stained transverse sections through the hindbrain of *Acipenser naccarii* at different developmental ages (the post-hatching day is indicated in each photograph). The figure at the bottom right shows at higher magnification the soma of a Mauthner cell in a 26-day-old larva. Magnification bars =200 μm. Abbreviations: *crcb*, cerebellar crest of the octavolateral area; *ma*; Mauthner cell; *mlf*, medial longitudinal fascicle; *slH*, sulcus limitans of His; *tc,* tela choroidea; *IV*, fourth ventricle; *Vm*, trigeminal motor nucleus

the ultrastructural characteristics of ampullary organs already correspond to those of adult animals (Camacho et al., 2007). Also, in a recent developmental study of the lateral line in the shovelnose sturgeon (*Scaphirhynchus platorynchus*) it was observed that the bony lateral line canals begin to form about 10 days posthatch (PH) and are fully formed by 36 days PH, while the ampullary organs are not fully developed until 27 days after hatching (Gibbs and Northcutt, 2004). In *A. naccarii* we have observed that by day 40 the dorsal and intermediate areas have already reached the normal adult pattern (Fig. 9.4). Also cytochrome oxidase histochemistry has shown that in juveniles the dorsal nucleus of the octavolateral area presents a high level of oxidative activity, which reveals the reception of a high proportion of excitatory inputs (Vázquez et al., 2002b).

The viscerosensory centres consist mainly of the projections from the VII, IX and X nerves. As is well known, the gustatory system is particularly prominent in the sturgeons. At the central level this difference is reflected in the size of the vagal lobe (Fig. 9.1), remarkably large in sturgeons, relative to other chondrosteans (Nieuwenhuys, 1998; Butler and Hodos, 2005), this indicates that in sturgeons the gustatory system plays a dominant role in detecting and selecting food. The adult pattern of the viscerosensory area is reached by day 90 in *A. naccarii* (Fig. 9.4).

The visceromotor nuclei are located in the basal plate. They are principally the vagal, the glossopharyngeal, and the facial motor nuclei. In sturgeons, all the visceromotor nuclei occupy a periventricular position. These nuclei are clearly evident by the time the larvae begin to feed, although considerable development is not achieved until reaching the adult pattern by day 40 (Fig. 9.4). It is well known that the efferent fibres from the nucleus of the solitary tract that go through the adjacent visceromotor nuclei are part of the reflex mechanisms to ingest and swallow food. Finally, among the somatomotor nuclei, the very conspicuous Mauthner cells appear bilaterally at the level of the entrance of the octavus nerve, transversally oriented and situated immediately below the ventricular surface (Fig. 9.4). These cells activate fast-start responses as seen in fishes eluding predatory attacks (Eaton et al., 2001). In *A. naccarii* the Mauthner cells are evident at hatching time but are not well myelinized until day 26 (Fig. 9.4). This age coincides with the complete differentiation of the caudal fin (Hernando et al., 2005b) and with a major improvement in the performance of the escape response, as the velocity increases and the time to complete the response decreases.

9.5 Development of the Cerebellum

The cerebellum typically reaches a notable size in chondrosteans (Fig. 9.1), and although the cytoarchitecture of this structure in Acipenser is very similar to that of the other vertebrates, it presents some morphological and cytoarchitectural characteristics that are exclusive to this group (Nieuwenhuys, 1967, 1998). In *A. naccarii* the central cerebellum region consists of an unpaired, massive body that protrudes within the ventricular cavity (Fig. 9.2). Its development begins with a paired anlage, clearly evident in the 4-day-old larvae (Vázquez et al., 2002a; Fig. 9.5), and its peculiarity is a consequence of the convergence of the cerebellar primordia during neural development and the fusing of the meningeal surfaces in the medial plane. This fusion process ends by PH day 10 (Fig. 9.5). In 26-day-old specimens the central body of the cerebellum already shows the three distinct zones characteristic of the adult. The rostral part forms the valvula cerebelli, which extends into the mesencephalic ventricle and becomes continuous with the mesencephalic tectum

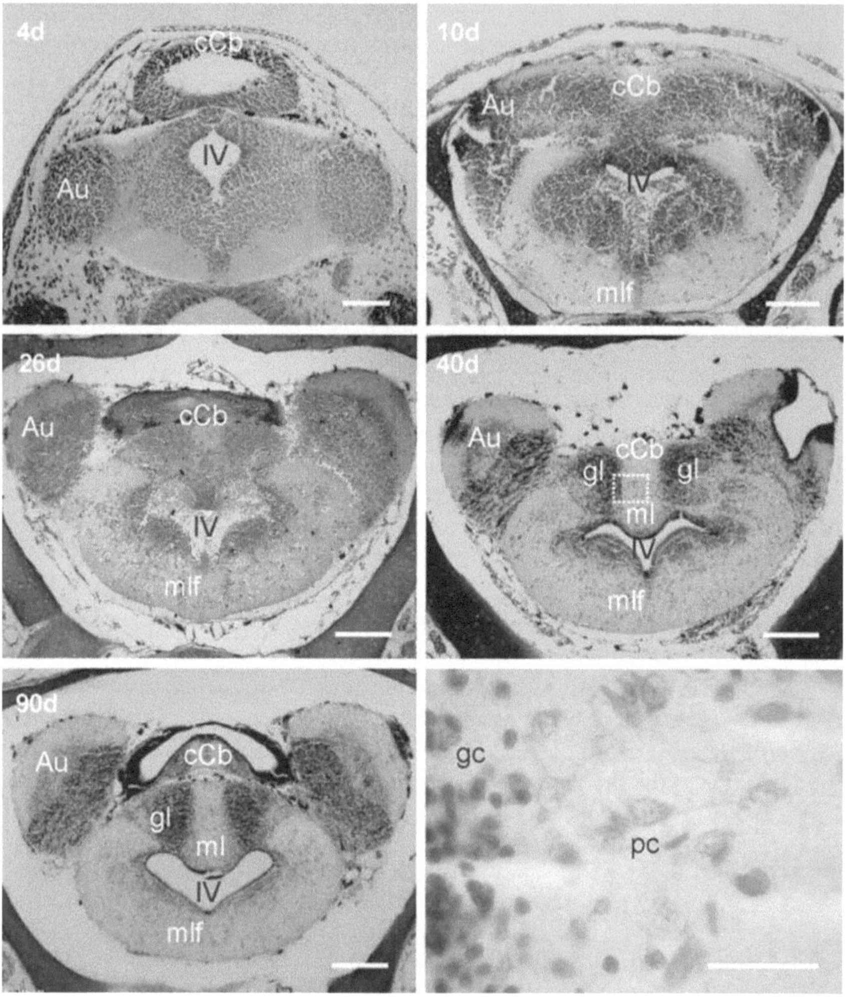

Fig. 9.5 Photomicrographs of Nissl-stained transverse sections through the cerebellum of *Acipenser naccarii* at different developmental ages (the post-hatching day is indicated in each photograph). The figure at the bottom right shows at higher magnification the molecular layer in a 40-day-old larva. Magnification bars =200 μm. Abbreviations: *Au*, auricula cerebelli; *cCb*, corpus cerebelli; *gc*, granule cells; *gl*, granular layer of the cerebellum; *mlf*, medial longitudinal fascicle; *ml*, molecular layer of the cerebellum; *pc*; Purkinje cells; *IV*, fourth ventricle

(Nieuwenhuys, 1998; Huesa et al., 2003). The corpus cerebelli, in the caudal zone, and the lateral zones, form the prominentia granularis, two granule cell masses that expand into the ventricle (Nieuwenhuys, 1998; Fig. 9.5). The stratum granulare of the corpus and valvula, and the eminentia granularis, contain large Golgi type II and Purkinje cells (Johnston, 1898a, 1901; Larsell, 1967; Nieuwenhuys, 1998) that are evident in 10-day-old larvae (Fig. 9.5). Later the dendrites of the Purkinje cells branch extensively and expand within the molecular zone. The normal cerebellar histological pattern of the adult is reached at 40 days PH in *A. naccarii* (Fig. 9.5).

9.6 Development of the Mesencephalon

In all chondrosteans, the mesencephalon is rather poorly developed, as it maintains the embryonic tubular morphology, which is particularly evident in sturgeons (Johnston, 1901; Nieuwenhuys, 1998; Butler and Hodos, 2005; Figs. 9.1 and 9.6). As in all vertebrates, the mesencephalon of the Acipenseridae presents two zones, the tectum, situated dorsally, and the tegmentum, ventrally. Although the sulcus limitans of His does not reach the mesencephalon, the subdivision between the tectum and the tegmentum appears during the initial post-hatching days (Fig. 9.6). During pre-hatching development, as a result of a process of evagination and folding of the tectum towards the interior of the dorsal roof, the mesencephalon of the chondrosteans has two bilateral lobes, although not clearly differentiated even in the adult. The scant tectal development, which *A. naccarii* shares with all other sturgeons, suggests that the visual system is less developed in this group than in other actinopterygian fishes. It is also important to note that the ventral, tegmental region develops much earlier than the tectum. Indeed, although the complete differentiation of the eye is achieved before the onset of exogenous feeding, the six-layered organization of the tectum appears only by post-hatching day 40, and the normal adult pattern is not reached until day 90 (Vázquez et al., 2002a,b; Fig. 9.6). By contrast, the tegmentum, which contains mainly two primary efferent centres (the trochlear and the oculomotor nuclei in a periventricular position), reaches the normal adult pattern by the 21-day-old stage.

9.7 Development of the Diencephalon

In sturgeons, the whole of the diencephalon and the pretectum can be subdivided into seven main regions: the preoptic area, the epithalamus, the dorsal and the ventral thalamus, the hypothalamus, the posterior tuberculum, the synencephalon, and the rest of the pretectum (Braford and Northcutt, 1983; Nieuwenhuys, 1998; Butler and Hodos, 2005). In an external dorsal view, the habenula, markedly asymmetric in Acipenser and Polyodon (Johnston, 1901; Hocke-Hoogenboom, 1929; Nieuwenhuys, 1998; Yáñez and Anadón, 1998), is the only visible zone in the diencephalon of sturgeons, as this brain subdivision is partially covered by the midbrain (Fig. 9.1).

In the diencephalon, the most dorsal area (epithalamus) differentiates very early during ontogenesis; for instance, the primordium of the epiphysis appears on day 1 PH but the caudal expansion of the hypothalamus, which forms the inferior hypothalamic lobes, occurs relatively late (Figs. 9.2 and 9.7). Although the thalamus

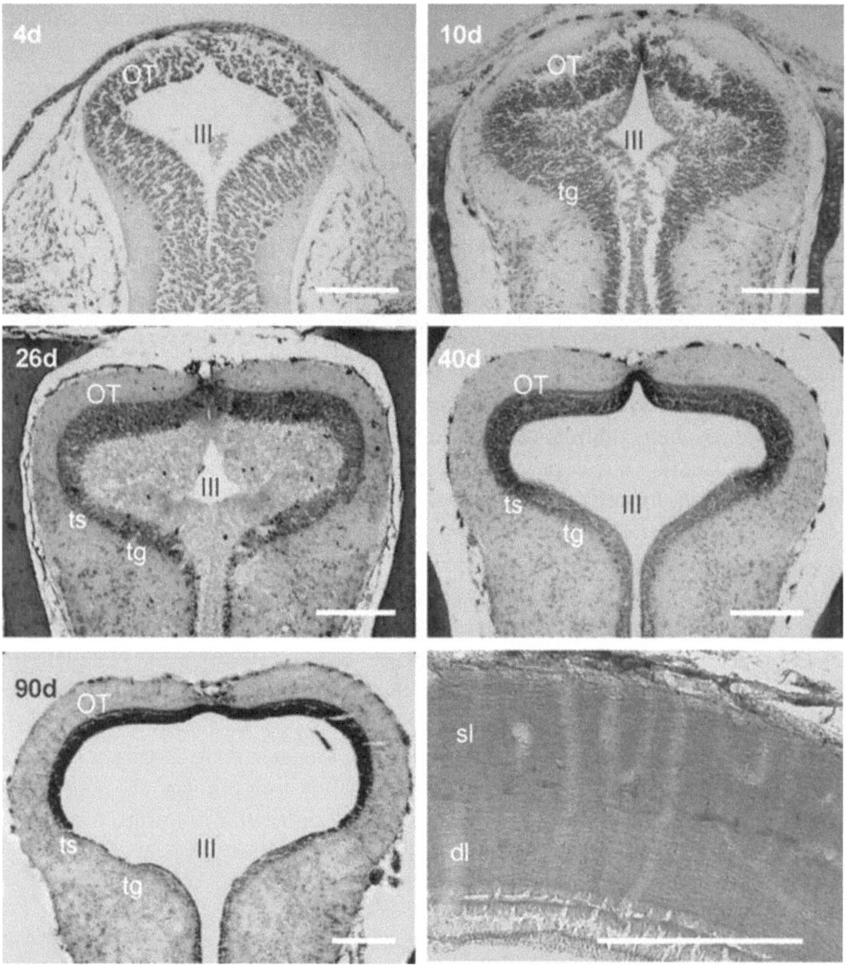

Fig. 9.6 Photomicrographs of Nissl-stained transverse sections through the mesencephalon of *Acipenser naccarii* at different developmental ages (the post-hatching day is indicated in each photograph). The figure at the bottom right shows a detail of heavy cytochrome oxidase labelling in the optic tectum of a juvenile (modified from Vázquez et al., 2002b). Magnification bars =200 μm. Abbreviations: *dl*, deep layers of the OT; *OT*, optic tectum; *sl*, superficial layers of the OT; *Tg*, tegmentum; *ts*, torus semicircularis; *III*, third ventricle

appears scarcely developed even in adult sturgeons, the hypothalamus reaches a considerable size. The thalamus consists of a zone of densely grouped periventricular cells (Repérant et al., 1982, Nieuwenhuys, 1998) that reaches its full development by the end of the second month of life.

The hypothalamus is the largest diencephalic structure in chondrosteans. This structure includes a large paired rostral region and a smaller, unpaired caudal one. The walls of the rostral area are evaginated, forming inferior lobes (Figs. 9.2 and 9.7).

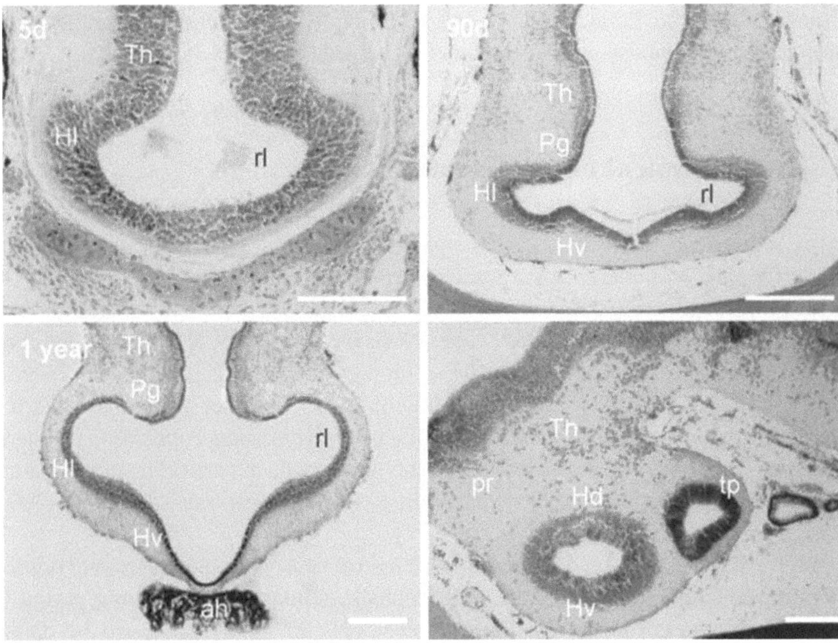

Fig. 9.7 Photomicrographs of Nissl-stained transverse sections through the diencephalon of *Acipenser naccarii* at different developmental ages (the post-hatching day is indicated in each photograph). The figure at the bottom right shows at higher magnification a mid-sagittal section through the hypothalamus in a 40-day-old larva. Magnification bars =200 μm. Abbreviations: *ah,* adenohypophysis; *Hd,* dorsal hypothalamus; *Hl,* lateral hypothalamus; *Hv,* ventral hypothalamus; *Pg,* preglomerular area; *pr,* pretectal area; *rl,* lateral recess of the third ventricle; *Th,* thalamus; *tp,* nucleus tuberis posterior

The inferior hypothalamic lobes receive visual afferents from the pretectum (Albert et al., 1999), and also important gustatory afferents (Nieuwenhuys, 1998), and thus the size of the lateral inferior lobe may correlate positively with the importance of the gustatory system in these species. In this area, the normal histological pattern of the adult is reached at 90 days PH (Fig. 9.7).

The hypophysis lies on the ventral surface of the hypothalamus. The adenohypophysis develops much earlier than the neurohypophysis, which does not begin its development until stage 80-day (Grandi and Chicca, 2004), and does not reach the adult morphology before 9 months of age. In the adult, the adenohypophysis is subdivided into a rostral pars distalis, a proximal pars distalis, and a pars intermedia deeply penetrated by neurohypophysial branches. A large central cleft separates the pars intermedia from the proximal one and divides the proximal pars into ventral and dorsal parts (Nieuwenhuys, 1998). The primordium of the adenohypophysis is visible at hatching (Fig. 9.2). In 40-day-old larvae the adenohypophysis enlarges but maintains its flat shape and its close contact with the diencephalon, although no nervous fibres appear to reach it from the hypothalamus (Polenov et al., 1983;

Grandi and Chicca, 2004; Figs. 9.2 and 9.7). Finally, the normal adult morphology of the adenohypophysis is observed in 5-month-old juveniles.

9.8 Development of the Telencephalon

In Acipenseridae, as in all actinopterygian fishes during embryonic development, the telencephalon undergoes an eversion process, which leads to the formation of solid hemispheres separated by a single telencephalic ventricle (Northcutt and Braford, 1980; Nieuwenhuys and Meek, 1990; Braford, 1995; Northcutt, 1995; Fig. 9.8). This peculiar morphology of the actinopterygian telencephalon has hindered not only the identification of possible homologies with other vertebrates, but has also contributed to the persistence of significant anachronisms concerning a general understanding of brain and behavioural evolution, and, in particular, to attributing fishes' poor learning and memory capabilities (for a review, see Salas et al., 2003; Rodríguez et al., 2006).

In *A. naccarii*, as in other sturgeons, the rostral neuropore (i.e. the rostral portion of the neural tube), persists in the telencephalic primordium for a long period of time. Nevertheless on day 3 PH three telencephalic subdivisions can be distinguished: a rostral part, consisting of very large, diverging paired olfactory bulbs with wide internal ventricles, and an intermediate and a caudal zones, which constitute the telencephalon proper (Johnston, 1911; Herrick, 1921; Nieuwenhuys, 1962, 1963, 1964). Whereas the small caudal telencephalic subdivision has remained uneverted (Yamamoto et al., 2007), the large intermediate zone, which forms the rostral telencephalon proper, shows the solid, everted lateral walls characteristic of the actinopterygian fish. The telencephalon proper is entirely covered by a wide tela ependymalis, attached rostrally to the dorsocaudal edge of the olfactory bulbs and laterally to the dorsolateral edges of the everted walls of the intermediate subdivision (Nieuwenhuys, 1998).

During the first week PH, the development of the most rostral zone of the forebrain does not differ from that of the remaining portions of the forebrain primordium. However, in the 14-day stage, the area of entrance of the olfactory nerves appears slightly thickened, showing a local inward bending or evagination that progressively increases to form a rostrocaudally directed evagination (Nieuwenhuys, 1964; Vázquez et al., 2002a). In sturgeons, most of the bulbar cells project to the intermediate and caudal telencephalon (Braford, 1995; Huesa et al., 2000, 2006). It bears noting that the neurons of the terminal nerve are not grouped in a distinct ganglion but rather appear distributed along the olfactory bulbs, the preoptic area and the ventromedial hypothalamus (Lepétre et al., 1993; Nieuwenhuys, 1998; Butler and Hodos, 2005). This structure is rich in gonadotropin-releasing hormone (GnRH) and is associated with reproductive behaviour.

The eversion of the telencephalic hemispheres begins in *A. naccarii* on PH day 4, with the sideways curving of the walls in the intermediate telencephalic subdivision (Vázquez et al., 2002a; Fig. 9.8). This process is clearly evident in 10-day-old

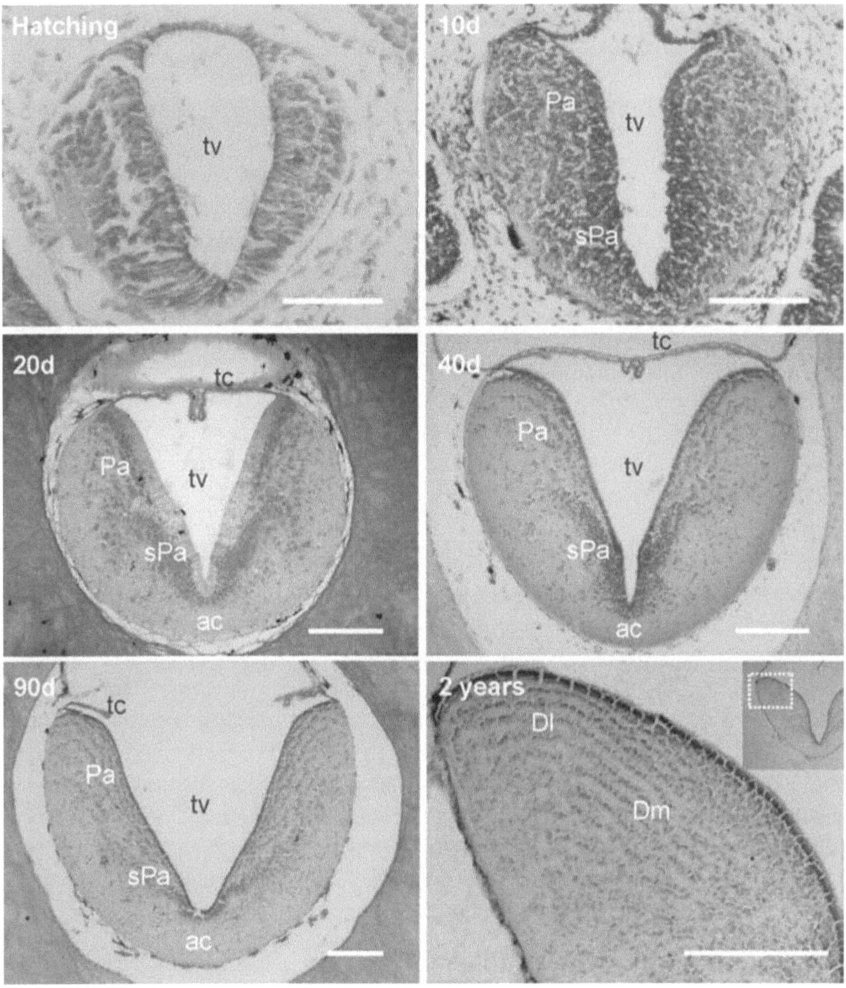

Fig. 9.8 Photomicrographs of Nissl-stained transverse sections through the telencephalon of *Acipenser naccarii* at different developmental ages (the post-hatching day is indicated in each photograph). The figure at the bottom right shows at higher magnification the dorsal pallium of the telencephalon in a 2-year-old juvenile. Magnification bars =200 μm. Abbreviations: *ac*, anterior commissure; *Dl*, lateral part of the dorsal telencephalon; *Dm*, medial part of the dorsal telencephalon; *Pa*, telencephalic pallium; *sPa*; telencephalic subpallium; *tc*, tela choroidea; *tv*, telencephalic ventricle

larvae, as a light dorsal divergence of the lateral walls of the neural tube. A few days later, this process has led to a considerable enlargement of the dorsal zone (Fig. 9.8). The shape of the telencephalic primordium changes relatively little during the following 60 days. The progressive thickening and intraventricular protrusion of the dorsal portion of the lateral plates lead to the elevation of the lateral recesses of the telencephalic ventricle (Figs. 9.2 and 9.8). It is noteworthy that the intraventricular expansion of the dorsal area of the telencephalic walls occurs later in the chondrostean fish than in holosteans and teleosts (Nieuwenhuys, 1964, 1998). Thus, in *A. naccarii*, the lateral walls of the telencephalon still appear very slightly everted on day 90, indicating that the distinctly everted morphological pattern of the adults is reached much later (Fig. 9.8).

In sturgeons, as well as in other actinopterygian fishes, the sulcus limitans telencephalic marks dorsally the zona limitans, an area with a low cell density that runs through most of the telencephalon and separates the pallium and the subpallium. The pallium is situated in a dorsal and lateral position relative to the subpallium and can be subdivided into three main areas—the lateral, dorsal and medial zones—and a central nucleus, although all of these appear in sturgeons less distinctly delimited and neurochemically and cytoarchitectonically more homogeneous (Nieuwenhuys, 1964, 1998; Northcutt and Bradford, 1980; Adrio et al., 2002; Huesa et al., 2006; Piñuela and Northcutt, 2007). In contrast to the situation in most actinopterygians, in sturgeons the dorsal area of the pallium shows a distinctly multi-layered appearance, containing large pyramidal neurons parallel to the ependymal surface. This cytoarchitectonic peculiarity is evident only in specimens older than a year (Fig. 9.8). Recent trace studies carried out in *Acipenser baerii* have shown that whereas the pallium presents extensive intratelencephalic connectivity, similar to other teleosts, the connectivity of the dorsal pallium with non-telencephalic structures appears to be more restricted in sturgeons than in other ray-finned fishes (Huesa et al., 2006). In particular, the thalamotelencephalic projections conveying visual information to the pallium have been shown to be quite less in *A. baerii*, *A. transmontanus* and *A. schenkii* (Albert et al., 1999; Huesa et al., 2006). This difference might be related to the fact that in sturgeons the gustatory system plays a dominant role in foraging behaviour.

Finally, the subpallium, situated in the ventral area of the telencephalon, consists of a number of nuclei, although these appear less differentiated in sturgeons, in terms of cytoarchitectonic and histochemical features (Nieuwenhuys, 1998; Huesa et al., 2006; Fig. 9.8). The subpallium develops faster than the pallium; in fact, the normal adult morphology of the subpallium is evident in 40-day-old larvae (Fig. 9.8). Recent studies, taking into account that the subpallium contains a high concentration of fibres positive to substance P, leu-enkephalin, serotonin and dopamine, have considered this area as homologous to the striatum of land vertebrates (Wulliman and Mueller, 2004; Adrio et al., 2005; Piñuela and Northcutt, 2007).

9.9 Conclusions

We have presented here recent data on the gross morphology and cytoarchitecture of the brain of the sturgeon *A. naccarii* during ontogenesis.

The overall organization of the central nervous system, consisting of prosencephalon, mesencephalon, rhombencephalon and spinal cord, is already defined in 3-day sturgeon embryos, and 22-day larvae already present the adult brain and spinal cord organization. Although *A. naccarii* shows marked differences in brain development, for instance, the adult pattern of the viscerosensory area (rhombencephalon) is reached on day-90 and the adult morphology of the adenohypophysis (diencephalon) appears as late as 5 months later, somatomotor differentiation occurs notably earlier. Thus, 30-day old larvae already show complete fin differentiation and well-myelinized fibres in the spinal cord, which coincide with the onset of the juvenile locomotion and exploration patterns, as well as with a major improvement in the efficacy of the escape response. Also in the cerebellum, the adult histological pattern is already present on PH day 40. It is important to note that the development of the telencephalon in sturgeons, described here in *A. naccarii*, confirms the eversion process proposed to explain the peculiar morphology of the telencephalon of Actinopterygian fishes. Although in sturgeons the rostral neuropore remains open for a long time, and the telencephalon continues growing and everting along life as in every actinopterygian, the eversion process of the telencephalic hemispheres is already evident in 10-day-old larvae of *A. naccarii.*

This work contributes with recent data to our knowledge on the development of the central nervous system in sturgeons. Given the strategic position that these species occupy in actinopterygian radiation, these results provide essential information not only concerning the neural development of *A. naccarii* but also help us to understand the evolution of the vertebrate brain and behaviour in general. Moreover, as chondrosteans comprise a number of endangered species, additional data on their normal development patterns is essential in order to determine the most adequate strategies for their conservation in natural habitats and for the achievement of optimal standards of culture in hatcheries and farms.

Acknowledgements We thank Gerardo Labrador for technical assistance. This work has been supported by grants AGL2000-0382-P4-05 and BFU2007-62228 from the Spanish Ministerio de Educación y Ciencia and CVI-242 from the Junta de Andalucía.

References

Adrio F, Rodríguez MA, Rodriguez-Moldes I. 2002. Distribution of tyrosine hydroxylase (TH) and dopamine β-hydroxylase (DBH) immunoreactivity in the central nervous system of two chondrostean fishes (*Acipenser baerii* and *Huso huso*). *J Comp Neurol* 448:280–297.

Adrio F, Rodríguez MA, Rodriguez-Moldes I. 2005. Distribution of galanin-like immunoreactivity in the brain of the Siberian sturgeon *Acipenser baerii. J Comp Neurol* 487:54–74.

Albert JS, Yamamoto N, Yoshimoto M, Sawai N, Ito H. 1999. Visual thalamotelencephalic pathways in the sturgeon *Acipenser*, a non-teleost Actinopterygian fish. *Brain Behav Evol* 53:156–172.

Bemis WE, Findeis EK, Grande L. 1997. An overview of Acipenseriformes. *Environ Biol Fish* 48:25–71.

Bergquist H. 1932. Zur Morphologie des Zwischenhirns bei niederen Wirbeltieren. *Acta Zool* 13:57–304.

Braford MR. 1995. Comparative aspects of forebrain organization in ray-finned fishes: touchstones or not? *Brain Behav Evol* 46:259–274.

Braford MR, Northcutt RG. 1983. Organization of the diencephalon and pretectum of the ray-finned fishes. In: Davis RE, Northcutt RG (eds.), *Fish Neurobiology, Vol. 2: Higher Brain Areas and Functions*. The University of Michigan Press, Ann Arbor, pp. 117–163.

Butler AB, Hodos W. 2005. *Comparative Vertebrate Neuroanatomy. Evolution and Adaptation*. Wiley, New Jersey.

Camacho S, Ostos Mdel V, Llorente JI, Sanz A, García M, Domezain A, Carmona R. 2007. Structural characteristics and development of ampullary organs in *Acipenser naccarii*. *Anat Rec* 290:1178–1189.

Chiba A, Honma Y. 1994. NeuropeptideY-immunoreactive structures in the telencephalon and the diencephalons of the white sturgeon, *Acipenser transmontanus*, with special regard to the hypothalamus-hypophyseal system. *Arch Histol Cytol* 57:77–86.

De la Herrán R, Robles F, Martínez-Espín E, Lorente JA, Ruiz-Rejón C, Garrido-Ramos MA, Ruiz-Rejón M. 2004. Genetic identification of western Mediterranean sturgeons and its implication for conservation. *Conserv Genet* 5:545–551.

Eaton RC, Lee RK, Foreman MB. 2001. The Mauthner cell and other identified neurons of the brainstem escape network of fish. *Prog Neurobiol* 63:467–485.

Garrido-Ramos MA, Soriguer C, De la Herrán R, Jamilena M, Ruiz Rejón C, Domezain A, Hernando JA, Ruiz Rejón M 1997. Morphometric and genetic analysis as proof of the existence of two sturgeon species in the Guadalquivir River. *Mar Biol* 129:33–39.

Garrido-Ramos MA, Robles F, de la Herrán R, Martínez-Espín E, Lorente JA, Ruiz Rejón C, Ruiz Rejón M. 2007. Analysis of mitochondrial and nuclear DNA markers in old museum sturgeons yield insights about the species existing in Western Europe: *A. sturio*, *A. naccarii* and *A. oxyrinchus* In: Carmona R. et al. (eds.), Biology, Conservation and Sustainable Development of Sturgeons. Springer. pp. 25–50.

Gibbs MA, Northcutt RG. 2004. Development of the lateral line system in the shovelnose sturgeon. *Brain Behav Evol* 64:70–84.

González GC, Belenky MA, Polenov AL, Lederis K. 1992. Comparative localization of corticotrophin and corticotrophin releasing factor-like peptides in the brain and hypophysis of a primitive vertebrate, the sturgeon *Acipenser ruthenus* L. *J Neurocytol* 21:885–896.

Goronowitsch N. 1888. Das Gehirn und die Cranialnerves von *Acipenser rhutenus*. *Morphol Jahrb* 30:427–574.

Grandi G, Chicca M. 2004. Early development of the pituitary gland in *Acipenser naccarii* (Chondrostei, Acipenseriformes): an immunocytochemical study. *Anat Embryol* 208:311–321.

Hernando JA, Domezain A, Zabala C, Domezain J, Cabrera R, Soriguer MC. 2005a. Desarrollo de las aletas pectorales durante la morfogénesis de *Acipenser naccarii*, Bonaparte 1836. In: *IX Congreso Nacional de Acuicultura. Abstract Book*. Junta de Andalucía, Cádiz, pp. 81–85.

Hernando JA, Domezain A, Zabala C, Domezain J, Cabrera R, Soriguer MC. 2005b. Morfogénesis de la aleta caudal de *Acipenser naccarii*, Bonaparte 1836. In: IX Congreso Nacional de Acuicultura. Abstract Book. Junta de Andalucía, Cádiz, pp. 109–113.

Herrick CJ. 1921. A sketch of origin of the cerebral hemispheres. *J Comp Neurol* 32:429–454.

Hocke-Hoogenboom KJ. 1929. Das Gehirn von Polyodon Folium Lacép. *Z Mikrosk Anat Forsch* 18:311–392.

Huesa G, Anadón R, Yáñez J. 2000. Olfactory projections in a chondrostean fish, *Acipenser baerii*. *J Comp Neurol* 428:145–158.

Huesa G, Anadón R, Yáñez J. 2003. Afferent and efferent connections of the cerebellum of the chondrostean *Acipenser baerii*. A carbocyanine dye (DiL) tracing study. *J Comp Neurol* 460:327–344.

Huesa G, Anadón R, Yáñez J. 2006. Topography and connections of the telencephalon of the chondrostean, *Acipenser baerii*. An experimental study. *J Comp Neurol* 497:519–541.

Ito H, Yoshimoto M, Albert JS, Yamamoto N, Sawai N. 1999. Retinal projections and retinal ganglio cell distribution patterns in a sturgeon (*Acipenser transmontanus*), a non teleost actinopterygian fish. *Brain Behav Evol* 53:127–141.

Ito H, Ishikawa Y, Yoshimoto M, Yamamoto N. 2007. Diversity of brain morphology in teleosts: brain and ecological niche. *Brain Behav Evol* 69:76–86.

Johnston JB. 1898a. Hind brain and cranial nerves of *Acipenser. Anat Anz* 14:580–602.
Johnston JB. 1898b. The olfactory lobes, fore-brain and habenular tracts of *Acipenser. Zool Bull* 1:221–241.
Johnston JB. 1901. The brain of Acipenser: a contribution to the vertebrate brain. *Zool Jahrb Anat Ontog* 15:59–260.
Johnston JB. 1911. The telencephalon of ganoids and teleosts. *J Comp Neurol* 21:489–591.
Kotrschal K, Krautgartner W-D, Adam H.1985. Distribution of aminergic neurons in the brain of the starlet, *Acipenser ruthenus* (*Chondrostei, Actinopterygii*). *J Hirnforsch* 26:65–72.
Larsell O. 1967. *The Comparative Anatomy and Histology of the Cerebellum from Myixinoids through Birds*. University of Minnesota Press, Minneapolis.
Lauder GV, Liem K. 1983. The evolution and interrelationships of the actinopterygian fishes. *Bull Mus Comp Zool* 150:95–197.
Leprêtre E, Anglade I, Williot P, Vandesande F, Tramu G, Kah O. 1993. Comparative distribution of mammalian GnRH (gonadotropin-releasing hormone) and chicken GnRH-II in the brain of the immature Siberian sturgeon (*Acipenser baerii*). *J Comp Neurol* 337:568–583.
McCormick CA. 1982. The organization of the octavolateralis area in actinopterygian fishes: a new interpretation. *J Morphol* 171:159–181.
Meek J, Nieuwenhuys R. 1998. Holosteans and teleosts. In: Nieuwenhuys R, Ten Donkelaar HJ, Nicholson C (eds.), *The Central Nervous System of Vertebrates*. Springer, Berlin, pp. 760–937.
Nieuwenhuys R. 1962. Trends in the evolution of the Actinopterygian forebrain. *J Morphol* 111:69–88.
Nieuwenhuys R. 1963. The comparative anatomy of the Actinopterygian forebrain. *J Hirnforsch* 6:171–192.
Nieuwenhuys R. 1964. Further studies on the general structure of the Actinopterygian forebrain. *Acta Morphol Neerl Scand* 6:65–79.
Nieuwenhuys R. 1967. Comparative anatomy of the cerebellum. *Prog Brain Res* 15:1–93.
Nieuwenhuys R. 1998. Chondrostean fishes. In: Nieuwenhuys R, Ten Donkelaar HJ, Nicholson C (eds.), *The Central Nervous System of Vertebrates*. Springer-Verlag, Berlin, pp. 701–757.
Nieuwenhuys R, Meek J. 1990. The telencephalon of sarcopteriginan fishes. In: Jones EJ, Peters A (eds.), *Cerebral Cortex*, Vol. 8A. *Comparative Structure and Evolution of the Cerebral Cortex*. Part 1. Plenum Press, New York, pp. 75–106.
Northcutt RG. 1995. The forebrain of gnathostomes: In search of a morphotype. *Brain Behav Evol* 46:276–318.
Northcutt RG, Braford MR. 1980. New observations on the organization and evolution of the telencephalon of actinopterygian fishes. In: Ebbeson SOE (ed.), *Comparative Neurology of the Telencephalon*. Plenum, New York, pp. 41–98.
Northcutt RG, Davis RE. 1983. *Fish Neurobiology*. University of Michigan Press, Ann Arbor.
Oka S, Honma Y, Iwanaga T, Fujita T. 1989. Immunohistochemical demonstration of urotensins I and II in the caudal neurosecretory system of the white sturgeon, *Acipenser transmontanus* Richardson. *Biomed Res* 3:329–340.
Piñuela C, Northcutt RG. 2007. Immunohistochemical organization of the forebrain in the white sturgeon, *Acipenser transmontanus*. *Brain Behav Evol* 69:229–253.
Polenov AL, Efimova NA, Konstantinova MS, Senchik YI. 1983. The hypothalamo-hypophyseal system in Acipenseridae. Formation of monoaminergic neurosecretory cells in the preoptic nucleus region during early ontogeny. *Cell Tissue Res* 232:651–667.
Repérant J, Vesselkin NP, Ermakova TV, Rustamov EK, Rio JP, Palatnikov GK, Peyrichoux J, Kasimov RV. 1982. The retinofugal pathways in a primitive actinopterygian the chondrostean *Acipenser güldenstädti*. An experimental study using radioautographic and HRP methods. *Brain Res* 251:1–23.
Rodríguez F, Broglio C, Durán E, Gómez A, Salas C. 2006. Neural mechanisms of learning in teleost fish. In: Huntingford F, Laland K, Krause J, Brown C (eds.), *Fish Cognition and Behaviour*. Blackwell Scientific, Oxford, pp. 243–277.

Rupp B, Northcutt RG. 1998. The diencephalon and pretectum for the white sturgeon (*Acipenser transmontanus*): a cytoarchitectonic study. *Brain Behav Evol* 51:239–262.
Salas C, Broglio C, Rodríguez F. 2003. Evolution of forebrain and spatial cognition in vertebrates: conservation across diversity. *Brain Behav Evol* 62:72–82.
Teeter JH, Szamier RB, Bennet MVL. 1980. Ampullary electroreceptors in the sturgeon *Scaphirhynchus platorynchus* (Rafinesque). *J Comp Physiol* 138:213–233.
Trabucchi M, Tostivint H, Lihrmann I, Sollars C, Vallarino M, Dores RM, Vaudry H. 2002. Polygenic expression of somatostatin in the sturgeon A*cipenser transmontanus*: molecular cloning and distribution of the mRNAs encoding two somatostatin precursors. *J Comp Neurol* 443:332–345.
Vázquez M, Rodríguez F, Domezain A, Salas C. 2002a. Development of the brain of the sturgeon *Acipenser naccarii. J Appl Ichthyol* 18:275–279.
Vázquez M, Rodríguez F, Domezain A, Salas C. 2002b. Cytochrome oxidase histochemistry in the brain of the sturgeon *Acipenser naccarii. J Appl Ichthyol* 18:359–364.
Von Kupffer K. 1893. *Studien zur vergleichenden. Entwicklungsgeschichte des Kopfe der Cranioten*, Vol. 1. Lehman, Munich, pp. 1–95.
Von Kupffer K. 1906. Die Morphogenie der Central nerven-systems. In: Hertwig O (ed.), *Handbuch der vergleichenden un experimentelien Entwicklungsgeschichte derWirbeltiere*, Vol. 2, part 3. Jena. Verlag von Gustav Fischer, pp. 1–272.
Wulliman MF, Mueller T. 2004. Teleostean and mammalian forebrains contrasted: evidence from genes to behavior. *J Comp Neurol* 475:143–162.
Yamamoto N, Yoshimoto M, Albert JS, Sawai N, Ito H. 1999. Tectal fiber connections in a non teleost actinopterygian fish, the sturgeon Acipenser. *Brain Behav Evol* 53:142–155.
Yamamoto N, Ishikawa Y, Yoshimoto M, Xue HG, Bahaxar N, Sawai N, Yang CY, Ozawa H, Ito H. 2007. A new interpretation on the homology of the teleostean telencephalon based on hodology and a new eversion model. *Brain Behav Evol* 69:96–104.
Yáñez J, Anadón R. 1998. Neural connections of the pineal organ in the primitive bony fish *Acipenser baerii*: a carbocyanine dye tract-tracing study. *J Comp Neurol* 398:151–161.

Chapter 10
Hormonal Induction of Ovulation In Vitro in Sturgeon Fishes

B.F. Goncharov, M.N. Skoblina, O.B. Trubnikova, and S.G. Vassetzky

Abstract It is shown that the capacity of progesterone to induce ovulation of the stellate sturgeon (*Acipenser stellatus* Pall.) oocytes depends on the osmolality of the culture medium. The use of 70% Ringer solution modified for sturgeons markedly increases the percentage of ovulated oocytes. However, the oocytes matured and ovulated under these conditions were characterized by low abilities of fertilization and subsequent development. The diluted (69%) Leibowitz medium proved to be much better but in this case also, the result depended to a great extent on the initial state of ovarian follicles. We achieved up to 90% of ovulated oocytes, of which 90% started development after insemination and about 60% reached the prelarval stage, and successfully passed through critical stages, such as transition to branchial respiration and exogenous feeding. Thus, it became possible for the first time to obtain viable progeny of sturgeons from the oocytes that had matured and ovulated in vitro.

Keywords Sturgeons, hormonal induction, ovulation, in vitro, development

10.1 Introduction

Hormonal induction of in vitro oocyte maturation is widely used for studying the mechanisms of this process in many animals, including sturgeons. The use of this model system also underlies the development of some methods that allow improvement of artificial reproduction of these economically valuable fish. They include the method of comparative quantitative estimation of the gonadotropic activity of pituitary preparations used to stimulate gamete maturation and the method of selecting sturgeon females for reproduction. In addition, monitoring of the state of ovarian follicles makes it possible to study the influence of the conditions of keeping mature fishes and of various hormonal treatments and, thereby optimizing their reproduction.

In all these studies, germinal vesicle breakdown (GVBD) was used as a criterion of oocyte maturation. However, oocyte maturation includes not only nuclear transformations, but also transformations of the cytoplasmic structures and, at the physiological level, leads to the formation of a mature egg capable of fertilization and development. Nevertheless, it was impossible to use the capacity for development as a criterion of normal in vitro maturation of the sturgeon oocytes, since either the matured oocytes do not ovulate, or the percentage of ovulation is much less than that of maturation. The available published data and our observations suggest that the use of an inadequate culture medium may cause the absence of in vitro ovulation. For example, it was reported that ovulation of the Pacific sturgeon oocytes was hardly stimulated either by gonadotropic preparations, or by progesterone (Lutes, 1985). On the contrary, in vitro ovulation in the presence of progesterone or gonadotropic preparations was observed when studying physiological changes of the Siberian sturgeon follicles of older generations. It was shown that the percentage of ovulation gradually increased as the spawning season approached (Goncharov et al., 1999). The discrepancy in these results could be explained by both different physiological states of the follicles and the use of different culture media. In our experiments, we often noted the discrepancy of the results of ovulation induction in vitro and in vivo: from complete ovulation in vivo under the influence of pituitary extract injection to the absence of ovulation in vitro in the follicles of the same females in the presence of the same gonadotropic preparation. This suggested that the absence of reaction in vitro is due to an inadequate culture medium, rather than to the absence of competence for ovulation, or inadequate hormonal treatment. Finally, when studying the influence of the composition of the culture medium on maturation of the sturgeon oocytes (Goncharov et al., 1997), we noted that the percentage of in vitro oocyte maturation in the presence of pituitary extract, given a certain physiological state of the follicles, increased with the culture medium osmolality, while the percentage of oocyte ovulation decreased.

The published data available and the results of our observations prompted us to study the influence of various culture media on the capacity of gonadotropic preparations and progesterone for induction of sturgeon oocyte ovulation. As a result, it was shown that under certain conditions a sufficiently high percentage of ovulated oocytes could be obtained and hormonal induction of in vitro oocyte maturation and ovulation proceeded quite normally, since the eggs obtained in vitro proved to be capable of fertilization and subsequent development.

10.2 Materials and Methods

At the Krasnodar sturgeon hatchery, studies were made on females of the stellate sturgeon *Acipenser stellatus* Pall. caught in nature during the spawning season and kept in basins for 2 to 3 weeks.

Ovarian follicles were isolated from the body using a special metallic probe. After isolation, the follicles were repeatedly washed by one of two media, in which they were then cultivated: Ringer solution for poikilotherms modified for sturgeon follicles (RMS) (Goncharov, 1978) containing an increased concentration (2 g/l) of sodium bicarbonate and Leibowitz medium (L-15) complemented by sodium bicarbonate as indicated in the results. Both media were also complemented by penicillin (500 U/l) and streptomycin (0.25 g/l).

Solutions and pituitary extracts were prepared just before the experiments. In order to obtain the pituitary extract, a suspension of acetonized pituitary powder was kept at room temperature for 1 h. The same stock powder was used in all experiments. Progesterone was introduced directly in Petri dishes as a concentrated alcohol solution. Its final concentration in the culture medium was 1 μg/ml and alcohol concentration was 0.1%.

The follicles were incubated in plastic Petri dishes (55 m in diameter) with 7.5 ml medium containing a tested preparation, 33 ± 2 follicles per dish. The period from isolation of the follicles to their transfer in the medium with hormonal preparations did not exceed 1 h. The follicles were incubated at ambient temperature. Ovulation was estimated by counting the follicle envelopes separated from oocytes. Some follicles were used to determine the percentage of oocyte maturation. For this purpose, the follicles were fixed by boiling and dissected by a safety blade under a dissection microscope to register the absence or of presence of germinal vesicle.

Insemination was performed by the semidry method currently used in sturgeon pisciculture (Dettlaff et al., 1993). Before insemination, the culture medium was poured out (in the case of L-15, the eggs were then rapidly washed by water) and then water and sperm were added as 1:25. The sperm was preliminarily checked for motility and sperm of high quality (more than four points according to the Persov, (1941), scale) only was used. The percentages of fertilization (stage 5–7) and of normal embryos at the small yolk plug stage (stage 17; Dettlaff et al., 1993) were determined by examination of the developing eggs under the dissection microscope. The percentage of normal hatched prelarvae and larvae on day 10 after transition to active feeding were also determined. The differences in frequency distribution were estimated using χ^2-test.

10.3 Results

10.3.1 Influence of Culture Medium Osmolality on Ovulation Induced by Sturgeon Pituitary Extract and/or Progesterone

In different stellate sturgeon females, the results of induction of in vitro ovulation by pituitary extract or progesterone in RMS may vary within wide limits. The data of three experiments on follicles of three females in each are presented in Fig. 10.1. The follicles of these females either did not respond to progesterone in 100% RMS, or ovulation was observed only in a small percentage of the cases (<10%). It can be seen that the percentage of ovulated oocytes increased, often significantly, with the RMS dilution. The highest response was observed in seven females in 70% RMS and in two females in 80% RMS. When RMS was diluted to 60%, the

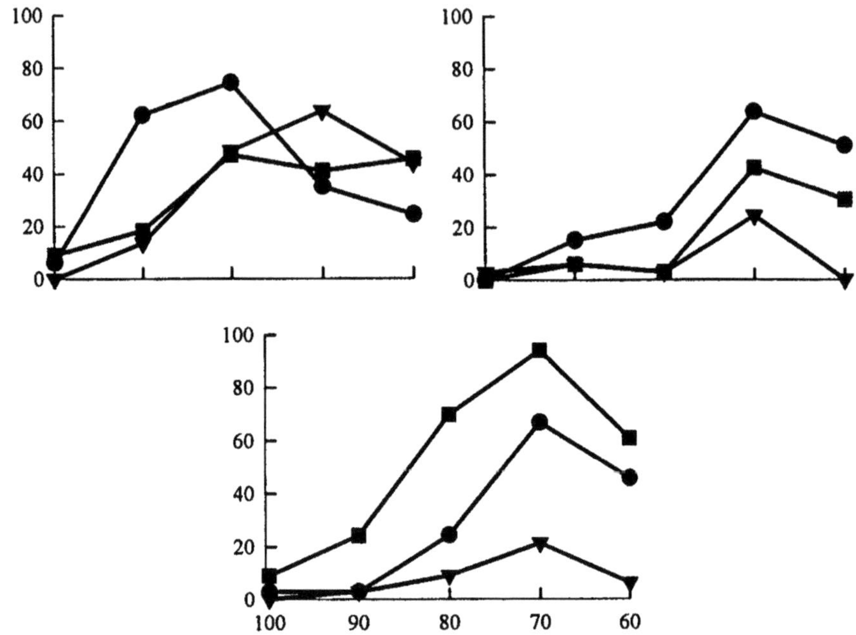

Fig. 10.1 Effect of RMS dilution of the capacity of progesterone to induce in vitro ovulation of the stellate sturgeon oocytes
Abscissa: RMS concentration, %. Ordinate: ovulated oocytes, %. In all three experiments, females are designated by different symbols (●, ■, ▲)

percentage of ovulated oocytes decreased in all females and when RMS was diluted to 50%, visible damage to the oocytes was noted (data not shown). In eight females, all oocytes matured in the presence of progesterone in all used RMS dilutions, and only in one female the percentage of matured oocytes fell sharply to 6.5% in 60% RMS.

In the second and third experiments, the follicles of the same females were incubated not only in the presence of progesterone, but also in the presence of pituitary extract (15 μg/ml) at the same RMS dilutions. In two females, no ovulation was observed in the presence of pituitary extract for all RMS dilutions, while in three females, the percentage of ovulated oocytes was low (<12%) and was registered only in two dilutions (100 and 90% RMS). Also, only in one female from the third experiment, in which more than 90% of oocytes ovulated in the presence of progesterone in 70% RMS, 30, 18, 15, 3, and 0% of oocytes ovulated in the presence of pituitary extract in the tested media: 100, 90, 80, 70, and 60% RMS, respectively. The level of oocyte maturation in the presence of pituitary extract was 100% (or nearly so) in 100 and 90% RMS. When more diluted media were used, the percentage of matured oocytes decreased, beginning from 80 to 60% RMS (in different females), varying from 0 to 100% in 70% RMS. Note that, in the absence of hormone, no ovulation was observed at any RMS dilution. Hence, the increase in the

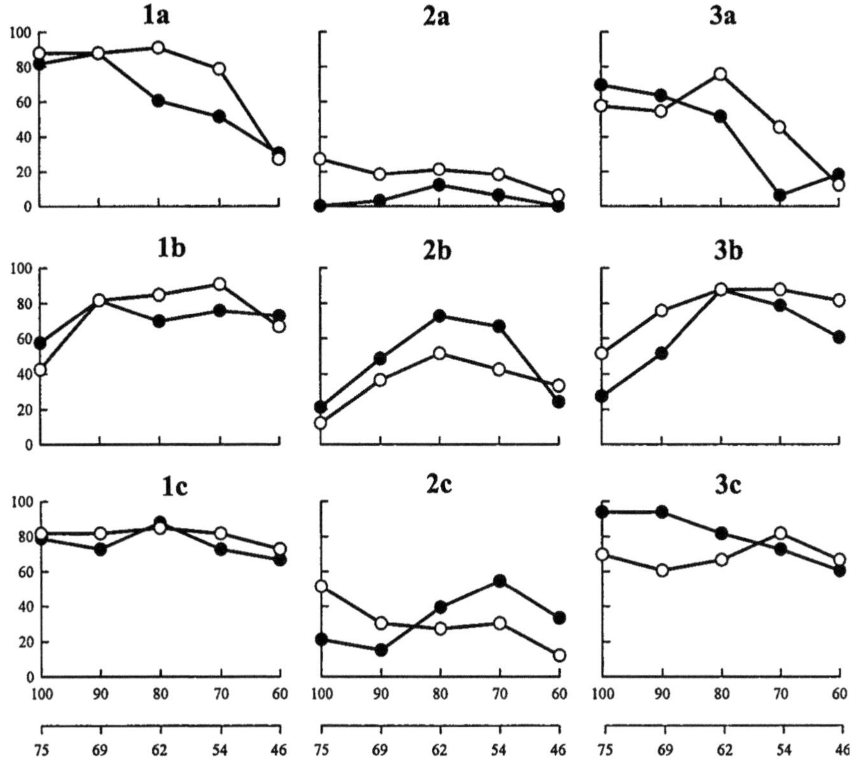

Fig. 10.2 Effect of culture medium (RMS and L-15) dilution on the capacity of progesterone and sturgeon pituitary extract to induce in vitro ovulation of the stellate sturgeon oocytes.
Abscissa: RMS concentration, %; on additional axis: L-15 concentration, %. Ordinate: ovulated oocytes, %. The results obtained in RMS and L-15 are designated as ● and ○, respectively. Variants with the use of sturgeon pituitary extract, progesterone, and mixture of two hormonal preparations are designated as **a**, **b**, and **c**, respectively; **1**, **2**, and **3** are ordinal numbers of females

percentage of ovulated oocytes with medium dilution was not related to the medium unspecific effect.

In the next experiment (Fig. 10.2), we compared different dilutions of two media: RMS and L-15. L-15 was diluted in such a way that its dilutions corresponded roughly to RMS dilutions by their osmolality. Besides, sodium bicarbonate (2 g/l) was added to 75% L-15 before dilution. The follicles of three stellate sturgeon females were used in this experiment; in all females, the percentage of oocytes ovulated in the presence of progesterone and, especially, pituitary extract in 100% RMS was higher than in the preceding experiments. Progesterone-induced ovulation was characterized by the same pattern as in the preceding experiments: as the culture medium was diluted, the percentage of ovulated oocytes increased and then decreased. This was true for both RMS and L-15. The fall in the percentage of oocytes ovulated in the presence of pituitary extract accompanying the culture medium (both RMS and L-15) dilution was confirmed (females 1 and 3). When two hormonal preparations (progesterone and pituitary extract) were used simultaneously, the percentage of oocytes ovulated at different medium dilutions varied to a much lesser extent than in the presence of each of these preparations separately. In this experimental variant, the percentage of ovulated oocytes was, on average, the same or somewhat higher than in the experiments with one hormonal preparation (females 1 and 3). In female 2, on the contrary, the mean percentage of oocytes ovulated in the presence of both preparations in RMS was somewhat lower than in the presence of progesterone alone. On the whole, there were no significant differences in the hormonal induction of ovulation between the corresponding dilutions of two media. The final data on the induction of ovulation are presented in Fig. 10.2. However, we recorded the ovulation level two more times: at 6.5 and 4.5 h before the end of the experiment. It was found that in all females, ovulation started in progesterone-containing media earlier than in those with pituitary extract. Note that in the variant with the treatment of follicles by both hormonal preparations, the percentage of ovulated oocytes was lower than in the presence of progesterone alone at the first and, in some cases, second time of observation. Another noteworthy observation concerns oocyte maturation in different dilutions of culture media. In this experiment, 100% of the oocytes from all three females matured in the presence of progesterone in all the media used, while in the presence of pituitary extract, 10% oocytes from two females (1 and 3) matured in 60% RMS whereas no oocytes from female 2 matured even in 70% RMS. On the contrary, in L-15, all oocytes from all females matured at all dilutions.

In the next experiment (Fig. 10.3), all the RMS used and the L-15 dilutions contained 2 g/l sodium bicarbonate, unlike the preceding experiments, when sodium bicarbonate (2 g/ml) was added to the initial medium and its concentration decreased with dilution. Only four dilutions of both media were tested in this experiment because of the lack of follicles. As in the preceding experiment, the percentage of oocytes ovulated in more concentrated media was rather high in two females and low in the third female. The percentage of ovulated oocytes in more diluted media containing progesterone increased, but remained unchanged or even decreased in the presence of pituitary extract. However, there was one clear distinction from the preceding experiment. In all females, especially in female 3, the response to all variants of hormonal treatment was more pronounced in all L-15 dilutions, as compared to RMS.

10.3.2 Developmental Capacities of Eggs Matured and Ovulated in Different Media in the Presence of Progesterone and Pituitary Extract

All oocytes ovulated in the experiment presented in Fig. 10.2 were inseminated and their development was followed until the larva stage (10 days after the beginning of exogenous feeding). Normal embryos were counted at the stages of four

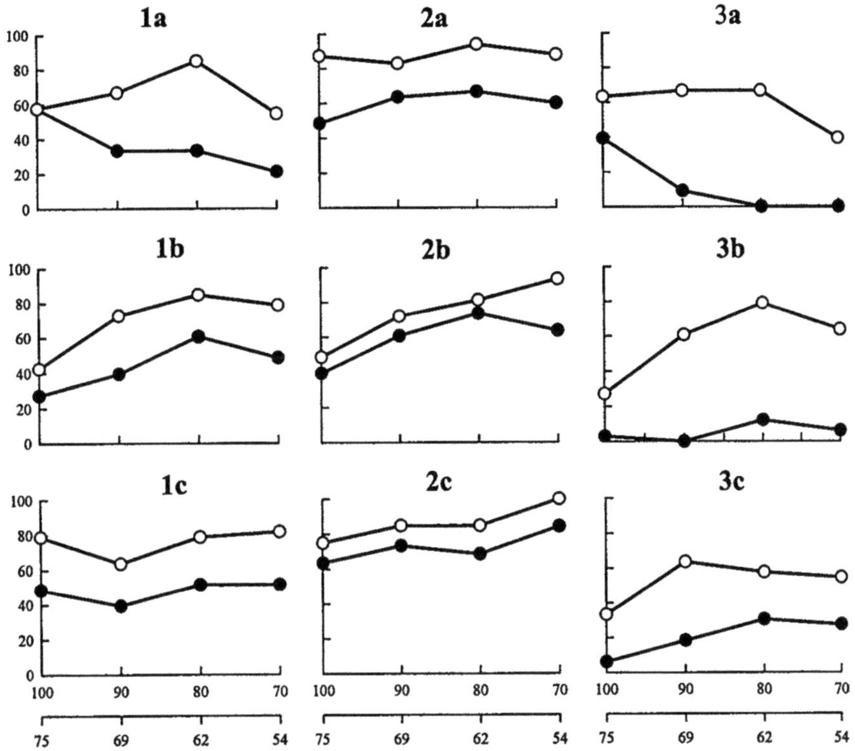

Fig. 10.3 Effect of culture medium (RMS and L-15) dilution on the capacity of progesterone and sturgeon pituitary extract to induce in vitro ovulation of the stellate sturgeon oocytes
For designations see Fig. 10.2. For differences in preparation of medium L-15 see text

blastomeres, small yolk plug, at hatching, and after transition to exogenous feeding. The results are provided only for females 1 and 3, as only a few oocytes of female 2 were fertilized. Analysis of the data presented in Fig. 10.4 allows the following conclusions to be confirmed by statistical processing. Medium L-15 is preferable to RMS for obtaining stellate sturgeon fry. Even when a rather high percentage of fertilization was found in RMS, the mortality of embryos was significant. The rate of larvae survival in L-15 was reliably higher for dilutions 69 and 62%, as compared to 75 and 54%, while in RMS the embryos survived only in two dilution variants: 90 and 100%. Regarding the influence of the inducers, the results proved to be somewhat different for two females. In female 1, the efficiency of progesterone and pituitary extract to induce maturation and ovulation of the oocytes capable of developing, after fertilization, into normal larvae was the same, while the joint application of both hormonal preparations gave worse results. In female 3, the best results (for all dilutions of both media) were in the experimental variant with pituitary extract, while the results of variants with progesterone or both preparations proved worse (the differences at the limit of significance) and did not differ from each other.

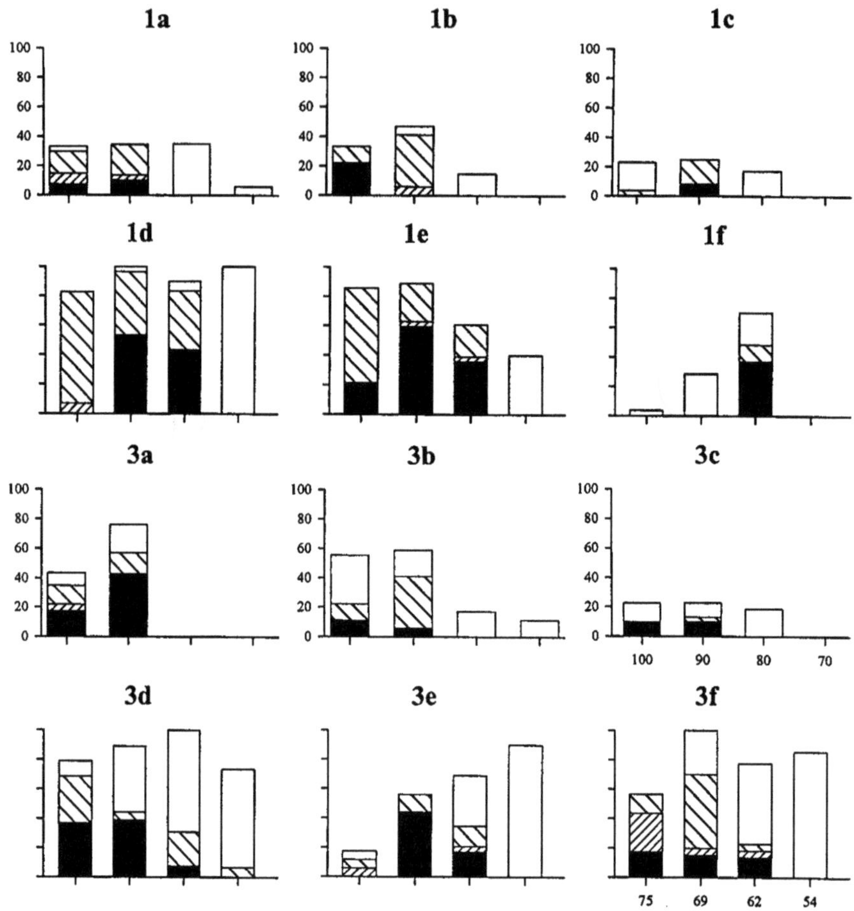

Fig. 10.4 Development of stellate sturgeon embryos obtained after artificial insemination of the oocytes matured and ovulated in vitro in different media and under the influence of different hormonal preparations

Abscissa: culture medium concentration, %. Ordinate: normal embryos, prelarvae, and larvae, %. **1** and **3**, ordinal number of females. Follicles were incubated in RMS (**a**, **b**, and **c**) or L-15 (**d**, **e**, and **f**) in the presence of sturgeon pituitary extract (**a** and **d**), progesterone (**b** and **e**), or mixture of these hormonal preparations (**c** and **f**). Survival and development were assessed at the four-cell stage (□), in the end of gastrulation (\\\), at the hatching stage (///), and within 10 days after transition to exogenous feeding (■)

10.4 Discussion

The possibility of in vitro hormonal induction of ovulation in sturgeons is essential for both fundamental research and for their artificial reproduction. The mechanisms of hormonal stimulation of ovulation have hardly been studied, since either this process could not be repeated on isolated follicles (Lutes, 1985), or stable reproducible results could not be achieved in in vitro experiments (our studies). In a single study of the effect that inhibitors of biosynthetic processes exert on ovulation, ovulation was triggered in vivo by the injection of pituitary extract and the follicles were then isolated within certain time intervals and all subsequent manipulations were performed in vitro (Trubnikova, 2003). Induction of in vitro ovulation opens possibilities for studying this process under controlled conditions. In addition, this is essential for studying oocyte maturation. If the conditions are found for not only maturation of the oocytes, but also for their ovulation, the normal course of these processes can be checked by insemination of the obtained eggs and analysis of their subsequent development. Obtaining embryos from the oocytes matured and ovulated in vitro is undoubtedly valuable for the improvement of sturgeon aquaculture. This approach opens possibilities for estimating the reproductive potential of a female even before obtaining mature gametes. The idea of extending the time of obtaining progeny at the expense of extracting a part of the follicles from the female body is not that unrealistic, if we take into account the easy and nontraumatic extraction of ovarian follicles from the body, high fecundity of sturgeon fishes, and capacity of gametes to retain their reproductive quality during long-term exposure of females to lowered temperatures. However, this approach may be especially useful for obtaining progenies from unique individuals (rare or disappearing species or females valuable for selection), when the errors in case of hormonal stimulation in vivo could prove to be fatal.

The results of our studies suggest that the percentage of oocytes ovulated in vitro in the presence of progesterone may be significantly increased in a culture medium with lowered osmolality. Note that for oocyte maturation the opposite results were found: the percentage of maturing oocytes declined with osmolality. This was especially pronounced when pituitary extract was used as a stimulus, but in this case another factor comes into force: sodium bicarbonate concentration. As the medium was diluted, its concentration decreased in all experimental variants, except the last. The concentration of bicarbonate ions is known to affect significantly the sensitivity of sturgeon follicles to gonadotropins (Goncharov et al., 1997). On the contrary, progesterone acts in a much wider range of concentrations in a minimum culture medium, such as RMS, where sodium bicarbonate acts as a pH regulator, or in the absence of sodium bicarbonate, for example in standard medium L-15. We would like to offer a few comments with respect to the opposite response of two processes, coupled in time and induced by the same stimulus, to changes in the culture medium osmolality. It was shown (Metallov et al., 1999) that when maturation and ovulation of the sturgeon oocytes are induced by pituitary extract in vivo, significant changes take place in the properties of the internal fluids of females and the blood osmolality is lowered by 15% by the time of ovulation. Since GVBD occurs much earlier than ovulation, it may well be that that these two processes take place in microenvironments with different osmolalities.

Obtaining a high percentage of mature ovulated oocytes in vitro is not the ultimate purpose. To solve the above-mentioned problems, it was important to find the conditions under which these eggs will be capable of subsequent development. In this work, we show that fully viable eggs could be obtained for some females. The best result corresponded to female 1 (Figs. 10.2 and 10.4). When the follicles were cultivated in 69% L-15 in the presence of progesterone or pituitary extract, 74 and 91% of oocytes, respectively, ovulated and, among these, 59 and 55% developed normally after fertilization, developing into larvae with no visible anomalies till the end of observations (10 days after transition to exogenous feeding). Note that in this female the highest percentages of ovulated oocytes were observed throughout the entire range of the medium dilutions used, including the most concentrated media. This agrees quite well with the data obtained when studying the correlation between physiological characteristics of the Siberian sturgeon (*Acipenser baerii*) follicles, as determined by their reaction to the hormones in vitro, and developmental potencies of the eggs obtained after injection of a hormonal preparation to females (Goncharov et al., 1999). The percentage of oocytes ovulated in vitro in the presence of pituitary extract proved to be the second best prognostic criterion of selection of females for artificial reproduction.

It is evident that the efficiency of obtaining sturgeon progenies by means of artificial insemination of the eggs matured and ovulated in vitro under the influence of hormones depends on three factors: the initial physiological state of the follicles, the conditions of hormonal stimulation, and the formation of a mature egg and its capacity for development, and character of hormonal treatment. For a desirable result to be achieved, further studies may be carried out along several lines: search for methods of determining the required physiological state of ovarian follicles, changes in their physiological state in a desired direction by means of influencing ecological or humoral factors on the female, further search for common conditions of cultivation and hormonal treatment, optimal for the follicles at different physiological states, or a search for different conditions of cultivation and hormonal treatments, adequate to different physiological states of the follicles.

Acknowledgements The authors express their sincere gratitude to Prof. M.S. Chebanov for his help in this investigation. This study was supported by the Russian Foundation for Basic Research, project no. 05-04-48433.

References

Dettlaff, T.A., Ginsburg, A.S., Schmalhausen, O.I. 1993. *Sturgeon Fishes. Developmental Biology and Aquaculture*. Springer Verlag, Berlin, 300 pp.

Goncharov, B.F. 1978. The influence of the culture medium composition on the capacity of the sturgeon follicles to mature in the presence of gonadotropic hormones. In: *Problems of Early Ontogenesis in Fish*. T.V. Dekhnik, V.N. Zhukinsky and L.S. Oven (eds.) Naukova Dumka, Kiev, pp. 77–78 (in Russian).

Goncharov, B.F., Polupan I.S., Williot, P., Le Menn, F. 1997. Influence of the culture medium composition on maturation of sturgeon oocytes induced by gonadotropic hormones and progesterone. *Russ. J. Devel. Biol.* 28, 55–64.

Goncharov, B.F., Williot, P., Le Menn, F. 1999. Morphological and physiological characteristics of the ovarian follicles of farmed Siberian sturgeon and their importance for predicting artificial reproduction success. *Russ. J. Devel. Biol.* 30, 46–54.

Lutes, P.B. 1985. Oocyte maturation in white sturgeon, *Acipenser transmontanus*: some mechanisms and applications. In: *North American Sturgeons: Biology and Aquaculture Potential*, E.P. Binkowski and S.I. Doroshov (eds.), Dr. W. Junk Publishers, Dordrecht, pp. 87–92.

Metallov, G.F., Lavrova, E.A., Aksenov, V.P., Geraskin, P.P. 1999. Physiological heterogeneity and ion homeostasis of internal fluids in the stellate sturgeon. *Toxicol. Herald* 1, 27–31 (in Russian).

Persov, G.M. 1941. Account of pisciculture studies with reference to the method of pituitary injections. In: *Method of Pituitary Injections and Its Role in Reproduction of Fish Reserves*. Leningrad State University, Leningrad, pp. 42–50 (in Russian).

Trubnikova, O.B. 2003. Effects of cycloheximide, aminoglutethimide, cytochalasin B, and colchicine on ovulation and the ultrastructure of the ovarian follicle wall in the stellate sturgeon *Acipenser stellatus* Pall. *Russ. J. Devel. Biol.* 34, 110–120.

Chapter 11
Dispermic Androgenesis as a Method for Recovery of Endangered Sturgeon Species

Anna S. Grunina, Alexander V. Recoubratsky, Vladimir A. Barmintsev, Ekaterina D. Vasil'eva, and Mikhail S. Chebanov

Abstract The possibility of the recovery of endangered sturgeon species is considered, with special reference to the production of androgenetic nucleocytoplasmic hybrids by dispermic androgenesis. The method of dispermic androgenesis, developed for sturgeon fishes, includes genetic inactivation of eggs, their insemination with concentrated sperm (to cause polyspermy), and heat shock to facilitate the fusion of male pronuclei. The restoration of the diploid state of androgenotes by fusion of two sperm nuclei allows androgenetic progeny to have a heterozygosity level similar to that in a regular crossing. Using this method, viable androgenetic progenies of several sturgeon species were obtained for the first time. Here we present the results of the first successful experiments in obtaining androgenetic hybrids in sturgeon fishes by dispermic androgenesis. Both the androgenetic hybrids, between stellate and great sturgeons and between Persian and Russian sturgeons, obtained in these experiments were fully viable. It was shown that androgenetic hybrids had the nuclear DNA of the paternal species and the mitochondrial DNA of the maternal species. The morphological analysis proved the androgenetic hybrids between Persian and Russian sturgeons to be completely identical to the paternal species at the age of 12 months, and those between stellate and great sturgeons by 3 years. The prolonged manifestation of matrocliny observed in both

A.S. Grunina
N.K. Koltsov Institute of Developmental Biology, Russian Academy of Sciences, Moscow, Russia

A.V. Recoubratsky
All-Russian Research Institute of Freshwater Fisheries, Moscow Province, Russia

V.A. Barmintsev
Russian Federal Research Institute of Fisheries and Oceanography (VNIRO), V. Krasnoselskaya, Moscow, Russia

E.D. Vasil'eva
Zoological Museum of the Moscow State University, Bolshaya Nikitskaya, Moscow, Russia

M.S. Chebanov
South Branch Federal Centre of Selection and Genetics for Aquaculture, Oktyabrskaya St., Krasnodar, Russia

R. Carmona et al. (eds.) *Biology, Conservation and Sustainable Development of Sturgeons*,

hybrids is most likely related to the extended life cycle of acipenserids. Our results suggest that endangered sturgeon fishes may be recovered by means of dispermic androgenesis if the sperm alone of a given species is available.

Keywords Androgenetic nucleocytoplasmic hybrids, dispermic androgenesis, DNA analysis, morphological analysis, sturgeons

11.1 Introduction

Induced androgenesis is a method to produce individuals whose development is controlled by the paternal chromosomes alone, without involving maternal nuclear heredity. To induce diploid androgenesis, it is necessary to eliminate the egg chromosome set and then cause diploidization of the male chromosomes.

Induced androgenesis can be employed in producing inbred strains and clones in fish, for sex regulation, for studying nucleocytoplasmic interactions, etc. This method attracts great attention because it allows the recovery of endangered fish species from the genetic material of spermatozoa alone (Veprintsev and Rott, 1979; Grunina and Neyfakh, 1997; Corley-Smith and Brandhorst, 1999). The importance of androgenesis is increasing because the technique of sperm cryopreservation is well developed (Suquet et al., 2000), whereas the problem of long-term storage of fish eggs and embryos remains to be solved (Lubzens et al., 1996; Chao and Liao, 2001). To recover a genotype from only the spermatozoa, when conspecific females are not available, the eggs from another species may be used, i.e. androgenetic nucleocytoplasmic hybrids will be produced (Grunina and Neyfakh, 1997).

Diploid androgenesis has been successfully produced in a number of teleostean fishes using the technique, which includes the inactivation of the female chromosomes by ionizing (X- or γ-rays) or UV irradiation of eggs and diploidization of the male chromosome set by the suppression of the first cleavage division in androgenetic haploid embryos (see reviews, Penman et al., 1996; Grunina and Neyfakh, 1997). Androgenetic sturgeons produced in this way, however, proved to be inviable (Grunina and Neyfakh, 1991). There is reason to believe that sturgeons are more sensitive than teleosteans to a high level of homozygosity arising from the suppression of the first cleavage.

At the same time, the investigations on induced androgenesis in sturgeons are of great importance as many sturgeon species and populations are endangered or close to extinction (Birstein et al., 1997b).

To avoid high homozygosity, we proposed that dispermic androgenesis be used. Androgenetic individuals, whose diploid state is restored through the fusion of the chromosome sets of two spermatozoa, have a heterozygosity level similar to that of a normal progeny. Two features of sturgeon eggs make dispermic fertilization possible: (1) Multiple micropyles allow several spermatozoa to penetrate the egg and (2) there are no intracellular mechanisms to exclude super-numerary spermatozoa from development (Ginsburg, 1968).

The method of dispermic androgenesis that we have developed for sturgeons includes genetic inactivation of eggs, their insemination with concentrated sperm (to cause polyspermy), and heat shock that facilitates the fusion of male pronuclei. Using this method, viable androgenetic progenies of several sturgeon species were obtained for the first time (Grunina et al., 1995; Recoubratsky et al., 1996; Grunina and Recoubratsky, 2005).

To obtain androgenetic nucleocytoplasmic hybrids, we applied the method of dispermic androgenesis to various hybrid combinations of sturgeons. Besides homozygosity, the viability of androgenetic hybrids can be unfavourably affected by nucleocytoplasmic incompatibility (Neyfakh and Radzievskaya, 1967; Grunina and Neyfakh, 1997; Neyfakh, 1999). We studied this phenomenon on sturgeon species that belonged to both diploid (approximately 120 chromosomes) and tetraploid (240–250 chromosomes) groups. We found that the extent of nucleocytoplasmic incompatibility depends not only on phylogenetic distance (that is obvious), but also on the differences in ploidy levels between the species used (Recoubratsky et al., 1998). For instance, diploid androgenetic hybrids obtained in reciprocal crosses between the Russian, *Acipenser gueldenstaedtii* (250 chromosomes) and stellate, *A. stellatus*, sturgeons (118 chromosomes) proved to be inviable and died at different stages of embryogenesis. Hence, sturgeon species of the same ploidy level are more promising for the production of viable androgenetic hybrids.

This paper describes the results of our first successful experiments in obtaining viable androgenetic nucleocytoplasmic hybrids in sturgeons. Two combinations were tested: (1) stellate sturgeon females and great sturgeon, *Huso huso*, males; and (2) Persian sturgeon (*A. persicus*) females and Russian sturgeon males. The first pair represents species with 120 chromosomes and the second, with 250 chromosomes (Vasil'ev, 1985; Birstein et al., 1997a). The data of molecular-genetic and morphological analyses of these hybrids are presented.

11.2 Material and Methods

The experimental work was carried out at the Krasnodar Sturgeon Hatchery (Russia) in 1999. Two experiments were performed. In experiment 1, stellate × great sturgeon and in experiment 2, Persian × Russian sturgeon androgenetic hybrids were produced. In experiment 1, a mixture of eggs from two females of the stellate sturgeon and two females of the great sturgeon were used. The sperm was collected from four males of stellate and four males of great sturgeons. The Persian sturgeon eggs were obtained from one female. Two males of Persian and four males of Russian sturgeons were used to collect sperm for experiment 2. The stellate, great, and Persian sturgeon females and Russian sturgeon males were caught in the Azov Sea. The Persian sturgeon sperm was collected from the males caught in the Caspian Sea. The great sturgeon sperm was obtained from the males that were reared in captivity at the Krasnodar Research Institute of Fishery.

11.2.1 Experimental Design

The progenies obtained in each experiment are listed in Table 11.1. In addition to diploid androgenetic hybrids, a number of control groups were produced: progenies from each parental species (except the Russian sturgeon, experiment 2), true hybrids, intraspecific diploid androgenotes, and haploid (unshocked) androgenotes. One more androgenetic group was produced in experiment 1 by fertilizing the enucleated stellate sturgeon eggs with a mixture of the stellate and great sturgeon sperm. In each experiment, to produce different androgenetic groups, one batch of eggs was irradiated and then divided into the appropriate number of portions.

11.2.2 Irradiation of Eggs and Embryo Manipulations

Eggs were X-irradiated at 220 Gy. This dosage was earlier shown to completely inactivate the maternal chromosome set both in the stellate and Russian sturgeon eggs (Recoubratsky et al., 1996, 1998). Weakly diluted (1:10 instead of the commonly used 1:100) sperm was used to cause polyspermic fertilization. According to our

Table 11.1 Results of embryo incubation and rearing of larvae obtained in experiments 1 and 2

		Yield of normal larvae		Yield of 1-month-old fingerlings		
Group	No. of fertilized eggs	No.	% From fertilized eggs	No.	% From larvae	% From fertilized eggs
Experiment 1						
(S) × H, shocked	2029	195	9.6	67	34.4	3.3
(S) × H, no shock	149	0	0.0	–	–	–
(S) × S, shocked	569	35	6.3	6	17.1	1.1
(S) × S, no shock	128	0	0.0	–	–	–
(S) × S+H, shocked	720	34	4.7	–	–	–
S × H	137	102	74.5	52	51.0	38.0
S × S	166	95	57.2	35	36.8	21.1
H × H	93	53	57.0	25	47.2	26.9
Experiment 2						
(P) × R, shocked	1799	451	25.1	179	39.7	9.9
(P) × R, no shock	120	0	0.0	–	–	–
(P) × P, shocked	392	71	18.1	35	49.3	8.9
(P) × P, no shock	41	0	0.0	–	–	–
P × R	305	140	45.9	105	75.0	34.4
P × P	135	76	56.3	57	75.0	42.2

Abbreviations: S, stellate; H, great; P, Persian; R, Russian sturgeons; S+H, a mixture of the stellate and great sturgeon sperm was used for insemination. The first capital letter in the group name designates the maternal species and the second capital letter in the group name designates the paternal species. A letter in brackets means that eggs of this species were X-irradiated previous (Grunina et al., 1995) and published (Ginsburg, 1968) data, the sperm dilution of 1:10 allows no more than two spermatozoa to penetrate most eggs. Embryos were incubated in Petri dishes at 19–20°C. The heat shock (37°C, 2.5 min) was applied 1.4 to 1.6τ_0 after insemination. This period is the most effective for the fusion of male pronuclei (Recoubratsky et al., 1996). τ_0 is a relative unit of developmental duration, which equals the length of one mitotic cycle during synchronous cleavage divisions. The age expressed in τ_0 is a temperature independent characteristic (Dettlaff et al., 1993).

11.2.3 Fish Rearing

The hatched prelarvae were transferred from Petri dishes to 1 L plastic basins in which they were kept until the beginning of exogenous feeding. Larvae were nursed in 10 L basins and fed on *Artemia salina* nauplei for a month. Further rearing was carried out in the tanks with constant water supply or in small ponds in summer. Fishes were fed on pelleted artificial and natural live diets. Androgenetic hybrids were raised under conditions identical to those of the progeny of their parental species and true hybrids.

A part of the experimental progenies at the larval stage were transferred to Kol'tsov Institute of Developmental Biology (Moscow). Each group of fish was reared in a separate aquarium under the same controlled conditions.

11.2.4 DNA Isolation

One-month-old specimens from all the groups were fixed by 96% alcohol. DNA was isolated from homogenized tissues using the standard chloroform–phenol extraction method. The concentration and purity of the isolated DNA were controlled using a spectrophotometer. The samples in which D260/280 ranged from 1.85 to 2.00 were selected for further analysis. Standard dilutions were prepared by collecting aliquots from the stock DNA solution to a final concentration of 20 ηg μL^{-1}

11.2.5 Random Amplification of Polymorphic DNA Polymerase Chain Reaction (RAPD-PCR)

RAPD–PCR was used to identify the nuclear DNA of androgenetic hybrids in comparison with the DNA of true hybrids and parental species. Reaction cocktail was prepared as follows: 15.8 µL deionized water, 2.5 µL 10× buffer, 2.0 µL dNTP (2.5 mM), 1.0 µL $MgCl_2$ (50 mM), 0.5 µL primer (5 OD mL^{-1}), 0.0625 µL Taq polymerase (5 U μL^{-1}), and 3.0 µL template DNA were added to 0.5 mL tube and covered with mineral oil. RAPD-PCR was performed using an amplifier "MJ Research PTC-150" and the procedure consisted of the following steps: the first denaturation –94°C, 3 min; recycling –94°C, 1 min, 36°C, 30 s, 72°C, 1 min (36 cycles total); additional synthesis –72°C, 10 min. A number of primers were tested and A07 primer (5′-GAA-ACG-GGT-G-3′) was found to be the most adequate to accomplish the task.

11.2.6 Mt-DNA PCR Assay

To identify the mitochondrial DNA of the great and stellate sturgeons, PCR assay was carried out using primer pairs specific to the diagnostic sequences of the cytochrome *b* gene of sturgeons (DeSalle and Birstein, 1997). The primer sequences and a protocol of the analysis were kindly granted by Dr. V. Birstein (USA) after permission by the patent supervisor, Dr. W. Burhenne, Executive Governor of the International Council of Environmental Law (Bonn, Germany).

11.2.7 Gel Electrophoresis

Gel electrophoresis of PCR products was performed in 6% polyacrylamide gel. DNA of λ phage restricted by *Pst*I was used as a positional marker.

11.2.8 Morphological Analysis

To compare morphological features of androgenetic hybrids with those of true hybrids and parental species, all experimental progenies were reared in laboratory aquaria under the same conditions up to 12-months of age. The low capacity aquaria we used allowed maintenance of only a restricted number of fishes. Hence, only a few fishes were analysed at the age of 6 and 12 months. Since experimental progenies of the stellate and the great sturgeons died at the early stages of development, the young of these species obtained from the Krasnodar Sturgeon Hatchery in 1999–2000 were used to compare the morphological features of androgenetic hybrids with those of the true hybrids and parental species. We tried to select numerous samples of specimens with the total body length *(TL)* similar to experimental fishes fixed at the age of 6 and 12 months. In addition, 3-year-old live androgenetic hybrids, as well as progenies of their parental species and true hybrids of the same age were investigated at the Hatchery under anaesthesia in May 2001. The total number of fishes examined in our study on the stellate × great sturgeon androgenetic hybrids is shown in Tables 11.2 and 11.4. In our investigation of the Persian × Russian sturgeon androgenetic hybrids we studied two 6-month-old and

12-month-old androgenotes, three 6-month-old, and five 12-month-old Persian sturgeons, four 12-month-old Russian sturgeons, and two and four true hybrids at the age of 6 and 12 months, accordingly (Vasil'eva et al., 2001).

All fishes were subjected to morphological analysis, which included generally accepted meristic and metric characters (Holčik et al., 1989), body pigmentation, relative size of dorsal scutes, shape of snout and mouth, diagnostic characters for genera *Acipenser* and *Huso*, and several indices of dermocranium were demonstrated to be of value for differentiation between different sturgeon species (Vasil'eva, 1999, 2004; Vasil'eva et al., 2001, 2005). The scheme of measurements used for craniological studies was published previously (Vasil'eva et al., 2001).

In statistical analysis of morphometric characters and craniological indices in different experimental groups and parental species, the standard univariate methods were used ($M \pm m$, t_{st}). After a preliminary comparison of samples, it was possible to select characteristics with the highest differences, with a hiatus or a minimum overlapping in the investigated samples of parental species of the same size. Comparative analysis of the androgenetic and true hybrids was performed using these characters (Vasil'eva et al., 2001) (Tables 11.2 and 11.4). Since it was shown (Vasil'eva, 1999) that a considerable number of the craniometric and morphometric characters in acipenserids varied with size, 6-month-old, 12-month-old, and 3-year-old sturgeons (and similar-sized specimens of indefinite age) were in all cases analysed separately.

The similarity level of the nucleocytoplasmic and true hybrids with the parental species was also estimated using the hybrid index (*HI*) calculated according the equation suggested by Verigin and Makeeva (1972): $HI = 2[100(M_h - M_f)/(M_m - M_f) - 50]$, where M_h is the average value of a character in the hybrid form, M_f is its average value in the maternal species, and M_m is its average value in the paternal species. A negative and positive value of *HI* mean that the hybrid deviates towards the maternal and paternal species, respectively.

11.3 Results

11.3.1 The Survival of Experimental Progenies

The results of incubation of embryos and rearing of androgenetic hybrids and appropriate control groups obtained in experiments 1 and 2 are presented in Table 11.1. In both experiments, all embryos from the irradiated but unshocked groups died before or during hatching. The heat shock caused the appearance of morphologically normal larvae that were capable of further development. Obviously, individuals from the irradiated unshocked groups were haploids and those from the irradiated shocked groups were androgenetic diploids.

In the stellate × great sturgeon combination, (S) × H (see footnote of Table 11.1 for abbreviations), 195 (9.6% of inseminated eggs) normal larvae were obtained and 67 (34.4%) of them were grown to the fingerling stage. The survival rate of the (S) × H androgenetic hybrids proved to be equal and even slightly higher than in the intraspecific (S) × S androgenotes.

In the Persian × Russian sturgeon combination, 451 (25.1% of inseminated eggs) normal androgenetic larvae hatched and 179 (39.7%) of them were grown to the fingerling stage (Table 11.1). Some androgenetic hybrids, both from the stellate × great sturgeon and Persian × Russian sturgeon combinations, have reached the age of 6 years.

11.3.2 Genetic Data

To confirm the androgenetic origin of (S) × H nucleocytoplasmic hybrids, RAPD-PCR was used to identify nuclear DNA of these hybrids in comparison with the DNA of true hybrids, S × H, and parental species taken from S × S and H × H groups (see Table 11.1). Mitochondrial DNA was identified in the same groups by the PCR assay. After the concentration and purity of the isolated DNA were controlled, the samples of DNA, obtained from 9 (S) × H androgenetic hybrids, 8 S × H true hybrids, 15 S × S stellate sturgeons, and 9 H × H great sturgeons were used for further analysis. All specimens from the (S) × H group had only nuclear DNA of the great sturgeon and the mitochondrial DNA of the stellate sturgeon (Figs. 11.1 and 11.2).

To confirm that the experimental conditions used provided for dispermic fertilization, one portion of the irradiated eggs was inseminated with a mixture consisting of equal parts of the great and stellate sturgeon sperm. Eight specimens from this (S) × S+H group were analysed and five of them had the RAPD-PCR patterns containing a combination of bands characteristic of both the stellate and great sturgeon nuclear DNA (Fig. 11.1, lanes 17–21). Two specimens had DNA of the great sturgeon and one specimen was identified as the stellate sturgeon.

In the Persian × Russian sturgeon combination, the RAPD-PCR analysis of larvae also confirmed their androgenetic origin (data not presented).

11.3.3 Comparative Morphology

When the characters specific for the genera *Huso* and *Acipenser* were analysed, all investigated (S) × H androgenetic hybrids were similar to the paternal species: they had semilunar mouth, their branchiostegal membranes were fused, not attached to the isthmus, and formed a free fold above it, and the barbels were flattened from the sides, with a foliar lobe. At the same time, in all true S × H hybrids the mouth was transverse, the branchiostegal membranes were attached to the isthmus, barbels were more or less cylindrical and without a foliar lobe, similar to the maternal species. The same situation was observed in the structural traits of the components and system of articulations of the palatoquadrate arch (Vasil'eva et al., 2005).

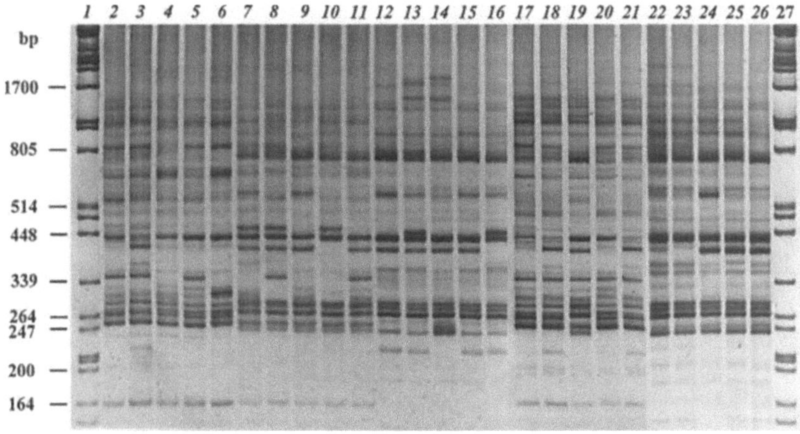

Fig. 11.1 RAPD-spectra of the total DNA obtained with the A07 primer. DNA was isolated from the great sturgeon, stellate sturgeon and their regular and androgenetic hybrids. The picture was compiled from photos of three polyacrylamide gel electrophoreses, no manipulations with the images being made

Lanes 1 and 27, positional markers (DNA of λ phage restricted by *Pst*I); lanes 2–6, stellate sturgeon [S × S]; 7–11, regular stellate × great sturgeon hybrids [S × H]; 12–16, great sturgeon [H × H]; 17–21, specimens produced by the insemination of enucleated stellate eggs with a mixture of the stellate and great sturgeon sperm [(S) × H+S]; 22–26, specimens produced by the insemination of enucleated stellate eggs with the great sturgeon sperm [(S) × H].

Specific band characteristics of the great sturgeon lie in the 750–805 bp area and those for stellate sturgeons are in the 164 bp area. Additional markers are also seen (zones of 810 bp, 370–460 bp, 245–340 bp, for example). These differences allow the great and stellate sturgeon DNA to be precisely distinguished without special calculations both in 'pure' and hybrid specimens obtained in this experiment. All markers specific both to the stellate and great sturgeons are available in regular hybrids (lanes 7–11), their intensity being approximately a half of that in 'pure' specimens. Specimens produced by the insemination of enucleated stellate sturgeon eggs with the mixture of the stellate and great sturgeon sperm (lanes 17–21) have DNA characteristic of regular hybrids. This suggests that these specimens originated from the fusion of two sperm nuclei (dispermic androgenesis). DNA of specimens produced by the insemination of the enucleated stellate eggs with the great sturgeon sperm (lanes 22–26) have markers specific to the great sturgeon only

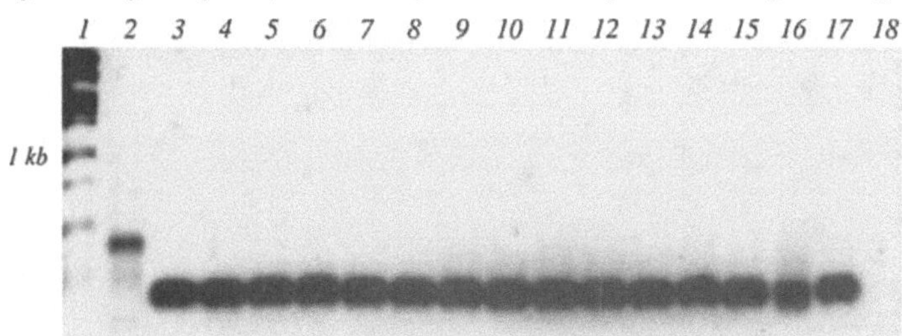

Fig. 11.2 PCR assay of mitochondrial DNA of the stellate and great sturgeons and their regular and androgenetic hybrids. Primer pairs specific to the stellate sturgeon (S2, S2an) and to the great sturgeon (B2, B2an) and a protocol were used as described by DeSalle and Birstein (1997). The same specimens analysed for the nuclear DNA on the Fig. 11.1 were used

Lane 1, positional marker; lane 2, the great sturgeon, H × H; lanes 3–5, stellate sturgeon S × S; lanes 6–8, regular stellate × great sturgeon hybrids, S × H; lanes 9–13, androgenetic hybrids produced using a mixture of the stellate and great sturgeon sperm, (S) × S+H; lanes 14–17, androgenetic stellate × great sturgeon hybrids, (S) × H; lane 18, Russian sturgeon.

Data presented show that eggs used for production of all types of hybrids belonged only to the stellate sturgeon females. Along with the RAPD-PCR spectra (Fig. 11.1) this confirms that hybrids developed from the irradiated eggs were of the androgenetic origin

Table 11.2 Ranges and means (in brackets) of morphometric characters in experimental groups of androgenetic hybrids ((S) × H) between the stellate and great sturgeons and true hybrids (S × H) of the same species at different ages and of juveniles of parental species similar in size

Progeny	*TL* (mm)	*Lsl*	*Lsr*	*a*	*aB*	*aM*	*barbi*	*mw*	*po*	*hc*	*lD*
6-month-old progenies											
(S) × H (*n* = 2)	166–201 (183.5)	35–39 (37.0)	36–38 (37.0)	43.8–48.2 (46.0)	25.3–25.7 (25.5)	48.5–49.7 (49.1)	20.8–21.3 (21.1)	31.1–37.1 (34.1)	40.4–46.1 (43.3)	34.7–36.2 (35.5)	10.1–10.4 (10.3)
S × H (*n* = 5)	148–205 (185.4)	30–38 (33.2)	31–38 (34.2)	49.2–53.0 (51.0)	30.0–33.7 (31.7)	54.3–59.5 (57.5)	14.7–19.7 (17.8)	27.6–29.0 (28.4)	37.9–42.0 (39.6)	32.7–36.0 (34.5)	8.3–10.5 (9.9)
S (*n* = 10)	179–215 (195.9)	33–37 (35.2 ± 0.39)	30–40 (35.6 ± 1.02)	51.5–56.0 (52.8 ± 0.42)	30.9–36.0 (32.6 ± 0.53)	52.9–62.9 (55.7 ± 0.87)	11.7–19.7 (16.9 ± 0.68)	23.0–30.3 (28.2 ± 0.67)	35.2–41.5 (39.0 ± 0.58)	35.4	7.2–10.1 (8.81 ± 0.28)
H (*n* = 8)	92–133 (111.4)	35–43 (39.8)	36–45 (38.9)	43.6–51.7 (47.6)	18.2–26.2 (22.1)	39.3–46.1 (43.1)	19.1–24.8 (23.2)	36.6–46.5 (41.5)	37.3–48.8 (42.9)	–	10.9–12.9 (11.8)
12-month-old progenies											
(S) × H (*n* = 2)	294–340 (317.0)	36 (36.0)	34 (34.0)	41.0–42.2 (41.6)	20.1–26.1 (23.1)	38.1–42.7 (40.4)	22.1–25.9 (24.0)	38.3–39.1 (38.7)	48.2–50.7 (49.5)	39.4–40.2 (39.8)	10.1–11.6 (10.9)
S × H (*n* = 4)	267–331 (296.7)	33–35 (34.0)	33–36 (34.3)	48.5–52.2 (50.6)	27.5–31.7 (30.1)	52.9–58.1 (55.5)	15.2–17.7 (16.4)	24.7–26.7 (25.7)	39.4–44.7 (41.9)	29.9–33.0 (32.0)	8.6–10.9 (9.6)
S (*n* = 14)	335–448 (404.1)	27–34 (30.4±0.58)	28–32 (30.7±0.41)	52.6–59.3 (56.4±0.46)	32.8–40.5 (36.9±0.48)	60.7–64.7 (63.2±0.29)	10.0–12.9 (11.8±0.24)	18.9–22.6 (19.9±0.25)	33.9–39.5 (36.3±0.43)	29.2–34.4 (31.5±0.40)	8.0–9.8 (8.8±0.14)
H (*n* = 3)	362–420 (398.7)	42–44 (43.3)	40–42 (41.3)	42.7–46.3 (44.2)	24.3–27.6 (25.8)	44.0–45.0 (44.5)	19.9–23.2 (21.4)	37.3–41.9 (39.1)	45.8–48.4 (47.1)	35.6–38.9 (37.6)	10.2–12.3 (11.5)

3-year-old progenies											
(S) × H (n = 4)	675–732 (697.8)	39–46 (42.5)	39–46 (42.5)	40.7–43.4 (42.5)	20.9–22.6 (22.1)	33.8–39.3 (37.3)	16.4–18.2 (17.5)	34.6–38.4 (36.5)	51.4–53.0 (52.1)	38.8–44.3 (41.6)	11.7–12.4 (12.1)
S × H (n = 5)	885–1100 (970)	30–41 (33.6)	30–38 (32.8)	52.1–55.9 (53.5)	31.9–34.1 (33.1)	57.5–58.8 (58.4)	13.6–15.1 (14.2)	20.8–22.4 (21.5)	39.1–42.6 (40.4)	34.4–38.2 (37.1)	8.1–9.5 (8.9)
S (n = 15)	603–845 (709.4)	27–34 (30.4±0.50)	25–34 (30.5±0.66)	49.1–58.8 (55.5±0.61)	31.4–39.2 (36.1±0.50)	55.8–65.1 (62.4±0.62)	9.6–13.8 (11.6±0.24)	18.0–20.0 (19.0±0.18)	35.4–41.0 (38.1±0.40)	31.7–37.0 (34.2±0.41)	7.8–10.4 (8.8±0.18)
H (n = 14)	592–860 (705.9)	34–46 (41.4±0.96)	34–51 (41.4±1.08)	39.4–45.7 (42.7±0.52)	20.5–26.8 (23.5±0.44)	35.5–42.9 (38.7±0.54)	15.9–22.2 (18.8±0.45)	36.3–40.7 (38.2±0.42)	49.3–55.4 (51.9±0.55)	32.6–45.5 (41.3±0.82)	9.5–13.9 (11.9±0.27)

Abbreviations of characters: *TL*, total body length; *Lsl*, lateral scutes number on the left side; *Lsr*, same, on the right; in % of head length: *a*, snout length; *aB*, distance from the snout tip to barbels; *aM*, distance from snout tip to upper lip; *barbi*, length of the inner barbell; mw, mouth width; *po*, postorbital distance; *hc*, head depth; in % *TL*: *lD*, length of dorsal fin base; abbreviations of species as in Table 11.1; *n*, number of specimens.

Table 11.3 Hybrid indices (HI) of morphometric characters in true (S × H) and androgenetic nucleocytoplasmic ((S) × H) hybrids between the stellate and great sturgeons at different ages

	Progenies and ages					
	(S) × H			S × H		
Characters	6 Months	12 Months	3 Years	6 Months	12 Months	3 Years
Lsl	−21.8	−13.2	120.0	−187.0	−44.2	−41.8
Lsr	−15.2	−37.7	120.2	−184.9	−32.1	−57.8
a	161.5	138.7	103.1	−30.8	−4.9	−68.8
aB	35.2	148.7	122.2	−82.9	22.5	−52.4
aM	4.8	148.1	111.8	−128.6	−17.7	−66.2
barbi	33.3	154.2	63.9	−71.4	−4.2	−27.8
mw	−11.3	95.8	82.3	−97.0	−39.6	−74.0
po	120.5	144.4	102.9	−69.2	3.7	−66.7
hc	–	172.1	108.5	–	−83.6	−18.3
lD	0	55.6	112.9	−26.7	−40.7	−93.5

Note: Designations of characters as in Table 11.2.

When morphometric characters differentiating the parental species were analysed, the 6-month-old (S) × H androgenetic nucleocytoplasmic hybrids were most similar to juveniles of the maternal species, the stellate sturgeon, in two characters: number of scutes in the lateral row and relative mouth width (Tables 11.2 and 11.3). At the same time, the true 6-month-old S × H hybrids manifested the highest similarity with juveniles of the maternal species in all these characters. At the age of 12 months, the true hybrids retained a high similarity to the maternal species in most characters, while the 12-month-old androgenetic hybrids retained similarity to the maternal species in only one character (number of lateral scutes). And at the age of 3 years, the androgenetic hybrids completely fitted in the range of the paternal species, great sturgeon, in all characters, and were separated with a hiatus from the variation range of the maternal species, while 3-year-old true hybrids manifested a high similarity to the maternal species in all nine characters (Tables 11.2 and 11.3).

When craniological indices differentiating the great and stellate sturgeons were examined, the 6-month-old (S) × H androgenetic hybrids manifested a similarity to the maternal species in only four characters (*Ri*/*Dc*, *Mr*/*Dc*, *W*/*Ri*, and *AM*/*Ri*), while true hybrids of the same age were similar to the maternal species in eight indices (Tables 11.4 and 11.5). At the age of 12 months, the true hybrids also manifested a high similarity to the maternal species, while androgenetic hybrids retained similarity to the maternal species in only one character (*Mr*/*Dc*). And at the age of 3 years, androgenetic hybrids were already indistinguishable from the paternal species, while true hybrids demonstrated a high similarity with the maternal species (Tables 11.4 and 11.5).

In the Persian × Russian sturgeon combination, the morphological studies demonstrated similar results: androgenetic nucleocytoplasmic hybrids acquired paternal (Russian sturgeon) characters at the age of 12 months itself (data have

Table 11.4 Ranges and means (in brackets) of craniological indices in experimental groups of androgenetic hybrids ((S) × H) between the stellate and great sturgeons and true hybrids (S × H) of the same species at different ages and of juveniles of parental species similar in size

Progeny	*TL*, mm	*Ri*/*Dc*	*Mr*/*Dc*	*Fds*/*Ri*	*Pds*/*Ri*	*W*/*Ri*	*B*/B_1	*B*/*Ri*	*Lb*/*B*	*Wm*/*W*	*Am*/*Ri*
6-month-old progenies											
(S) × H (*n* = 2)	166–201 (183.5)	70.6–73.0 (71.8)	62.2–64.4 (63.3)	85.6–90.6 (88.1)	58.2–58.6 (58.4)	57.7–57.8 (57.8)	77.5–87.0 (82.3)	39.6–45.9 (42.8)	97.2–100.6 (98.9)	81.6–84.1 (82.9)	77.8–87.3 (82.6)
S × H (*n* = 5)	148–205 (182.4)	68.5–76.4 (72.4)	51.2–64.2 (60.7)	82.7–96.7 (87.0)	54.7–62.8 (58.8)	50.9–54.2 (52.1)	119.2–151.8 (134.7)	44.7–51.5 (48.0)	79.7–89.0 (84.0)	42.9–68.5 (53.6)	84.8–92.8 (88.6)
S (*n* = 10)	179–215 (195.9)	70.1–77.2 (73.3±0.66)	49.9–68.4 (61.1±2.15)	67.5–87.5 (79.3±1.75)	40.9–59.5 (51.1±1.71)	48.5–58.1 (53.3±1.14)	98.3–152.5 (129.4±5.33)	48.3–53.1 (50.5±0.56)	68.6–87.7 (82.4±1.71)	35.7–63.6 (52.3±2.32)	82.1–92.5 (86.5±0.99)
H (*n* = 10)	92–190 (127.1)	65.3–71.4 (68.2±0.58)	49.1–63.9 (54.9±1.42)	84.9–102.0 (92.2)[a]	55.2–68.4 (60.6±1.47)	65.6–91.8 (77.6±3.09)	74.3–138.5 (98.8)[a]	32.9–46.0 (39.0±1.40)	84.3–93.7 (89.2)[a]	83.3–146.0 (107.2)[a]	69.4–79.1 (74.4)[a]
12-month-old progenies											
(S) × H (*n* = 2)	294–340 (317.0)	68.3–72.6 (70.5)	63.5–66.8 (65.2)	91.9–114.8 (103.4)	56.1–63.9 (60.0)	61.5–67.5 (64.5)	68.0–94.2 (81.1)	36.6–46.8 (41.7)	105.7–112.0 (108.9)	84.7–128.8 (106.8)	69.4–76.8 (73.1)
S × H (*n* = 4)	267–331 (296.7)	73.4–74.8 (74.4)	56.2–71.1 (64.6)	86.0–93.2 (89.9)	51.2–58.6 (55.0)	45.5–51.5 (49.4)	112.8–129.4 (118.4)	44.7–49.6 (47.1)	80.7–89.3 (83.6)	49.5–58.8 (55.2)	81.4–87.2 (83.9)
S (*n* = 26)	255–448 (381.0)	71.5–88.6 (77.5±0.58)	64.2–74.0 (69.2±0.62)	61.3–88.8 (67.5±1.39)	38.9–67.9 (48.1±1.07)	30.4–48.5 (42.8±0.74)	107.9–142.0 (125.6±2.54)	43.5–61.8 (53.9±0.86)	58.6–70.5 (64.3±0.86)	26.6–38.2 (31.9±0.92)	89.6–93.9 (91.4±0.32)
H (*n* = 15)	224–414 (310.7)	63.4–69.6 (66.4±0.56)	49.1–67.4 (57.5±1.24)	92.1–96.9 (94.3)[a]	55.2–65.3 (59.0±0.91)	54.0–75.9 (66.8±1.44)	86.1–133.6 (103.7)[a]	42.3–52.2 (46.3+0.74)	93.8–98.7 (96.0)[a]	72.5–93.0 (85.8)[a]	76.3–78.4 (77.2)[a]
3-year-old progenies											
(S) × H (*n* = 4)	675–732 (697.8)	62.7–64.4 (63.3)	62.2–64.7 (63.5)	86.6–97.3 (93.2)	–	68.3–80.4 (73.6)	93.8–98.0 (95.0)	33.2–44.1 (37.5)	93.9–101.5 (98.2)	72.5–81.6 (77.9)	69.1–76.5 (73.6)

(continued)

Table 11.4 (continued)

Progeny	*TL*, mm	*Ri*/*Dc*	*Mr*/*Dc*	*Fds*/*Ri*	*Pds*/*Ri*	*W*/*Ri*	B/B_1	*B*/*Ri*	*Lb*/*B*	*Wm*/*W*	*Am*/*Ri*
S × H (*n* = 5)	885–1100 (970.0)	68.0–69.8 (68.8)	57.1–69.0 (63.4)	81.8–89.9 (85.3)	52.2–63.0 (56.7)	44.7–47.8 (45.9)	111.2–118.3 (114.5)	52.3–55.2 (53.6)	73.2–79.5 (76.2)	40.0–44.4 (43.1)	93.5–96.4 (94.8)
S (*n* = 15)	603–845 (709.4)	67.6–75.9 (71.7±0.59)	59.2–69.1 (64.6±0.58)	63.8–98.6 (81.2±2.42)	–	40.4–49.5 (44.0±0.66)	107.0–131.1 (119.0±1.79)	49.5–57.0 (54.1±0.48)	57.8–68.0 (64.9±0.72)	25.8–38.7 (32.0±0.86)	91.2–95.4 (93.5±0.28)
H (*n* = 14)	592–860 (705.9)	61.3–66.6 (63.9±0.44)	52.1–75.7 (64.5±1.43)	86.6–103.8 (96.8±1.26)	–	66.9–76.9 (74.5±0.96)	88.2–110.8 (99.9±2.07)	43.4–50.3 (45.6±0.52)	100.0–109.8 (104.2±0.81)	68.5–96.9 (80.1±2.44)	71.8–81.8 (75.4±0.69)

Abbreviations of measurements used to obtain indices (in %): *Am*, distance from rostrum tip to upper edge of upper lip; B, distance from anterior end of rostrum to the base of inner pair of barbels; B_1, distance from base of barbels to mouth cartilage; *Dc*, length of dermocranium from rostrum end to posterior edge of dermosupraoccipitale; *Fds*, distance from anterior end of frontale to posterior edge of dermosupraoccipitale; *LB*, length of the inner barbel; *Mr*, length of the part of dermocranium occupied by the complex of bones of rostralia-medialia; *Pds*, distance from anterior end of parietale to posterior edge of dermosupraoccipitale; *Ri*, distance from rostrum tip to posterior angle of infraorbitale accessorium; *W*, dermocranium width at the level of posterior angle of infraorbitale accessorium; *Wm*, mouth width; other designations as in Table 11.2.

[a]The number of investigated specimens is less than *n*.

Table 11.5 Hybrid indices (*HI*) of craniological indices in true (S × H) and androgenetic nucleocytoplasmic ((S) × H) hybrids between the stellate and great sturgeons at different ages

	Progenies and ages					
	(S) × H			S × H		
Characters	6 Months	12 Months	3 Years	6 Months	12 Months	3 Years
Ri/Dc	–41.2	26.1	115.4	–64.7	–44.1	–25.6
Mr/Dc	–171.0	–31.6	nc	–87.1	–21.4	nc
Fds/Ri	36.4	167.9	53.9	19.4	67.2	–46.2
Pds/Ri	53.7	118.4	–	62.1	26.6	–
W/Ri	–63.0	80.8	94.1	–109.9	–45.0	–87.5
B/B_1	207.8	306.4	151.3	–134.6	–34.3	–52.9
B/Ri	33.9	221.1	290.6	–56.5	79.0	–88.2
Lb/B	385.3	181.4	69.5	–52.9	21.8	–42.5
Wm/W	11.5	177.9	90.9	–95.3	–13.5	–53.9
Am/Ri	–35.5	157.8	119.9	–134.7	5.6	–114.4

Note: Designations of characters as in Table 11.4; nc, the value of the index was not calculated as there was no difference between the progenies of parental species.

been presented in details in previous publication - Vasil'eva et al., 2001). At the same time we found some retardation in the formation of paternal characters in the androgenetic (P) × R hybrids in comparison with the true P × G hybrids, which we explained by manifestation of nucleocytoplasmic incompatibility (Vasil'eva et al., 2001).

11.4 Discussion

Thus, as a result of our experiments, viable androgenetic nucleocytoplasmic hybrids were obtained, both between stellate and great sturgeons and between Persian and Russian sturgeons. This was the first case of dispermic interspecific androgenesis in vertebrates. Previously, dispermic hybrid androgenotes were obtained only in the silkworm (Astaurov and Ostryakova-Varshaver, 1957). Evidently, the viability of our androgenetic hybrids was due to successful dispermic androgenesis. Dispermic fertilization allows production of heterozygous individuals, which is especially important in the case of sturgeons (see above).

The androgenetic origin of nucleocytoplasmic hybrids obtained in experiments 1 and 2 has been confirmed by DNA analysis: all specimens from the irradiated shocked groups, both (S) × H and (P) × R, had nuclear DNA of only the paternal species and mitochondrial DNA of maternal species (Figs. 11.1 and 11.2). Some of the androgenotes obtained by fertilization of the irradiated stellate sturgeon eggs with the mixture of sperm from stellate and great sturgeons [(S) × S+H] had nuclear DNA of both paternal species. The experimental conditions used, therefore, provided for dispermic fertilization.

Our morphological studies revealed a pronounced maternal effect in both (S) × H and (P) × G androgenetic hybrids at the age of 6 months, through some craniological and morphometric characters. But at the age of 12 months, the maternal effect of morphometric and craniological characters in the androgenetic (S) × H hybrids was significantly decreased and was retained only in a few characters, while in 12-month-old (P) × G androgenetic hybrids, as well as in 3-years-old (S) × H androgenotes, any manifestation of matrocliny disappeared completely. By this age, the androgenetic hybrids were identical to the paternal species in all characters. The prolonged manifestation of matrocliny in the androgenetic hybrids is most likely related to the extended life cycle of acipenserids (more extended for the pair stellate–great sturgeons) and relatively late development of bony covers in these fish (Vasil'eva et al., 2001, 2005).

Therefore, our results show that in principle, genotypes of endangered and even extinct species can be restored from spermatozoa alone. Fresh or cryopreserved sperm of an endangered species and eggs of another species which is more abundant but closely related to the paternal species, can be used to produce dispermic androgenetic hybrids. This is especially important for such endangered group of fishes as sturgeons (Birstein et al., 1997b).

In connection with the possibility of using dispermic androgenesis for the restoration of endangered sturgeon species, the question of sex composition in androgenetic progeny is of great importance. Sex composition will depend on male homo- or heterogamety in sturgeons, if sex is determined by sex chromosomes. In male heterogamety (XY male and XX female) which is inherent in many fish species (Vasil'ev, 1985; Devlin and Nagahama, 2002), dispermic androgenesis allows production of bisexual progeny consisting of regular XY males and XX females, as well as exotic YY males. Male homogamety (ZZ males and WZ females) results in unisexual androgenetic all-male (ZZ) progeny. In this case, the regular reproduction of the fishes restored by androgenesis is impossible and additional manipulations (e.g., sex inversion) are to be applied.

However, there are many small chromosomes in sturgeon karyotypes (Vasil'ev, 1985) and, therefore, direct identification of sex chromosomes in these fishes is not feasible. In such situations, the sex composition of gynogenetic or androgenetic progenies with either maternal or paternal heredity alone may be examined for studying the mechanism of sex determination. Two studies based on induced gynogenesis have been carried out on acipenserid fishes. In these studies, male heterogamety was proposed for the American paddlefish (*Polyodon spathula*) (Mims et al., 1997) and female heterogamety for the white sturgeon (*Acipenser transmontanus*) (Van Eenennaam et al., 1999).

For studying the mechanism of sex determination in sturgeons, we carry investigations of both androgenetic and gynogenetic progenies. Besides androgenesis, we have already obtained meiotic gynogenesis in several sturgeon species (Recoubratsky et al., 2003). Comparative analysis of gynogenetic and androgenetic progenies in the same species can provide us with valuable information concerning the sex determination mechanism in sturgeons.

Acknowledgements The authors are extremely grateful to the Organizing Committee of the 1st International workshop on the biology, conservation and sustainable development of sturgeons in southern Europe (Bioresturgeons 2005) and its sponsoring entities for making it possibile for us to attend this meeting. These studies were supported by the Russian Foundation for Basic Research (project nos. 06–04–49637-a and 07–04–00219-a), as well as by RAS Programs 'Biodiversity and genofond dynamics', 'Biodiversity', and 'Biological resources of Russia'. We are very thankful to our colleagues who participated at different stages of our studies, as well as to the Krasnodar Research Institute of Fishery staff, who helped us in our experiments. We would like to thank Prof. S.G. Vassetsky, N.K. Koltzov Institute of Developmental Biology, for assistance in our work and helpful comments on the manuscript.

References

Astaurov BL, Ostryakova-Varshaver VP. 1957. Production of full heterospermic androgenesis and interspecific hybrids in silkworm: experimental analysis of relative roles of nucleus and cytoplasm in development and heredity. Izv Akad Nauk SSSR. *Ser Biol* 2:154–175.

Birstein VJ, Hanner R, DeSalle R. 1997a. Phylogeny of the Acipenseriformes: cytogenetic and molecular approaches. *Environ Biol Fish* 48:127–155.

Birstein VJ, Bemis WE, Waldman JR. 1997b. The threatened status of acipenseriform fishes: a summary. *Environ Biol Fish* 48:427–444.

Chao N-H, Liao ICh. 2001. Cryopreservation of finfish and shellfish gametes and embryos. *Aquaculture* 197:161–189.

Corley-Smith GE, Brandhorst BP. 1999. Preservation of endangered species and populations: a role for genome banking, somatic cell cloning, and androgenesis? *Mol Reprod Devel* 53:363–367.

DeSalle R, Birstein VJ. 1997. Method and compositions for identification of species origin of caviar. Appl. No. 97107809.2. European Patent Application Bulletin, p. 47.

Dettlaff TA, Ginsburg AS, Schmalhausen OI. 1993. *Sturgeon Fishes. Developmental Biology and Aquaculture*. Springer-Verlag, Berlin, 300 pp.

Devlin RH, Nagahama Y. 2002. Sex determination and sex differentiation in fish: an overview of genetic, physiological, and environmental influences. *Aquaculture* 208:191–364.

Ginsburg AS. 1968. Oplodotvorenie u ryb i problema polispermii (*Fertilization in Fish and Problem of Polyspermy*). Nauka, Moscow, 358 pp (in Russian).

Grunina AS, Neyfakh AA. 1991. Induction of diploid androgenesis in Siberian sturgeon *Acipenser baerii* Brandt. *Ontogenez* 22:53–56 (in Russian with English summary).

Grunina AS, Neyfakh AA. 1997. Induced diploid androgenesis. *Physiol Gen Biol Rev* 12:73–103.

Grunina AS, Recoubratsky AV. 2005. Induced androgenesis in fish: obtaining viable nucleocytoplasmic hybrids. *Russ J Dev Biol* 36:208–217.

Grunina AS, Recoubratsky AV, Neyfakh AA. 1995. Induced diploid androgenesis in sturgeons. *Sturgeon Quart* 3:6–7.

Holčik J, Bănărescu P, Evans D. 1989. General introduction to fishes. In: Holčik J (ed.), *The Freshwater Fishes of Europe*. 1. Pt. 2. AULA-Verlag, Wiesbaden, pp. 18–147.

Lubzens E, Pekarsky I, Magnus Y, Ar A. 1996. Term and prospect for long term storage of teleost ova and embryos. In: *Proceedings of the Conference of Refrigeration and Aquaculture*, Bordeaux, pp. 491–501.

Mims SD, Shelton WL, Linhart O, Wang C. 1997. Induced meiotic gynogenesis of paddlefish *Polyodon spathula. J World Aquacult Soc* 28:334–343.

Neyfakh AA. 1999. Nucleo-cytoplasmic incompatibility of androgenetic hybrids in sturgeons. *J Appl Ichthyol* 15:318–319.

Neyfakh AA, Radzievskaya VV. 1967. On morphogenetic nuclear function in standard and androgenetic goldfish/loach hybrids (*Carassius auratus* × *Misgurnus fossilis*). *Genetika* 12:80–88 (in Russian with English summary).

Penman DJ, Myers JM, McAndrew BJ. 1996. Restoration of diploid genotypes by androgenesis. In: *Proceedings of the Conference Refrigeration and Aquaculture*, Bordeaux, pp. 469–474.

Recoubratsky AV, Grunina AS, Minin AA, Duma LN, Neyfakh AA. 1996. Dispermic androgenesis in *Acipenser stellatus*. *Sturgeon Quart* 4:12–14.

Recoubratsky AV, Grunina AS, Myuge NS, Neyfakh AA. 1998. Production of androgenetic nucleocytoplasmic hybrids in sturgeon fish. *Russ J Dev Biol* 29:224–229.

Recoubratsky AV, Grunina AS, Barmintsev VA, Golovanova TS, Chudinov OS, Abramova AB, Panchenco NS, Kupchenko SA. 2003. Meiotic gynogenesis in Stellate and Russian sturgeons and Sterlet. *Russ J Dev Biol* 34:92–101 (in Russian).

Suquet M, Dreanno C, Fauvel C, Cosson J, Billard R. 2000. Cryopreservation of sperm in marine fish. *Aquacult Res* 31:231–243.

Van Eenennaam AL, Van Eenennaam JP, Medrano JF, Doroshov SI. 1999. Evidence of female hetero-gametic genetic sex determination in white sturgeon. *J Hered* 90:231–233.

Vasil'ev VP. 1985. *Evolutionary Karyology of Fishes*. Nauka, Moscow, 300 pp (in Russian).

Vasil'eva ED. 1999. Some morphological characteristics of Acipenserid fishes: considerations of their variability and utility in taxonomy. *J Appl Ichthyol* 15:32–34.

Vasil'eva ED. 2004. Morphological data corroborating the assumption of independent origins within octoploid sturgeon species. *J Ichthyol* 44 (Suppl. 1):63–72.

Vasil'eva ED, Grunina AS, Recoubratsky AV. 2001. The pattern of manifestation of some morphological characters in androgenetic nucleo-cytoplasmatic hybrids of the Persian *Acipenser persicus* and Russian *A. gueldenstaedtii* sturgeons in the postlarval period. *J Ichthyol* 41:454–460.

Vasil'eva ED, Grunina AS, Recoubratsky AV, Pavlinov IYa. 2005. Manifestation pattern of some morphological characters in androgenetic nucleocytoplasmatic hybrids between stellate sturgeon *Acipenser stellatus* and great sturgeon *Huso huso* (Acipenseridae) during postlarval ontogeny. *J Ichthyol* 45:465–478.

Veprintsev BN, Rott NN. 1979. Conserving genetic resources of animal species. *Nature* 280:633–634.

Verigin BV, Makeeva AP. 1972. Hybridization of Carp with *Aristichthys nobilis*. *Genetika* 8(7): 55–64 (in Russian).

Chapter 12
Influence of Temperature on the Sterlet (*Acipenser ruthenus L.*) Ovarian Follicles State

B.F. Goncharov, M.N. Skoblina, O.B. Trubnikova, and M.S. Chebanov

Abstract We studied the effects of changes in the temperature of water in which the sterlet (*Acipenser ruthenus* L.) females were kept on morphometrical and physiological characteristics of the ovarian follicles: size of oocytes, their polarization index, and their capacity to mature and ovulate in vitro in the presence of progesterone or purified preparation of gonadotropic hormone of the sturgeon pituitary, as well as 'spontaneously' in the culture medium. When the temperature was lowered from 13°C, the optimal for spawning, to 6°C, the morphometrical parameters remained unchanged, but the indices of the physiological state underwent changes. After the temperature was gradually elevated to the initial value, some indices returned fully or partially to the initial state, while others continued to change in the same direction.

Key words Sturgeons, oocytes, maturation, ovulation, in vitro, temperature

12.1 Introduction

Temperature is the main factor that regulates the seasonal character of reproduction in sturgeon females. Rich experience of the national sturgeon aquaculture based on utilization of spawners caught in nature made it possible to establish some patterns of the influence of this factor on the state of gonads and artificial reproduction. For example, it was shown that keeping females even for a short time (several days) at the spawning temperature may lead to the irreversible loss of the capacity to respond to hormonal stimulation in the production of high-quality eggs. It was proposed long ago (Kazanskii, 1963) to change the temperature regime for spawners to preserve their reproduction capacity. The temperature could be lowered by

---

B.F. Goncharov, M.N. Skoblina, and O.B. Trubnikova
Kol'tsov Institute of Developmental Biology, Russian Academy of Sciences, Moscow, Russia

M.S. Chebanov
South Branch Federal Centre of Selections and Genetics for Aquaculture, Krasnodar, Russia

R. Carmona et al. (eds.) *Biology, Conservation and Sustainable Development of Sturgeons*,

3–4°C below the threshold of spawning temperature and maintained until a given moment and then gradually elevated to the optimal one. Under these conditions, the females could be used in the second cycle of artificial reproduction. This method was successfully used at sturgeon hatcheries.

With the advent of keeping sturgeons for a long term or full-cycle rearing, the problem of controlling seasonal reproduction became even more acute. It was shown that, given a certain temperature regime, mature gametes could be obtained in practically any season of the year (Chebanov et al., 2002). Nevertheless, the results of artificial reproduction are often far from ideal and sometimes even disappointing. The matter is that fish populations may be heterogeneous, especially fish reared in captivity, in which asynchronous gametogenesis may be rather pronounced. In this respect, the problem of the criteria for the selection of females suitable for reproduction remains urgent.

A method was proposed (Goncharov, 1993) for selection, based on the determination of the time of in vitro maturation of 50% of oocytes in the presence of progesterone (T50). For the stellate sturgeon (*Acipenser stellatus* Pal.) females caught in nature, this method allowed identification and removal from artificial reproduction of a group of fishes that were not capable of producing mature high-quality eggs in this state and under standard conditions of hormonal stimulation. Later (Goncharov et al., 1999), a similar study was made on Siberian sturgeon (*Acipenser baerii* Brandt) females reared in captivity. The seasonal changes in some morphometrical and physiological characteristics of follicles were monitored and their prognostic value was determined. It was seen that all the characteristics studied underwent changes during the observation period from December until April, and the oocytes ovulated in vitro under the influence of hormonal preparations had the best prognostic value. The application of these criteria allowed us to increase the efficiency of artificial reproduction but none of them could ensure unmistakable sorting of females by the quality of gametes that they could produce after hormonal stimulation. Hence, it was necessary to continue the studies of factors affecting the physiological state of follicles. The seasonal changes in the reaction of follicles to hormones in vitro suggest their ontogenetic character, but the information we obtained led us to hypothesize that the competence of maturation and ovulation of oocytes is maintained by homeostatic mechanisms. This implies not only unidirectional changes of these characteristics, but also the possibility of their transient return to the preceding state. Here, we study the influence of a gradual decrease in the temperature at which sterlet females are reared in captivity and an increase in the initial value on some morphometrical and physiological characteristics of follicles, finding evidence in favor of the above hypothesis.

12.2 Materials and Methods

Studies were carried out at the Krasnodar (Adygean) Sturgeon Hatchery on sterlet females reared in ponds at ambient seasonal temperatures at the South Branch Federal Centre of Selection and Genetics for Aquaculture.

The females were transferred from a pond, where the temperature reached 13°C, into large concrete basins with regulated temperature in a closed room. The temperature was gradually lowered to 6°C within 5 days, maintained at this level for another 5 days, and then gradually elevated to the initial level within 6 days.

Studies were carried out on two groups (experiment and control) of 10 females in each. In the experimental group, the follicles were sampled at the beginning of the experiment (0 days), within 10 days, and at the end of the experiment (16 days). In the control group, which remained under the same temperature regime, the state of the follicles was estimated only at the end of the experiment (control to stress related to the sampling of follicles).

Ovarian follicles were taken using a special metallic probe and placed in a medium with hormonal preparations within 1 h. The follicles were repeatedly washed by Ringer solution for poikilotherms modified for sturgeons (RMS) (Goncharov, 1978) containing sodium bicarbonate at an increased (1.5 g/L) concentration. The follicles were incubated in the same medium in 55 mm plastic Petri dishes, 36 ± 3 follicles in 7.5 mL per dish, at 16 ± 1°C.

Oocyte maturation and ovulation were estimated with 38–42τ_0 (τ_0 is a dimensionless unit proposed by T.A. Dettlaff for measuring the development duration; see Dettlaff et al., 1993). Ovulation was estimated by counting the follicle envelopes separated from the oocytes. Maturation was estimated according to the germinal vesicle breakdown (GVBD). For this purpose, the oocytes were fixed by boiling and cut through with a safety blade under a dissection microscope.

Progesterone was introduced directly in Petri dishes as concentrated alcohol solution to a final concentration of 1 μg/mL (alcohol concentration 0.1%). Gonadotropic hormone of stellate sturgeon (aciGTH) was purified as described elsewhere (Burzawa-Gerard et al., 1975), with slight modifications. The lyophilized aciGTH powder was dissolved to a concentration of 2 μg/mL in the culture medium directly before the experiment.

The following characteristics of the follicles were determined in the course of the experiment: oocyte diameter, oocyte polarization index (OPI), percentage of oocytes capable of maturation and ovulation in vitro under the influence of progesterone, aciGTH, or culture medium, and duration (in the number of τ_0) of maturation of 50% oocytes in vitro in the presence of progesterone (T50).

Thirty follicles were fixed immediately after their isolation from the female and envelopes were removed with sharpened forceps. The diameter (mean of two measurements: between the poles and perpendicular one) and shortest distance between the GV surface and oocyte membrane were measured using ocular micrometer to a precision of 0.03 mm. The ratio of the distance from GV to the oocyte surface to the mean diameter (expressed in %) was designated as OPI.

In order to determine T50, maturation was monitored within 12τ_0 after the beginning of incubation by fixing 12 to 15 oocytes with an interval of 1τ_0. From the moment of the appearance of the first mature oocytes, double numbers of oocytes were fixed with the same interval until all or almost all oocytes matured. T50 was determined using the graphic method after transformation of percentages into probits (Finney, 1971).

To estimate the differences between the mean quantitative indices, the criteria for randomization for paired samples (for studying the changes in the follicles of females of the experimental group) and independent samples (for comparison of follicles of females of the experimental and control groups) were used (MegaStat software for Excel).

12.3 Results

The mean values of the morphometrical and physiological characteristics of the follicles and data for individual females are given in the Table 12.1 and Fig. 12.1, respectively.

At the beginning of the experiment, practically 100% of oocytes of the experimental females matured in the presence of progesterone or aciGTH. This was only partly due to the effect of the hormones, since in different females 17 to 97% of oocytes matured in a hormone-free medium. The percentage of oocytes ovulated in the presence of these hormonal preparations was much lower and progesterone was a more effective inducer. Ovulation was practically absent in the hormone-free medium.

Within 10 days of the beginning of the experiment, some parameters of the follicles underwent changes. The mean T50 value increased by approximately $6\tau_0$. The mean percentage of ovulated oocytes decreased approximately twice in the experiments with progesterone and somewhat more with aciGTH. The mean percentage of oocyte maturation in the hormone-free medium decreased as expected. On the contrary, the lowered temperature did not notably affect the percentage of oocyte maturation in the presence of progesterone or aciGTH, nor did it alter the mean morphometrical indices, such as oocyte diameter and OPI.

After the temperature rose to the initial level, some indices were fully or partly restored. The mean values of T50 and percentage of oocyte ovulation in the presence of aciGTH practically returned to the initial level, while that in the presence of progesterone was only partially restored. On the contrary, the percentage of oocyte maturation in the hormone-free medium continued to decrease. The morphometrical parameters and percentage of oocyte maturation in the presence of hormones remained at the same level.

Comparison of the state of follicles of the experimental and control females at the end of the experiment did not reveal statistically significant differences for any of the morphometrical or physiological characteristics studied.

12.4 Discussion

The efficiency of artificial reproduction of sturgeons is determined by a number of factors. The obtaining of mature eggs capable of normal development after insemination depends on two main factors. First, the follicles of the older generation should

Table 12.1 Changes in morphometrical and physiological characteristics of sterlet follicles after lowering and subsequent elevation of the temperature of keeping of the females

Time after the beginning of experiment (days)	Physiological characteristics of ovarian follicles incubated *in vitro* in the presence of							Morphometric characteristics of oocytes	
	Progesterone (1 µg/mL)			aciGTH (2 µg/mL)		Without hormones			
	GVBD,%	Ovulated oocytes, %	T50 (τ_0)	GVBD (%)	Ovulated oocytes (%)	GVBD (%)	Ovulated oocytes (%)	Diameter (mm)	OPI (%)
0	100 (a)	42.11 (a)	13.66 (a)	98.71 (a)	28.09 (a)	54.83 (a)	0.76 (a)	1.99 (a)	11.71 (a)
10	99 (a)	20.27 (b)	19.60 (b)	94.46 (b)	10.17 (b)	30.32 (b)	0.16 (a)	1.97 (a)	11.65 (a)
16	100 (a)	29.72 (c)	13.96 (a)	99.51 (a)	24.79 (a)	15.8 (b)	0 (a)	1.99 (a)	11.07 (a)
16[a]	100 (a)	29.35 (c)	15.84 (a)	99.57 (a)	27.19 (a)	17.29 (b)	0 (a)	2.02 (a)	8.98 (a)

Numerals designate the mean values. Identical letters within the column designate the absence of differences between the mean values at $p < 0.05$.
[a]Control group.

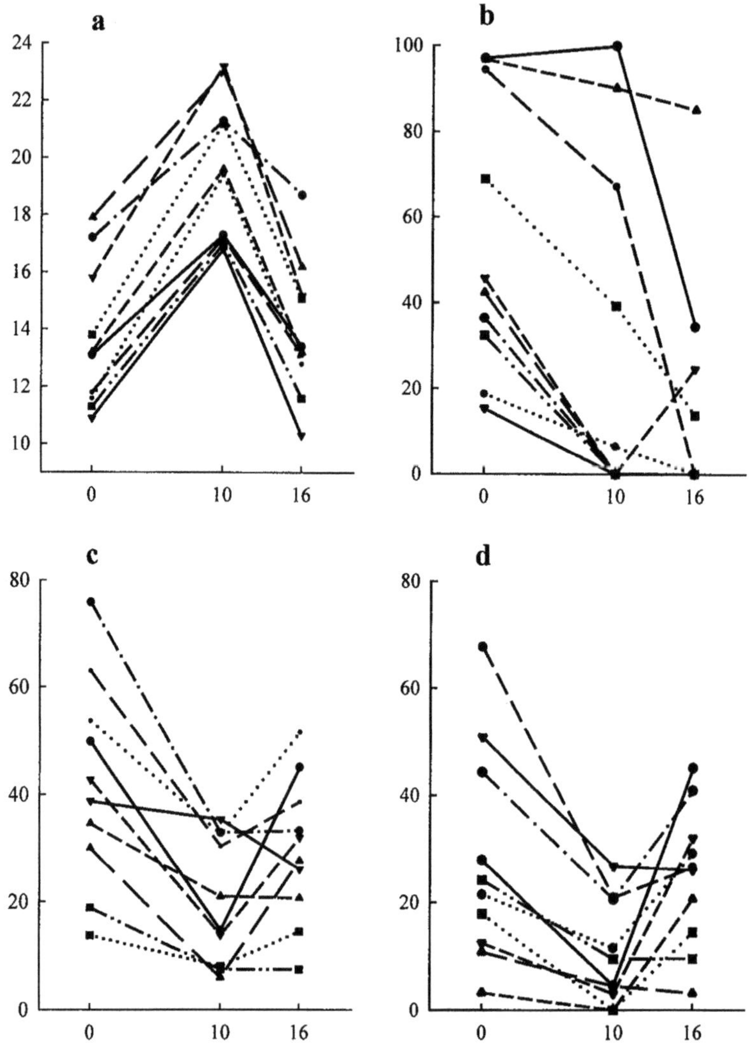

Fig. 12.1 Changes in physiological characteristics of the sterlet ovarian follicles (a, T50; b, percentage of oocyte maturation in hormone-free medium; c, percentage of oocyte ovulation in the presence of progesterone; d, percentage of oocyte ovulation in the presence of aciGTH) after lowering and subsequent raising of the temperature of keeping females. Abscissa, time after the beginning of experiment, days; ordinate, number of τ_0 (a), GVBD, % (b), ovulated oocytes, % (c and d). Identical lines and marks represent the data for the same female

not be damaged and should be capable of responding to the hormonal preparations through oocyte maturation and ovulation. Second, an adequate hormonal treatment should be given to the females. As shown by the experience of artificial breeding of sturgeon fishes, the first factor is of paramount importance. In the beginning of the era of sturgeon aquaculture in the USSR, there was no problem concerning the production of high-quality gametes from the broodfish, as the fish used for reproduction were caught in nature during anadromous migration, when the state of their gonads was close to spawning. Under these conditions, a single injection of pituitary suspension led to the desirable result in the vast majority of cases. However, as the ecological situation in the seas and rivers worsened, the efficiency of artificial reproduction was reduced and the problem of selection of broodfish became very acute. The numbers of sturgeon individuals were then catastrophically decreased, forcing many hatcheries to resort to repeated utilization and, recently, to utilization of broodfish reared in captivity. When the fishes are held or reared in captivity, the conditions are different from the natural ones: in the feed and often in the temperature regime. This may lead to disturbances of oogenesis and/or its desynchronization. Hence, the problem of selection of females for reproduction becomes especially acute.

It is known that as the oocytes grow and the reproduction season approaches, certain changes take place in the oocytes and, specifically, GV is displaced to the animal pole. This parameter can be rapidly and relatively easily estimated. It was proposed long ago (Trusov, 1964; Kazanskii et al., 1978) to use it for selecting the females suitable for reproduction, but its efficiency turned out to be far from always effective. Later another approach was proposed on the basis of estimating the physiological state of follicles according to their reaction to hormonal preparations in vitro (Goncharov, 1993). Among the indices studied, the most suitable were the duration of in vitro oocyte maturation in the presence of progesterone (T50) and, next, the percentage of in vitro oocyte ovulation in the presence of gonadotropic preparations (Goncharov et al., 1999). A correlation was found between T50 and the quality of eggs. It proved to be highly significant, but weak. The use of this criterion can increase the efficiency of artificial reproduction on a large scale, but cannot guarantee results for individual females. When studying the in vitro reaction of the Siberian sturgeon follicles to hormones, it was found that the pattern of seasonal changes in physiological parameters correlated quite well with seasonal temperature changes. Hence, it was confirmed that as the spawning season approached, the sensitivity of the gonads to hormones increased, the reaction to these hormones became more complete, and it occurred at a higher rate. All these data suggest unidirectionality of the processes in question. However, data accumulated suggesting their changes in the reverse direction as well. It was also reported (Dettlaff and Davydova, 1974, 1979) that the sensitivity of follicles to gonadotropic hormones in vitro could be lost after their sharp short-term cooling in vivo and in vitro and could be restored after injection of triiodothyronine to females, while such treatment did not affect the reaction of oocytes to progesterone. However, two conditions were not fulfilled in this work, which are essential for correct comparison of the results of different experiments on the induction of in vitro oocyte maturation by gonadotropic preparations. First, strict identity of the culture medium is necessary, but it could not be ensured because the medium pH was brought to a certain value by sodium bicarbonate. The data on the important role of the sodium bicarbonate concentration for the sensitivity of follicles to gonadotropins were obtained later (Goncharov, 1978). Second, the use of the hormonal preparation of standard activity is necessary.

The present work was undertaken to test whether the follicles may return to a less advanced state and, if so, whether this return is an initial sign of pathological changes or, under certain conditions, progression to the prespawning state may be restored. We chose temperature as a natural factor, essential for sturgeons and, possibly, capable of altering the physiological state of follicles, and we switched the direction of its changes to the opposite with reference to the seasonal changes. It is believed that the range of spawning temperatures for the sterlet lies within the limits of 9 to 20°C (cf. Dettlaff et al., 1993). Hence, our experimental females were at the temperature conditions favourable for spawning and we might have expected that their gametes were ready for maturation. Actually, a study of the initial state showed that this was the case. In all the females, practically 100% of oocytes were capable of responding to both progesterone and aciGTH by maturation, the level of in vitro ovulation in the presence of hormones was sufficiently high, while T50 and OPI were rather low. The work on artificial reproduction of the sterlet at the hatchery was under way at that time, which also suggested that the experimental females were ready for reproduction.

After the decrease in temperature, OPI remained unchanged, and some physiological parameters, such as percentage of oocyte maturation in the presence of progesterone and aciGTH, also remained unchanged and others were displaced to a less advanced state. The mean T50 increased from 13.7 to $19.6\tau_0$ and in 6 females out of 10 it exceeded $18\tau_0$. It was shown for the stellate sturgeon (Goncharov, 1993), Siberian sturgeon (Goncharov et al., 1999), and sterlet (B.F. Goncharov et al., unpublished data) that the females with $T50 \geq 18\tau_0$ are not capable of producing high-quality mature gametes. The transformation of the data on T50 for the Siberian sturgeon became possible after the dependence of τ_0 on temperature was studied (Gisbert and Williot, 2002). The percentages of oocyte ovulation in the presence of progesterone and aciGTH and in the hormone-free medium were also predictably decreased.

To determine whether these changes in the state of follicles are a normal reaction to the lowered temperature or whether they reflect the initial stages of pathological change, we gradually elevated the temperature so that, by the end of the experiment, the females were again under the initial temperature conditions. Analysis of the follicles sampled at this time showed that the characteristics that were invariable at the first stage of experiment (temperature lowering) again remained unchanged. These characteristics include the percentage of oocyte maturation in the presence of progesterone and aciGTH, as well as oocyte diameter and OPI. The mean T50 returned almost to the initial value. It was noted that, despite different initial levels of this parameter in different females, the pattern of changes proved to be similar: increase and then decrease to the initial level. The mean percentage of oocyte ovulation in the presence of progesterone increased, but did not reach the initial level. In this case, however, the pattern of changes at the level of individual females was not entirely similar, especially during the second stage of experiment. If at the lowered temperature, this parameter decreased in all females, the reaction to temperature elevation was more variable: from return to the initial level to the absence of changes or further slight decrease. Similar heterogeneity was also observed in the case of oocyte ovulation in the presence of aciGTH. At a certain state, the oocytes placed in RMS are capable of maturing in the absence of hormonal preparations. This phenomenon is called 'spontaneous maturation' and has been described for some fishes and amphibians. Spontaneous maturation is increasingly expressed as the spawning season approaches. Goncharov et al. (1999) demonstrated that this characteristic also bears some prognostic significance for the selection of females. In our study, this characteristic diminished in practically all the females both during temperature lowering and the subsequent raising to the initial level.

The aim of this work was to study the influence of temperature changes on the state of follicles, but we know that the parameters in question may be affected also by stress related to manipulation with the broodfish (Goncharov and Polupan, 1997). To identify the changes that may be related to stress influences, we subjected a group of 10 females to the same temperature regime, but the state of their follicles was assessed only at the end of the experiment and, hence, they were not subjected to stress influences connected with double sampling of follicles from the females. Although there were no statistically significant differences for any of the characteristics studied, T50 in the control group was somewhat higher and OPI somewhat lower. It cannot be ruled out that in a larger sample, the differences in T50 could be significant. This suggestion is based on the fact that the procedure of follicle extraction may lead to a decreased T50. In all likelihood, OPI is a less labile parameter. In many of our experiments on different sturgeon species we found no correlation between these parameters and T50 had the best prognostic significance, while OPI had the lowest.

The results suggest that the characteristics of the follicles studied combine the features of the ontogenetic process, as there is, undoubtedly, a vector of their changes over the course of oogenesis. However, the maintenance of their state at any given moment appears to be determined by a homeostatic mechanism permitting differently directed deviations under the influence of different factors. Both vitellogenesis and terminal stages of oogenesis, oocyte maturation and ovulation, are regulated by gonadotropic hormones. It can be proposed that the observed return of the follicles to an earlier ontogenetic state is related to a decreased level of gonadotropins and/or sensitivity of the follicle to them, but, as was shown in the present study, this deviation may be fully or partly reversible.

Unfortunately, the circumstances did not allow us to test the capacities of all these females to reproduction but, according to the evidence of fish culturists, who worked with these females after our departure, these temperature influences did not lead to significant losses, and high-quality gametes were obtained from at least some of these females.

Acknowledgements This study was partially supported by the Russian Foundation for Basic Research, project no. 05-04-48433.

References

Burzawa-Gerard, E., Goncharov, B.F., Fontaine, Y.A. 1975. L'hormone gonadotrope hypophysaire d'un poisson chondrostéen, l'Esturgeon (*Acipenser stellatus* Pall.). 1. Purification. *Gen. Comp. Endocrinol.* 27, 289–295.

Chebanov, M.S., Karnaukhov, G.I., Galich, E.V., Chmir Yu, N. 2002. Hatchery stock enhancement and conservation of sturgeon, with an emphasis on the Azov Sea populations. *J. Appl. Ichthyol.* 18, 463–469.

Dettlaff, T.A., Davydova, S.I. 1974. Effect of triiodothyronine on oocyte maturation in the stellate sturgeon after the action of low temperature and reservation of females. *Ontogenez* 5, 454–461 (in Russian).

Dettlaff, T.A., Davydova, S.I. 1979. Differential sensitivity of cells of follicular epithelium and oocyte in the stellate sturgeon to unfavorable conditions and correcting influence of triiodothyronine. *Gen. Comp. Endocrinol.* 39, 236–243.

Dettlaff, T.A., Ginsburg, A.S., Schmalhausen, O.I. 1993. *Sturgeon Fishes. Developmental Biology and Aquaculture*. Springer Verlag, Berlin, 300 pp.

Finney, D.J. 1971. *Probit Analysis*. Cambridge University Press, Cambridge, 333 pp.

Gisbert, E., Williot, P. 2002. Duration of synchronous egg cleavage cycles at different temperatures in Siberian sturgeon (*Acipenser baerii*). *J. Appl. Ichthyol.* 18, 271–274.

Goncharov, B.F. 1978. The influence of the culture medium composition on the capacity of the sturgeon follicles to mature in the presence of gonadotropic hormones. In: *Problems of Early Ontogenesis in Fish*. T.V.Dekhnik, V.N.Zhukinsky and L.S.Oven (eds.) Naukova Dumka, Kiev, pp. 77–78 (in Russian).

Goncharov, B.F. 1993. Duration of oocyte maturation time in vitro as a criterion for selecting sturgeon spawners for breeding. In: Dettlaff, T.A. et al. (eds.), *Sturgeon Fishes. Developmental Biology and Aquaculture*, Appendix B. Springer Verlag, Berlin, pp. 218–219.

Goncharov, B.F., Polupan, I.S. 1997. Stress affects the physiological state of sturgeon ovarian follicles and female reproductive potential. In: *Abstracts of the Third International Symposium on Sturgeon*, Piacenza, Italy, 8–11 July 1997.

Goncharov, B.F., Williot, P., Le Menn, F. 1999. Morphological and physiological characteristics of the ovarian follicles of farmed Siberian sturgeon and their importance for predicting artificial reproduction success. *Russ. J. Devel. Biol.* 30, 46–54.

Kazanskii, B.N. 1963. Obtaining of different season progeny of fishes in order to provide for repeated fish culture cycle (sturgeon taken as an example). In: Pavlovskii, S.N. (ed.), *Sturgeon Culture in Water Bodies of the USSR*. Izdatel'stvo Acad. Nauk SSSR, Moscow, pp. 56–64 (in Russian).

Kazanskii, B.N., Feklov, Yu. A., Podushka, S.B., Molodtsov, A.N. 1978. Express method for determination of degree of gonadal maturity in sturgeon spawners. *Rybn. Khoz.* 2, 24–27 (in Russian).

Trusov, V.Z. 1964. Method of estimation of degree of gonadal maturity in sturgeon females. *Rybn. Khoz.* 1, 26–28 (in Russian).

Chapter 13
On Nutrition and Feeding Studies as the Basis for the Culture of Different Sturgeon Species

M. García-Gallego, A. Domezain, M. de la Higuera, M.C. Hidalgo, M. Furné, A.E. Morales, and A. Sanz

Abstract Any attempt to save the endangered sturgeon species will depend on the availability of a sufficient stock of the fish in question, as aquaculture is a very suitable tool and has proved feasible for several sturgeon species and, additionally, it could contribute to the development of a new productive activity.

The breeding of fish in captivity depends on a series of determining factors, among which the design of optimized feeding practices for the different growing phases is fundamental.

In the present review we shall focus on the two species on which the most systematic studies have been performed: *Acipenser transmontanus* (mainly in the USA) and *Acipenser baerii* (primarily in France), and our experience with *Acipenser naccarii*.

Keywords Sturgeon culture, Sturgeon feeding, *Acipenser transmontanus*, *Acipenser baerii*, *Acipenser naccarii*

13.1 Introduction

There is broad agreement on the serious risk facing natural populations of many sturgeon species, and the resolve to improve this situation by defining and neutralizing as far as possible the threats underlying this regrettable situation, in addition to the formulation of intervention policies, where needed, involving the reintroduction of endangered species. It is here that the culture of sturgeons can make a decisive contribution.

M. García-Gallego, M. de la Higuera, M.C. Hidalgo, M. Furné, A.E. Morales, and A. Sanz
Research group 'Fish Nutrition and Feeding', Dpt. Biología Animal, Universidad de Granada, Granada, Spain

A. Domezain
Dpt. R + D Piscifactoría Sierra Nevada S.L., Riofrio, Granada, Spain

The success of any attempt to save any fish species depends largely on the availability of sufficient stock of the fish in question at different growth stages and sizes, well typified and with optimum health. For the production of these stocks, breeding in captivity (aquaculture) has been demonstrated to be essential and, thus, fish culture aimed at repopulation is now developing, besides that specifically, or mainly, devoted to commercial purposes.

The breeding of healthy fish in captivity depends on a series of determining factors, among which the design of optimized feeding practices is fundamental. This task includes the definition of adequate nutritional composition for each age and the selection of raw materials (**what** has to be delivered to fish as food) as well as the design of what is known generically as a 'feeding strategy' (**how** to do it). This in turn requires information from fish-farmers' previous experience with related species in addition to specifically designed and conducted research, both in specialized centres as well as in the corresponding departments of fish farms.

Several research teams have made an effort in this regard, on different species that have varying conservation status in their natural environments, but which can all provide useful results that can be applied at least in part to other closely related species. Although feeding experiments with artificial diets have been made in numerous sturgeon species and their hybrids, in the present review (focusing on the last 20 years, using the databases most accessible in our geographic setting), before describing our experience with *Acipenser naccarii*, we would like to refer exclusively to the two species with which the most systematic studies have been performed: *Acipenser transmontanus* (mainly in the USA) and *Acipenser baerii* (primarily in France).

13.2 Studies on *A. transmontanus*

Once the reproduction technology was established by the team at the University of California at Davis headed by Dr. S.I. Doroshov for white sturgeon (*A. transmontanus*) in captivity (see Doroshov et al., 1994), it was proposed that this culturing approach be extended to the regeneration of dwindling natural stocks of the species as well as for the production of flesh and caviar, as it is a very attractive species in terms of a high growth rate and edible quality.

Initially, the sturgeon growers used salmonid feed for their fish, but this practice proved inadequate because, over a long term, it resulted in poor growth and abnormal symptoms suspected to be related to nutritional deficiencies. Therefore, the development of specific suitable commercial feed for sturgeons became a target of top priority.

Faced with the absence of specific data on the nutritional requirements of these species, phylogenetically so distant from other cultured fish (mainly teleosts), the research team of S.S.O. Hung at the University of California at Davis undertook an approach that had been developed over the years in a long series of noteworthy work, complemented afterwards by other research groups. After the analysis of the composition of the edible part of this fish in culture and the effect of age on this parameter (Hung et al., 1987a), it was decided as the first step to design purified diets that could be used as a tool for this type of study (Hung et al., 1987b). After comparing this with other formulas, including salmonid feeds, the researchers decided that the diet that they called SPD-C (sturgeons purified diet with lecithin and without cholesterol) was a suitable formula for the study, at least for juvenile specimens (20–30 g). This was a diet based on casein, wheat gluten, white egg, and corn starch, which also included soy lecithin and oil mixture (cod-liver oil, corn oil, and lard) and the corresponding vitamin and mineral premixes. The proximate composition was, in rounded-off values, 44% protein, 15.5% ether extract and 4.2% ash with a moisture content of about 7.5%.

With this diet as the base, it was proposed initially to determine the protein needs of the juveniles of this species, as protein was perhaps the prime determinant of the growth rate and doubtless the most expensive ingredient in that diet. They tested eight experimental diets (Moore et al., 1988) that contained 20 to 53% protein (all with the same amount of fat, ca. 13%, and with decreasing levels of dextrin) in an experiment that lasted 8 weeks and for which they used sturgeons about 6 months old (145–300 g). With growth rate as a criterion, the researchers determined that, for these animals, the recommended proportion of protein in the diet should be set at around 40%. This percentage of protein is among those provided by the experimental diets containing 38.4 and 43% protein and that incorporate 30.4 and 24.6% dextrin, respectively. As might be expected, the utilization of the protein for growth (PER) or for protein retention (PPV) was somewhat better for the diets having a lower protein content (higher in dextrin).

Clearly, the establishment of a suitable level of protein in a diet depends on the quality of the protein (digestibility, content in essential amino acids, etc.) and therefore it is advisable to verify whether the optimal protein level proposed for the commercial feed should be maintained or altered if the protein sources change. In 1989 (Stuart and Hung, 1989) the results of testing eight different protein sources were published, involving both animal derivatives (casein, white egg, gelatine, herring, shrimp) and vegetable (soybean concentrate, zein), in comparison to control (casein, wheat gluten meal, white fish meal). All the diets were formulated as isonitrogenous (about 6%, i.e. 38% crude protein) and theoretically isoenergetic (13% lipid, 33–43% dextrin in all diets). Given the significant differences found in body composition and weight in the final test, the authors considered the percentage of retained energy to be the most sensitive index for assessing the quality of dietary proteins. Based on this index, apart from the control mixture, casein proved to be the most efficient source, followed by shrimp meal, herring meal, and soybean concentrate; white egg was in an intermediate position, while gelatin and zein were clearly the worst. The authors, without dismissing other possible causes such as other nutrients, digestibility, palatability and/or possible presence of anti-nutritional/toxic factors in the respective protein source, point towards the possible role of the different profile in presumed indispensable amino acids, an issue for which no information was available with respect to this species.

This was the aim of another work (Ng and Hung, 1995) which, by using the indirect method of measuring daily deposition in fed fish versus the daily mobilization of different amino acids in starved fish, estimated a dietary indispensable amino acid pattern for young growing white sturgeon that included (g/100 g protein): arginine 4.8, histidine 2.3, leucine 4.3, isoleucine 3.0, lysine 5.4, methionine (plus cystine) 2.2, phenylalanine (plus tyrosine) 5.3, threonine 3.3, tryptophan 0.3 and valine 3.3. These data were compiled on the basis of 67 g sturgeons on a diet of 40% protein. Previous studies of the amino acid composition of whole body, egg, and several tissues (Ng and Hung, 1994) were complemented by this one, providing information on the amino acid profile that should be taken into account when considering the nature of the protein provision in a commercial diet for this species.

In this sense, the replacement of full protein in a diet by a mixture of free amino acids, while delivering the same amounts of individual amino acids, proved negative when considering both diet acceptance and its utilization for growth (Ng et al., 1996). The latter could be due to quicker and unbalanced absorption in relation to the capacity of protein anabolism of the fish, which translates as greater renal and branchial excretion of the amino acids when free in the diet.

The lowest-cost energy source in the feed of farmed animals is carbohydrates. However, many carnivorous fish have shown poor utilization, at both the digestive and metabolic levels, at least when carbohydrates were provided in a crude form and at high amounts. It remains to be determined whether this general situation is also applicable to white sturgeon.

In a former study (Hung et al., 1989a), different carbohydrates were tested in 25 g fingerlings, by including them at 27% in several other experimental diets. The best results on retained energy were found with maltose and glucose, intermediate values with raw corn starch, dextrin, and sucrose, and the lowest ones with fructose, lactose, and cellulose. This was attributed essentially to differences in the efficiency of the corresponding digestion/absorption machinery (reduced alpha-amylase, saccharase, or lactase activity and deficient fructose transport). In a later assay (Herold et al., 1995) with larger specimens (824 g) it was confirmed that while the apparent digestibility of the monosaccharides exceeded 99%, that of disaccharides was between 36% (lactose) and 57% (fructose). Dextrin was digested at 75% and crude corn starch at only 32%, while the cellulose did not appear to be digested at all. Given the short duration of the test in relation to the expected lifespan of the fish, the authors did not rule out possible ontogenetic changes in these values. In relation to the above, changes were detected in certain metabolic pathways, especially lipogenesis, or in the hepatic storage rate of glycogen as well as some histological changes, both in the liver as well as in the intestine. In addition, it was concluded that this species had good homeostatic capacity to control glycaemia.

One technique that combined oesophageal intubation, dorsal aorta cannulation, and urinary catheterization, enabled the simultaneous monitoring of the glycaemic and glucosuric responses after the administration of diverse carbohydrates (Deng et al., 2001). Although the glycaemic peak appeared in all cases some 3–6 h after administration of the corresponding dosage, it was more pronounced after the administration of glucose and maltose than in any other case. The glycaemic index that takes into account the magnitude of the overall hyperglycaemic response (amplitude and duration) was clearly greater for these two simple sugars than for the polymers dextrin and starch—from corn and potato—that were assayed. In all cases, the glycaemic index was higher for dextrins than for native starches. These data correlated quite well with the digestibility data previously available.

The hydrolysed potato starch proved to be an effective source of carbohydrates, at least up to 15% of the diet, providing better indices than those previously found with other polysaccharides in the native state (Deng et al., 2005).

With respect to the possible responses to the growing levels of carbohydrates (Fynn-Aikins et al., 1992), it can be noted that the inclusion in the diets of progressively higher levels of D-glucose (0–35%), promoted metabolic changes in the liver, especially of lipogenic enzymes, glycogen accumulations, a higher hepatosomatic relation, higher levels of circulating glucose and triglycerides, as well as—from levels of 7% on—greater growth and a higher percentage or retained energy due at least in part to a certain body fattening. From this percentage on, the changes in growth and body composition were not very significant, although those of the aforementioned hepatic enzymes were. Afterwards, with the multicannulation system, the effects of ascending dosages of glucose administrated by oesophagus probe (0 to 1500 mg/kg body weight) were tested with regard to glycaemic responses (Gisbert et al., 2003). The highest hyperglycaemia was found 3 to 4 h after the administration of the two highest doses. In almost all cases, it took between 12 and 14 h for the basal glycaemia values to be restored. The intestinal absorption rate of glucose was found to depend on the dose supplied, but with saturation for the highest dose. Clearance increased at these doses, while the gluconeogenic rate did not appear to be significantly inhibited at high doses, a result that could explain, according to the authors, the maintenance of hyperglycaemia.

The good capacity of this species to utilize dietary carbohydrates was shown in a comparative study with tilapia hybrids (*Oreochromis niloticus* × *Oreochromis aureus*) (Lin et al., 1997), which reflected the clear benefit of continued feeding versus a regime of two feedings per day with respect to growth and diet utilization. This positive effect was not as marked for the type of carbohydrate used (glucose as opposed to cornstarch, both at 30%). On the other hand, in the tilapia hybrid, the feeding regime had less effect, while in this species starch proved to be better than glucose for growth. In any case, keeping other factors constant, carbohydrates was used better by the white sturgeon than by the other species, considered as omnivorous, so this fact could reflect a relation with the habitual diet in the wild of this sturgeon species.

The data concerning lipid nutrition of this species are notably more scarce, both in terms of their role as a source of essential fatty acids as well energy (Gawlicka et al., 2002). Regarding the former, experiments have been performed but the data are hardly conclusive (being based on the use of synthetic triglycerides that trigger poor growth in animals) but appear to indicate that the species requires both *n*–3 as well as *n*–6 fatty acids (Deng et al., 1998), though it appears capable of elongating and desaturating linolenic and linoleic to HUFA (highly unsaturated fatty acids) of both series. As in other fish species, the accumulation of 20:3n9 in hepatic phospholipids may be indicative of a dietary deficit of essential fatty acids.

In relation to the advisable level of total lipids in the feed, an assay with high-lipid/high-energy diets showed that the juveniles (110 g) of this species are well suited to efficient utilization of these diets, and thus there were no significant variations in growth rate and feed efficiency when the dietary-protein content fell from 53.5 to 45% with a concomitant rise in the lipid content (fish origin) from 26 to 36% (this involving an increase in gross energy from 23.6 to 25.3 MJ/kg and a decrease in the crude protein/gross energy ratio from 22.7 to 17.8 g/MJ). An increase in the dietary fat content to 40.2% negatively influenced growth and feed utilization (Hung et al., 1997).

As in other fish species, the fatty acid composition of the dietary lipids influenced the various tissues of the fish itself. Although the sensitivity of the response varied according to the type of body lipids (phospholipids proved more conservative than the triglycerides) and the tissue (the brain being less sensitive than the muscle or the liver). If the diet includes HUFAn3-rich fish oil, the accumulation of these fatty acids in bodily tissues, including the edible part (mainly muscle), is greater than when the dietary fat is based on vegetable oils, which may have repercussions from a commercial standpoint (Xu et al., 1993, 1996).

With respect to the strategy of feed distribution, the results of some tests have been published. The effect of continuous feeding, as opposed to only two feeds per day in the use of diets with different carbohydrate sources, has already been discussed. In a broader study with juveniles (8 g), several systems of continuous feeding were tested (24 h/day, 12 h daytime, 12 h nighttime) as well as various intermittent systems (2, 4 and 6 feedings per day) (Cui et al., 1997). Again, continuous feeding over the entire day proved the most effective in terms of growth, followed by the regime of six meals per day. There were no significant differences between continuous daytime or nighttime regimes. Given the slow feeding habit of these fish, the most continuous feeding possible proves to be the most effective food-distribution system, at least for juveniles.

With reference to the most adequate daily ration size, 2% of body weight in juveniles (30–100 g) was found to be best, while 1.5 to 2% of body weight proved adequate in larger fish (250–500 g) (Hung and Lutes, 1987; Hung et al., 1989b). When, in addition to ration size, water temperature was varied, it was found that 23°C was more favourable than 26.5°C, and that the optimal daily ration slightly increased at the highest temperature (2.5–3% body weight/day vs. 2.0–2.5% body weight/day) (Hung et al., 1993).

From these data, in 1995 (Cui and Hung, 1995) a prototype feeding–growth table was developed to predict growth rate from optimum feeding rate, body weight and water temperature. Such models may need a greater number of data to support and extend their practical validity.

13.3 Studies on *A. baerii*

According to Barrucand et al. (1979), as a result of an agreement with the former USSR, in May 1976 an acclimation programme began in France for this species, considered one of the most promising for aquaculture, given its plasticity (concerning temperature and salinity) as well as certain hybridization possibilities,. The first tests were carried out at two different locations: one in the INRA facility in Donzaq (at low temperatures of 16–18°C) and the other at the CTGRES, in Gironde (at high temperatures of 18–25°C) where preliminary feeding tests were performed with composite pelleted feeds based on trout models, giving results that, though improvable in terms of feed-conversion indices, proved promising. Thus, the growth rate in the cold-water installations raised from 1.72 g/day during the first year to a 3.63 g/day the second, data that approximated those of other temperate water species (trout, carp). The growth rates were even higher in the warm-water facilities, reaching up to 5.3 g/day, within the test period. In this group, the optimal temperature was found to be ca. 20°C, growth diminishing when the temperature fell in winter to below 15°C or when it rose in summer to 25–30°C.

In 1985, the INRA (St. Pée-sur-Nivelle, S. Kaushik and team) began testing the feeding of larvae of this species, studying diverse diets, natural and artificial, evaluating survival, growth, and protein as well as energy metabolism, in addition to amino acid absorption. As a general conclusion, it was verified as feasible to replace the live food of the initial diet.

In 1991, the results of the first study on the specific needs in protein and essential amino acids for this species (Kaushik et al., 1991) were published . This study was carried out with juveniles of approximately 20 g of initial weight maintained at 18°C for 45 days. Diets were tried with protein content (based mainly on fish meal with some soy and wheat meal) which went from 29 to 53% of the diet in dry matter (25–45% digestible protein according to estimates by the authors), maintaining these theoretically isoenergetic (20 mJ/kg dry matter) on lowering the fat content (fish oil) and carbohydrates (starch from crude and gelatinized corn) as the protein content was raised.

The break-point analysis showed that this occurred at a digestible-dietary-protein level of 40 ± 2%, which, expressed in terms of unit weight gain, was about 300 g/kg body weight gain, a figure comparable to data for other fish. Based on the data of this experiment the optimum dietary protein:energy ratio for this age of Siberian sturgeons ranged from 23 to 25 mg protein/kJ.

Data for essential amino-acid (EAA) requirements were based on the analysis of daily EAA increments, these being generally within those attributed to other fish species (in % of dietary protein: arginine 2.8, lysine 6.3, histidine 1.1, isoleucine 2.1, leucine 3.2, valine 2.3, phenylalanine 1.5, threonine 2.2). The authors recognize that corroboration on this information requires other approaches to the problem, such as an evaluation of the losses during starvation, to estimate maintenance needs, or studies based on growth or on biochemical and metabolic responses.

The effects of different protein sources on growth performance, apparent digestibility coefficients, amino acid availabilities and plasma free amino acid were posteriorly measured (Kaushik et al., 1994) showing that casein and/or soy bean proteins were digested better than fish meal.

The possibility of including carbohydrates in the diet for this species was considered in an experiment testing four diets containing 36–42% protein and different carbohydrate sources such as crude starch, gelatinized starch, extruded starch and extruded whole corn, at an amount of 380 g/kg of dietary mix. The data for growth and feed efficiency showed clearly better utilization of the pretreated starches than of crude ones, in line with data for other carnivorous species. The rise in the protein level to 42% proved detrimental, by which the authors interpret that the needs would be closer to 36% than 42%. The diets that induced the best growth also provoked hepatic accumulation of glycogen, which, for the duration of the study, did not prove unfavourable for protein utilization (Kaushik et al., 1989).

A later study involved juveniles (ca. 50 g initial weight) fed during 8 weeks on two diets that, having the same quantity of protein (51%) and energy (22 MJ/kg dry matter), differed in carbohydrate and lipid levels; one contained 200 g/kg of gelatinized starch and only 50 g/kg of fish oil, while the other had more fat (148.3 g/kg of fish oil) and less carbohydrate (98.9 g/kg crude starch was the only supply of this nutrient). In this way, according to the calculations of the authors, while the first covered 23% of their energy supply with lipids and 21% with carbohydrate, in the second these figures went from 38 to 7%, respectively. In both cases, protein digestibility, though low, was similar, but that of fat fell from 90 to 68% on increasing its presence in the diet, so that the overall energy fell from 73.4 to 55.5% (the presence of crude starch in this diet could have something to do with this finding, and in any case the authors themselves mentioned the need to verify the figures with new experiments). Although the diet richest in fat was slightly better accepted by the fish, the specific growth rate and therefore feed efficiency were somewhat lower. Nevertheless, from the body-composition data (retained energy) and N-excretion rates, the authors point out that the lipids appeared to be used more efficiently than carbohydrates at the metabolic level and also had a better protein-sparing action (Medale et al., 1991).

However, the same authors had observed that in this species the age of the sturgeons could be a determinant in the utilization of dietary energy. Thus, when this parameter was studied at three ages (3, 10, and 24 months) using a commercial trout diet (50% crude protein, 11% fat, 21.4 kJ/g dry matter gross energy), the oldest fish proved most effective in terms of energy retention, which at this age was mainly in the form of lipids, while at the other two ages protein retention predominated. The good digestive utilization of fat and the comparatively poorer utilization of protein are significant (Medale and Kaushik, 1991).

With larvae of the same origin and at the same facilities, Salin and Williot (1991), studying the endogenous N-excretion rate in specimens subjected to starvation, reported low values, perhaps indicative of a reduced basal metabolic rate and/or a greater dependence during this period on fat as the energy source. Also, in the Cemagref of Bordeaux, examining the daily pattern of N excretion in sturgeons differing in weight, these authors found a steady reduction of this parameter as fish weight increased which might indicate both a fall in the metabolic rate as well as a progressively greater dependence on fat as the source of energy.

More recently, this same research group (Gisbert and Williot) has published broad and informative work on the improvement of raising larvae of this species, describing the ontogenetic development, feeding behaviour, and response to different diets (Gisbert et al., 1998a,b; Gisbert and Williot, 2002; Gisbert and Ruban, 2003).

Researchers from other teams have made recent contributions on the feeding of *A. baerii* such as the possibility of partially replacing animal protein by the microalgae *Spirulina* or by soy supplemented or not with lysine and methionine. Although the results are good in terms of growth, changes were detected in the fatty acid profile of the edible portion with a reduction of HUFAn3 with respect to a control diet rich in fish meal (Palmegiano et al., 2005).

13.4 Studies on *A. naccarii*

Knowledge of the physiology of sturgeons in general and of *A. naccarii* in particular is still rudimentary. The research group Nutrición y Alimentación de Peces (Fish Nutrition and Feeding) of the University of Granada (Spain), has collaborated for more than 10 years with the RD Department of the Piscifactoría Sierra Nevada (Spain) to study the basic biology of this species and its farming on a large scale. After the ground-breaking studies in Italy, where the species was successfully bred in captivity and the various key elements for intensive farming were defined (by Giovannini, Arlati, Bronzi and others), the company proposed the farming of this fish in Spain, initially for meat and caviar, though other products derived from this sturgeon are being marketed. In addition, the recognized autochthonous character of the species in the Iberian Peninsula (Garrido-Ramos et al., 1997; De la Herrán et al., 2004) adds the advantage of possible restocking of this species in its ancestral distribution area and, in any case, would provide abundant, well-characterized specimens for a systematic approach to the knowledge of this species.

Along with the general interest in a new species of fish to farm, the sturgeon deserves special attention as an animal from which everything can be used: skin, cartilage, meat, and, of course, the roe (caviar). Furthermore, the Acipenseriformes in general and sturgeons in particular have notable peculiarities in many aspects of their biology, especially in their physiology and metabolism, with respect to other fish groups. In large measure, these peculiarities are ascribed to their notable phylogenetic age (these fish are considered as authentic 'living fossils').

Few of the early work on this species focused on the characterization of certain haematic and biochemical parameters (erythrocytes, glucose, amino acids, total lipids), biochemical analyses of sturgeons at different stages of the culture cycle (6, 18, 36, and 66 months old) or biometric studies (evolution of whole-animal weight and length as well as of different visceral organs) including the composition of several body fractions (muscle, digestive tract, liver, gonads, whole body). All this was concentrated on acquiring better knowledge of the different aspects of the species that might serve as a basis for nutritional experiments (see some results on Tables 13.1 and 13.2).

The nutritional assays were all performed in the Sierra Nevada fish farm, where some facilities were especially adapted to the experimental designs and objectives (animal and lot size, number of replicates, and so on). The nutritional assays were based on the following scheme:

Table 13.1 Some biometrical and blood parameters of *Acipenser naccarii* in culture at different ages

	Age (months)			
	6	18	30	66
Weight (g)	138.6	637.2	2275.0	6137.5
Total length (cm)	32.3	52.4	78.5	100.7
Nutritional index[a]	0.41	0.45	0.47	0.60
Eviscerated wt/body wt (%)	91.1	92.4	92.6	85.1
Liver wt/body wt (%)	4.4	3.7	2.9	1.8
Digestive wt/body wt (%)	4.5	3.3	3.1	1.2
Gonads wt/body wt (%)	n.d.	0.6	1.3	11.7
Haematocrit (%)	26.0	31.5	32.0	22.0
Red blood cells count (thousands/μm^3)	778	657	845	870
Haemoglobin (g/100 mL)	5.5	7.70	8.85	10.25
MCH[b] (pg)	73.8	117.3	104.69	118.2
MCV[c] (μ^3)	348.6	480.7	378.4	258.8
MCHC[d] (%)	22.8	24.44	27.7	52.3

n.d., no detected.
[a]Weight (g)/length^3 (cm) × 100.
[b]Mean corpuscular haemoglobin.
[c]Mean corpuscular volume.
[d]Mean corpuscular haemoglobin concentration.

Design of experimental dietary formulas, selection of the raw materials (first with purified ingredients, afterwards with commercial ingredients, Table 13.3), and manufacture of these diets and test with replicates of lots of sturgeons (a) of different ages and (b) at different scales (pilot, semi-industrial, and industrial, characterized by the lot sizes and the duration of the experiments).

In all cases, the response of the animals was measured in terms of (a) growth, (b) diet-utilization indices, (c) body composition and (d) repercussions on the physiology and metabolism of fish (see Table 13.4 for main indices determined).

After specific mineral and vitamin complements were defined (based on commercially available ones but slightly modified in certain components), a series of four different experiments were made to test several experimental diets formulated by varying the proportion of the different energy-providing nutrients (protein, carbohydrates, and fats). This enabled estimates of the most appropriate total dietary energy content at different ages as well as the most favourable protein/energy ratio (P/E) for the utilization of the normally more costly ingredient (protein) in addition to the possible influences on the metabolism and physiology of the sturgeon.

For the first experiment (pilot scale, purified or laboratory ingredients, sturgeons of age 1 year+: 475 g), we formulated and manufactured five experimental diets and used a sixth one (commercial) as an external control (RC.1). As reflected in Table 13.5, the five formulated diets were isoenergetic (21 MJ/kg) but with different protein/energy ratios (P/E). Of these, the one we called LC.1 or laboratory control had a high P/E ratio, while the other four substituted protein by carbohydrates (diets 1.1

Table 13.2 Main fatty acid composition of total lipids (%) of different body fractions of *Acipenser naccarii* in culture at different ages

	Muscle		Liver		Gonads	
Age (months)	6	66	6	66	30	66
Fatty acid						
14:0	4.92	3.47	4.32	4.13	3.68	4.19
14:1n5	0.29	0.25	0.34	0.22	0.25	0.20
15:0	0.45	0.33	0.35	0.34	0.32	0.33
16:0	16.76	20.07	17.18	19.15	17.73	19.74
16:1n7	5.34	8.22	6.30	8.20	7.74	8.29
17:0	0.27	0.20	0.19	0.22	0.00	0.27
18:0	2.42	1.50	3.42	2.30	2.59	1.65
18:1n9	20.34	27.47	21.94	33.48	31.35	28.75
18:1n7	2.68	1.34	2.83	2.96	2.82	2.89
18:2n6	6.66	6.53	5.33	6.32	5.86	6.96
18:3n3	1.03	0.91	0.76	0.73	0.79	0.90
20:0	0.09	0.30	0.07	0.00	0.00	0.05
20:1n9	8.78	3.37	5.56	4.92	4.15	3.79
20:2n6	0.42	0.27	0.38	0.27	0.28	0.23
20:3n3	0.27	0.16	0.33	0.18	0.14	0.15
20:4n6	0.64	0.76	0.87	0.48	0.70	0.96
20:5n3	5.30	5.38	3.62	2.59	3.69	4.46
22:1n9	4.84	1.53	3.34	2.68	1.98	2.01
22:2n6	0.33	0.17	0.23	0.00	0.00	0.24
22:5n3	1.19	1.32	1.01	0.88	1.02	1.01
22:6n3	9.03	8.02	7.71	3.79	4.92	4.27
24:1n9	0.10	0.00	0.63	0.00	0.00	0.11
Total saturated	24.99	26.13	26.29	26.14	24.52	26.60
Total monoenes	42.38	42.18	40.95	52.46	48.29	46.03
Total polyenes	24.88	23.53	20.23	15.23	17.40	19.17
Total HUFA	16.50	15.65	13.44	7.73	10.33	10.93
Total HUFAn3	15.52	14.72	12.34	7.26	9.64	9.74
Total n6	7.30	7.29	6.19	6.80	6.56	7.92
Total n3	16.55	15.64	13.10	7.98	10.42	10.63
n3/n6 ratio	2.27	2.14	2.12	1.23	1.56	1.33

and 1.2) or by fat (diets 1.3 and 1.4) at two different levels, and thus had lower P/E relationships (19 and 16.5 g/MJ, respectively). After a period of adaptation to the diets by the fish, the experimental period in the strict sense lasted 60 days.

All the macronutrients were well digested by the animals: values higher than 90% digestibility for the protein, higher than 80% for the fat, and 60 to 80% for the carbohydrates (mainly dextrins) were found. No influence of the protein or fat level was detected in their digestive utilization, and thus levels of up to 20% of fat were well digested (83.4%), and the protein was digested at a similar proportion, either at low dietary levels or as high as 48%. The digestibility of carbohydrates varied

Table 13.3 Ingredients used in the experimental diets

Experiment #	Ingredients
1 (pilot scale)	Casein, lysine, methionine, fish oil, sunflower oil, corn dextrins, vitamin premix, mineral premix, choline chloride, vitamin C, betain, carboxy-methyl-cellulose (binder), etoxiquin (antioxidant), cellulose (filler)
2, 3, and 4 (semi-industrial and industrial scales)	Fish meal, fish oil, sunflower oil, pregelatinized corn starch, breadcrumbs (only in experiments 3 and 4), vitamin premix, mineral premix, choline chloride, vitamin C, vitamin E, carboxy-methyl-cellulose (binder), etoxiquin (anti-oxidant), cellulose (filler)

Table 13.4 Main parameters determined in nutritional studies

	Experiment #
In blood samples	
Red blood cells count, haematocrit, haemoglobin content, related indexes	1
Total free amino acids	1
Glucose	1, 4
Total lipids	4
Biometry	
Liver/body weight ratio	1
Digestive/body weight ratio	1
Fillet/body weight ratio	1
Body, fillet and liver composition	1
Diets performance	
Growth rate (absolute, percentage, standard growth rate)	1, 2, 3, 4
Apparent digestibility coefficients	1
Food conversion	1, 2, 3, 4
Protein efficiency ratio	1, 2, 3, 4

among diets, presumably because of the cellulose content used as filler in making the diets isoenergetic. In fact, we found an inverse relationship between this cellulose content and the digestibility values of total carbohydrates.

On considering production indices, we found that the substitution of protein by fat or carbohydrates did not significantly reduce the growth rate of the sturgeons, perhaps for better utilization of the remaining protein for growth (as indicated by the PER values), which in turn should provoke a lower predictable release of nitrogen waste into the medium, resulting also in a lower diet cost, given the lower proportion of the most expensive ingredient, protein. On replacing protein with other energy-yielding nutrients, the fats proved somewhat more effective than carbohydrates.

On the basis of the results in assaying these diets, five other diets were designed (Experiment #2, Table 13.5), this time tested with fish of two different ages, 1+ and 2+ years of age under semi-industrial conditions (many of greater sizes, diets based on commercial ingredients). Two of the diets (LC.2 and CoC) presented a high P/E

Table 13.5 Basic formulation of the diets used in the different nutritional assays

	Diet	Protein (% dm)	Fat (% dm)	CHO[a] (% dm)	Energy[b] (MJ/kg)	P/E ratio[c] (g/MJ)
Experiment #1						
	RC.1	50	9	20	18.7	26.7
	LC.1	48	15	22	20.9	22.9
	1.1	40	15	33	21.0	19.1
	1.2	35	15	40	21.0	16.7
	1.3	40	16	30	20.8	19.2
	1.4	35	20	30	21.3	16.5
Experiment #2						
	LC.2	48	13	22	20.2	23.8
	CoC	48	13	22	20.2	23.8
	2.1	40	13	32	20.0	20.0
	2.2	40	17	26	20.6	19.5
	2.3	33	22	26	20.9	15.8
Experiment #3						
	RC.3	Closed formula				
	3.1	40	17	32	21.6	18.5
	3.2	35	20	32	21.6	16.2
Experiment #4						
	RC.4	Closed formula				
	4.1	30	13	40	19.0	15.8
	4.2	30	23	25	20.4	14.7
	4.3	40	10	32	18.8	21.3
	4.4	40	17	26	20.6	19.5

RC, commercial diets used as a control.
[a]Digestible carbohydrates.
[b]Calculated by using the following coefficients (kJ/g): 23.4 for proteins, 39.7 for fats, 17.1 for CHO.
[c]Protein/energy ratio.

relationship (23.8 g/MJ) prompted by a high protein content and the other three with a lower P/E relationship achieved by different ways: two diets, with a P/E of about 19.5 g/MJ, with the substitution of protein by carbohydrates (diet 2.1) or by carbohydrates and, mainly, fat (diet 2.2) and a third (diet 2.3), with the lowest P/E of 16 g/MJ, where the protein level also fell while the fat and carbohydrate levels rose. All the diets were approximately isoenergetic (20–21 MJ/kg). The two with the greatest protein content (LC and CoC) differed in protein composition, one containing exclusively fish meal (LC) and the other 2% protein in the form of collagen, to investigate whether the greater proline and hydroxyproline supply by this protein source would be beneficial in a species with such a high content of body cartilage.

The production results proved to be dependent on age. Thus, the 1year + fish registered better utilization indices of the diets with a lower P/E relationship, the best being for the fat substitution of protein. However, the 2 years + fish showed better utilization of the diets with the higher P/E relationship and therefore a higher protein level. This was surprising as the protein needs of fish are known to decline with age parallel to the specific growth rate and therefore the demand for structural proteins. One explanation might be the biological cycle of the species in which at a certain age (still unknown in this species), the individual begins migration to more saline waters (this being an anadromous species), which would be preceded by a preparatory stage that should include an increase in the fatty reserves that could be achieved more easily, at least with lower metabolic cost, from non-protein macronutrients. This would make sense if this preparatory process were nearer the age 1+ than age 2+. A higher growth rate in 2+ age, as sustained by the biometrical data of previous assays, and so, a higher proteins requirement could contribute to this different situation according to age.

The incorporation of collagen as a source of amino acids (proline and hydroxyproline) that in some way could complete the dietary protein did not prove effective. This might be because either the fish meal (main protein source in these diets) was sufficiently complete for sturgeon needs or else the quantity of supplement used (2% of the total protein) was not enough to promote a differential effect in the period tested (2 months).

To date, the assays indicate that the substitution of protein by fat was beneficial, particularly in the younger fish. To test this result, the following experiment (#3, Table 13.5) was designed, in which three diets were formulated to be assayed again at two different ages (1+ and 2+) at the industrial scale. Two of the experimental diets were isoenergetic and were differentiated in that one presented a higher quantity of protein (diet 3.1) and the other a greater content in fat that replaced the protein (diet 3.2), maintaining the carbohydrate level high and constant. The third diet referred as RC.3, based on a private formula made on the fish farm itself, presented a higher protein/lower CHO level than the experimental ones.

When we compared the results of assaying the diet with a higher fat level against that of a higher protein level, both being isoenergetic, we again found it was effective to lower the protein level and raise the fat level. On the other hand, the two experimental diets presented a higher level of carbohydrates than the reference diet and also gave better results. Therefore reduction in the protein component, accompanied by increase in carbohydrates also proved to be advantageous, especially for animals of 1year + in age.

The results in successive assays indicated repeatedly that, on one hand, a reduction of the dietary protein could prove advisable, so long as the protein was replaced by fat and/or carbohydrates, and that, on the other hand, this measure was more effective in animals of 1year + of age than in those of 2years +.

In a final experiment (#4, Table 13.5), we tested the results of the nutritional assays on an industrial scale (100 kg of fish/lot) and for a longer time period (4 months). For this, four isoenergetic diets were formulated; two of these had the same protein percentage, one with a higher quantity of carbohydrates (diet 4.1) and the other of fat (diet 4.2) with a P/E ratio of 16 and 15 g/MJ, respectively; the other two diets had a higher protein percentage, but also, as in the first two, raising the carbohydrate level in one (diet 4.3) and fat in the other, with a P/E relationship of 21 and 20 g/MJ, respectively; a fifth diet (RC) with a higher protein and energy level was used as control. Once again, the diets were assayed in sturgeons of 1year + and 2 year + ages. The results confirmed previous findings, indicating that:

1. It may be beneficial in the formulation and manufacture of commercial diets for the sturgeon *A. naccarii* to reduce the protein level to below the level considered optimum for a strictly carnivorous fish, adding the corresponding amount of non-protein macronutrients (carbohydrates and/or fat).
2. Sturgeons of 2years + in age appear to require approximately 5% more protein than those of 1year +.

Thus, we find that this fish, from the standpoint of macronutrient needs, is closer to fish considered omnivorous than to those strictly carnivorous, as we found also for the European eel (*Anguilla anguilla*). In fact, the eel shows better utilization of diets where especially the carbohydrates, but also fats, replace the protein, having protein needs of almost 10% lower than those of strictly carnivorous farmed fish. Both species are possibly in a situation of needing to store energy reserves for the migration process (Sanz et al., 1993).

In addition to these nutritional studies, we have undertaken others with the same species that include aspects of digestive physiology, metabolic pathways involved in the utilization of the nutrients and the study of different markers of oxidative stress. All these are focused on acquiring better knowledge of the nutritional physiology of these fish.

In short, the studies on digestive physiology (Llorente et al., 2005) indicate that:

1. Activities of digestive enzymes, α-amylases and proteases, are present in the earlier phases of life, from the fertilization of the oocyte, and are probably related to the rupture of the embryonic sac.
2. The detected enzymatic activities become particularly evident at different moments of the development of the fish and probably have different origins and functions, as indicated in microscopy studies. It should be emphasized that at the beginning of the exogenous feeding, though the pancreas appear developed, there is an absence of zymogen granules in the stomach, and thus the peak of enzymatic activity detected, mainly acidic, should have a major exogenous component. That is, we cannot rule out the digestive contribution of the enzymes present in the live feed of the fish in this transition phase.
3. At 1 month of life, counting from the moment of fertilization of the egg, the values of the digestive α-amylase and protease activities stabilize; indicating functionally developed digestive enzymatic apparatus. This circumstance was confirmed on making histological, histochemical, and ultrastructural studies of the gastrointestinal tract.
4. In the adult stage, the results indicate that the sturgeon would be prepared not only to digest the fat and the protein like any carnivorous fish, but also that the amylase pool would help it to digest carbohydrates at levels characteristic of an omnivore. It is known that the quantity of proteases present in the digestive tract of fish is not related to the feeding habit and that, on the contrary, the secretion of carbohydrases is greater in omnivorous and herbivorous fish than in carnivo-

rous ones. Specifically, in work carried out by our team with species of fish from different feeding habits (Hidalgo et al., 1999), on comparing the digestive pool of amylase against proteases, we found higher values in carnivores as trout (*Oncorhynchus mykiss*), than in omnivores and herbivores such as carp and tench. A subsequent comparative study with trout indicated that sturgeon had a lower protease/amylase ratio (Furné et al., 2005a).

Thus, the studies on the digestive physiology reinforce the previous findings in the nutrition assays, that the characteristic closest to the omnivory of the sturgeon *A. naccarii*.

In addition, a posterior study, made to ascertain the evolution of the activities of digestive enzymes during a period of food deprivation in this species of sturgeon and also in trout, indicated a similar pattern in both species both after starving for 72 days as well as after 60 days of subsequent re-feeding. The activity of digestive enzymes began to be reduced in both species from the second week of starvation, amylase activity being affected before protease activity. Nevertheless, after a month of starvation, both species continued to be prepared to digest the protein in an effective way. When the starvation was prolonged to 72 days, the digestive apparatus of both species were noticeably altered in their capacity to digest macronutrients. After 60 days of re-feeding following this period of prolonged starvation, both species had recovered the capacity to digest proteins, though that corresponding to carbohydrates and lipids was still depressed. In practical terms, for better utilization of the feed supplied and greater performance of the farmed fish, the feed in both species should be slowly reintroduced after a prolonged starvation period if this event occurs.

Another type of study we are involved in refers to the antioxidant defences of the sturgeon *A. naccarii* in comparison to the trout *O. mykiss* (Trenzado et al., 2006) by the determination in different tissues (gills, heart, digestive tract, liver, muscle, skin, red blood cells and swimming blader) of the activities of the enzymes superoxide dismutase (SOD), catalase (CAT), glutathione peroxidase (GPx), and glutathione reductase (GR) as well as of the lipid peroxidation index. These determinations have shown that:

1. The sturgeon has a great antioxidant activity in the liver. In the rest of the tissues, this antioxidant activity, except for CAT, is greater in the trout.
2. In both species, the lipid peroxidation index shows the highest values in the digestive tract.
3. The high lipid content and the relatively high proportion of HUFAn3 (García-Gallego et al., 1999), together with the good state of these as manifested by the low lipid peroxidation index, confer the liver and muscle of this sturgeon with great advantages from the standpoint of human consumption.

In addition, we are now performing comparative studies on the intermediate metabolism in the sturgeon *A. naccarii* and the rainbow trout *O. mykiss* (Furné et al., 2005b), based on the determination of the activities of some key enzymes such as fructose 1,6-bisphosphatase (FBPase), pyruvate kinase (PK), glucose 6-phosphate dehydrogenase (G6PDH), malic enzyme (EM), glutamate pyruvate transaminase (GPT), glutamate oxalacetate transaminase (GPT), citrate synthase (CS), 3-hydroxyacil

CoA dehydrogenase (HOAD), glycerol kinase (GK), acetoacetil CoA thiolase (ACoAT), and β-hydroxybutyrate dehydrogenase (β-OHBDH). These studies have revealed that:

1. Lipid oxidation takes place both in the liver and in extrahepatic tissues, especially in the heart, in both species. At this level, sturgeons would have a metabolic profile similar to that of teleosts and different from that of elasmobranchs.
2. The hepatic lipogenic and glyconeogenic capacities are greater in the sturgeon.
3. The trout exhibits a more active muscle metabolism..
4. No differences have been found in the branchial metabolism in the two species.

Finally, we must mention other, though not many, studies published by other authors more or less directly related with these issues, especially several aspects of fat and fatty acids and vitamin E utilization and accumulation (Agradi et al., 1993; McKenzie et al., 1995, 1997; Badiani et al., 1996; Cataldi et al., 1999; Vaccaro et al., 2005).

This research team is now involved in the improvement of fish diets to be utilized in an organic aquaculture production. So, data from previous assays are being used to design organic feeding practices that include the utilization of raw materials obtained only from organic production and/or specifically authorized by legislation, and practices of manufacturing, storage and delivery of fish according to the principles of sustainable aquaculture. As a result of this cooperation, the fish farm is now marketing meat and caviar from *A. naccarii* with the 'organic product' label awarded by the CAAE (Andalusian Committee of Organic Agriculture).

13.5 Conclusion

We trust that all these results, and those soon to be forthcoming in similar research, will expand our knowledge of the biology, not only of this species but, by extension, of sturgeons as a group, enabling us at the same time to improve culture conditions. Farming allows access to animals which were previously found only in their natural environment and it opens immense study possibilities, for promoting the advancement of ichthyophysiology in general and for sturgeon farming in particular, both in terms of production as well as in relation to any recovery attempt of these endangered wild populations.

References

Agradi E, Abrami G, Serrini G, McKenzie D, Bolis C, Bronzi P. 1993. The role of dietary *n*–3 fatty acids and vitamin E supplements in growth of sturgeon (*Acipenser naccarii*). *Comparative Biochemistry and Physiology 105*(A), 187–195.

Badiani A, Anfossi P, Fiorentini L, Gatrta P, Manfredini M, Nanni N, Stipa S, Tolomelli B. 1996. Nutritional composition of cultured sturgeon (*Acipenser* spp.). *Journal of Food Composition and Analysis 9*(2), 171–190.

Barrucand M, Ferlin P, Lamarque P, Sabaut JJ. 1979. Alimentation artificielle de l'esturgeon *Acipenser baerii*. In: Halver JE and Tiews K (eds.), *Finfish Nutrition and Fishfeed Technology*, Heenemann Verlag, Berlin. *Proceedings of the World Symposium on Finfish Nutrition and Fishfeed Technology*, Hamburg, 20–23 June, 1978, Vol. I, pp. 411–421.
Cui Y, Hung SSO. 1995. A prototype feeding-growth table for white sturgeon. *Journal of Applied Aquaculture 5*(4), 25–34.
Cui Y, Hung SSO, Deng D, Yang Y. 1997. Growth performance of juvenile white sturgeon as affected by feeding regimen. *Progressive Fish-Culturist 59*(1), 31–35.
De la Herrán R, Robles F, Martínez-Espín E, Lorente JA, Ruiz Rejón C, Garrido-Ramos MA, Ruiz Rejón M. 2004. Genetic identification of western Mediterranean sturgeons and its implication for conservation. *Conservation Genetics 5*, 545–551.
Deng DF, Hung SSO, Conklin DE. 1998. White sturgeon (*Acipenser transmontanus*) requires both n–3 and n–6 fatty acids. *Aquaculture 161*, 333.
Deng DF, Refstie S, Hung SSO. 2001. Glycemic and glycosuric responses in white sturgeon (*Acipenser transmontanus*) after oral administration of simple and complex carbohydrates. *Aquaculture 199*(1–2), 107–117.
Deng DF, Hemre GI, Storebakken T, Shiau S Y, Hung SSO. 2005. Utilization of diets with hydrolyzed potato starch, or glucose by juvenile white sturgeon (*Acipenser transmontanus*), as affected by Maillard reaction during feed processing. *Aquaculture 248*(1–4), 103–109.
Doroshov S, Van Eenennaam J, Moberg G. 1994. Reproductive management of cultured *white sturgeon (Acipenser transmontanus)*. In: MacKinlay DD (ed.), *High Performance Fish, Proceedings of an International Fish Physiology Association*, Vancouver, BC, Canada.
Furné M, Hidalgo MC, López A, García-Gallego M, Morales AE, Domezain A, Domezain J, Sanz A. 2005a. Digestive enzyme activities in Adriatic sturgeon *Acipenser naccarii* and rainbow trout *Oncorhynchus mykiss*. A comparative study. *Aquaculture 250*, 391–398.
Furné M, García-Gallego M, Sanz A, Hidalgo MC, Domezain A, Domezain J, Morales AE. 2005b. Estudio comparado del metabolismo intermediario en el esturión *Acipenser naccarii* y la trucha *Oncorhynchus mykiss*. X Congreso Nacional de Acuicultura, Gandía, Valencia, Spain.
Fynn-Aikins K, Hung SSO, Liu W, Li H. 1992. Growth, lipogenesis and liver composition of juvenile white sturgeon fed different levels of D-glucose. *Aquaculture 105*(1), 61–72.
García-Gallego M, Sanz A, Domezain A, De la Higuera M. 1999. Age-size influences on tissues-lipid quality of the sturgeon *Acipenser naccarii* from intensive culture. *Journal of Applied Ichythyology 15*, 261–264.
Garrido-Ramos M, Soriguer MC, de la Herrán R, Jamilena M, Ruiz Rejón C, Domezain A, Hernando J, Ruiz Rejón M. 1997. Morphometric and genetic analysis as proof for the existence of two sturgeon species in the Guadalquivir River. *Marine Biology 129*, 33–39.
Gawlicka A, Herold MA, Barrows FT, de la Nouee J, Hung SSO. 2002. Effects of dietary lipids on growth, fatty acid composition, intestinal absorption and hepatic storage in white sturgeon (*Acipenser transmontanus* R.) larvae. *Journal of Applied Ichthyology 18*(4–6), 673–681.
Gisbert E, Ruban GI. 2003. Ontogenetic behaviour of Siberian sturgeon, *Acipenser baerii*: a synthesis between laboratory test and field data. *Environmental Biology of Fishes 67*(3), 311–319.
Gisbert E, Williot P. 2002. Advances in the larval rearing of Siberian sturgeon. *Journal of Fish Biology 60*, 1071–1092.
Gisbert E, Rodriguez A, Castello-Orvay F, Williot P. 1998a. A histological study of the development of the digestive tract of Siberian sturgeon (*Acipenser baerii*) during early ontogeny. *Aquaculture 167*, 195–209.
Gisbert E, Williot P, Castelló-Orvay P. 1998b. Morphological development of Siberian sturgeon (*Acipenser baerii*, Brandt) during pre-larval and larval stages. *Rivista Italiana di Acquacoltura 33*, 121–130.
Gisbert E, Sainz RD, Hung SS. 2003. Glycemic responses in white sturgeon after oral administration of graded doses of D-glucose. *Aquaculture 224*(1–4), 301–312.
Herold MA, Hung SSO, Fynn-Aikins K. 1995. Apparent digestibility coefficients carbohydrates for white sturgeon. *Progressive Fish-Culturist 57*(2), 137–140.

Hidalgo MC, Urea E, Sanz A. 1999. Comparative study of digestive enzymes in fish with different nutritional habits. Proteolytic and amylase activities. *Aquaculture 170*, 267–283.

Hung SSO, Lutes PB. 1987. Optimum feeding rate of hatchery-produced juvenile white sturgeon (*Acipenser transmontanus*): at 20 degree C. *Aquaculture 65*(3–4), 307–317.

Hung SSO, Lutes PB, Conte FS. 1987a. Carcass proximate composition of juvenile white sturgeon (*Acipenser transmontanus*). *Comparative Biochemistry and Physiology 88B*(1), 269–272.

Hung SSO, Moore BJ, Bordner CE, Conte FS. 1987b. Growth of juvenile white sturgeon (*Acipenser transmontanus*) fed different purified diets. *Journal of Nutrition 117*(2), 328–334.

Hung SSO, Fynn-Aikins K, Lutes PB, Xu R. 1989a. Ability of juvenile white sturgeon (*Acipenser transmontanus*) to utilize different carbohydrate sources. *Journal of Nutrition 119*, 727–733.

Hung SSO, Lutes PB, Conte FS, Storebakken T. 1989b. Growth and feed efficiency of white sturgeon (*Acipenser transmontanus*) sub-yearlings at different feeding rates. *Aquaculture 80*(1–2), 147–153.

Hung SSO, Lutes PB, Shqueir AA, Conte FS. 1993. Effect of feeding rate and water temperature on growth of juvenile white sturgeon (*Acipenser transmontanus*). *Aquaculture 115*(3–4), 297–303.

Hung SSO, Storebakken T, Cui Y, Tian L, Einen O. 1997. High-energy diets for white sturgeon, *Acipenser transmontanus* Richardson. *Aquaculture Nutrition 3*(4), 281–286.

Kaushik SJ, Luquet P, Blanc D, Paba A. 1989. Studies on the nutrition of Siberian sturgeon, *Acipenser baeri*. 1. Utilization of digestible carbohydrates by sturgeon. *Aquaculture 76*, 97–107.

Kaushik S, Breque J, Blanc D. 1991. Requirements for protein and essential amino acids and their utilization by Siberian sturgeon (*Acipenser baerii*). In: Williot P., editor. Actes du Premier Colloque International sur l'Esturgeon, Bordeaux, France. Cemagref Publ., Antony, France. pp. 25–39.

Kaushik SJ, Breque J, Blanc D. 1994. Apparent amino acid availability and plasma free amino acid levels in Siberian sturgeon (*Acipenser baeri*). *Comparative Biochemistry and Physiology (A) 107*, 433–438.

Lin J, Cui Y, Hung SSO, Shiau S. 1997. Effect of feeding strategy and carbohydrate source on carbohydrate utilization by white sturgeon (*Acipenser transmontanus*) and hybrid tilapia (*Oreochromis niloticus* × *O. aureus*). *Aquaculture 148*(2–3), 201–211.

Llorente JI, Hidalgo MC, García-Gallego M, Furné M, Morales AE, Carmona R, Ostos MV, Domezain A, Domezain J, Sanz A. 2005. Enzimas digestivas durante el desarrollo ontogénico del esturión, *Acipenser naccarii*, Bonaparte 1836. X Congreso Nacional de Acuicultura, Gandía, Valencia, Spain.

McKenzie DJ, Piraccini G, Steffensen JF, Bolis CG, Bronzi P, Taylor EW. 1995. Effects of diet on spontaneous locomotor activity and oxygen consumption in Adriatic sturgeon (*Acipenser naccarii*). *Fish Physiology and Biochemistry 14*, 341–355.

McKenzie DJ, Piraccini G, Papini N, Galli C, Bronzi P, Bolis CG, Taylor EW. 1997. Oxygen consumption and ventilatory reflex responses are influenced by dietary lipids in sturgeon. *Fish Physiology and Biochemistry 16*, 365–379.

Medale F, Kaushik SJ. 1991. Energy utilization by farmed Siberian sturgeon (*Acipenser baerii*) from 3 age classes. In: Williot P., editor. Actes du Premier Colloque International sur l'Esturgeon, Bordeaux, France. Cemagref Publ., Antony, France. pp. 13–23.

Medale F, Blanc D, Kaushik SJ. 1991. Studies on the nutrition of Siberian sturgeon, *Acipenser baeri*. 2. Utilization of dietary non-protein energy by sturgeon. *Aquaculture 93*, 143–154.

Moore BJ, Hung SSO, Medrano JF. 1988. Protein requirement of hatchery-produced juvenile white sturgeon (*Acipenser transmontanus*). *Aquaculture 71*(3), 235–245.

Ng WK, Hung SSO. 1994. Amino acid composition of whole body, egg and selected tissues of white sturgeon (*Acipenser transmontanus*). *Aquaculture 126*(3/4), 329–339.

Ng WK, Hung SSO. 1995. Estimating the ideal dietary indispensable amino acid pattern for growth of white sturgeon, *Acipenser transmontanus* (Richardson). *Aquaculture Nutrition 1*(2), 85–94.

Ng WK, Hung SSO, Herold MA. 1996. Poor utilization of dietary free amino acids by white sturgeon. *Fish Physiology and Biochemistry 15*(2), 131–142.

Palmegiano GB, Agradi E, Forneris G, Gai F, Gasco L, Rigamonti E. 2005. Spirulina as a nutrient source in diets for growing sturgeon (*Acipenser baeri*). *Aquaculture Research 36*, 188–195.

Salin D, Williot P. 1991. Endogenous excretion of Siberian sturgeon *Acipenser baeri* Brandt. *Aquatic Living Resources 4*, 249–253.

Sanz A, Suárez MD, Hidalgo MC, García-Gallego M, De la Higuera M. 1993. Feeding of the European eel (*Anguilla anguilla*). III. Influence of the relative proportions of the energy yielding nutrients. *Comparative Biochemistry and Physiology 105A*(1), 177–182.

Stuart JS, Hung SSO. 1989. Growth of juvenile white sturgeon (*Acipenser transmontanus*) fed different proteins. *Aquaculture 76*(3–4), 303–316.

Trenzado C, Hidalgo MC, García-Gallego M, Morales AE, Furné M, Domezain A, Domezain J, Sanz A. 2006. Antioxidant enzymes and lipid peroxidation levels in sturgeon *Acipenser naccarii* and trout *Oncorhynchus mykiss*. A comparative study. *Aquaculture 254*, 758–767.

Vaccaro AM, Buffa G, Messina CM, Santulli A, Mazzola A. 2005. Fatty acid composition of a cultured sturgeon hybrid (*Acipenser naccarii* × *A. baerii*). *Food Chemistry 93*(4), 627–631.

Xu R, Hung SSO, German JB. 1993. White sturgeon tissue fatty acid compositions are affected by dietary lipids. *Journal of Nutrition 123*(10), 1685–1692.

Xu R, Hung SSO, German JB. 1996. Effects of dietary lipids on the fatty acid composition of triglycerides and phospholipids in tissues of white sturgeon. *Aquaculture Nutrition 2*(2), 101–109.

Part III
Recovery and Conservation

Chapter 14
Restoration of Adriatic Sturgeon (*Acipenser naccarii*) in Italy: Situation and Perspectives

Giovanni Arlati and Lidia Poliakova

Abstract This work is a report of the activities concerning the conservation of the cobice sturgeon endemic to Italy.

In addition to restocking in Lombardy rivers from 1991, and in the rivers of Veneto from 1999, brood fish of wild origin (P) have been genetically characterized. This offers the possibility of verifying without microchip signal, whether the specimen caught comes from brood fish stock of wild origin (P) maintained in captivity.

It is necessary to continue this activity, especially monitoring and identification of natural areas for the conservation of the species.

Keywords *Acipenser naccarii*, recovery plan, genetic investigation, restocking, adriatic sturgeon

14.1 Introduction

This chapter gives a brief history of the situation and activities related to sturgeon conservation in Italy. The work has been done with the species *Acipenser naccarii* (Bonaparte, 1836), which is generally called Italian or adriatic sturgeon; in Italy it is known as 'Storione del Naccari' or 'Storione cobice' or 'Storione Italiano' or 'Storione dell'Adriatico'.

The 'cobice' sturgeon has been reported in Italy from the rivers of the Po River basin (Sesia, Terdoppio, Agogna, Ticino, Adda, Oglio, Mincio) and from rivers of the Veneto region (Adige, Brenta, Bacchiglione, Livenza, Piave and Tagliamento). It was most common in the Po River and migrated into the Piedmont part of the Po River as far as Turin. Once relatively abundant in the Po River and estuary, the

G. Arlati and L. Poliakova
ERSAF—Regione Lombardia, Struttura Valorizzazione delle Risorse Ittiche,
via Copernico, Milano, Italy

R. Carmona et al. (eds.) *Biology, Conservation and Sustainable Development of Sturgeons*,

'cobice' sturgeon underwent a dramatic decline in abundance and became rare by the early twentieth century. Sturgeon numbers are decreasing mainly because of:

1. The building of dams and barriers which prevent migration of spawning adults
2. Extreme fishing and poaching of the largest fish over time and thus endangering natural reproduction
3. Environmental pollution, reducing the availability of adequate reproduction areas

In brief, the cobice sturgeon (D'Ancona, 1924; Paccagnella, 1948; Tortonese, 1989; Rossi et al., 1991; Bronzi et al., 1994) was at risk of extinction.

The first experiments in breeding young *A. naccarii* at Orzinuovi (near Brescia) from fishes caught in the delta Po, Adda, Adige, Piave and Oglio rivers were conducted in 1977. Since June 1988, it has been possible to reproduce this species with a technique based on the hormonal manipulation of broodfish and non-surgical collection of gametes (Fig. 14.1) (Arlati et al., 1988).

The method of using a synthetic LHRH-analogue (luteinizing hormone-releasing hormone) (Poliakova et al., 2003) has been used to induce ovulation in sturgeons either maintained in farming conditions or caught from the wild (Table 14.1).

For fish-farm sturgeons, there is a particular management method of the brood stock with special care given to both the feeding and the physicochemical water conditions.

From June 1988, every year, controlled reproductions have been carried out, always with different females and males (P) to preserve the biodiversity for restocking.

Although the cobice sturgeon is a long-lived species (in nature it lives for more than 50 years) and comes to sexual maturity (under the experimental conditions

Fig. 14.1 Eggs stripping from female sturgeon *Acipenser naccarii*

Table 14.1 Main parameters of controlled reproduction for several species of sturgeons

Species	Age of broodfish (years)	Weight (kg)	Total dosage LHRH (μg/kg)	Temperature of reproduction (°C)	Temperature of incubation (°C)	Time of incubation (day degrees)
A. naccarii[a]	14–25	16–70	22–35	18–23	15.5	116–130
A. naccarii f1	11	24–38	22–35	22–23	15.5	116–130
A. baerii f3	8–9	9–12	20–24	14–18	15	90–110
A. ruthenus[a]	6–7	1.5–2.5	15	15–18	15	100–110
A. gueldenstaedtii f1	10–11	16–22	20	15–18	15	100–110
A. oxyrinchus[b]	18	83	25	18–21	19	90–105
A. brevirostrum[b]	9	4–9	20	13–16	14	100–110
A. transmontanus	11	72	35	18–20	15.5	105–110

[a]Sturgeons caught in the wild and reared in captivity.
[b]Sturgeons caught in the wild.

used) at 14–15 years, it has been planned since 1994 to reproduce F1 specimens obtained from primary stock of wild origin, both for aquaculture and for restocking and faunal recovery.

Concerning the biotechnological aspects, experience has led to:

1. The use of an innovative method for a controlled reproduction—'from the laparatomy to the stripping'—on the different species of sturgeons reared in Italy (Sterlet, Siberian, White, Russian)
2. Better technology of controlled reproduction, so that nowadays a hormone posology of 1 μg/kg is used for the females, while semen from the males is got naturally
3. An in depth study of the hygienic and sanitary needs and feeding necessities during weaning

14.2 Recovery Plan of Adriatic Sturgeon

The possibility of breeding this species with non-surgical bloodless eggs has led to careful and targeted action for restocking the main rivers in the Lombardy region and in other rivers of the Veneto and Emilia-Romagna regions.

The work carried out until now has been based on a specific plan of which the main actions are:

1. Genetic control of the broodstock and F1 produced, to guarantee that all sturgeons used for experimental restocking are from different parents
2. Weaning aimed to obtain a fully guaranteed healthy product ready for restocking in streams

3. Identifying the sites and choosing the most appropriate size for restocking
4. Proper management of brood stock and refinement of the breeding biotechnology to safeguard the wild caught–original stock (P) of cobice sturgeon

14.2.1 Genetic Investigations

In continuation of the research regarding classical cytogenesis (Fontana et al., 1997; Lanfredi et al., 1999), a genetic characterization programme of brood fish of wild origin (P) was begun, with the collaboration of the University of Ferrara (Congiu et al., 2001; Fontana et al., 2001) to:

1. Create a gene bank (genomic library) and a database of brood fish both wild and F1 (120 subjects)
2. Try to identify a method of genetic tagging to use on stocks in nature for recapture. In fact, for practical application, we used only A. naccarii specimens of wild origin (at that time nearly extinct from the rivers)
3. Investigate genetic variability of stock from which ichthyogenic specimens can be obtained for restocking

Until now, brood fish (P and F1) have been characterized with genetic survey for microsatellites, AFLP (amplified fragment lenght polymorphism), and mitochondrial loci (12) and from these specimens we can obtain individuals for restocking (Zane et al., 2002). This is very important because it makes it possible, through biological sampling, to identify the origin (wild or from P/F1specimens), of those specimens captured during monitoring, which do not show the presence of the 'transponder'.

14.2.2 Restocking

The Lombardy region, in cooperation with the provinces, started a faunistic restocking and recovery plan in 1991 (Arlati et al., 1999, 2003). The stocking areas involved were the most important rivers, in particular the Adda and Oglio. Till now, restocking activity has involved the re-introduction of 438,633 individuals (Table 14.2).

Of these 75.93% are 2.5 cm; 10.76% 8–18 cm; 3.00% 18–40 cm; 7.26% 40–60 cm and 3.05% 60–90 cm.

It is important to emphasise that 12,390 subjects (40–60 cm) were marked with PIT (passive integrated transponder) tags or transponder (Fig. 14.2a,b) for individual recognition of the specimens captured during monitoring (Fig. 14.3a,b).

14.2.3 Life COBICE

LIFE Natura COBICE 'Conservation and Breeding of Italian Cobice Endemic sturgeon', a 3-year project began on 1 October 2004.

Table 14.2 Restocking actions from June 1988 to April 2007

Size (cm)	Subjects	Rivers	Region (Province)
2.5	333,050		
	283,050	Adda, Oglio	Lombardy (Brescia, Cremone, Lecco, Lodi, Mantova, Milan)
	44,000	Channel Lore, Adiga, Brenta, Sila, Piave, Tagliamento	Veneto (Padova, Rovigo, Treviso, Venice, Verona)
	6,000	Po	Emilia-Romagna (Piacenza)
8–18	47,200		
	47,200	Ticino, Adda, Lake Iseo, Mincio	Lombardy (Brescia, Cremone, Lecco, Lodi, Mantova, Milan)
18–40	13,162		
	13,162	Tardoppio, Agogna, Ticino, Adda, Oglio, Mincio[a/b]	Lombardy (Brescia, Cremone, Lecco, Lodi, Mantova, Milan)
40–60	31,341		
	20,800	Tardoppio, Agogna, Ticino, Adda, Oglio, Lake Iseo, Mincio	Lombardy (Brescia, Cremone, Lecco, Lodi, Mantova, Milan)
	8,041	Po, Scolo Caresolo, Adige, Bacchiglione, Brenta, Sile, Piave. Livenza, Lemene	Veneto (Padova, Rovigo, Treviso, Venice, Verona)
	3,000	Po	Emilia-Romagna (Piacenza)
60–90	13,380		
	9,680	Tardoppio, Agogna, Ticino, Adda, Oglio, Mincio	Lombardy (Brescia, Cremone, Lecco, Lodi, Mantova, Milan)
	3,700	Adige, Piave	Veneto (Padova, Rovigo, Treviso, Venice, Verona)
	438,633		

[a]Of which 16,850 from LIFE COBICE project.

[b]12,390 with PIT tags (rivers Ticino, Adda, Oglio, Mincio, Adige, Sile, Piave, Livenza, Po, Brenta, Bacchiglione).

The ultimate aim of the project is the cobice sturgeon's recovery and conservation—a sturgeon protected by the European Community Habitat Directive as a priority species—in its geographical distribution area that includes the main rivers of northern Italy (Po Valley -Venetian District).

The chief objective is the natural increase of the species to restore a live population that is able to sustain itself.

The activities of the various partners, whether before or during the project, was related to the characterization of each particular area. In fact, every area shows different peculiarities, for example: availability of protected areas suitable for the reproduction or the need to create them; measures of environmental improvements; different degrees of public awareness; the presence of fishing-boat fleets useful for monitoring the situation, etc. These features will enable us to carry out effective action plans that can also be adopted by the Emilia-Romagna and Lombardy Regions and by Veneto Province, all very interested in the current project, according to their respective ichthyological regulations.

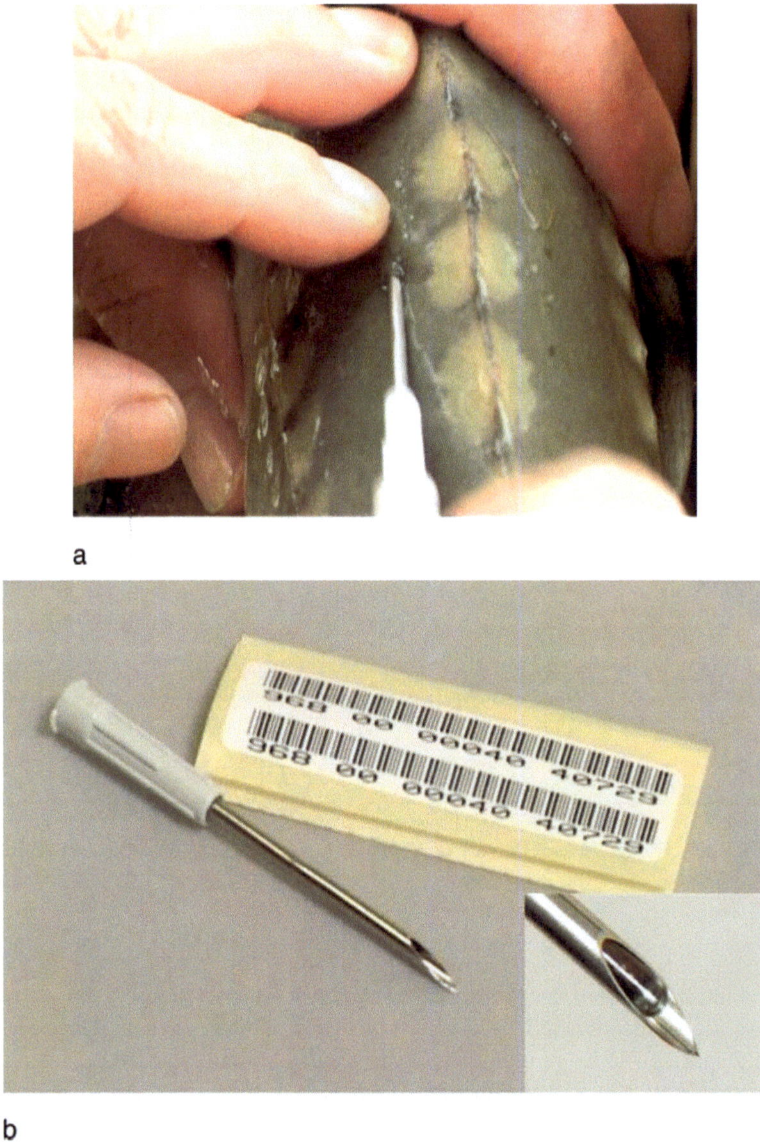

Fig. 14.2 (a, b) Graft of transponder in exemplar of *Acipenser naccarii*

14.3 Perspectives

In particular, the main actions involve:

1. Several restocking actions in various rivers
2. Wide partnership establishment

a

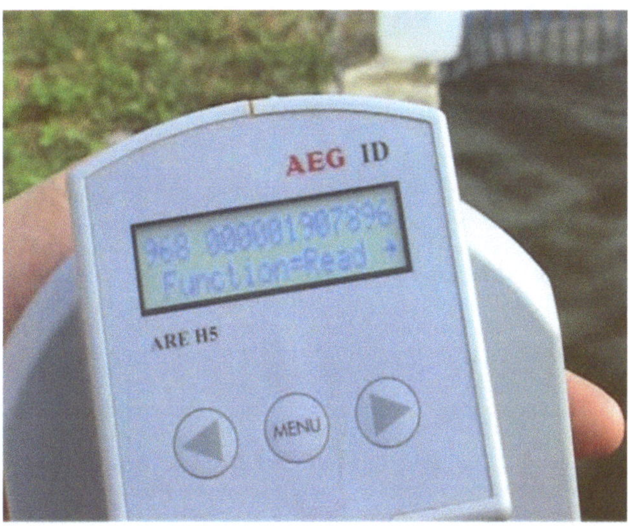

b

Fig. 14.3 (a, b) Use of transponder reader

3. Genetic characterization of broodstock (F1)
4. Parental genetic control of specimens captured during monitoring, which do not show the presence of the transponder
5. Monitoring net both in fresh and marine waters (provincial guards, volunteers, sport anglers, professional fishermen)
6. Information and outreach actions (towards the fishing industry, as well as towards schools)

14.4 Conclusion

The repopulation of the adriatic sturgeon has two main objectives:

1. Increase of natural population in the whole area
2. Guidelines for the correct management of the species

Moreover, a national study is needed to establish special conservation areas for *A. naccarii*, a strict regulation, and continue with monitoring and genetic research.

The project of the cobice sturgeon's recovery will not end with the Life Project but will continue with the commitment of all partners and of all involved provinces to help this native species, threatened with extinction, to take up again an important naturalistic, faunal and economic role.

References

Arlati G., Bronzi P., Colombo L. and Giovannini G. 1988. Induzione della riproduzione nello storione italiano (*Acipenser naccarii*) allevato in cattività. *Riv. Ital. Acquac.* 23: 94–96.

Arlati G., Grassi A. and Granata A. 1999. Restocking *Acipenser naccarii* in the Lombardy region. *J. Appl. Ichthol.* 15: 298 (abstract).

Arlati G., Poliakova L. and Granata A. 2003. Restocking of the autochthonous sturgeon cobice (*Acipenser naccarii*) in the Lombardy Region. In: *Proceedings of 4th International Symposium on Sturgeon*, Oshkosh, WI, USA.

Bronzi P., Arlati G., Cataudella S. and e Rossi R. 1994. Sturgeon presence and distribution in Italy. Presented at The International Conference on Sturgeon Biodiversity and Conservation, July 28–30, New York.

Congiu E., Dupanloup I., Patarnello T., Fontana F., Rossi R., Arlati G. and Zane L. 2001. Identification of interspecific hybrids by amplified fragment length polymorphism: the case of sturgeon. *Mol. Ecol.* 10: 2355–2359.

D'Ancona U. 1924. Contributo alla biologia degli storioni nelle acque italiane. Ministero dell'economia nazionale. Direzione generale dell'agricoltura, Divisione V. Osservatorio di pesca di Fiumicino. Roma, Libreria dello Stato, 58 pp.

Fontana F., Rossi R., Lanfredi M., Arlati G. and Bronzi P. 1997. Cytogenetic characterization of cell lines from three sturgeon species. *Caryologia* 50 (1): 91–95.

Fontana F., Tagliavini J. and Congiu L. 2001. Sturgeon genetics and cytogenetics: recent advancements and perspectives. *Genetica* 111: 359–373.

Lanfredi M., Rossi R., Arlati G., Bronzi P. and Fontana F. 1999. Cytogenetic characterization of cell lines from six sturgeon species. *J. Appl. Ichthyol.* 15: 282–283.

Paccagnella B. 1948. Osservazioni sulla biologia degli storioni del bacino padano. *Archivio Oceanogr. Limnol. di Venezia* 5 (1/3): 141–154.

Poliakova L.A., Arlati G. and Giovannini J. 2003. Use of the synthetic hormone LH-RH [D-ala6] for reproductive control of wild-caught and captivity-raised sturgeons. In: *Proceedings of 4th International Symposium on Sturgeon*, Oshkosh, WI, USA.

Rossi R., Grandi G., Trisolini R., Franzoi P., Desfuli B.S. and Veccietti E. 1991. Osservazioni sulla biologica e la pesca dello storione cobice, (*Acipenser naccarii* Bonaparte) nella parte terminale del fiume Po. *Atti Soc. Ital. Sci. Nat.* 132: 121–142.

Tortonese E. 1989. *Acipenser naccarii* Bonaparte, 1836. In: *The Freshwater Fishes of Europe. 1/II. General Introduction to Fishes. Acipenseriformes*, Holcik J. (ed.), Aula-Verlag, Wiesbaden, pp. 285–293.

Zane I., Patarnello T., Ludwig A. and Fontana F. 2002. Isolation and characterization of microsatellites in the Adriatic sturgeon (*Acipenser naccarii*). *Mol. Ecol. Notes* 2: 586–588.

Chapter 15
Acipenser sturio Recovery Research Actions in France

P. Williot, E. Rochard, T. Rouault, and F. Kirschbaum

Abstract The European Atlantic sturgeon *Acipenser sturio*, formerly present throughout Europe, is currently represented by a very critically endangered population in the Gironde-Garonne-Dordogne basin in France. In spite of its protected status in France since 1982, the relict population has continued to decline. Better knowledge of its biology and ecology has been acquired with regard to potential spawning grounds, migrations both downstream into the estuary and on the continental shelf, strength of the last representative cohorts in the estuary, the inhabited area and feeding habits in the Gironde estuary, adaptation of stocked fish to the wild, and characteristics of wild brood fish (frequency, date, sex, reproductive status). Due to the dramatic decline in brood fish by-catch, only four artificial reproductions were successful between 1981 and 2006, the last taking place in 1995. This was the first opportunity to grow larvae and to mark fingerlings for stocking. From the early 1990s, brood-stock building appeared to be the only way to potentially produce fingerlings on a regular basis. As there were no references in either husbandry or related ecophysiology standards, a precautionary strategy was adopted. The most critical environmental factors are salinity, temperature, and light regime. Some encouraging results (semen with motile spermatozoa) were recorded from reconditioned adult males held in brackish water. Some females initiated only partial ovarian development and therefore we introduced some changes, especially in the temperature regime. Complementary investigation and data have been forthcoming through European cooperation on husbandry, feeding, genetics, and endocrinology and this will be developed. Further actions are briefly presented.

Keywords *Acipenser sturio*, endangered species, relict population, recovery plan, ex situ measures, gonad maturation

P. Williot, E. Rochard, and T. Rouault
Cemagref, Estuarine Ecosystems and Diadromous Fish, Cestas cedex, France
[1]4 rue du Pas de Madame, 33890 Audenge, France

F. Kirschbaum
Leibniz-Institute of Freshwater Ecology and Inland Fisheries, Department of Biology and Ecology of Fishes, Müggelseedamm, Berlin, Germany

R. Carmona et al. (eds.) *Biology, Conservation and Sustainable Development of Sturgeons*,

15.1 Introduction

The European Atlantic sturgeon, *Acipenser sturio* Lacépède 1758, formerly present in the whole of Western Europe (Magnin, 1962; Holcik et al., 1989), has been significantly declining since the nineteenth century to such an extent that the species is now present in France only in the Gironde–Garonne–Dordogne basin (Fig. 15.1), with a very critically endangered population (Williot et al., 2002a,b).

The species has disappeared from all other river basins without any recovery attempts being made, with the exception of Germany, where reintroduction measures were launched in the mid 1990s (Kirschbaum et al., 2000), and more recently in the River Rhône (Mediterranean basin, France), where preliminary investigations into reintroduction projects were initiated (Brosse et al. com.pers.).

In France, scientists have been drawing attention to the risks to the population and recently to the species itself (Roule, 1922; Vibert, 1953; Rochard et al., 1990). A survey focusing on estuarine fisheries (CTGREF, 1973) has been the signal for a small group from the Cemagref Institute to initiate an *A. sturio* restoration

Fig. 15.1 Map of main French river drainages including the last river system (Gironde-Garonne-Dordogne) inhabited by the sturgeon *Acipenser sturio*

programme, as the species, because of its rarity, was no longer an economic resource for fishermen. .

By the late 1970s, a four-point recovery strategy was set up: (1) additional biological and ecological knowledge on the species was needed; (2) stocking would most probably be needed to sustain the threatened population; (3) know-how was to be acquired about another sturgeon species (the Siberian sturgeon, *Acipenser baerii*) as a surrogate to avoid any further adverse effects on the relict population (Williot et al., 2004); and (4) necessary to inform stakeholders and the public of the status of the species and the risk of it disappearing. At that time, stocking was expected to be achieved by rearing larvae obtained from spawning wild-originated brood fish, as developed by the former USSR in the Caspian and Azov Seas (Charlon and Williot, 1978). As the status of the species continued to deteriorate, it became necessary to change the preservation strategies.

The aim of this paper is to provide an overview of past and present recovery actions on the last French sturgeon population, focusing on: (1) preliminary recovery actions, (2) current status of the species, and (3) special emphasis on ex situ measures.

15.2 Preliminary Recovery Actions

15.2.1 Relict Population

In the early 1980s, preliminary investigations on the biology, ecology and species of the the relict population (Trouvery et al., 1984) and the first 3-year tagging survey (Castelnaud and Trouvery, 1984)were initiated. . These preliminary investigations were helpful in promoting the fishing ban on the species in 1982 and later the protection of a well-known spawning ground in the river Garonne in 1985. Available potential spawning grounds were further mapped and characterized (Jego et al., 2002). Some methods had to be updated, e.g. age determination (Rochard and Jatteau, 1991). Over time, better knowledge of the ecology of the species was acquired: growth (more rapid than that reported by Magnin (1962)), estuarine and marine migrations, and strength (a few hundreds to a few thousands) of most recent cohorts present in the estuary in summer from 1984 to 1988 (Castelnaud et al., 1991). Moreover, due to the absence of new recruitment, the age of the fish had increased and the number of cohorts present in the lower estuary had declined (Castelnaud et al., 1991; Williot et al., 1997). Marine distribution was shown to be restricted to the continental shelf and fish were mainly found (71%) between −10 and −40 m (Rochard et al., 1997a). Feeding of juveniles in the estuary was shown to be essentially composed of small polychaetes (Brosse et al., 2000). Downstream juvenile migration (Rochard et al., 2001), as well as juvenile habitats in the estuary (Taverny et al., 2002) were documented. This allowed us to attract the attention of policy-makers on the risks of gravel extraction in the estuary (Lepage et al., 2000).

15.2.2 Wild Brood Fish

Sustaining a population by stocking requires mastery in spawning and fingerling production.

The basic procedure for artificial spawning of the model species was set up in 1981 (Williot and Rouault, 1982) and has been considerably improved since (Williot et al., 1991; Williot, 1997). Very soon after the initial success in 1981, we asked fishermen to contact the Cemagref in case of incidental catches of *A. sturio* brood fish migrating upstream in the Rivers Garonne and Dordogne during the lamprey and shad fishing season (Williot et al., 1997). Records of by-catches are reported in Fig. 15.2. The number of specimens was extremely low up to the late 1980s and later was only anecdotic. Moreover, the two sexes were rarely caught simultaneously and brood fish were not often of a suitable physiological status (over- or under-mature) for artificial reproduction.

Therefore only four artificial reproductions were performed between 1981 and 2006 (1981 (1), 1985 (2) and 1995 (1)) (Williot et al., 2002a, 2007). In addition, there were indications after 1985 that the reproductive potential of the males was deteriorating as some produced non-motile spermatozoa in the presence of water (Williot et al., 2002a). From the last spawning, some thousands of larvae were obtained (Williot et al., 2000) which allowed us to perform the first successful larval rearing with nauplii of *Artemia salina* and chironomids (Williot et al., 2005). Hatchlings (2000) and later fingerlings (5000 ~1 g in June and 2000 ~6.5 g in August 1995) were released in equal numbers into the Garonne and Dordogne Rivers in 1995 (Williot et al., 2005). Both batches of fingerlings were marked. Assessment of survival in the 2 years post release (Rochard et al., 1997b), suggested that the survival rate was about 3–4% which is very promising for restocking strategy. Moreover, the growth of the released fish has proved to be similar to that of their wild counterparts (Lochet et al., 2004), thus proving how well the survivors had instead adapted.

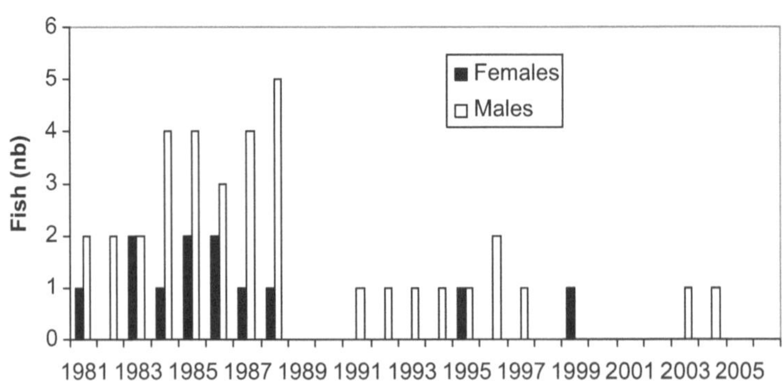

Fig. 15.2 Distribution of brood fish of *Acipenser sturio* by-catch during the experimental period (1981–2006) (updated from Williot et al., 2002a)

15.3 Current Status of the Species

The species has been totally protected since 1982 in France and 1998 in Europe. At present, the species is also protected by several conventions. It is classified in Appendix I of the Convention on Migratory Species (CMS, Bonn), which lists endangered migratory species). It is listed in annex. 1 of the Convention on International Trade in Endangered species (CITES, Washington), and as a strictly protected faunal species (Annex II) of the Convention on the Conservation of European Wildlife and Natural Habitats (Bern convention). It is also protected by the Convention for the Protection of the Marine Environment of the North-East Atlantic (OSPAR convention). In European Community Law, especially the Habitat Directive, the species is listed among the animals of Community interest (Annex II) whose conservation requires the designation of Special Areas of Conservation (SAC). Finally, in 2005, the species became, on paper, one of the priorities of French policy for biodiversity conservation. The establishment of a European Action Plan for the preservation of the species within the Bern convention is in progress.

Despite its protection status in France since 1982, the species has continued to be threatened (Rochard et al., 1990) and the last spawns in the wild were recorded in 1984 and 1988. Poaching from the seashore has continued (Lepage and Rochard, 1997) and recent landings in the UK, the Netherlands and France in 2004 were quasi-auctioned, illustrating the ineffectiveness of the protection status (Guth and Laurent, 2004).

In the early 1990s, due to the deterioration in the status of the species, it was decided, as a precaution, to build up an ex situ brood stock (Williot et al., 1997; Williot et al. 2007).

15.4 Ex Situ Measures

15.4.1 Main Lines

This ex situ measure was initiated when Cemagref started to operate its own new facility.

One of the major difficulties was that the conditions that should be provided for the fish to enable them to develop a normal gametogenesis were unknown. Moreover, there were no physiological data to assess the health status of the fish in relation to environmental rearing conditions. We therefore had to face the challenge of investigating the appropriate rearing conditions by carrying out research into husbandry and different fields of biology while providing the few farmed specimens with presumably the best possible conditions. The following strategy has thus been set up, drawing on the documented experience gained from other sturgeon species, the known ecology of the species, a step-by-step procedure, non-invasive practices, and conservative decisions, as much as possible.

Among the most critical rearing factors are water salinity and temperature, light regime, feeding, and current management. When the brood-stock building objective was established, all rearing successes with sturgeon species were achieved in freshwater, whatever the ecology of the species (this is still the case) (Williot et al., 1997; Williot et al. 2008). The only way to finally judge the efficiency of the entire management scheme was to perform artificial reproduction, and the corresponding data had to be analysed both quantitatively and qualitatively. Therefore, surveys on sexual maturation have to be carried out on a regular basis.

15.4.2 Installing a Future Brood Stock

All data regarding the origin and size of the fish have been considered, to increase the rate of rearing success and prevent maturation delay (Williot et al., 2007).

With regard to the primary environmental conditions, we initially postulated the possibility of rearing the species in freshwater (as mentioned above) and therefore tested the immediate acclimation to freshwater. Our first attempts to adapt wild-originated juveniles caught in brackish water (low estuary) to freshwater were unsuccessful except when fish were ≥105 cm in length (Williot et al., 1997). Survival, delay in resuming food intake, and therefore growth, were considerably improved by reducing all kinds of fishing stress and acclimation to brackish water (Williot et al., 2007). Thereafter, as a precaution, large juveniles and adults were held in brackish water throughout the year. Most of the fish lost weight for months and sometimes years before recovering their initial weight. Growth (as well as food intake) was irregularly cyclic (Williot et al., 2007). There was no statistical influence of salinity level (0 and 15 ‰) on growth in young wild-originated fish caught in the upper or mid estuary during a 5.5 year experimental period (from ~1 year-old fish to ~6.5 year-old fish). A new salinity-growth replicated comparison (with a new seawater batch, as fish spend most of their adult phase in ocean waters) was set up some months later using all available juveniles (Fig. 15.3). With the exception of a replicate in a freshwater batch (due to poor rearing conditions of this subsample in the first year of life), the growth curves, which fit well in linear regression, do not differ significantly within the 3-year experimental period (2001–2004) whatever the salinity level. It is worth noting that the growth trend of the two batches in seawater was in the lower range, as illustrated by the lowest slopes (Fig. 15.3).

Up to late summer 2006, the summer water temperature in the holding tanks could reach 25°C or even higher, depending on air temperature, as demonstrated by the brackish water tank (Fig. 15.4). It bears noting that growth (weight curve) resumed when the water temperature decreased (Fig. 15.4).

Consumed food and weight were plotted for the same tank in brackish water (Fig. 15.5). The experimental period was divided into five phases. Phases *c* and *e* were similar, as growth either remained stable and consumed food displayed sudden

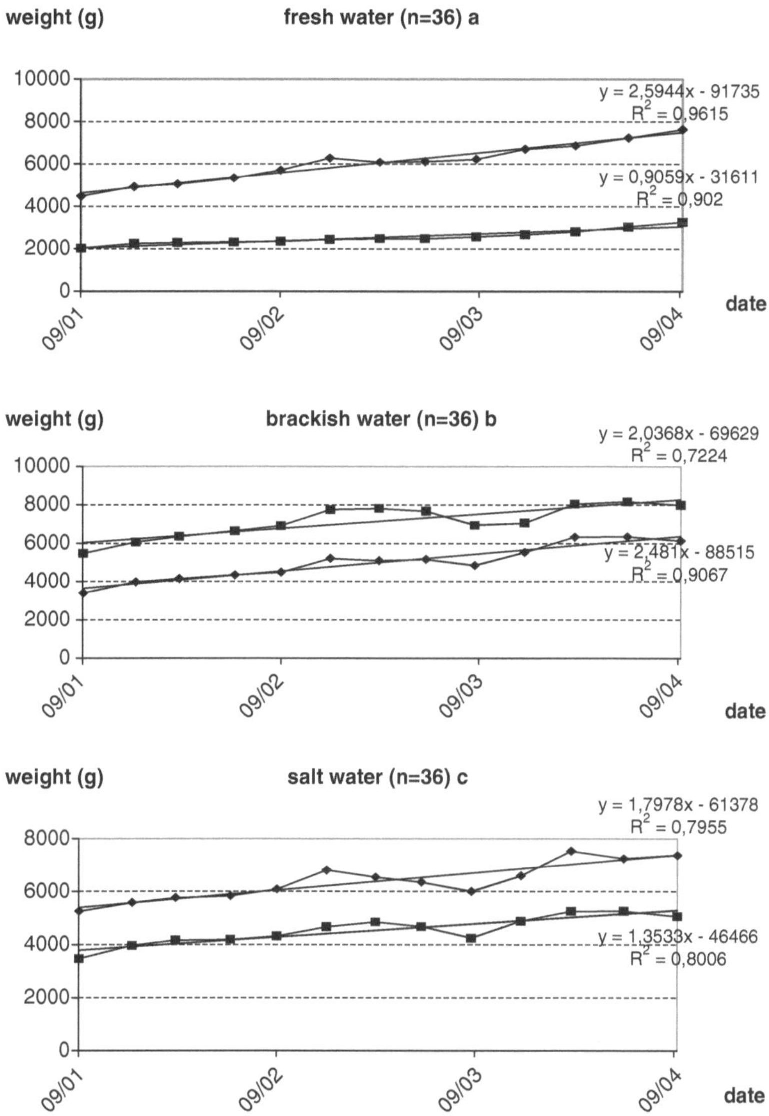

Fig. 15.3 Growth of juvenile *Acipenser sturio* according to salinity and holding tanks. Weights represent median values (g)

changes around a fairly stable trend. Phases *b* and *d* were high growth periods, with a corresponding increase in consumed food.

They were observed in autumn and winter during 2002–2003, i.e. in the period when the temperature was rather low. Phase *a* corresponded to moderate growth while consumed food decreased slowly. These observations suggested different types of metabolic activity in the species.

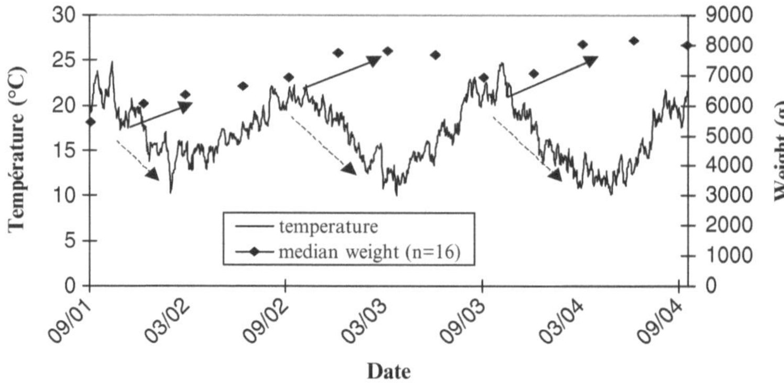

Fig. 15.4 Growth of juvenile *Acipenser sturio* and corresponding water temperature in tank 'b' in brackish water. Weights are median values (g) and temperatures (°C) are daily values. Solid arrows underline increasing growth and dotted arrows signal corresponding decreasing water temperature

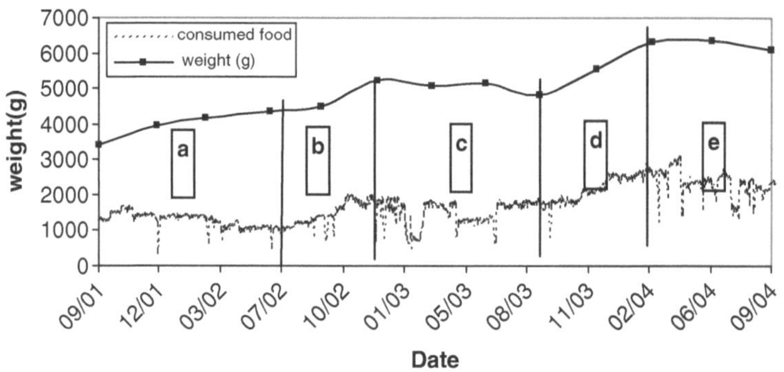

Fig. 15.5 Growth of juvenile *Acipenser sturio* in brackish water tank 4 and consumed food. Weights are the median (g). Consumed food is the daily difference between quantities of food delivered and recovered (g). Curves are divided into five successive phases (a–e) exhibiting fairly homogenous growth trends

These two experiments concerning the effects of salinity on growth ended when one of the batch died due to human error (Table 15.1). The fish from the last massive loss (55 specimens were full siblings) exhibited a balanced sex ratio.

Thanks to the database, it has been possible to reconstruct the growth of the dead fish from 1996 onwards (Fig. 15.6). Growth was similar regardless of sex, apparently indicating that intense ovary growth (or vitellogenic phase) did not start for those ~7-year-old fish under the experimental conditions.

The remaining stock is shown in Table 15.2.

As shown in Fig. 15.4, winter temperature fell to ~10°C. This was achieved by opening the gate of the building where the tanks were located to take advantage of the low winter temperatures. Most sturgeon species need this yearly cycle to

Table 15.1 Loss of *Acipenser sturio* during the period 1993–2006

Origin	Birth year	Number (period)	Causes (number)
Wild	1984–1988	6 (1997–2001)	Toxicity (3)
			Unknown (3)
	1994	3 (late spring)	Unknown
		11 (summer 2001)	Human error and/or insecure water system
Artificial breeding	1995	55 (late 2004)	Toxicity due to non-watertight closed gate
		3 (mid 2006)	Unknown
Total		78	

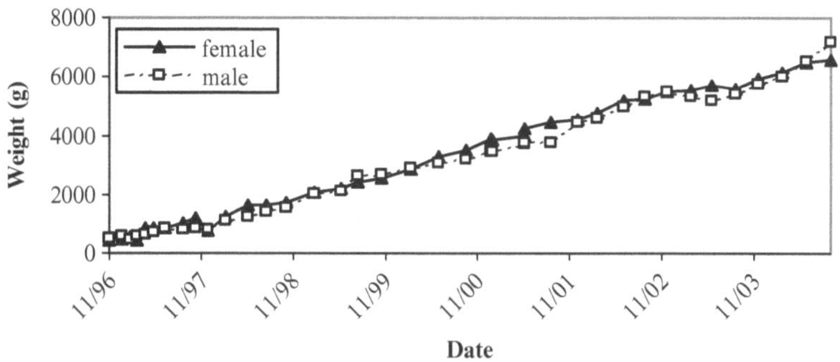

Fig. 15.6 Growth of juvenile male and female *Acipenser sturio* produced in the hatchery in 1995 which died accidentally in November 2004. Dead fish numbered 55, i.e. the present freshwater fish-group (n = 36) and 19 specimens held in three separate freshwater tanks. The database allowed us to reconstruct their growth from late 1996 onwards

Table 15.2 Cultured specimen of *Acipenser sturio* in France (December 2006)

Origin	Birth year	Number (sex)	Weight (min–max) kg
Wild	1984–1989	7 (6 ♂), 1 ♀	10.4–22.8
	1994	24 (8 ♂, 12 ♀,4 ?)	3.9–18.7
	?	10 (5 ♂, 5 ♀)	12–33.3
Artificial breeding	1995	40 (?)	5.7–13.8
Total		81	

encourage good maturation. The normal photoperiod was applied. The intensity of the light remains to be clearly determined as sturgeons prefer shadowy environments rather than exposure to direct light. Large fish were held in tanks that were either 1 or 2 m deep, and 4 m in diameter; the deeper tanks allowed better growth in the 2–3 months post transfer period (Williot et al., 2007).

With regard to food, sturgeons fed ad libitum (judged by refused food) on frozen shrimp, and by preferance on the common prawn *Palaemon longirostris* rather than the common shrimp *Crangon crangon*. Though costly and risky (from a health standpoint, as proved by some mortalities where shrimp food was involved), these were the only accepted food items over a long period of time. A small batch of larvae born in 1995 were weaned onto artificial compound diet with moderate survival success (~30%). They continued to exhibit fairly good growth (currently fed the artificial diet) up to 1.5 years, an age at which some fish were bending and finally died some months later. Total loss was higher than that observed in fish fed natural food items. The causes remain mostly unknown as some vitamin deficiencies and oxidized lipids provoked no or some abnormalities in 2 month-old Siberian sturgeons (Fontagné et al., 2006).

Current management includes daily care for fish and tanks (individual closed water systems), i.e. feeding, cleaning, checking of water temperature, oxygen, and pH; monthly checking of the nitrogen cycle, and a 3-month control (appearance, tagging, weight, sexual maturity advancement, and blood sampling for endocrine measurements).

In the mid 1990s, a contract was established with Germany, to focus on brood-stock building, and 40 juveniles born in our hatchery in 1995 were sent to Berlin (Leibniz-Institute of Freshwater Ecology and Inland Fisheries, IGB) in 1996. The activity pattern of young juveniles in the context of different water temperatures was investigated. There are indications of a preference for the range 12.6–18.5°C and a greater activity at night (Staaks et al., 1999). The study rapidly extended to other fields such as feeding and growth (Kirschbaum et al., 2000; Hensel et al., 2002; Kirschbaum et al., 2006), genetics (Ludwig et al., 2004; Tiedemann et al., 2006), and the whole restoration programme (Kirschbaum et al., 2004; Williot et al., 2007; Kirschbaum et al. chapter 24). As the best management technique has still not been determined, the French and German brood stocks are reared in slightly different ways, e.g. with sampling frequency fortnightly in Germany and quarterly in France.

15.4.3 Brood Stock Functioning

As the aim is to build a functional brood stock, the setting up of a long-term management scheme is needed—that is, it is necessary to define the means and methods for dealing with fish management in order to have a self-sustainable brood stock. The first step is to get the fish to spawn and to acquire maximum knowledge of their reproductive physiology, especially endocrinology.

The oldest available fish are mainly male (Table 15.3).

They were reconditioned and exhibited cyclical maturation according to the pattern shown in Fig. 15.7, with a peak in two successive years, each with very different status at the individual level. However, the present findings suggest limited sexual activity for males and a great inter-individual variation, substitute (one male matured 7 times out of the 8 year-period, one male matured 4 times, two males matured 3 times, one male matured 2 times, and one male matured only once).

As mentioned above, these fish were held for the entire year in brackish water, so it has been necessary to develop a procedure to manage the mature fish in an appropriate way to allow them to provide good-quality gametes. As judged by the

Table 15.3 Maturation in female *Acipenser sturio*

Experimental year	Name (birth year)	Preliminary observations spring time	Observations posthormonal treatment
2002	DN (1988)	Abdomen not swollen Difficulties in sampling ovf[a] Unusual drawings at oolema level Irregular Gv[b] migration GVBD[c] = 0%	Priming GVBD = 79% Whole injection Poor ovulation No fertilization
2005	DN (1988)	Abdomen not swollen Difficulties in sampling ovf Unusual drawings at oolema level Irregular Gv[b] migration GVBD = 28%	Priming GBVD = 0%
	Karine (1994)	Abdomen not swollen Difficulties in sampling ovf Unusual drawings at oolema level Irregular Gv[b] migration GVBD = 30%	Priming GVBD = 84% No ovulation
	Alice (1994)	Abdomen not swollen Difficulties in sampling ovf Unusual drawings at oolema level Irregular Gv[b] migration GVBD = 5%	Priming GVBD = 10%
2006	Karine (1994)	Abdomen not swollen GVBD = 10%	Priming GVBD = 97% Whole injection No ovulation
	Alice (1994)	Abdomen not swollen GVBD = 56%	Priming GVBD = 97% Whole injection No ovulation
	Lucie (1994)	Abdomen not swollen GVBD = 10%	Priming GVBD = 100% Whole injection No ovulation

[a]ovf, ovarian follicle.
[b]Gv, germinal vesicle or nucleus.
[c]GVBD, germinal vesicle break down.

quantity of semen and more importantly by the density and potential activity of spermatozoa (motility in the presence or absence of water), a 1-week holding in freshwater (simulation of an upstream migration in freshwater) followed by current hormonal injection (either carp pituitary powder or analogues of GnRH) have been producing good-quality semen with 50% success (Williot et al., 2007).

With regard to the females, the synthesis of recorded sexual maturation (at least partial) is shown in Table 15.3. In spite of very few available fish, some females initiated sexual maturation. As for males, the methods used are those developed on Siberian sturgeon as a model species (Williot, 2002). It is noteworthy that the oldest fish displayed two maturations at 3-year intervals and the fish born in the wild in

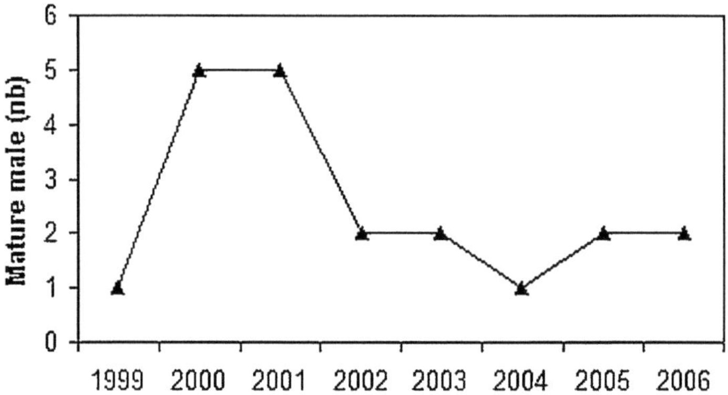

Fig. 15.7 Yearly mature male *Acipenser sturio* among the oldest fish born in the wild in 1984 and 1988 (see Table 15.2). Until 2001, seven fish were alive and six later on

1994 exhibited signs of maturation from their 11th year onwards. In all these cases, fish did not display a swollen abdomen, which indicates that the ovaries were not well developed and therefore only a very limited number of ovarian follicles had developed. Moreover, there were signs of disturbed growth of these cells, especially in 2005. It was even possible to obtain a fairly high germinal vesicle break down (GVBD) rate (in vitro maturation-competence test of ovarian follicles) by priming hormonal injection in some fish; however, no ovulation (or very poor ovulation) was recorded after the complete hormonal injection. The most likely interpretation is that one (or many) environmental factor(s) has a negative impact on the normal gametogenesis of the fish.[1]

The second step required for brood-stock functioning is related to genetics. This is of great importance in the case of a threatened species such as *A. sturio*, where the number of available specimens is very limited (Table 15.2). A preliminary survey 'analysed the genetic structure of Gironde sturgeon population including both genetic variability in the entire population and the degree of relatedness between single individuals' (Ludwig et al., 2004). Six non-specific microsatellites and a fragment of mtDNA were used. All specimens shared the same mitochondrial haplotype, and the allelic richness was low compared with other sturgeon species. It is likely that the fish born in the wild in 1994 (Table 15.2) were produced by a single mating pair. The allelic distribution provides indications for the most valuable mating with regard to genetic variability (Ludwig et al., 2004). New approaches with sturio-specific microsatellites and nuclear markers are in progress (L. Congiu and R. Tiedemann, respectively, personal communication).

The third group of actions that had to be taken into account to ensure brood-stock functioning covers various fields. Together with the establishment of the brood stock, an endocrine survey of reproductive steroid hormones (testosterone,

[1] By the late June 2007, some thousands of larvae hatched. Three fish (one female (1994) and two males (1984 and 1994)) held in brackish water except for the last week, produced gametes.

11-ketotestosterone and estradiol) was performed (Davail, 2006). This kind of investigation should now be generalized. The high levels of estradiol enabled us to find the two young maturing females in 2005 (Table 15.3). When working with endangered species, sperm preservation should be set up on a regular basis. A preliminary attempt was carried out as soon as it proved possible (Kopeika et al., 2000), with some success. New, easier methods are now available (Horvath, personal communication) and should be used to perform sperm cryopreservation each time semen is produced. Some extreme safeguarding solutions, such as androgenesis and tissue conservation, have also been explored, so far without success.

15.5 Conclusions and Perspectives

Ex situ brood-stock building is the only chance to save this species, which is teetering on the verge of extinction.

15.5.1 Further Investigations

The aim of the ex situ measures is to define the best husbandry management strategies on a long-term basis (including the production of identifiable fingerlings for stocking); the main fields of investigation are therefore physiology (reproduction, exchange-regulation, health status), feeding, genetics, behaviour, and brood-stock management (Rochard and Williot, 2006). This document also includes the investigations that need to be performed as soon as stocking is effective. A key point is cooperation, including at the international level (there is at least European cooperation at present) to multiply the approaches, increase the efficiency, benefit from available skills, and finally enhance the potential recovery of the species in a complete programme. The German batch held in freshwater (Kirschbaum et al. chapter 24) and used to replace the French counterparts lost from late 2004 onwards illustrates this point (see Section 15.2).

In the short-term, maintaining the fish and providing them with the best possible conditions are the priorities. There are some encouraging signs, such as the reconditioning of some adult males, some of which have been able to provide semen. Additionally, females have initiated incomplete maturation; it has been hypothesized that potential causes might be the high summer temperatures. A thermoregulation system was set up in late summer 2006, the effects of which will probably be evident from 2008 onwards because of non-yearly oogenesis. Current light regime and food deprivation prior to spawning should also be considered as important parameters.

In addition to the many uncertainties with regard to the best rearing conditions, it has been shown that the definition of such conditions is complicated by the great inter-individual variation in sturgeons (growth, feed intake, maturation, etc.).

15.5.2 Lobbying Actions

Both the Research Institutes, Cemagref[2] and IGB,[3] have been participating in various lobbying actions led by WWF, France. The objectives are to draw the attention of both the French and European authorities to the dramatic situation of the species, despite its legal protection, and to help to save the species from extinction. The targeted bodies were the European Council, the Environmental and Fisheries department of the EU, some European Conventions (see protection status section), and the French Ministry for Ecology. The following have contributed to these lobbying actions: the regional public body EPIDOR,[4] both the French and German scientific institutes (Cemagref and IGB), the Society to Save the sturgeon (D), the Society in Favour of Sturio (F), the World Society for the Conservation of Sturgeons (established in D), the Society for Diadromous Fish in the River Rhône and adjacent Mediterranean coast (F).

At present, WWF (2005) France, has taken charge of setting up a European Action Plan for the recovery of the species within the Bern Convention. The draft, drawn up mainly by French and German parties, will be proposed very soon to other European partners.

15.5.3 Recovery Projects in Europe

At present, most French and German research actions are concerted as much as possible, as illustrated by the common studies (e.g. Kirschbaum et al., 2004; 2006; chapter 24; Williot et al. 2007). Concrete recovery projects are currently in progress in France and Germany on the Atlantic cost and in the North Sea, respectively. Due to the former distribution of the species, some other European countries are interested in our findings (Spain, Poland, Italy, the Netherlands, Belgium, UK and some Central European countries), and some have already asked for biological material to initiate re-introduction programmes (Poland, Italy). Part of the second Life contract was devoted to preparing a questionnaire for other European Environmental agencies to analyse the state of the art in terms of possibilities for recovery plans (Rochard, 2002).

Acknowledgements The authors are grateful to Marcel Pelard, Daniel Mercier and Louis Jacobs for taking care of the fish on a daily basis, to Mario Lepage and Charlie Roqueplo for their help in capturing the fish held in 2-m deep tanks, to Vincent Bouju for securing the database and running some analyses, and to Paul Gonthier for his constant support. Investigations and investments were partly funded by two life contracts B4-3200/94/754 and B-3200/98/460.

[2]Cemagref = Agricultural and Environmental Engineering Research Institute

[3]IGB = Leibniz Institute of Freshwater Ecology and Inland Fisheries

[4]EPIDOR = Etablissement Public Interdépartemental Dordogne

References

Brosse L, Rochard E, Dumont P, and Lepage M. 2000. Premiers résultats sur l'alimentation de l'esturgeon européen, *Acipenser sturio* Linnaeus, 1758 dans l'estuaire de la Gironde et comparaison avec la macrofaune estuarienne présente. *Cybium* 24:49–61.

Castelnaud G and Trouvery M. 1984. Premiers resultants de trois années de marquage de l'esturgeon *Acipenser sturio* dans le bas estuaire de la Gironde. Etude n° 18, Série esturgeon n°2. CEMAGREF de Bordeaux, Division A.L.A/Agedra, 25 pp.

Castelnaud G, Rochard E, Jatteau P, and Lepage M. 1991. Données actuelles sur la biologie d'*Acipenser sturio* dans l'estuaire de la Gironde. In: Williot P, editor. *Acipenser*. Cemagref Publication, Antony, pp. 251–275.

CTGREF, 1973. Rapport sur la Pêche en Gironde. *Ctgref Bordeaux, Aménagements Littoraux et Aquaculture*, 24 pp.

Charlon N and Williot P. 1978. Elevage d'esturgeons de repeuplement et de consommation en URSS. *Bull Centr Etud Rech Sci Biarritz* 12 (1):7–156.

Davail B. 2006. Contrôle de la reproduction des esturgeons: suivi hormonal et développements d'outils moléculaires. *Rapport scientifique FEDER, convention 2003–235, LPTC/UMR-5472 CNRS*, 32 pp.

Fontagné S, Bazin D, Brèque J, Vachot C, Bernade C, Rouault T, and Bergot P. 2006. Effects of dietary oxidized lipid and vitamin A on the early development and antioxidant status of Siberian sturgeon (*Acipenser baeri*) larvae. *Aquaculture* 257:400–411.

Hensel E, Kirschbaum F, Williot P, Wirth M, and Gessner J. 2002. Restoration of the European sturgeon, *Acipenser sturio* L., 1758 in Germany: effect of different feed items on specific growth rates of large juvenile fish. *Int Rev Hydrobiol* 87:539–551.

Holcik J, Kinzelbach R, Sokolov LI, and Vassilev VP. 1989. *Acipenser sturio* Linnaeus, 1758. In: Holcik J, editor. *The Freshwater Fishes of Europe*. Aula Verlag, Wiesbaden, pp. 367–394.

Guth MO and Laurent JL. 2004. Retour d'expérience sur la capture et la vente illicite d'un esturgeon en criée aux Sables d'Olonne (Vendée). *Rapport de l'inspection générale de l'environnement No. IGE/04/034. Paris:* Ministère de l'Ecologie et du Développement Durable, 30 pp.

Jego S, Gazeau C, Jatteau P, Elie P, and Rochard E. 2002. Spawning grounds available for the European sturgeon *Acipenser sturio* L. 1758 in the Garonne-Dordogne basin. Methods used, present status and prospects. *Bull Fr Pêch Piscic* 365/366:487–505.

Kirschbaum F, Gessner J, and Williot P. 2000. Restoration of *Acipenser sturio* L., 1758 in Germany: growth characteristics of juvenile fish reared under experimental indoor conditions. *Bol Inst Esp Oceanogr* 16:157–165.

Kirschbaum F, Ludwig A, Hensel E, Würtz S, Kloas W, Williot P, Tiedemann R, Arndt GM, Anders E, Norheim Hv, and Gessner J. 2004. Status of the project on protection and restoration of Atlantic sturgeon in Germany: background, current situation, and perspectives. In: Gessner J and Ritterhoff J, editors. Species differentiation and population identification in the sturgeons Acipenser sturio L. and Acipenser oxyrhinchus. *Bundes Naturschutz* 101:36–53.

Kirschbaum F, Wuertz S, Williot P, Tiedemann R, Arndt GM, Bartel R, and Gessner J. 2006. Prerequisites for the restoration of Atlantic sturgeons, *Acipenser sturio* and *A. oxyrinchus,* in Germany – Report on the twelve-year preparatory period. Voraussetzungen für die Wiedereinbürgerung der Atlantischen Störe, *Acipenser sturio* und *A. oxyrinchus,* in Deutschland – Bericht über die 12-jährige Vorbereitungsphase. *Verhandl Ges Ichthyol* 5:79–93.

Kopeika E, Williot P, and Goncharov B. 2000. Cryoconservation of Atlantic sturgeon *Acipenser sturio* L., 1758: first results and associated problems. *Bol Inst Esp Oceanogr* 16:167–173.

Lepage M and Rochard E. 1997. Estimation des captures accidentelles d'*Acipenser sturio* réalisées en mer. In: Elie P, coordinator. *Restauration de l'esturgeon européen Acipenser sturio. Contrat Life rapport final du programme d'exécution.* Cemagref Bordeaux étude n°24, pp 377–381.

Lepage M, Rochard E, and Castelnaud G. 2000. Atlantic sturgeon *Acipenser sturio* L., 1758 restoration and gravel extraction in the Gironde estuary. *Bol Inst Esp Oceanogr* 16:175–179.

Lochet A, Lambert P, Lepage M, and Rochard E. 2004. Growth comparison between wild and hatchery-reared juvenile European sturgeons *Acipenser sturio* (Acipenseridae) during their stay in the Gironde estuary (France). *Cybium* 28:91–98.

Ludwig A, Williot P, Kirschbaum F, and Lieckfeld D. 2004. Genetic variability of the Gironde population of *Acipenser sturio*. In: Gessner J & Ritterhoff J, editors. Species differentiation and population identification in the sturgeons Acipenser sturio L. and Acipenser oxyrhinchus. *Bundes Naturschutz* 101: 54–72.

Magnin E. 1962. Recherches sur la systématique et la biologie des Acipenséridés. *Ann Stat Centr Hydrobiol Appl* 9:7–242.

Rochard E, coordinator, 2002. Restauration de l'esturgeon européen *Acipenser sturio*, *Rapport scientifique Contrat LIFE n° B - 3200/98/460. Etude Cemagref n°80*, Groupement de Bordeaux, 224 pp.

Rochard E and Jatteau J. 1991. Amélioration de la méthode de détermination de l'âge de l'esturgeon commun *Acipenser sturio* et premières applications. In: Williot P, editor. *Acipenser*. Cemagref Publication, Antony, pp. 193–208.

Rochard E and Williot P, coordinators, 2006. Actions de recherches proposées pour contribuer au plan international de restauration de l'esturgeon européen *Acipenser sturio. Etude Cemagref groupement de Bordeaux n° 103*, 51 pp.

Rochard E, Castelnaud G, and Lepage M. 1990. Sturgeon (Pisces: Acipenseridae); threats and prospects. *J Fish Biol* 37:123–132.

Rochard E, Lepage M, and Meauzé L. 1997a. Identification and characterisation of the marine distribution of the European sturgeon *Acipenser sturio. Aquat Living Resour* 10:101–109.

Rochard E, Lepage M, Gazeau C, and Lambert P, 1997b. Tableau de bord de la population. Estimation de l'abondance des différentes classes d'âge. In : Elie P, coordinator. *Restauration de l'esturgeon européen Acipenser sturio. Rapport final d'exécution Contrat Life N° B4–3200/94/754, Etude Cemagref n° 24*. p 349–374.

Rochard E, Lepage M, Dumont P, Tremblay S, and Gazeau C, 2001. Downstream migration of juvenile European sturgeon *Acipenser sturio* L. in the Gironde estuary. *Estuaries* 24: 08–115.

Roule L. 1922. Etude sur l'esturgeon du golfe de Gascogne et du basin girondin. *Off Sci Tech Pêches Marit, Notes et Mémoires N°20*, 12 p.

Staaks G, Kirschbaum F, and Williot P. 1999. Experimental studies on thermal behaviour and diurnal activity rhythms of juvenile European sturgeons (*Acipenser sturio*). *J Appl Ichthyol* 15:243–247.

Taverny C, Lepage M, Piefort S, Dumont P, and Rochard E. 2002. Habitat selection by juvenile European sturgeon *Acipenser sturio* in the Gironde estuary (France). *J Appl Ichthyol* 18:536–541.

Tiedemann R, Moll K, Paulus KB, Scheer M, Williot P, Bartel R, Gessner J, and Kirschbaum F, 2006. Atlantic sturgeons (*Acipenser sturio, Acipenser oxyrinchus*): American females successful in Europe. *Naturwissenschaften* 5 p. DOI 10.1007/s0014–006–0175–1.

Trouvery M, Williot P, and Castelnaud G. 1984. Biologie et Ecologie d'*Acipenser sturio. Etude de la pêcherie. Cemagref, Etude n°17, « Série Esturgeon n° 1 »*, 79 p.

Vibert R. 1953. Les poissons migrateurs dans l'économie piscicole du Sud Ouest (écrit en 1944). *Bul Fr Piscic* 136 :121–135.

Williot P. 1997. Reproduction de l'esturgeon sibérien (*Acipenser baeri* Brandt) en élevage : gestion des génitrices, compétence à la maturation in vitro de follicules ovariens et caractéristiques plasmatiques durant l'induction de la ponte. *Thèse n°1822, Université Bordeaux I*, 227 p.

Williot P. 2002. Reproduction des esturgeons. In: Billard R, coordinator. *Esturgeons et caviar. Lavoisier Tech et Doc*. p: 63–90.

Williot P and Rouault T. 1982. Compte rendu d'une première reproduction en France de l'esturgeon sibérien *Acipenser baeri. Bul Fr Piscic* 286:255–261.

Williot P, Brun R, Rouault T, and Rooryck O. 1991. Management of female breeders of the Siberian sturgeon, *Acipenser baeri* Brandt: First results. In: Williot P, editor. *Acipenser*, Cemagref Publ, Antony. p 365–379.

Williot P, Rochard E, Castelnaud G, Rouault T, Brun R, Lepage M, and Elie P. 1997. Biological characteristics of European Atlantic sturgeon, *Acipenser sturio*, as the basis for a restoration program in France. *Environ Biol Fishes* 48:359–370.

Williot P, Brun R, Pelard M, and Mercier D. 2000. Induced maturation and spawning in an incidentally caught adult pair of critically endangered European sturgeon, *Acipenser sturio* L. *J Appl Ichthyol* 16:279–281.

Williot P, Rouault T, Brun R, Pelard M, and Mercier D. 2002a. Status of caught wild spawners and propagation of the endangered sturgeon *Acipenser sturio* in France: a synthesis. *Intern Rev Hydrobiol* 87:515–524.

Williot P, Arlati G, Chebanov M, Gulyas T, Kasimov R, Kirschbaum F, Patriche N, Pavlovskaya L, Poliakova L, Pourkazemi M, Kim Yu, Zhuang P, and Zholdasova IM. 2002b. Status and management of Eurasian sturgeon: an overview. *Intern Rev Hydrobiol* 87:483–506.

Williot P, Rouault T, Rochard E, Castelnaud G, Lepage M, Gonthier P, and Elie P. 2004. French attempts to protect and restore *Acipenser sturio* in the Gironde: Status and perspectives, the research point of view. In: Gessner J and Ritterhoff J, editors. *Bundes Naturschutz* 101: 83–99.

Williot P, Brun R, Rouault T, Pelard M, and Mercier D. 2005. Attempts at larval rearing of the endangered western European sturgeon, *Acipenser sturio* L. (Acipenseridae), in France. *Cybium* 29: 381–387.

Williot P, Rouault T, Pelard M, Mercier D, Lepage M, Davail-Cuisset B, Kirschbaum F, and Ludwig A. 2007. Building a broodstock of the critically endangered sturgeon *Acipenser sturio* L.: problems associated with the adaptation of wild-caught fish to hatchery conditions. *Cybium* 31: 3–11.

WWF France, 2005 (July). STURIO Project. *Towards a European Action Plan for Sturgeon Conservation*. 20 p.

Chapter 16
Conservation of the Sturgeon Fish in Lower Volga

S.A. Maltsev

Abstract The river Volga is the largest in Europe. Its length is about 3600 km. The difference in height between the source and the mouth is 256 m. The mid-annual drainage is about 250 km^3; water flow is more than 7000 m^3/s; during spring high water it is 26,000–28,000 m^3/s. The basic supply to the Volga is provided by snow (60% of annual drainage), ground (30%) and rain (10%) water. In midsummer (July) the temperature of the water in the river reaches 23–25°C.

Eight hydroelectric power stations and a water-divider have been erected on the Volga.

On the Lower Volga, fishery is traditionally well developed. About 70 species of fish, including six species of sturgeon, are found. All the sturgeon fish, except the sterlet, are diadromous. Nowadays the general annual catch reaches 40,000 tons.

It is well known that catching sturgeon fish began in this area long ago. As far back as the middle of the seventeenth century, catches of sturgeon here reached 50,000 tons a year. There were rather big catches in the nineteenth– and the beginning of the twentieth centuries. However, due to the extremely high intensity of fishing around 1930, the catch of sturgeon decreased to 13,500 tons, and by 1946to 4,900 tons. Owing to the accepted measures of preservation of the stock, the catch of these fish increased during the subsequent period, and in 1968 reached 11,600 tons. In the 1970s and 1980s the catch of sturgeon reached 20,000–25,000 tons, and black caviar production reached 2000 tons.

The short period of sufficient stocks ended in the late 1980s, when there was a reduction in the catch. Within 15 years, the catch fell ten times, and in 2004 the catch of the sturgeon by fishermen of the Russian Federation in the Volga-Caspian basin was about 400 tons, including 153 tons of sturgeon fish caught for research purposes and reproduction.

In these circumstances the preservation of sturgeon fish in Russia, and, in particular, on the Lower Volga, has become especially significant.

Keywords Volga, beluga, sturgeon, reproduction, regulation, protection

---

S.A. Maltsev
Ecology Expert of Federal Hydrogeneration Company, the Assistant Professor of Volgograd State Technical University, Cand. Sc. (Biol.). Volgograd Region, Voljsky, Russia

R. Carmona et al. (eds.) *Biology, Conservation and Sustainable Development of Sturgeons*,

16.1 Introduction

The river Volga is the largest in Europe, with a length of about 3600 km. The difference in height between the source and the mouth is 256 m. The mid-annual drainage is about 250 km^3; water flow is more than 7000 m^3/s, and, during spring high water, 26,000–28,000 m^3/s. The basic supply to the Volga is provided by snow (60% of annual drainage), ground (30%), and rain (10%) water. In midsummer (July) the water temperature reaches 23–25°C.

Eight hydroelectric power stations and a water-divider have been erected on the Volga.On the Lower Volga, fishing is traditionally well developed. About 70 species of fish, including six species of sturgeon, are found here. All the sturgeons, except the sterlet, are diadromous. As of now, the general annual catch reaches 40,000 tons.

It is well known that the catching of sturgeons began in this area long ago. As far back as the middle of the seventeenth century, catches of sturgeon here reached 50,000 tons a year. There were rather big catches in the nineteenth and the beginning of the twentieth centuries. However, the extremely high intensity of fishing towards 1930 decreased the sturgeon catch to 13,500 tons, and by 1946 to 4,900 tons. Owing to the accepted measures of preservation of the stock, the catch increased during the subsequent period, and in 1968 reached 11,600 tons. In the 1970s and 1980s the catch of the sturgeon rose to 20,000–25,000 tons, with black caviar production reaching 2000 tons.

The short favourable period for the stocks ended in the late 1980s, when a need to reduce the catch became obvious. Within 15 years the catch fell ten-fold, and in 2004 the sturgeon catch by fishermen of the Russian Federation in the Volga-Caspian basin was about 400 tons, including 153 tons of sturgeon caught for research purposes and reproduction.

Under the circumstances the preservation of sturgeons in Russia, and, in particular, in the Lower Volga, has become especially significant.

16.2 Measures to Preserve the Sturgeon Fish

To preserve the sturgeon fish on the Lower Volga, a 3-point programme has been devised, consisting of reproduction (natural and artificial), protection of fish stocks and regulation of fishing.

16.2.1 Natural Reproduction

Before the construction of the hydroelectric power stations, 3390 ha of natural spawning area was available near the river Volga. It was located at a significant distance from the mouth of the river, and spawning of all species took place there. It helped in the preservation of the genetic fund of sturgeon populations, their genetic heterogeneity, multi-aged structure of the shoal and large numbers (Milshtein and Pashkin, 1971).

The construction of a hydroelectric power station at Volgograd (1958) caused the loss of 187 sites, 2869 ha in area, or 85% of the total spawning ground in the upper reaches. The spawning grounds of the beluga were reduced almost by 100%, of the sturgeon by60%, and the stellate sturgeon by 40%.

The spawning areas near the Central stadium of Volgograd may be regarded as an example of a typical Volga spawning area for sturgeon fishes. It is located in the bed of the Volga, 480–485 km from Astrakhan and 15 km from the dam of the Volgograd hydroelectric power station, the total area comprising 58 ha.

The spawning ridge stretches along the right side of the river Volga. The main bank is high, steep, and formed of loams and sandy loams. The ridge is made up of the spring-flooded area and the river bed, which is constantly under water. The total length of the ridge is 4700 m, the width is 100–200 m, the speed of the current changes from 1.0 to 1.2 m/s.

Fragments of flag, bottle and pebble, with an admixture of coarse-grained quartz sand and mussels, characterise the spawning substratum. Sturgeon fishes used the spawning ground intensively during spring high water, with the temperature of the water being 9–11°C. The average density of eggs was 620 pieces/m^2.

From 1975 to 1980, because of the large numbe of spawners accumulated at the dam of the Volgograd hydropower unit (up to 1–1.5 million specimens) and the high density of the ovum mass, which reached 4000–5000 eggs/m^2, and the adverse hydrological conditions as a result of significant changes in the level and speed of the river, a reduction in natural reproduction efficiency became inevitable. On the spawning grounds at the dam of the hydroelectric station the survival rate did not exceed 30%.

At the end of the 1970s and the beginning of the 1980s, scientists recommended constructing artificial spawning areas and clearing the ridges on the Volga as one of the measures to rectify the damage to natural reproduction, and to the environment, caused by hydro constructions and anthropogenic factors.From 1966 to 1988 six experimental ridges, spread over 65 ha in area, were constructed, and clearing of 27 ha of the natural spawning area was carried out.

Long-term research has shown that fish productivity in the constructed spawning areas was not lower than that of the natural ones. In the years of average water level, the yield of fish per hectare has been 13 tons. In the cleared areas, the spawning of sturgeon fishes (the sturgeon and the white sturgeon) increased 2.7 times and the amount of live eggs of embryonic development for the period increased by 18%. The data testified to the efficiency of artificial spawning grounds and cleared natural ones.

Now the spawning ground on the Volga consists of 22 spawning places.

The calculations of the capacity of the spawning areas have shown that for their optimum one-time filling, the number of spawners should be 131,000,including the sturgeon(50,000), the stellate sturgeon(73,000), and the beluga(8,000).

However, the number of sturgeon spawners coming to the spawning grounds of the Volga does not exceed 50,000(in the year 2000), which is obviously not enough for optimum natural reproduction.

Beluga (*Huso huso* L.) is the largest sturgeon fish in the Caspian basin. The reason for the decrease of the beluga population is the sharp reduction in its spawning ground, which causes difficulties in providing sturgeon fish-breeding factories with the necessary quantity of spawners. In 1991–1995 the number of spawners was 950 (Novikova and Khodorevskaya, 2000; Khodorevskaya et al., 2000). Since 2000, natural spawning of the beluga and of larvae within the limits of the upper spawning areas have not been found.

The Russian sturgeon—Acipenser guldenstadti Brandt: The stocks of the sturgeon migrating to the Volga have reduced more sharply than those of other sturgeon fishes. From 1966 to 1995 the number of spawners of the summer sturgeon migrating to the Volga spawning areas in early spring increased from 15,000 to 44,000; in 1997–1999 there were no more than 2000. The total number of spawners of the summer and winter sturgeon which passed through to the spawning areas in 1999 was reduced to 20,100, which is obviously not enough for its high-grade natural reproduction.

The stellate sturgeon—Asipenser stellatus Pall: From 1986 to 1990, on an average, 230,000 stellate sturgeons migrated to the spawning areas; in 1996 –1997—the number was 58,000. The reduction in the number of migrating stellate sturgeons has caused a change in the qualitative structure of the population. The number of the stellate sturgeon after spawning has reduced sharply of late. In 1990–1993 the relative index of catadromous migrants of the stellate sturgeon decreased 2.1 times in comparison with the period from 1986 to 1990; in the last 2 years it has reduced 50 times.

Barbel sturgeon (ship)—Acipenser nudiventris Lov.: Nowadays spawning of the barbell sturgeon in the river Volga is not noticed.

Sterlet—Acipenser ruthenus L.: It is a fresh-water fish, living in the river Volga. The stock of the sterlet in the river Volga and, in particular, below the dam of the Volgograd hydroelectric power station, has been constantly reducing. In the last 5 years its number in this area has decreased more than five times (Nizhnevolgryzhvod; Annual Statistics of Fishery, 2003).

Natural spawning of all sturgeon fishes has decreased catastrophically. In the 1970s and 1980s more than 100 million larvae were found in the spawning areas located in the lower pool of the Volgograd hydroelectric unit (they are only a part of the available sturgeon spawning areas); but now, the number of registered larvae which will be found downstream does not exceed two million.

Taking into account the importance of natural spawning for the maintenance of sturgeon stock, certain measures are being taken to increase the number of spawners. In the lower reaches of the Volga, special rules have been introduced to reduce fishing of sturgeon, fish way channels have been cleared, protective measures have been intensified, and other necessary work has been carried out.

One of the interesting projects that met with approval, and is introduced into the federal programme, is the construction of an adjustable artificial spawning area on the river Volga for sturgeon. The spawning ground of 5 ha can serially support natural spawning of all the sturgeon species; after spawning, the spawners will be returned to fish-breeding factories for maintenance and preparation for the following spawning.

16.2.2 Reproduction of Sturgeon Fish

Sturgeon breeding in Russia is more than 100 years old.

Up to the middle of the 1920s, experimental work on sturgeon fish breeding was conducted regularly, though on a small scale: annually from 10,000 to 150,000 larvae were let out into the Volga.

In the 1930s, sturgeon fish breeding was started on an industrial scale. The breeding points were located mainly in spawning areas or near them. The final output was released into the river as 1 or 2-day-larvae. In the Volga the release of larvae increased from 1.4 (1923) to 33.8 (1937) million.

To get mature reproductive products, the method of hypophysial injections developed by N.L. Gerbilskii began to be applied in 1938. It led to a significant increase in sturgeon fish reproduction (Milshtein, 1971).

Industrial sturgeon breeding in Russia began in the second half of the twentieth century. Eight fish-breeding factories were constructed on the Lower Volga. Annually they produce about 70 million sturgeon fingerlings of various species.

The sturgeon fish-breeding factories on the Lower Volga are designed, constructed and operated according to the technology developed in the 1960s which provided for:

1. Annual preservation of spawners from commercial catches
2. Hormonal stimulation for maturing
3. Short-term breeding of larvae in ponds with natural feed
4. Release of fingerlings into natural water reservoirs

This scheme of sturgeon farming did not include the live preservation of spawners after reception of eggs and sperm. As a result, artificial reproduction was completely dependent on the catch of spawners from natural reservoirs.

At present, there are serious problems at the fish-breeding factories about catching the necessary number of male and female specimens during spawning migration. This has resulted in reducing fingerling release by half. An urgent need to form the brood stock and provide for a practical reserve for factories has sprung up in this connection.

This challenge is being met in two ways. One is the formation of brood stock from eggs. It is a long process, besides demanding a large productive area and significant financial expenses. The other is the formation of brood stock of immature fish caught in the river and sea, and their further maintenance in appropriate reservoirs with regular feeding and providing normal mature reproductive stock.

Thus, special ponds and areas for the maintenance of brood stocks have been introduced into the standard working routine of sturgeon fish-breeding factories.

The indispensable condition for artificial reproduction of these valuable fishes is the preservation of spawners after reception of the reproductive products, for the purpose of their repeated use in the reproductive process.

At the fish-breeding enterprises, the widely known methods of egg reception from sturgeon females, described by Burtsev (1969) are followed. Both methods are being successfully applied in sturgeon breeding and have contributed significantly to the preservation of brood stocks of the sturgeon.

The fecundity of a she-sturgeon, the fertility rate and other parameters are not affected by these methods. It has also been shown that their genetic structure is not broken (Burtsev et al., 2002).

In all the stages of the process, certain biological norms or standards are specified for fish-breeding enterprises, to assess the quality of work handled and the output product. According to the existing biological specifications the weight of the released baby sturgeons, under the terms of feeding in ponds during 25–60 days, is to be not less than 2 g. The yield of baby fishes from the hatchery is accepted as 0.55%, the weight being 2 g, and 3%—when the weight is 15 g. From one she-sturgeon not less than 17,000 baby fish of standard weight must be received.

Of late, fish-breeding factories have carried out work on the reception and release of baby fish from 3 to more than 10 grams. The monitoring of baby fish of increased weight on a river home range of 100 km showed that their survival rate after release from ponds exceeds the survival rate of baby fishes of 2–3 g. However, the change-over to more modern standards of released baby fish demands additional survey and research as it is necessary to make sure that baby fish will not lose the instinct of rolling down into the sea.

Now, all fish-breeding factories are under reconstruction. The production capacities have been adapted to new technologies, including maintenance of brood stocks, lifetime reception of reproductive products and so forth. Significant attention is given to feedstuff production. In practice, production of feedstuff, which is available in the market, does not fully satisfy the requirements of a spawning school. There are some problems concerning guaranteed reception of qualitative posterity under industrial fish-breeding conditions. The use of natural feed or its elements mixed with those offered on the market for feeding of spawners and baby fish, if applied, will raise the quality coefficients at all stages of breeding.

One such factory is the Volgograd sturgeon fish-breeding factory. Constructed in 1961 as a compensation for the damage caused by the construction of the Volgograd hydroelectric power station, it annually produces more than four million baby sturgeon fish. In 2000 a fish-breeding complex for the maintenance of a brood stock with a production capacity of 35 million eggs or seven million baby fish was built in cooperation with the Volgograd hydroelectric power station.

The fish-breeding complex is located in the body of the dam and takes advantage of the difference in water level between the top and tail pools for water supply. The closed-circuit system of water supply equipped with trickling filters helps to hold the temperature of water at not less than 15°C during winter. Under such conditions, repeated maturing of the Russian sturgeon spawners takes place every 2–3 years.

In the opinion of the scientists, the optimum output for a sturgeon farm of 100 million baby fish can be achieved, if breeding technology is improved and the reconstruction of the existing fish-breeding enterprises is carried out.

16.2.3 Fishery Regulation and Fish Stocks Protection

One of the important elements of sturgeon industry management is control of sturgeon fishing and fish stock protection.

Until the administrative reform which began in the Russian Federation in March 2004, fishery management and fish stock protection were organized according to the federal structural division that had control and protection over the water area, covering, as far as possible, a home range of basic water species.

Nowadays a territorial principle is applied, according to which fishery control and protection are exercised within the limits of a part of the Russian Federation. A protection service of fish stock has federal permission and financing and works in cooperation with the regional authorities. The norms of regulation for commercial fishing are established by federal legislation, and the norms of sports and amateur fishing are fixed by local legislation.

The protection of fish stocks and regulation of fishing on the Lower Volga and Northern Caspian Sea are executed by state inspectors of a special federal service of the Ministry of Agriculture. Besides this, frontier troops and special divisions of the militia (in the zone of Northern Caspian Sea) participate in protection work. Annually more then 1000 infringements, connected with illegal catch of sturgeon fish, are registered in the Lower Volga, within the limits of the Volgograd and Astrakhan regions. If an illegal catch is committed, a "sturgeon case" is tried by a nature protection Office of Public Prosecutor and referred to the court. Punishment for illegal catch of sturgeon fish is a fine upto 500,000 rubles or 3 years of imprisonment, and confiscation of the equipment and vehicles used during the catch, as well as compensation for the damage of the specimens.

However, the measures taken are obviously insufficient. As one of the additional security measures, the Russian Federation government is considering the formation of a uniform security service and introduction of state monopoly in the sturgeon industry.

While constructing industrial projects which may affect the conditions of sturgeon habitation, state supervision bodies are to make demands for measures to be taken to avoid any negative influence, or insist on obligatory construction of units which could compensate the loss of the stocks. Thus, all fish-breeding factories on the Lower Volga have been constructed as damage compensating enterprises, and the construction of these factories has outpaced that of the main units.

Fish passes have been constructed with a view to protect the route of migration at all dams of the hydroelectric power stations in the Lower Volga. Besides their basic function of transfering migrating sturgeon spawners, fish elevators work as registration stations, where the whole process of fish migration is recorded with absolute accuracy: the number, specific structure and so forth. In some years, up to 70,000 specimens of sturgeon have passed through the fish way of the Volgograd hydroelectric power station.

To preserve the conditions for natural reproduction, some other measures have also been taken, such as a ban to withdraw soil from the river in the places of sturgeon spawning and obligatory equipping of all pump stations with fish protecting devices.

In the Lower Volga, fishery regulation and, accordingly, all normative documents are basically aimed at the preservation of sturgeons and similar kinds of fishes:

1. A ban has been imposed on sturgeon fishing in the Caspian Sea.
2. In case a sturgeon is caught in the equipment of those fishermen who have no sanction to catch these species of fish, the caught fishes should be let out alive into the reservoir. If fishermen are caught in the act of illegal fishing, they can be deprived of their licenses and criminal prosecution can be initiated against them.
3. Using equipment that can injure a fish and catch baby sturgeon is prohibited.
4. Catching fish for scientific purposes and reproduction is carried out on the basis of licensed governmental decisions and the order of a corresponding federal department and is under its control.
5. Amateur fishermen are forbidden to catch sturgeon.. The fishcaught accidentally by a fisherman should be released alive into a reservoir; otherwise the person is subject to criminal prosecution.

With the view of protection, reproduction and regulation of sturgeon fishing in the Russian Federation, a number of federal Acts have been passed, such as the Fishery Law, and the concept of preservation and augmentation of sturgeon fish. The work on the Law on sturgeon fish and other Acts is almost complete, and will enable the regulation of sturgeon protection and turnover of sturgeon products. It will also determine the priorities in the field of sturgeon breeding to direct the increase of their number, and by that, help to preserve and restore sturgeon fish stocks in the Caspian basin and the Lower Volga.

References

Annual Statistics of Fishery. 2003. Regional Department of Fish Stocks Protection and Reproduction, Nizhnevolzhrybvod.

Burtsev, I. A. 1969. Obtaining offspring from intergeneric hybrid between beluga, *Huso huso* (L.), and the sterlet, *Acipenser ruthenus L.* In: *Genetic, Selection and Hybridization of Fishes*. Ed. Science, Moscow, pp. 232–242 (in Russian).

Burtsev, I. A., Nikolaev, A. I., Maltsev, S. A., and Igumnova, L. V., 2002. Formation of domesticated broodstocks as guarantee of sustainable hatchery reproduction of sturgeon for sea ranching. Russian Federal Research of Fisheries and Oceanography (VNIRO), and Regional Department of Fish Stocks Protection and Reproduction, Nizhnevolzhrybvod.

Khodorevskaya, R. P., Dovgopol, G. F., and Zhuravleva O. L. 2000. Commercial sturgeon stocks dynamics in the Volga-Caspian region, Caspian Fisheries Research Institute (KaspNIRKh), Astrakhan.

Milshtein, V. V. 1971. Sturgeon-breeding century. Proceeding of the Central Research Institute of Sturgeon-breeding (ZNIORH), Astrakhan.

Milshtein, V. V. and Pashkin, L. M. 1971. Sturgeon fishes reproduction on the regulated Volga. Central Research Institute of Sturgeon-breeding (ZNIORH), Volgograd Department, V. P. Shilov, Saratov Department.

Novikova, A. S. and Khodorevskaya, R. P. 2000. Opportunities and actual state of natural reproduction of sturgeon fishes within a non-regulated area of the river Volga. Caspian Fisheries Research Institute (KaspNIRKh), Astrakhan.

Chapter 17
Experience of Conservation of *Acipenser naccarii* in the Ticino River Park (Northern Italy)

Cesare M. Puzzi, Adriano Bellani, Stefania Trasforini, and Alessandra Ippoliti

Abstract Started in October 2003 and concluded in September 2006, the project has been realized thanks to the financial support of the European Community and the co-financial support of Lombardy Region—Environment Quality Department, with the participation, as a partner, of South Oglio Park.

The objective of the project was the conservation of the adriatic sturgeon population found in the Ticino River and in the middle reach of the Po River, blocked there by the Isola Serafini dam on the Po river that prevents the fish from migrating to and from the Adriatic sea.

Sporadic but reliable records provide evidence of the presence in the Ticino River of a spawning population that passes its life cycle in freshwater, and so is called a *landlocked* population.

This population, with little information about its real size and status, could actually represent one of the last groups, if not *the* last, of the original population of *Acipenser naccarii* that once inhabited the Po Basin.

That is why it is necessary to have a conservation project that would define the best strategy to manage and protect the species, and carry out the activities necessary for its preservation.

Keywords Biotelemetry, land-locked population, Po River, repopulation, river fragmentation, *Silurus glanis*, Ticino River

---

C.M. Puzzi, S. Trasforini, and A. Ippoliti
GRAIA srl, via Repubblica, Varano Borghi (VA), Italy

A. Bellani
Consorzio Parco Lombardo della Valle del Ticino—via Isonzo, Pontevecchio-Magenta (MI), Italy

R. Carmona et al. (eds.) *Biology, Conservation and Sustainable Development of Sturgeons*,

17.1 Introduction

The adriatic sturgeon (*Acipenser naccarii*) is the only native sturgeon that still inhabits Italian water courses (Bronzi et al., 2005). It is classified as a 'priority species' in the 'Habitats Directive' (92/43/CEE) and 'threatened by extinction' in the IUCN (International Union for Conservation of Nature) Red List.

The certified presence of the adriatic sturgeon in the Ticino River dates back to the end of the nineteenth century, when the species was common. Since the 1960s, however, the population began to decline progressively, because of the high fishing pressure, but especially because of the construction of the Isola Serafinidam on the Po River at Monticelli d'Ongina (PC), which caused the end of the migrations of the adriatic sturgeon from the Adriatic Sea to the middle reach of the Po river and its affluents. As a consequence, a nucleus of adriatic sturgeon remained isolated from the sea, settling down in the middle reach of the Po and in the middle-low reach of the Ticino, creating a so-called 'land-locked' population, the presence of which has been confirmed by many records of its reproductive activity (Bernini and Nardi, 1989; Bernini and Nardi, 1990b). The conservation status of this population is currently worrying, in view of its isolation, its reduced dimensions and the numerous factors that threaten its survival, and predict its extinction (Fig. 17.1) (Bronzi et al., 2005).

The adriatic sturgeon is an anadromous species, endemic to the high Adriatic Sea and originally present in the larger and deeper rivers of the high Adriatic Basin, such as the Po, Adige, Brenta, Piave, and Tagliamento, where it lived in proximity of the river bottom; in the sea it lives near the river mouths, on the muddy and sandy bottom, at a depth of 10–40 m, reaching freshwater for the reproductive period. The upstream migration to Italian rivers occurs in the first months of the year and the spawning takes place between April and June (Gandolfi et al., 1991; Rossi et al., 1992). The species feeds on macrobenthic invertebrates like annelids, crustaceans, gastropods and bivalve molluscs, and, in minor amounts, also small fish. According to the studies carried out by Bernini and Nardi (Bernini and Nardi, 1990a), gammarids represent the most important food source in the diet of the Ticino sturgeon, with a contribution of about 50%, while the Po population receives an alimentary contribution from Diptera (90%).

Fig. 17.1 A specimen of the adriatic sturgeon

Aiming at protecting and, if possible, increasing the landlocked population in the middle-low reach of the Ticino and in the middle reach of the Po, the Ticino Park decided to promote a conservation project in favour of the adriatic sturgeon, financially supported by the European Union.

17.2 The Project

The project, a LIFE-Natura 2003 Project, is called 'Conservation of *Acipenser naccarii* in the River Ticino and in the middle reach of the River Po (Life03nat/it/000113)'. It started in 2003 and ended in October 2006. It was carried out by the Lombardy Ticino Park, which coordinated with the Lombardy Region - Environment Quality Department (as co-financial supporter), South Oglio Park (as a partner, responsible for the activities carried out in the territory of the South Oglio Park), local fisher Associations, and GRAIA srl (as external consultant).

17.2.1 The Project Area

The project area includes the fluvial reach of the Ticino River from Vigevano (PV) to the Po confluence (about 40 km) and a reach of the Po River about 110 km upstream of the Isola Serafini dam. The project also planned a comparison between the Ticino River and the Oglio River, for which continuity to the sea has never been interrupted, but in which the adriatic sturgeon has long been extinct, for causes obviously not related to the fluvial fragmentation but not still well defined. The stretch of the Oglio chosen is within the territory of the South Oglio Park, approximately 50 km long, to the confluence with the Po River (Fig. 17.2).

17.2.2 The Project Aims

Pursuing the objective of species conservation in the Ticino River, the project is intended to achieve at first the following general aims:

1. To increase the local population of *Acipenser naccarii*
2. To improve the knowledge concerning the ecology of this species, in particular its reproduction
3. To define the best management strategies for the species
4. To conserve the natural habitats
5. To counteract poaching
6. To sensitize public opinion

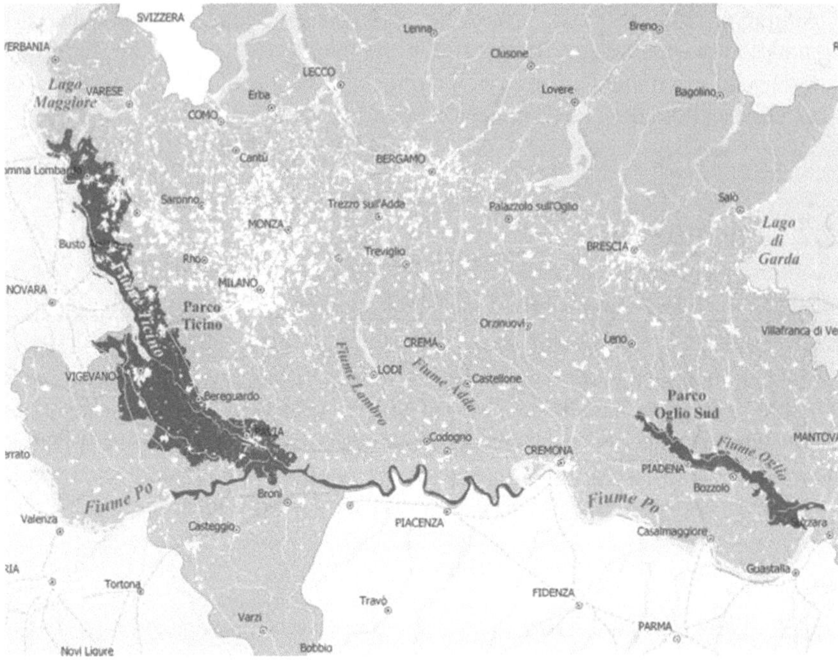

Fig. 17.2 The study area; in black the stretches of rivers studied

The first step in the Life project planning has been the identification of the main factors that menace the adriatic sturgeon's survival in the Ticino River. Five big threats have been spotted, as described below.

17.2.2.1 Loss and Degradation of the Available Habitat

The destruction or degradation of the habitat that can be colonized by the sturgeon during each different stage of its life cycle can reduce the possibility of survival of the species. Above all, the loss of reproduction sites can negatively affect the status of the landlocked population.

17.2.2.2 Habitat Fragmentation

Habitat fragmentation is still an active threat to the adriatic sturgeon population because the fish pass at Isola Serafini dam on the Po River has not yet been made. The isolation of this population represents a strong threatening factor because they are unable to reproduce with other sturgeons that are free to migrate, and this reduces the genetic flow among the different populations, strongly limiting the viability of the species.

17.2.2.3 Genetic Drift

Considering the probable small size of the population, the risk of inbreeding is high. Genetic drift can result in dramatic changes in allele frequency, determining moreover the selection of hereditary characteristics, no longer through natural selection but through casualty, causing the loss of 'evolutionary success' of the population and local extinction.

17.2.2.4 Spread of the Sheat-Fish

A Life2000Natura Project carried out by GRAIA srl and the Ticino Park, named 'Conservation of *Rutilus pigus* e *Salmo marmoratus* in the Ticino River', and previous research on the Ticino River fish community showed the increasing numbers of the sheat-fish (*Silurus glanis*) in the entire Po Basin, representing a strong menace to the survival of all autochthonous species (GRAIA srl, 1999, 2004). The sheat-fish became dominant in terms of both density and biomass in many reaches of the Ticino River; the ecology study on the Ticino sheat-fish population showed that it is able to occupy all the available habitats, and confirmed its intense activity of hunting, mainly nocturnal, and its opportunistic diet, based mostly on fish during the adult stage and mostly on gammarids during the juvenile phase. According to these results, the sheat-fish seems to represent a real obstacle to the recovery of the adriatic sturgeon, both by direct predation and by trophic competition (Fig. 17.3).

Fig. 17.3 A specimen of sheat-fish (*Silurus glanis*)

On the whole, the reasons that make the Sheat-fish a strong enemy of the native fish fauna of the Ticino River are summarized as follows:

1. Rapid growth that allows the species to reach a length of 50–70 cm and a weight of a kilogram in only 3 years
2. Its almost ichtyophagic diet from the first years of life and completely opportunist predation, particularly effective because of the large size of the fish and its habit of nocturnal hunting
3. The capacity to occupy the best covers, chasing away other fish
4. The long reproductive period that lasts for many months, from late spring to late summer, assuring successful reproduction

17.2.2.5 Fishing Pressure

As of now, even though fishing activity cannot be considered the first cause of the reduction of the adriatic sturgeon in the Po basin, catching even a few specimens can strongly damage the population, because of its small size. Although angling activity has been forbidden for many years, poaching is still frequent.

17.2.3 The Project Structure

The project has been structured as many 'Actions', aimed at offsetting the limiting factors that threaten the survival of the species.. All the actions are summarized in Fig. 17.3.

To offset habitat reduction, many measures have been carried out (Action A.1), which are given below:

1. Characterization of the area in which the adriatic sturgeon lives
2. Assessment of the size and status of the population in the project area
3. Definition of the relationships with other species in the river
4. Assessment of its distribution

All these actions are aimed at realizing a management plan of the species and the environment, called 'Action Plan of the *A. naccarii* management' (Action A.1), that would define the best conservation strategies for the species, with special emphasis on spawning-habitat conservation, restocking activities and fishing management.

To offset the fragmentation of the distribution area of the species, a survey has been carried out along the stretch of the Po River between the Isola Serafini dam and its mouth, to characterize other possible impediments that prevent the migration of sturgeon (Action A.3).

To counter the threat of genetic drift, the repopulation (Action C.1) of the river has been undertaken with sturgeons coming from the same river basin (Action A.2; Action C.3) and, as shown from the genetic analysis with molecular markers (Action A.4), with levels of genetic variability compatible with those shown by

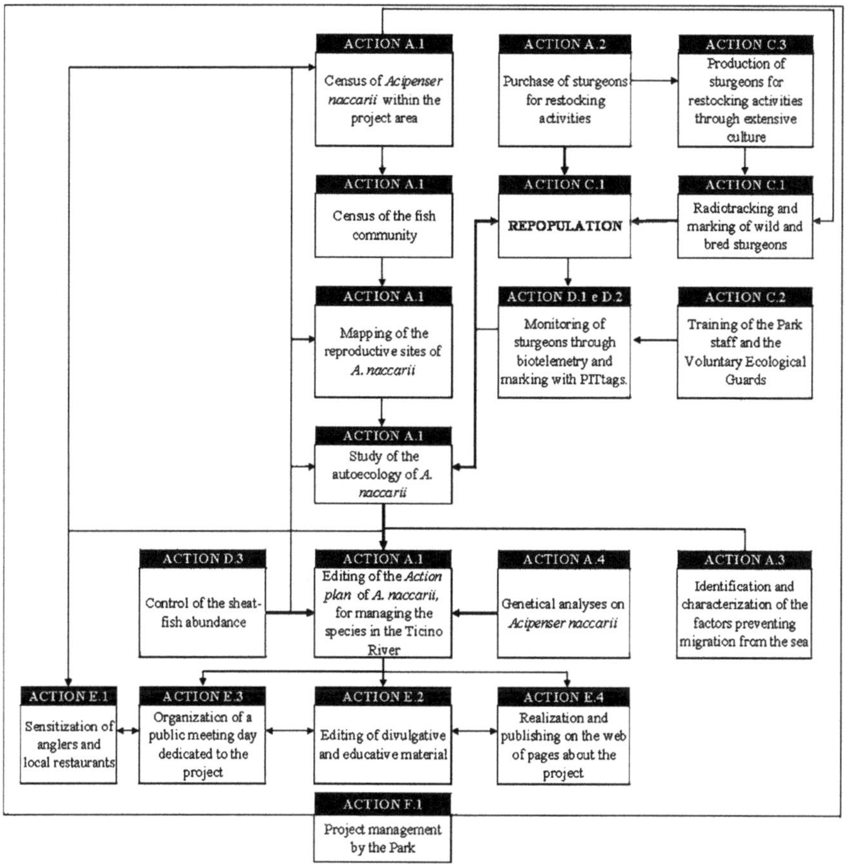

Fig. 17.4 Chart of the project structure

other natural populations. All the specimens introduced into the river have been marked with a microchip for recognition in case of re-capture (Action C.1).

To combat the threat from sheat-fish, control of the invasive species (Action D.3) has been carried out, with campaigns of diurnal and nocturnal electro-fishing, started by the Park during the previous Life-Nature plan of conservation. In addition, the study of sheat-fish biology in the Ticino River, started during the previous project, is being continued.

To stem the threat from fishing pressure, i.e. from 'poaching', considering the impossibility of maintaining tight and constant surveillance of an area as big and complex as that of the Ticino River, besides the creation of a nucleus of vigilance, constituted by two patrols charged with the surveillance of the river and the safeguard of the fish fauna, the Plan raised the awareness of the fishermen and the people. A series of initiatives have been promoted, such as meetings with the Fishing Associations, distribution of informative pamphlets, creation of a web-site devoted to the project and creation of a group of local restaurants cooking sturgeon, which will guarantee the safety of the species, through a label attesting the exclusive capture of bred fish (Fig. 17.4) (Action E.1; E.2; E.3; E.4).

17.3 Material and Methods

The Action Plan will be prepared on the basis of the investigations on the autoecology of *A. naccarii*, by means of electrofishing, net fishing, hydroacoustic surveys, underwater observation, PIT (Passive Integrated Transmiter) tagging, biometric studies, and habitat analysis.

17.3.1 Action A.1

To define the current 'status' of the landlocked population of the adriatic sturgeon in the project area, a census was carried out 60 times in 3 years through electrofishing and more than 10 underwater explorations, for direct observation of the preferred habitats of the sturgeon. The Ticino River and the middle reach of the Po River have been subdivided into fluvial segments, each comprising many sampling units, and characterized on the basis of the environmental characteristics of the river, which make them recognizable ecological units of the type of fluvial habitat and fish community.

Pamphlets to raise awareness have been distributed to fishermen, including a sheet to be given back, after filling without giving the name, for eventual recording of sights or captures. Moreover, the traditional spawning areas of the adriatic sturgeon have been investigated during the reproductive period, to look for some mature adults in the reproductive age.

Three wild sturgeons have been captured in the Ticino River; they have been measured, weighed, sampled for genetic analyses, marked with a microchip or radiotransmitter and set free.

17.3.2 Action A.3

To characterize the impediments between the Adriatic Sea and the middle reach of the Po, in September 2004, the Po River and the last reaches of its main affluents were explored, using an aluminium boat with a hydrojet motor, to navigate even the low waters. The exercise started at Bereguardo (PV) and involved the last 30 km of the Ticino River and the stretch of the Po between the confluence of the Ticino and Isola Serafini; this lasted 2 days. The total length covered by the exploration was about 420 km.

17.3.3 Action A.4

The study of the genetic variability, at both the individual and population levels, has been carried out through the analysis of microsatellites (nuclear markers). These are short repeated nucleotidic sequences that follow a Mendelian hereditary transmission. The number of repeats is variable and constitutes the allelic characteristic that is transmitted from parent to offspring. The microsatellites are characterized by a high mutation rate per generation, and therefore by high variability, making them especially useful for the study of the genetic variability of isolated and small populations. In this case, the genetic analyses are particularly complex, because the sturgeon is polyploid—that is, it has a high number of chromosomes and consequently they can have more than two alleles per locus, complicating the statistical analysis. The genetic analyses have been carried out using six species-specific microsatellite markers, the sequences of which are described in the literature. The Ticino Park gave 100 bred sturgeons, analysed together with three wild specimens caught in the Ticino, probably part of the original population.

17.3.4 Action D.1 and D.2

The sturgeons have been monitored with ultrasound biotelemetry. This technique consists of attaching to the animals a transmitter a few centimetres long, with a battery that converts electric power into acoustic waves of a frequency between 20 and 300 kHz, not audible to humans. The battery (which lasts 3 years) produces electrical impulses that, applied to a quartz, are transformed into mechanical vibrations with ultra-sound frequencies. These run in the water and can be picked up by a receiver dipped in water, called hydrophone, which contains a transducer that absorbs the mechanical vibrations and converts them into impulses, to be translated into sounds. This type of marking technique has many advantages—first of all the possibility to control the activity and movements of the fish, without causing them excessive stress (Kieffer and Kynard, 1993; Collins et al., 2000).

The transmitters have been inserted through a laparotomy operation into the abdominal cavity of 30 sturgeons that are being monitored, which allowed us to assess the maturation level of the gonads. After 1 month, during which the sturgeons have been kept under observation in the breeding tanks, we verified that they were in good health. Then they were set free at different points of the Ticino River between Vigevano and Pavia. From the moment of release, the movements of the sturgeons have been monitored, at first daily, then weekly, through biotelemetry monitoring with a hydrophone from a boat. In addition, at Pavia, near the covered bridge and near the floating centre of the Association for the Defence of Nature and Environment (ADNA), a fixed hydrophone has been set to pick up and record the signals of the possible passings of the sturgeons towards the sea (Fig. 17.5).

17.3.5 Action D.3

The sheat-fish that have spread throughout the project area of the Ticino and the middle reach of the Po have been controlled through periodic campaigns of both nocturnal and diurnal electrofishing, in the Ticino between Abbiategrasso and the

Fig. 17.5 Insertion of the radio transmitter in the adriatic sturgeon abdomen and radiotelemetry activity

mouth, in the Po between the confluence of Agogna stream and the dam of Isola Serafini, and in the Oglio River within the South Oglio Park.

On the whole, more than 200 km of the rivers have been investigated, in 128 attempts, of which 86 were in the Ticino River (78) and affluents, 29 were in the Po River (20) and affluents and 13 were in the Oglio River.On the whole, 112 attempts through diurnal electrofishing, 14 through nocturnal electrofishing during the reproductive period and 2 campaigns through underwater fishing in large pools have been undertaken.

17.3.6 Action C.1

Another specific objective of the project was the increase of the landlocked population of adriatic sturgeon through repopulation activities. This activity has been conducted in different steps:

1. Purchase of selected sturgeons for repopulation
2. Temporary enclosure of sturgeons in intensive rearing tanks

3. Pre-adaptation to the life in nature
4. Release in the river

17.3.6.1 Purchase of Selected Sturgeons for Repopulation

In 2004 and 2005, about 1300 1-year-old sturgeons (age class 1+) and more than 70 adults or sub-adults were bought from the farm VIP of Giovannini Giacinto at Orzinuovi (BS), the first and only Italian producer of *A. naccarii*, who in 1988 started an intensive production of adriatic sturgeon meant for repopulation activities, starting from a nucleus of wild reproducers captured in the Po basin.

In addition to this group of young, adult and sub-adult sturgeons, in 2005 a group of 2500 yearlings (age class 0+), were bought and transferred to the tanks of the Fagiana Hatchery building, the property of the Park during the previous LIFE-Nature project for the conservation of marble trout and pigo.

17.3.6.2 Temporary Enclosure of Sturgeons in Intensive Rearing Tanks

All the sturgeons, bought at different times, were at first transferred to the concrete tanks at the Fish Farm of Marano Ticino (NO), by Piero Fantinato. Here, each stock was enclosed for about a month, during which the sturgeons were fed intensively. On their arrival, all the animals were measured, weighed, and marked with an under-skin microchip, called PITtag, similar to those used for dogs, containing a readable numerical code (Nielsen, 1992). The 0+ sturgeons, too small for being marked with microchip, were instead tattooed with the Pan Jet, blue ink injected under the skin (using a type of gun) that lasts at least 2 years (Fig. 17.6).

17.3.6.3 Pre-Adaptation to Life in Nature

Before releasing them in the river, the sturgeons were transferred, in small groups (30–80 specimens, depending on the size), to the semi-natural breeding tanks at Cassolnovo (PV), belonging to the Park. Here the sturgeons found an environment very similar to the natural one, improved by renaturalization activities and by bed excavation to create deep pools, where they were able to feed on their own, hunting the macroinvertebrates at the bottom, through a gradual reduction of the artificial food supplied, until its complete suspension in approximately one month.

17.3.6.4 Release in the River

After the phase of pre-adaptation to natural life, the sturgeons were gradually set free in the Ticino and Oglio Rivers.

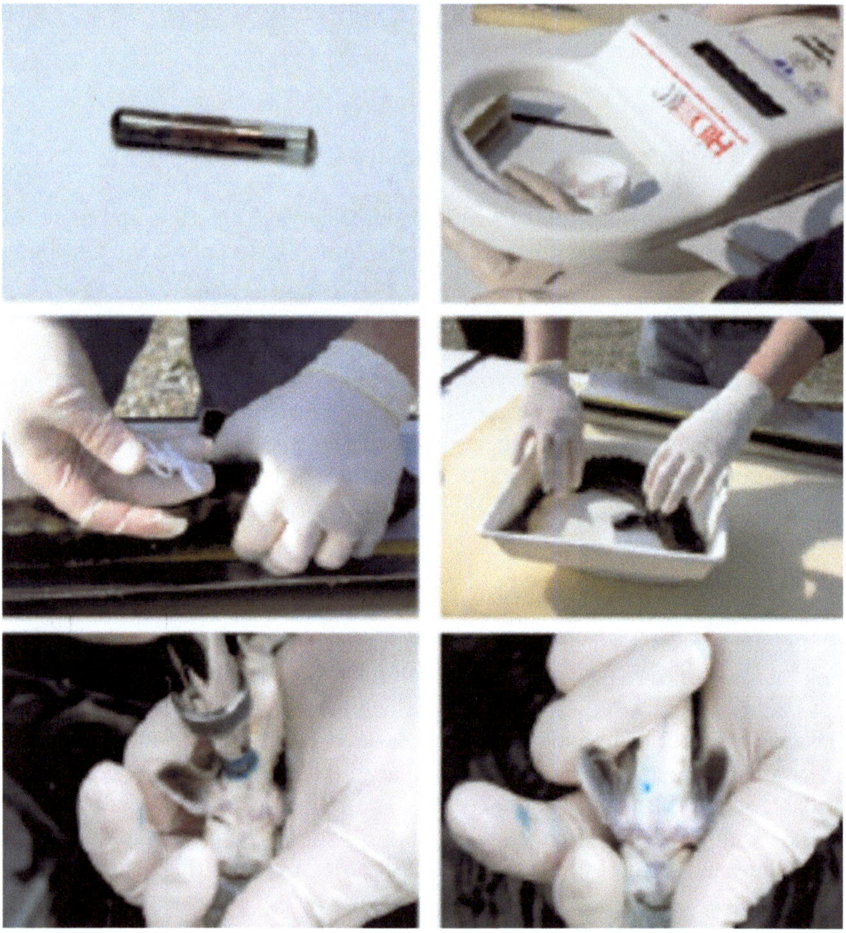

Fig. 17.6 Pit-tagging

17.3.7 Action C.3

The artificial reproduction of adriatic sturgeon, in two reproductive seasons (2005–2006), using as breeders some of the adults bought from the farm at Orzinuovi, was carried out in steps.

17.3.7.1 Monitoring of the Maturation Level of the Male and Female Gonads

To control the maturation level of the gonads, a sample of gametes was taken from adult subjects near the reproductive period. The sperm was examined in response to water activation, to determine the motility of spermatozoa. The maturation level of the oocytes was checked by boiling and cutting them longitudinally to examine the position of the germinative vesicle. When the germinative vesicle was near the animal pole, revealing complete maturation, and when, at the same time, the test of activation in water of the sperm revealed optimal motility, the artificial reproduction was performed.

17.3.7.2 Hormonal treatment

The hormonal induction, undertaken using commercially available inducers (e.g. carp ipophysis) at the recommended doses, led the reproducers to the optimal conditions for reproduction.

17.3.7.3 Abdominal-Pressure Technique and Egg Fertilization

The abdominal pressure on the females, after anaesthetic bath, was applied repeatedly, a few hours between sessions, because the maturation of the sturgeon ovary is progressive and the spawning occurs in a serial way, starting from the eggs of the distal area. The males instead were treated once and the sperm produced was conserved in a refrigerator till the successive fecundation. The eggs produced were fertilized with sperm diluted with water 1:50 (this activates the spermatozoa that remain in such a state for 3 min, after which the motility decreases until stopping 5 min after activation). After a few minutes, a maize–starch watery emulsion was added to the fertilized eggs, to prevent the formation of an adhesive film that would have made them extremely sticky, causing 'death by asphyxia'.

17.3.7.4 Incubation of Eggs and Rearing of the Alevins

The eggs were transferred to the Fagiana Hatchery and the Abbiategrasso Hatchery, run by the District of Milan Administration, in Zug Bottles that, maintaining the eggs in constant suspension, prevent them adhering to each other. The embryonic development was monitored daily.

17.4 Results

17.4.1 Study of the Landlocked Population of Adriatic Sturgeon in Ticino River (Action A.1; D.1; D.2)

17.4.1.1 The Current Status of the Population

The results of data collection describe the current status of the landlocked population of adriatic sturgeon, summarized in the following points:

1. After 3 years of constant work, using a variety of survey methods, only 29 records of the presence of the adriatic sturgeon in the Ticino River have been collected
2. After a first reliable record, in July 2003, of a young specimen, no young individuals, coming from natural reproduction, have been recorded
3. For many years, no natural reproductive activities of the adriatic sturgeon were observed anywhere in the project area and the investigations carried out at the sites normally used by the species during the spawning period, gave no positive outcome

On the whole, therefore, the information collected indicates the presence of a rarefied and declining population that is finding it difficult to carry on reproduction activities. On the other hand, all the specimens caught or recorded showed optimal health status, confirming the suitability of the Ticino River, the habitats and the food found in it, for the survival of this species

Furthermore, no record of wild sturgeons was collected from the Po River, indicating the tendency of the adriatic sturgeon to settle in the Ticino River, possibly because of the better environment and quality of the water. It might also be because of the difficulties in investigating the Po River, which is much larger and deeper than the Ticino, even in its middle reach.

17.4.1.2 Sturgeon Behaviour

The graph (Fig. 17.7) shows the data on the movements of sturgeons recorded between March 2005 and September 2006. The data also shows the behaviour of the released sturgeons and their environmental preferences, confirmed by the observations made for the wild specimens. These data clearly show the tendency of some sturgeons to occupy precise zones of the river, in which they have found the ideal conditions for living, and from which they donot stray; such zones are mostly big pools, many metres deep, in which the sturgeons stay near the bottom in the deepest points, or downstream at the deepest point where there is laminar flow. Often, specimens were recorded in reaches where the flow was quite high, probably for feeding.

The biotelemetric monitoring of adult, sub-adult and young individuals allowed the study of the environmental preferences of the different age classes, leading to the result that they are similar. The preference for large pools, as places of long-term stay, is shown both by the young and adults, while most of the young specimens tended to go downstream quite immediately, perhaps because of the attraction of the sea.

A point-by-point census of all the great pools present in the study area was carried out. Each pool was characterized, and the dimensions, maximum depth, bathymetric profiles and location were recorded. On the whole, 21 large pools were characterized. For some pools, subjected to a complete bathymetry through ecoscanning, the 3D shape was also reconstructed; in the zones of calm water, the

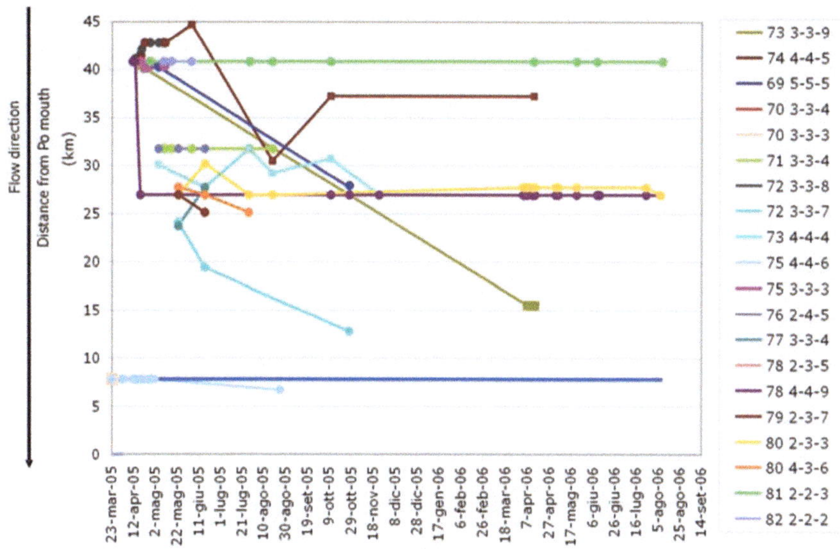

Fig. 17.7 Biotelemetric records

sturgeon typically occupied the deepest point of the pool; where the pool was placed after a reach of high flow and there were whirlpools in the deepest point, the sturgeon occupied the zone of laminar flow immediately downstream, near the gravel bottom. Analysing their main morphologic characteristics, we found that the large pools are in general very wide, averaging about 2 ha, with maximum depths of 13 m, with medium values of 8 m and an average water transparency of approximately 2 m, depending on the wind. The substrate was mostly gravel and pebbles.

17.4.1.3 The Availability of Reproductive Habitat

One of the aims of the Life Project was to locate and characterize the areas of the river chosen by the sturgeon for reproduction activity. For this, there were fish samplings, underwater observations, inspections of the river during the reproductive period and interviews with anglers. However, no evidence of real reproductive activity of the sturgeon in the study area has been found. This means that there may be some environmental factors, not yet understood, that affect this essential phase of the biological cycle of the species or that it is difficult to find an area in which the spawning activity is happening. This possibility becomes even rarer, considering the typical characteristics of the spawning areas of the adriatic sturgeon: gravely and sandy bottom in long, deep reaches with laminar flow.

Because of the failure to follow the reproduction in nature, a point-by-point characterization was made of the entire fluvial habitat of the Ticino River currently inhabited by the adriatic sturgeon, with the aim of spotting, mapping, and therefore quantifying the availability of potential reproduction areas.

Through an examination of the river, by boat from Vigevano to the Po confluence, the different types of meso-habitat were located and measured (on the basis of flow speed, turbulence of water, type of flow, and type of substrate): the large pools, as stated above; the zones with turbulent flow and low water (riffle); the zones with laminar, but fast flow and low depth (fast run); the zones with low or very slow or very slow flow and deep water (slow run). The latter ones, having the hydraulic, morphological, and substrate characteristics typical of the preferred reproduction areas of the adriatic sturgeon, have been classified as potential reproductive areas for the landlocked population. Their georeferencing enabled the drawing of a map of the potential spawning sites. Such areas constitute approximately 39% of the study area, with more than 230 ha, and indicate the great availability of habitat during this vital phase for the adriatic sturgeon (Fig. 17.8).

17.4.1.4 Relationships of the Adriatic Sturgeon with the other Fish Species of the River

All the information acquired was organized into a database connected to a georeferencing system. The first data set, collected experimentally within this project, comes from 46 episodes of electrofishing sampling or, in some case, from records

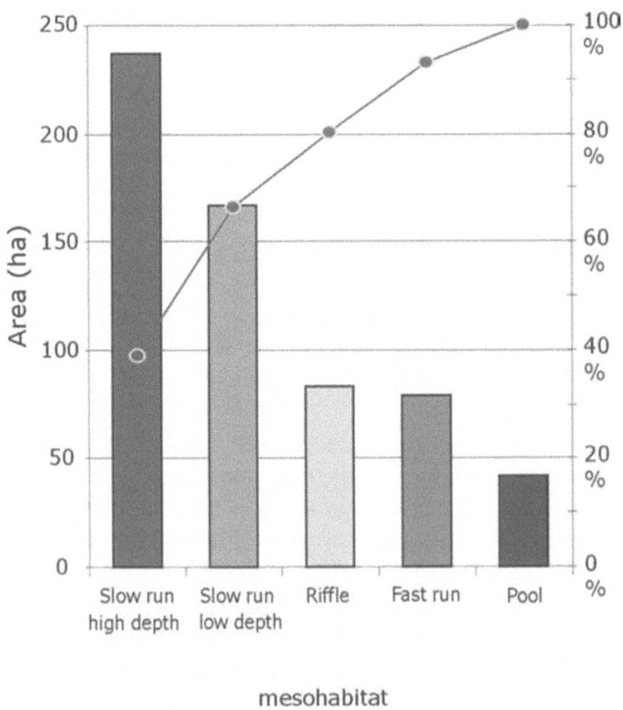

Fig. 17.8 Reproductive habitat availability

or capture made by fishermen and reported to the park authorities. The second set of data instead refers to the period 1993–2006 and comes from different studies and fish-management projects conducted in the Ticino River prior to or at the same time as the present project, by or on behalf of the Ticino Park or other local administrations, such as those of Milan or the Pavia Districts. Regarding the main reach of the Ticino River and the stretch of the Po included in the project area, 109 surveys were carried out, with more than 1500 recordings of the presence and estimate abundance of the different fish species.

The description of the fish community inthe collected data shows the existence in the Ticino and the middle reach of the Po of a great variety of fish species; 47 fish species were recorded, 21 of which were exotic. Unfortunately, such a variety does not come from the natural processes of evolution and conservation of the native populations, but rather from historical and profound alterations of the fish community caused by humans during the last century.

There are remarkable overlaps in the distribution areas of the different species. This is true in particular for the adriatic sturgeon, which, being distributed in the reaches with the highest densities and the greatest biodiversity, has to share the habitat with many native and exotic species, i.e. with nearly all the species currently present in the Ticino River.

Considering that the adriatic sturgeon has locally co-evolved with the other native species of the river, the forced cohabitation with newly introduced species could represent a threat, not from the exotic species found occasionally or those introduced ages ago (e.g. the Crucian Carp and the Carp), but rather those recently acclimatized with consistent and increasing populations. Comparing the auto-ecological characteristics of the exotic species with those of the adriatic sturgeon, it is possible to create competition and predation mechanisms. There are, in fact, many species that share the same habitat of the sturgeon or have the same feeding preferences, or that choose similar reproductive areas, with a temporal overlap. Considering the results from the observations and the surveys on the sturgeon (through biotelemetry) and on the fish community structure carried out during the project, we can affirm that:

1. There is not a real habitat competition between the adriatic sturgeon and other species. Many of the sturgeons monitored through biotelemetry, and also those recorded during the nocturnal samplings, showed in fact sedentary habits, occupying the same pool for a long time (several months), where the fish community is particularly diversified and abundant.
2. There is not even a real feeding competition with the exotic species, because the macrobenthos (and in particular the food preferred by the adriatic sturgeon, such as Gammarids, Lumbricids) is very abundant in the river.
3. The most important disturbance to the reproduction of the Adriatic Sturgeon is probably linked to eggs predation; however, this can be contrasted to the adoption by the sturgeon of an *r* reproductive strategy, with the production of a huge number of eggs.
4. The most important threat is the predation on larvae and yearlings (much more difficult in the adults) by ichtyophagic species.

In relation to the results, it appears that the exotic species currently most widespread in the project area, the sheat-fish, the asp, and the exotic barbel, can be considered the most threatening for the adriatic sturgeon.

Regarding the sheat-fish, even though no case of direct predation on sturgeons has been recorded during the control activities, the results on diet and feeding behaviour confirm that the spread of this species can really represent a strong menace to the adriatic sturgeon, particularly in relation to possible predation pressure. The data collected over more than 5 years of study on this species, in a previous Life-Nature project of conservation of two other species—marble trout and pigo—and this Life project, point out the biological and ecological characteristics that make this species particularly invasive.

17.4.2 The River Fragmentation (Action A.3)

The Ticino River, from the Bereguardo (PV) to the confluence with the Po: this fluvial reach of about 30 km presents no impediments, even partial ones, that could prevent the free migration of fish.

The Po River, from the Ticino confluence to the dam of Isola Serafini: this reach of about 90 km has no significant impedidments. Downstream of the thermoelectrical plant of Piacenza, there is an underwater weir of huge stones, which does not represent an obstacle to the free migration of fish. The dam, instead, constitutes a complete interruption of the fluvial corridor, preventing fish from moving upstream. The inactivity of the navigation basin prevents even the hypothetical minimum passage of fish due to the transit of boats.

Particularly significant from an environmental point of view is the alteration of the river near the dam of Isola Serafini, which for a few kilometres transforms the river into a lake-like environment, with a very slow speed flow. This represents another serious perturbation to the fluvial ecosystem, which leads to a loss of environmental and biological diversity, besides the fragmentation of the Po River.

Adda River: It has many interruptions along its course, such as the weir at Maccastorna (LO), located few hundreds metres upstream of the Po confluence; which limits the possible passage of the Po in flood. It is a small weir that in periods of low flow can represent an obstacle to fish migration. At Pizzighettone (CR) there is a big weir, in two steps, with a total gap of 7–8 m, which is insurmountable. At the old bridge of Lodi, there is another weir without a fish pass, which creates a level difference of about 3.5 m, which also is insurmountable. Therefore, the Adda River, even though it attracts sturgeon and other species in migration from the sea, is unfortunately fragmented by many impediments that prevent the free movement of fish.

Oglio River: The only impediment to fish migration in the entire fluvial course, within the South Oglio Park (a reach of about 50 km), is a control weir at the street bridge of Isola Dovarese, made up of two steps about 2.5 m high, insurmountable during low flow periods. During floods, they are underwater, restoring, only in a

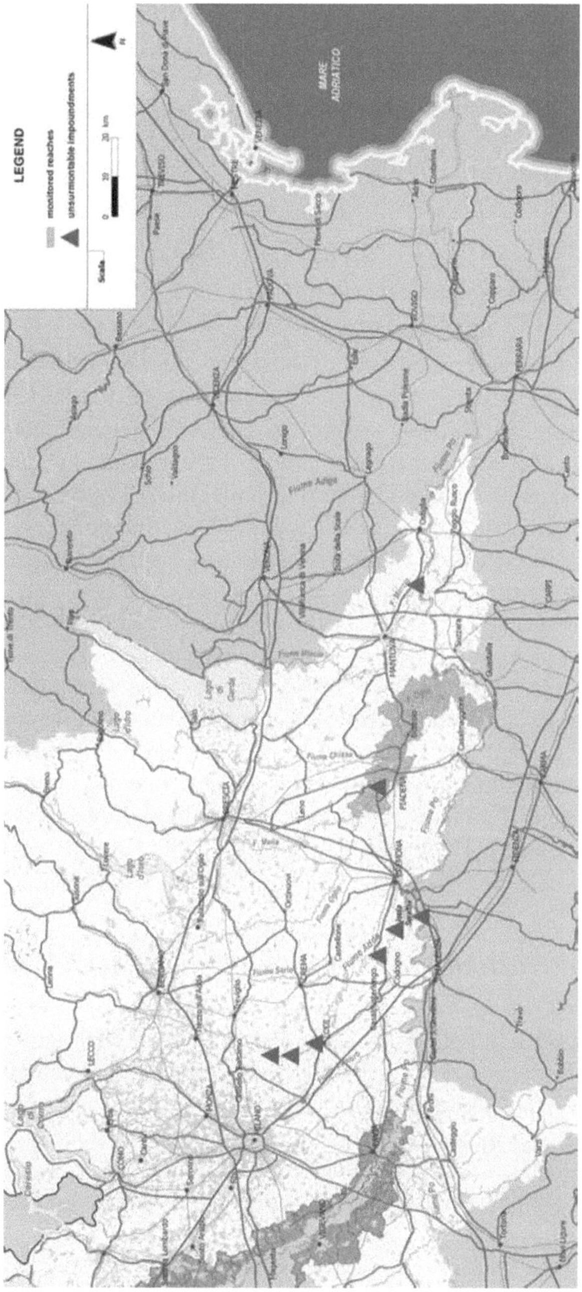

Fig. 17.9 Insurmountable obstacles within the project area

theoretical way, the free flow of the river. Upstream of this obstruction, the river guarantees free movement until a water diversion at Roccafranca (BS), which supplies water to the Carosa Channel, for agricultural use.

Mincio River: It allows free migration upto the dam of Governolo (MN), 2.5 km from the Po confluence. Such an impediment is totally insurmountable, except during exceptional floods (Fig. 17.9).

17.4.3 Genetic Analysis (Action A.4)

The study of the genetic variability is a fundamental tool for the management and conservation of small populations. In such populations, it is possible that the effects of genetic drift and inbreeding, i.e. crossings between consanguineous individuals, can reduce the genetic variability of the population and endanger its potential for adaptation, compromising the possibility of survival. The level of genetic variability, expressed by the heterozygosis and inbreeding rate, is used to estimate the real dimensions of the population in the bred sturgeons used for repopulation activities.

The results show that the level of genetic variability of bred individuals is similar to that of the wild population formerly studied in the literature. Unfortunately, it has not been possible to estimate the level of variability of the wild population, due to the very small number of samples (three specimens). However, the results showed that the three wild sturgeons, at the six loci analysed, have all the alleles present in the bred sturgeons, although it has not been possible to compare the frequencies. This result, although preliminary, supports the use of bred sturgeons to reinforce the natural population in the Ticino River.

17.4.4 Counteraction to the Spread of Sheat-Fish (Action D.3)

The sheat-fish control activity has been carried on for three consecutive years, from 2003 to 2006. The diurnal electrofishing has proved the most effective technique of hunting, allowing the capture of many fish of all the sizes. The nocturnal electrofishing, on the other hand, was particularly effective in summer, allowing the capture of adult subjects during reproductive activity.

Underwater fishing in large pools was particularly effective in the capture of adults of large size, but less effective when considering the catch per unit effort and therefore abandoned in favour of a greater number of electrofishing surveys.

On the whole, 4300 sheat-fish, weighing 9.55 tons, have been collected: (1) 3359 specimens, weighing 7.7 tons, in the Ticino River, (2) 880 specimens, weighing1.5 tons, in the Po River and (3) 88 specimens, weighing350 kg, in the Oglio River.

The data collected show that more than 180 specimens have been captured in each survey, sometimes reaching a catch of more than 400 kg.

17.4.5 The Study of the *S. glanis* Autoecology (Action D.3)

Data on growth rate and reproductive biology of the species have been collected.

The weight growth rate of the sheat-fish population was:

$\ln P = -11.28 + 2.9 \ln \mathrm{LT}$ $r^2 = 0.96$

The length growth rate is expressed by the following curve elaborated through the application of the Von Bertalanffy model (Fig. 17.10):

$L_t = 1900\ (1 - \exp^{(-0.098\ (t + 0.74))})$ $r^2 = 0.96$

The results show a very fast growth rate: after a year, the species reaches 30 cm and after 5 years, 80 cm.

The first sexual maturation is reached by the females at 4–5 years of age, after a process of maturation of the gonads that lasts about 2 years.

Reproduction lasts from June (sometimes even from May) to August. The long duration of the reproductive period results in the presence within the same age class of animals of very different size. Among the individuals that were 1 year old in spring, there were some that were double the size of the others. Such diversity tends to become lower through the years, though very slowly; it is therefore possible to find individuals not yet sexually mature but of large size and individuals considerably smaller but already mature and in reproduction.

The fecundity data show that the species, besides giving parental care till the first growth stage, has adopted a reproductive strategy that consists of producing a great number of eggs. According to the count of the eggs in subsamples of gonads ($N = 69$), the absolute fecundity is 17,000–380,000 eggs produced by each female (average of 75,000 eggs) and the relative fecundity is 5,000–25,000 eggs/kg of fish.

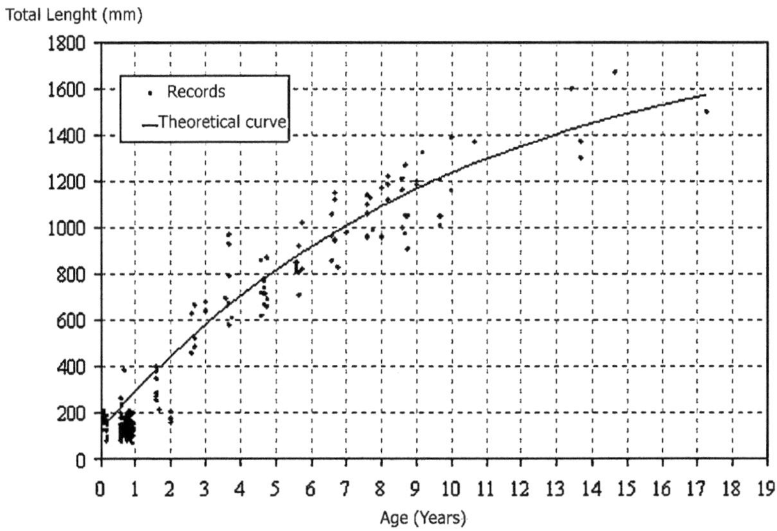

Fig. 17.10 Linear growth rate of *Silurus glanis*

17.4.6 Repopulation Activity (Action C.1)

To monitor the nutrition state and the growth rate of the sturgeons, both before they were transferred to the Cassolnovo tanks, and at the moment of their release in the river, the fish were all measured and weighed and, thanks to the PIT tagging, it has been possible to reconstruct the individual curve of growth in weight and length for all specimens. Comparing the growth rate recorded in the intensive and extensive breeding tanks, a decrease in the growth in weight, and a loss in weight, have been recorded after the period of auto-feeding in the semi-natural tanks; however, the constant increase of length recorded, even though slower in the second phase, demonstrates that the chosen duration of the pre-adaptation phase is enough to make the sturgeons ready to live in the wild. Also the 0+ sturgeons were transferred to the tanks at Cassolnovo for the phase of pre-adaptation, but the presence of a small unknown passage allowed them to exit and reach the open water.

After the phase of pre-adaptation to the natural life, a total number of more than 1200 specimens of sturgeons were gradually set free in the Ticino and Oglio Rivers.. In the Ticino River, the fish were set free in the reach having the greatest concentration of the adriatic sturgeon, that is, between Vigevano and Pavia.

17.4.6.1 Tests of Artificial Reproduction (Action C.3)

As part of the project, the first experiments in breeding the adriatic sturgeon were carried on, as described. The embryonic development was monitored daily and dead eggs were removed. After about 24 h from fertilization, gastrulation occurred; after 96 h the embryos were in an advanced stage of neurulation; after 120 h the embryos were developed and wrapped in the yolk sac, inside the membrane of the egg; after approximately 180 h the hatching started (Fig. 17.11). The larvae were transferred in a glass-reinforced plastic tank and the alevins were fed with *Artemia salina*, when the resorption of the yolk sac was nearly completed. In both cases, the hatching yield was quite low, probably because of the incomplete maturation of the female gonads. On the whole, a few thousand sturgeon alevins were produced.

17.4.7 Action Plan for the Management of *A. naccarii*

The experience and knowledge acquired during the project constitute the background for the formulation of management and conservation proposals in the Action Plan. They provide hope for the restoration, and therefore for conservation, of the adriatic sturgeon in the Ticino River and in the middle reach of the Po River. The Plan consists of a series of activities aimed at the conservation of the species:

1. Monitoring the state of the adriatic sturgeon population
2. Protecting and restorating the natural environment in the course of land management

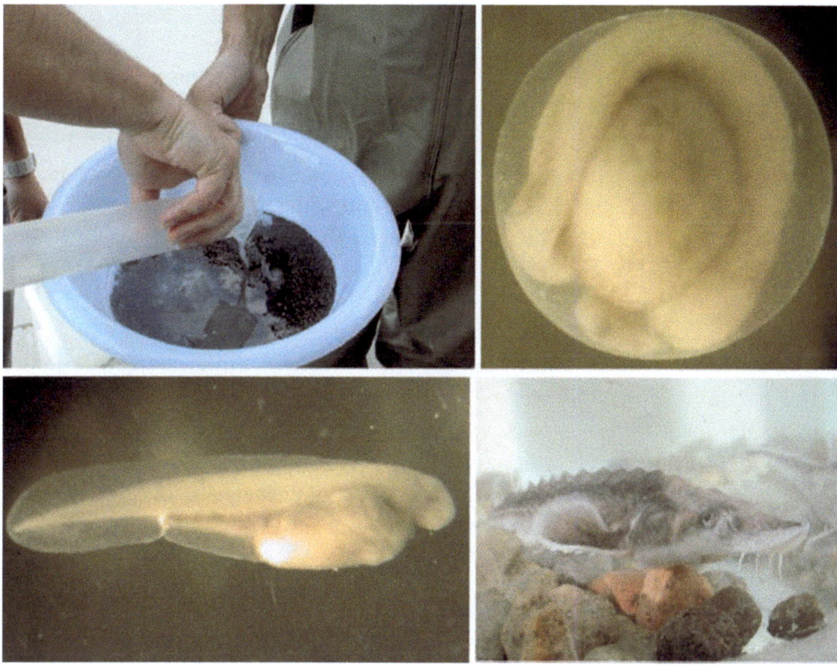

Fig. 17.11 Artificial reproduction of the adriatic sturgeon; from left to right: fecundation of the eggs; an egg 120 h after fecundation; a larva; a yearling

3. Counteraction to the spread of invasive exotic species, in particular *S. glanis* and *Aspius aspius*
4. Sensitization and active involvement of the local human population
5. Breeding and restocking activities
6. Activation of legislative instruments: proposal of insertion of *A. naccarii* in Appendix I of the Bonn Convention; proposal of reclassification of *A. naccarii* in the Red List of IUCN as 'EN-endangered'; proposal of the promotion of conservation plans for sturgeons in the Mediterranean Sea, within the Barcelona Convention
7. Auditing and monitoring of natural reproduction in the river
8. Reinforcement of surveillance and assessment of the opportunity of sanctioning further non-legal behaviour associated with the possession of an underwater gun

17.5 Conclusions

After 3 years of study and experimental activities of restoration of the adriatic sturgeon in the Ticino River and the middle reach of the Po River, important goals have been achieved and a series of conclusions can be drawn, as summarized below:

1. The 'Action Plan of management of the species' has been formulated and published, containing not only the guidelines for its management and conservation, but also a study of the biology and autoecology of the Adriatic Sturgeon, at least for all the aspects that could be analysed.
2. The important activity of repopulation of the Ticino River (and marginally of the Po River and the Oglio River) has been completed with bred sturgeons coming from the Po Basin, genetically compatible and with a sufficient degree of genetic variability, as reflected in the genetic analyses carried out.
3. The demographic control of the sheat-fish population has been carried out successfully; this fish is a very invasive species, dangerous for the native ichthyofauna, as found in previous studies.
4. Through many public meetings and the numerous outreach publications, a steady campaign has been carried out to raise the awareness of the people, in particular local fishermen and young people.
5. Thanks to this research, a large amount of information is available on the fish community of the river, defining the best management guidelines of the fluvial ecosystem to maintain its high level of naturalness and biodiversity.

References

Bernini F. & Nardi P.A. 1989. Caratteri morfometrici e meristici del genere *Acipenser* L. (Osteichthyes, Acipenseridae) nel tratto pavese dei Fiumi Po e Ticino. *Boll. Mus. Reg. Sci. Nat. Torino*, 7(2): 321–340.

Bernini F. & Nardi P.A. 1990a. Regime alimentare di *Acipenser naccarii* Bp. (Osteichthyes, Acipenseridae) nel tratto pavese dei fiumi Po e Ticino. *Boll. Mus. Reg. Sci. Nat. Torino*, 8(1): 429–439.

Bernini F. & Nardi P.A. 1990b. Accrescimento di *Acipenser naccarii* Bp. (Osteichthyes, Acipenseridae) nel tratto pavese dei Fiumi Po e Ticino. *Boll. Mus. Reg. Sci. Nat. Torino*, 8(1): 159–172.

Bronzi P., Vecsei P., & Arlati G. 2005. Threatened fishes of the world: *Acipenser naccarii* Bonaparte, 1836 (Acipenseridae). *Environmental Biology of Fishes*, 72: 66.

Collins M.R., Smith T.I., Post W.C., & Pashuk O. 2000. Habitat utilization and biological characteristics of adult Atlantic sturgeon in two South Carolina Rivers. *Transactions of the American Fisheries Society*, 129: 982–988.

Gandolfi G., Zerunian S., Torricelli P., & Marconato A. 1991. *I pesci delle acque interne italiane.* Ministero dell'Ambiente—Unione Zoologica Italiana, Istituto Poligrafico e Zecca dello Stato, Roma, 616 pp.

GRAIA srl. 1999. Ricerca sulla Fauna Ittica del Fiume Ticino. Parco del Ticino, Magenta, 500 pp.

GRAIA srl. 2004. Conservazione di *Salmo marmoratus* e *Rutilus pigus* nel Fiume Ticino, (Life00nat/it/7268). Rapporto tecnico consegnato alla Commissione Europea.

Kieffer M.C. & Kynard B. 1993. Annual movements of Shortnose and Atlantic Sturgeons in the Merrimack River, Massachusetts. *Transactions of the American Fisheries Society*, 122: 1088–1103.

Nielsen L.A. 1992. *Methods of Marking Fish and Shellfish.* American Fisheries Society, Special Publication 23, 208 pp.

Rossi R., Grandi G., Trisolini R., Franzoi P., Carrieri A., Dezfuli B.S., & Vecchietti E. 1992. Osservazioni sulla biologia e la pesca dello storione cobice *Acipenser naccarii* Bonaparte nella parte terminale del Fiume Po. *Atti della Società Italiana di Scienze Naturali e del Museo Civico di Storia Naturale di Milano*, 132(10): 121–142.

Chapter 18
Identification of Sturgeon Caviar Using DNA Markers

Manuel Ruiz Rejón, Francisca Robles, Roberto de la Herrán, Manuel Garrido-Ramos, and Carmelo Ruiz Rejón

Abstract The meat and, above all, the caviar from sturgeons have been gastronomic emblems of delicacy from time immemorial. When sturgeons were abundant, caviar was marketed primarily from three species (*Huso huso, Acipenser gueldenstaedtii* and *A. stellatus*) mostly from the Caspian Sea. However, these are not the only species of sturgeons used traditionally to produce quality caviar, nor is the Caspian its only geographic origin. For example, caviar from sturgeons belonging to the species *A. sturio* or *A. naccarii* captured in the Guadalquivir River (southern Spain) gained renown in the 1960s, winning prizes at world fairs. Now, when sturgeons are almost disappearing all over the world, the caviar trade has become far more complex. In fact, practically all sturgeon species can produce quality caviar, not only in the wild (in steady decline) but also in fish farms. This scenario makes it necessary to monitor the sale of caviar at a worldwide level, and for this the application of molecular techniques based on DNA, together with others, can be of great use. In this chapter, we present two types of DNA markers for this purpose: some, especially nuclear-DNA markers, indicate qualitative differences between the different species (presence/absence of these markers); and others, particularly mitochondrial-DNA, indicate differences in certain bases of nucleotide sequences. In addition, sturgeons produce interspecific hybrids. Consequently, nuclear markers along with mitochondrial ones are needed for accurately identifying the species responsible for a given caviar.

Keywords Sturgeon, caviar, molecular markers, genetic identification, hybridization

M.R. Rejón, R. de la Herrán, M. Garrido-Ramos and C.R. Rejón
Departamento de Genética, Facultad de Ciencias, Universidad de Granada, Granada, Spain

F. Robles
Present address: Institute for Zoo and Wildlife Research, Department of Evolutionary Genetics, Berlin, Germany

R. Carmona et al. (eds.) *Biology, Conservation and Sustainable Development of Sturgeons*, 299

18.1 Introduction: The Problem of Sturgeons and Their Caviar

Sturgeon and paddlefish meat, and especially their eggs sold as caviar, are symbols of gastronomic delicacy and have been consumed from time immemorial. Sturgeons and paddlefish constitute the order Acipenseriformes, a group of fish comprising 27 species in six genera: two genera of paddlefish, each with one species (*Polyodon*, *Psephurus*); and four genera of sturgeons (*Acipenser* with 17 species, *Huso* with two, and *Pseudoscaphirhynchus* and *Scaphirhynchus* each with three species) (Rochard et al., 1991; Birstein, 1993; Bemis and Kynard, 1997; Birstein and Bemis, 1997). Sturgeons are normally anadromous fish (i.e. they spawn in the freshwater of rivers, migrate to the salt water of the sea, and later return to reproduce in the same rivers where they were born). They live in rivers and seas of different geographical regions. Today, in North America, there are one species of paddlefish (*Polyodon spathula*) and eight species of sturgeons (*Scaphirhynchus albus*, *S. platorhynchus* and *S. suttkusi*, *Acipenser oxyrinchus*, *A. brevirostrum*, *A. medirostris*, *A. transmontanus*, and *A. fulvescens*); there are two sturgeon species in Europe (*A. sturio* and *A. naccarii*); seven live in Eurasia (*A. gueldenstaedtii*, *A. baerii*, *A. nudiventris*, *A. stellatus*, *A. ruthenus*, *A. persicus*, and *Huso huso*), and, finally, one species of paddlefish (*Psephurus gladius*) and eight species of sturgeons (*Pseudoscaphirhynchus kaufmani*, *P. hermanni*, *P. fedtschenkoi*, *Huso dauricus*, *Acipenser mikadoi*, *A. dabryanus*, *A. sinensis*, *A. schrenckii*) live in the rivers and seas of Asia.

This group of fish appeared more than 200 Myr ago (Gardiner, 1984) but now, due to different factors including the contamination and changes undergone by the rivers and seas in which they live (in all the cases sturgeons are native to the Northern Hemisphere; Billard and Lecointre, 2000) and above all to overfishing for meat and caviar, they are one of the most threatened groups of vertebrates (McDowald, 1999). In fact, at present, most of the sturgeon species are considered threatened and some in danger of extinction by the International Union for Conservation of Nature (IUCN, 2004; www.iucnredlist.org). For this reason, at present, it is prohibited to trade or traffic in some species (among these, *A. sturio*, a species normally present in rivers and seas of Western Europe). In other cases, sturgeons are under control (see below), and recently it has even been recommended to prohibit the international trade of caviar from sturgeons of some areas, such as the Caspian Sea, which are subject to great international demand.

When sturgeons were abundant in rivers and seas, caviar was marketed primarily from three species: *H. huso* (beluga caviar), *A. gueldenstaedtii* (osetra) and *A. stellatus* (sevruga), in descending order of quality and price (Fig. 18.1). Most of these are from the Caspian Sea and regions such as Russia and Iran. However, these are not the only species of sturgeon used traditionally to produce quality caviar, nor is the Caspian its only geographical origin. For example, caviar from sturgeons captured in the Guadalquivir River (*A. sturio* or *A. naccarii,* see below) gained world renown in the 1960s, winning prizes at world fairs (Fig. 18.2). Also, caviar from other species and regions has been sold with the above names, depending on the one they most resembled.

Fig. 18.1 Commercial examples of caviar sold from *Huso huso* (beluga caviar), *Acipenser gueldenstaedtii* (osetra) and *A. stellatus* (sevruga)

Fig. 18.2 Caviar produced by the commercial brand 'Hijos de Ybarra' in the factory of Coria del Río (Seville), Spain

In recent years, and particularly at present, given that sturgeons are disappearing, the situation regarding their meat and caviar trade has become far more complex. The black market has increased and caviar of almost any quality is sold under almost any name, regardless of the species that produced the caviar. For example, caviar can be bought under the label *H. huso* when in reality it is *A. gueldenstaedtii, A. stellatus, H. dauricus* (kaluga) or some other species. In addition, caviar is being sold from other sturgeon species in the wild (in America, for example, from *A. oxyrinchus*) or from fish farms: e.g. from *A. naccarii* farmed in Spain (Riofrío, Granada, Spain) from *A. baerii* in France or Italy or from *A. transmontanus* in Italy and America (Fig. 18.3). In fact, practically all sturgeon species can produce quality caviar.

Finally, taking advantage of the natural tendency of sturgeons to produce viable and fertile interspecific hybrids in nature, breeders are rearing interspecific hybrids, in some cases marketing the caviar incorrectly labelled as 'pure' of one species. For example, the hybrid between *H. huso* (as female parent) and *A. ruthenus* (as male parent), known as 'bester', has been marketed as pure *H. huso* when it is actually a mixture, with the nuclear genes from the two species, *H. huso* and *A. ruthenus* and the mitochondrial genes from only *H. huso*. In this respect, it should be borne in mind that it is the female vertebrate (in this case sturgeon) that transmits the mitochondria to the offspring through the cytoplasm of the ovule; the spermatozoid, on the other hand, contributes practically no cytoplasm to the fertilized ovule and therefore contributes no mitochondrial DNA to the offspring.

In an attempt to correct these marketing problems, the Convention on International Trade in Endangered Species of Wild Fauna and Flora (CITES) adopted a mandatory guideline (CITES Resolution Confs. 11.3 and 12.7-Rev. Co P13) for the label, according to which, caviar sold in any form, regardless of species or country of origin, must carry a given code. For exports and re-exports without re-packaging, the label or mark must include a formula that provides the following information:

Fig. 18.3 Caviar obtained from *Acipenser naccarii* in Spain and from *Acipenser transmontanus* in California

Table 18.1 Standard three-letter species code

Species	Code	Species	Code	Species	Code
Acipenser baerii	BAE	*A. oxyrinchus*	OXY	*H. dauricus*	DAU
A. brevirostrum	BRE	*A. persicus*	PER	*Scaphirhynchus platorhynchus*	PLA
A. dabryanus	DAB	*A. ruthenus*	RUT	*S. albus*	ALB
A. fulvescens	FUL	*A. schrenckii*	SCH	*S. suttkusi*	SUT
A. gueldenstaedtii	GUE	*A. sinensis*	SIN	*Pseudoscaphirynchus kaufmani*	KAU
A. medirostris	MED	*A. stellatus*	STE	*P. hermanni*	HER
A. mikadoi	MIK	*A. sturio*	STU	*P. fedtschenkoi*	FED
A. naccarii	NAC	*A. transmontanus*	TRA	*Polyodon spathula*	SPA
A. nudiventris	NUD	*Huso huso*	HUS	*Psephurus gladius*	GLA

1. Standard three-letter species code, as indicated in Table 18.1. Hybrids should be identified using the codes for the male and female parents, separated by the letter '×'. For example: GUE × BAE.
2. One-letter source code of the specimen.
3. Two-letter International Organization for Standardization (ISO) code of the country of origin.
4. Four-digit year of harvest.
5. Processing-plant code (representing the plant producing the caviar or the exporter). This code can be letters or numbers, or a combination, as determined by the labeller.
6. Lot identification number (corresponds to the caviar-tracking system used by the processing or repackaging plant). It does not have to be four digits.

These measures are appropriate, although the problem of possible mislabelling persists (Raymakers, 2002), as when caviar is sold without this label or when caviar protected by a particular label is stated to be of a certain species when it is in reality another, and the verification difficulties remain. For these reasons, the need arises for techniques that can verify as accurately as possible the majority of items included in the guideline of CITES, and above all the species that in fact produced the caviar.

In this chapter, we review the attempts to develop DNA markers intended to identify sturgeon species and their caviar. These genetic markers can also be useful to some extent for verifying other items of the guidelines such as the country or the population of origin, although they are not valid to verify some others such as the manner or year of obtaining the caviar. These latter problems require different techniques such as gas–liquid chromatography, which determines the fatty-acid composition of caviar (to identify the source, wild or farm, Wirth et al., 2002) or microbiological and food-control methods (to determine the age of the caviar).

18.2 Techniques for Caviar Identification

Traditionally, the caviar species has been identified by morphological traits (size and colour of the eggs) or by its flavour. For example, beluga caviar from *H. huso* is purported to be composed of bigger, blacker, and more flavourful eggs than of those the other species. But clearly, these characteristics, being highly variable and subjective, are not trustworthy. In fact, recently there have been some attempts to make these aspects objective through scientific analysis of oocyte morphology of different species, i.e. by scanning electron microscopy (SEM) (Debus et al., 2002). However, these attempts have not been successful because all the species have similar oocytes and because caviar processing blurs the possible diagnostic differences between species.

Therefore, in the last few years, efforts have shifted to applying techniques and markers that are more scientifically reliable. These include techniques from molecular biology, which seek to identify the caviar of the different sturgeon species based on the differences in very basic biological characteristics of these species: proteins and DNA.

With respect to the possibility of distinguishing caviar by analysing its proteins, two techniques have been employed: isoelectric focusing (IEF) and gel electrophoresis. However, IEF has given weak differentiation between species and is very sensitive to the processing method and to the sampling of the caviar (Rehbein, 1985; Keyvanfar et al., 1988 and Chen et al., 1996). Different authors have undertaken protein-electrophoresis studies of enzymatic and non-enzymatic proteins in sturgeons, providing some worthwhile results for differentiating species, populations and even sexes; nevertheless, the techniques are difficult to apply in caviar used for trade because these techniques can be applied only to live or very fresh material.

Finally, techniques based on DNA markers have been successfully used in many problematic issues for identifying biological material. The use of molecular markers based on DNA is supported by several facts. First, DNA and the specific markers within this material are present in all biological material (in this case meat, eggs, and the remains of sturgeon, though at times deteriorated). On this point, it should be mentioned that organisms have two sources of DNA. On the one hand, they have a great quantity of DNA from the nucleus of the cell (in the case of sturgeons this ranges from 2000 million bp up to 6500 million bp, given the high number of chromosomes that they present—120, 240, 500 chromosomes, depending on the sturgeon species—as compared with the human species, which has 46 chromosomes and 3000 million bp in its genome). On the other hand, there is a small quantity of DNA in the mitochondria, the machines that produce energy for the cell and which are located in the cytoplasm of the cells; this DNA is normally inherited from the mother. In the case of sturgeons, as in other vertebrates, mitochondrial DNA is a molecule that does not reach 17 kb. However, the fact that there are several DNA molecules per mitochondria, and particularly that the cytoplasm can contain up to thousands of mitochondria, offsets the initial difference regarding the quantity of the DNA of the nucleus and therefore it is often more feasible to analyse mitochondrial than nuclear DNA.

In addition, DNA, after being obtained from biological material, can be cloned and amplified (in bacteria) and sequenced relatively easily, providing information on the nucleotide composition—that is, the base composition (adenine, thymine, guanine, cytosine, normally in double chains in those that form A-T,G-C pairs). It should be taken into account that, after the DNA is obtained and before or after cloning or sequencing, it can be amplified by techniques such as polymerase chain reaction (PCR), the technique consisting of the in vitro amplification of a specific DNA fragment.

Finally, the idea to use the molecular markers based on DNA to identify caviar is based on the fact that the nucleotide sequences of DNA of different sturgeon species are normally more differentiated than the populations, forms or subspecies of the same species, and, obviously, on the fact that nucleotide sequences of DNA of different sturgeon species are normally more differentiated between distant species than between closely related species.

However, before considering the use of such markers for identifying caviar, we should take into account three facts currently known about the molecular biology of sturgeons. Firstly, sturgeons are living fossils in which the rate of evolutionary change appears to be very low, not only within vertebrates in general but even within fish (Krieger et al., 2000; De la Herrán et al., 2001; Krieger and Fuerst, 2002; Robles et al., 2004). This characteristic to some extent hampers the use of molecular data for inter- and intraspecies differentiation of sturgeons.

Secondly, it should be borne in mind that the molecular data (both nuclear and mitochondrial) point to phylogenetic relationships within the Acipenseriformes which appear to differ from traditional classifications based on morphological analyses. Thus, the molecular data—using mitochondrial and nuclear data—indicate that the two sturgeon species grouped under *Huso* would be more closely related to species belonging to the genus *Acipenser* than are *A. sturio* and *A. oxyrinchus*. However, it is not completely clear whether these two latter species or the species of *Scaphirhynchus* and *Pseudoscaphirhynchus* are the basal lineage in the sturgeon phylogeny, with the paddlefish—*Polyodon* and *Psephurus*—as a sister group (Fig. 18.4) (Artyukhin, 1995; Birstein and DeSalle, 1998; Tagliavini et al., 1999; De la Herrán et al., 2001; Ludwig et al., 2001; Robles et al., 2004). In fact, Birstein et al. (2002), have recently suggested that *Pseudoscaphirhynchus* can be clustered within *Acipenser*, as a species very closely related to *A. stellatus*.

Thirdly, an issue that further complicates the use of molecular data for identifying species producing caviar is the existence not only of artificial hybrids between species, as mentioned above, but also of natural hybrids or forms resulting from introgression. Such hybrids and forms have arisen after processes such as historical and natural movements of distribution areas of sturgeons and in some cases of uncontrolled restocking with exotic species. In these cases sturgeons can have different mitochondrial and nuclear genetic characteristics. i.e. having mitochondrial characteristics of one species but nuclear characteristics of another one, making it necessary to analyse both the markers (mitochondrial and nuclear) in order to identify the caviar.

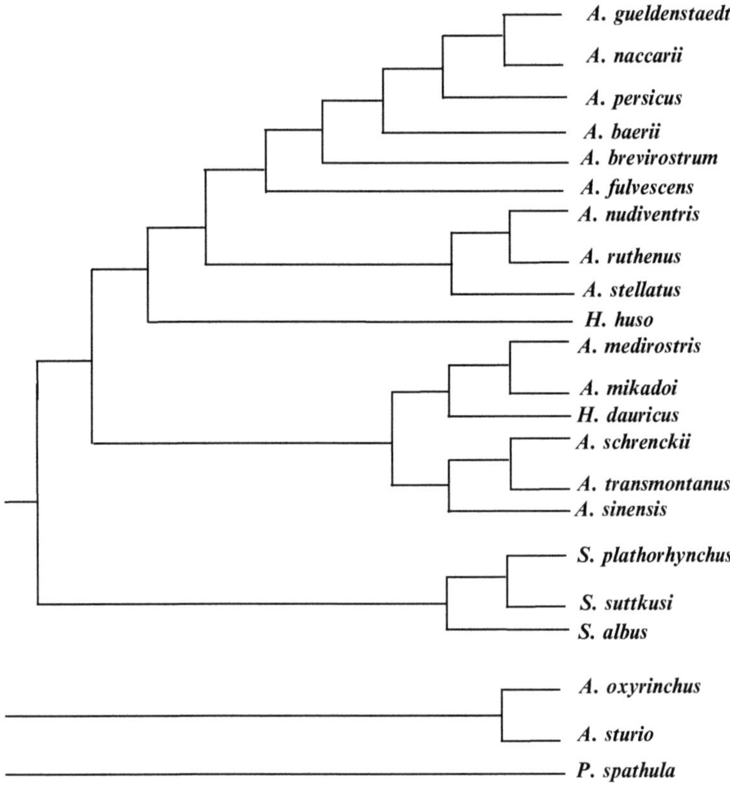

Fig. 18.4 Phylogenetic tree of sturgeon species inferred from mitochondrial DNA sequences according to Ludwig et al. (2001)

With these three characteristics, in this chapter, we review the efforts made in the last few years to develop DNA markers for differentiating sturgeon species in general and specifically their caviar. These markers are based not only on the differences existing in various regions of the mitochondrial DNA, but also on the differences found in certain nuclear-DNA sequences, such as the microsatellites, amplified fragment length polymorphisms (AFLPs), random amplified polymorphic DNAs (RAPDs) and satellite DNAs. We analyse in detail the possibilities of applying these markers in order to differentiate the various sturgeon species and their caviar at two levels:

1. Searching for genetic characteristics, present in some sturgeon species and absent in others, which can be used as qualitative genetic markers which clearly differentiate species and their caviar.
2. Searching for differences in diagnostic nucleotide positions between species for genetic characteristics in all of these, and in this latter case separating the techniques that use DNA sequencing and its subsequent amplification by PCR (nested or semi-nested PCR) from those that use another step, cleaving the amplified DNAs with restriction enzymes (PCR-restriction fragment-length polymorphisms [RFLPs]).

18.3 Identification of Caviar Using DNA Markers

18.3.1 DNA Markers Giving Species-Specific Characteristics

Up to now, the only DNA markers that could give 'qualitative' differences to discriminate different sturgeon species and caviar have been nuclear markers: satellite and microsatellite DNAs, and RAPDs and AFLPs.

18.3.1.1 Satellite DNAs

Satellite DNAs are highly repetitive sequences found in all eukaryotic genomes studied so far. A satellite-DNA family is composed of millions of copies of short tandem repeats (about 180–200 bp in length) forming the heterochromatic parts of the genomes usually at centromeres and telomeres. These types of sequences do not evolve independently but in concert, from the intra-specific homogenization of new variants arising by mutation among repeats and their subsequent fixation in a population (Dover, 1986; Ugarković and Plohl, 2002). Thus, concerted evolution leads to inter-specific sequence identity among repeat units of a satellite-DNA family (Dover, 1986; Charlesworth et al., 1994). On the other hand, at the inter-specific level, concerted evolution results in rapid divergence between repeat sequences of different species and usually leads to species-specific satellite DNAs with repeats that are completely different in sequence (Dover, 1986; Charlesworth et al., 1994). However, there are exceptions in which a group of species shares the same satellite-DNA family. For example, a group of very closely related species commonly conserves the same satellite-DNA family (Wijers et al., 1993; Pons et al., 2002; Pons and Gillespie, 2003) and, less common, this preserved satellite DNA occurs in a whole family of species (Garrido-Ramos et al., 1999). Strikingly, a broad group of species such as a whole order could share a satellite DNA family because of a presumed functionality (Fanning et al., 1993) or because of extremely slow rates of evolutionary change (Grétarsdóttir and Arnason, 1992; Arnason et al., 1992). In these cases, thanks to the high rate of concerted evolution of these types of DNA, repeat sequences of different species can be compared for species-specific diagnostic nucleotide sites. Species-diagnostic mutations in satellite-DNAs are expected to be useful then for genetic identification as well as for systematic purposes (De la Herrán et al., 2001).

However, in sturgeon, we have found one of the most noteworthy exceptions to the rules of the evolutionary dynamics of satellite DNAs (De la Herrán et al., 2001; Robles et al., 2004). The *Hind*III and the *Pst*I satellite DNAs were exceptional in that they are ancient repetitive DNA conserved in many species of the order Acipenseriformes, having an antiquity of about 90 Myr. Slow rates of evolution as well as slow rates of sequence homogenization have been found for the *Hind*III and *Pst*I repeats. It appears that gene flow between species by hybridization and historic introgression has reduced genetic differences between species and has made it difficult to find specific diagnostic mutations (Robles et al., 2004). Nevertheless, satellite-DNA sequences have proved useful for certain taxonomic identification purposes (De la Herrán et al., 2004), since divergence between repeat sequences has led to the high differentiation between the *A. sturio* and *A. oxyrinchus* group and the rest of sturgeon species for these two nuclear repetitive-DNA markers (Robles et al., 2004). Diagnostic differences have been also found when comparing species belonging to two different phylo-biogeographic groups of the four described by Ludwig et al. (2001). Specifically, the *Hind*III satellite is found in sturgeon species belonging to the genera *Huso*, *Scaphirhynchus*, and *Pseudoscaphirhynchus* but in the case of the genus *Acipenser*, it is found in all of them except in *A. sturio* and *A. oxyrinchus*. This finding has enabled the differentiation of biological material of these two species, including their caviar, from the rest of the species: all biological material of these two species lacks this satellite DNA, which is present in the other sturgeon species.

In fact, the existence of this qualitative molecular marker may have, and in fact has had, many useful applications (Fig. 18.5). The presence of the *Hind*III satellite

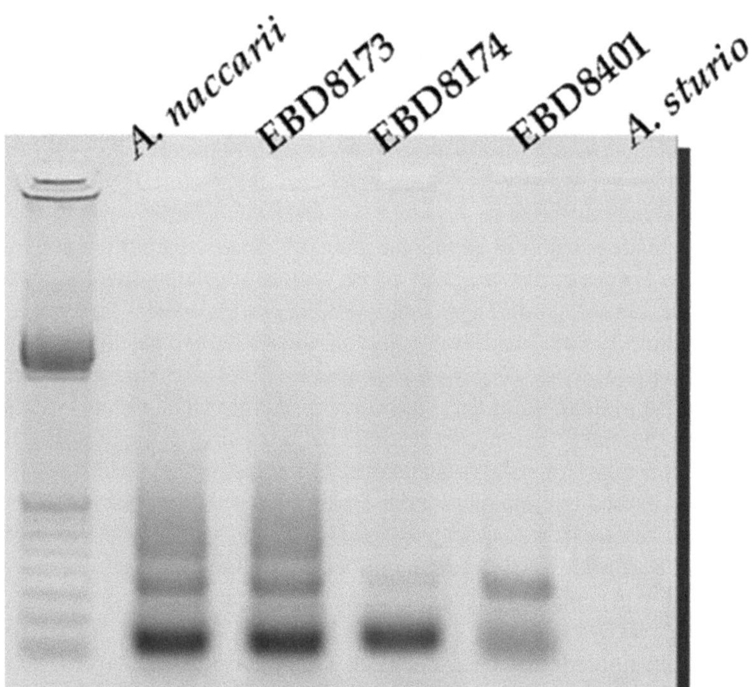

Fig. 18.5 PCR assay demonstrating the presence of *Hind*III satellite-DNA sequences in the genome of three sturgeon specimens kept at the Doñana Biological Station (Seville, Spain) and caught in the Guadalquivir river during last century. This satellite-DNA family is present in the genome of *Acipenser naccarii* but absent from the genome of *Acipenser sturio*, a species that do not show PCR amplification with specific primers for this marker

in some sturgeons captured in the Guadalquivir River and other Mediterranean rivers of France and Italy (in this case not the Adriatic) and preserved in museums (Garrido-Ramos et al., 1997; Ruiz Rejón et al., 2000) provided the first data (confirmed afterwards with the sequencing of the satellite itself and of other nuclear and mitochondrial markers; De la Herrán et al., 2004) indicating that the species *A. naccarii* has a distribution area that is larger than previously suspected (originally thought to be restricted to the Adriatic). In fact, our data demonstrate that *A. naccarii* has inhabited rivers of Western Europe such as the Guadalquivir River. This has important implications for the genetic conservation of this species and programmes for the recovery of this sturgeon (see Chapter 2).

With respect to the use of this marker for identifying caviar, although *A. sturio* is not farmed and its existence and recuperation in nature is problematic, the same is not true of *A. oxyrinchus*. From this latter species, caviar was obtained in the United States a few years ago and there is discussion of reintroducing it into Germany (see Chapter 24) and Poland, on the basis of its close kinship with *A. sturio* (sometimes considered the same species). *A. oxyrinchus* is the species normally found in North America, and even in some periods such as the Little Ice Age it was present in the Baltic in the wild. Therefore, the caviar from *A. oxyrinchus* produced by wild sturgeons, as occurs in North America, or what can be obtained after its reintroduction in Germany, or even what will result from its farming, can be distinguished in two steps from that produced by the rest of the species. In principle, the caviar of *A. oxyrinchus* and also in theory from *A. sturio* (by the absence of this satellite DNA), can be qualitatively identified. After the confirmation of this, a species assignment can be made using the other molecular markers mentioned below.

18.3.1.2 Microsatellites

Other molecular markers, for which qualitative differences have been found between sturgeon species and therefore are useful for caviar identification, are microsatellites. These make up another part of the repetitive sequences present throughout the nuclear genomes of eukaryotic organisms. In these cases, the monomers repeated in tandem are shorter, normally 4–6 bp going from one or two nucleotides, such as CA, to longer monomers such as $(AG)_{12}(CG)(AG)$; $(CG)(AG)_{40;}$ $(ATT)_2(ACT)_{10}$; $(ATTT)_5(ATTC)$, and the degree of repetition is lower, from tens to hundreds. The fact that each monomer motif could be flanked by a characteristic and unique nucleotide sequence enables analysis by means of PCR using primers in these characteristic sequences. In this case, the variability in the number of repetitions of each microsatellite in each individual characterizes the populations and even species for one microsatellite or a combination of microsatellites. In fact, the microsatellites with different numbers of repetitions of the monomer motif are considered alleles of each locus that are inherited in a Mendelian way. This opens the possibility, which is the focus of this chapter, that some populations or even some species can display unique alleles for some microsatellites, and this in turn opens the possibility of distinguishing populations or species using these diagnostic alleles (Fig. 18.6).

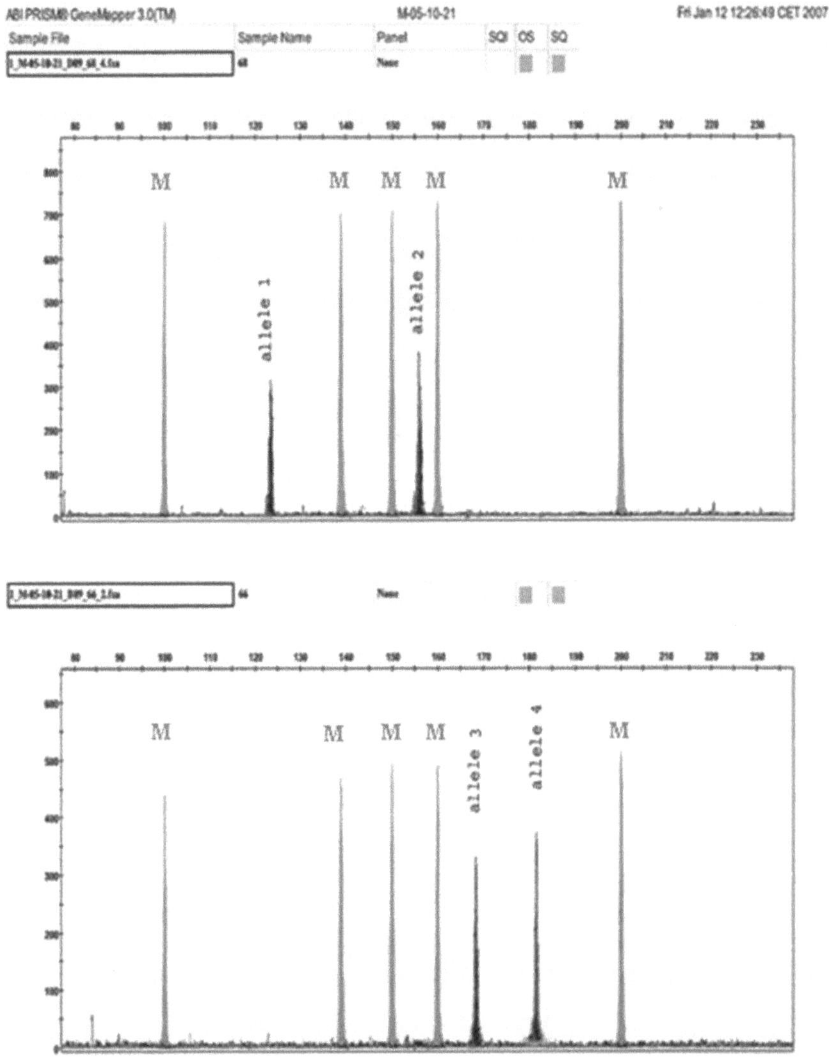

Fig. 18.6 Microsatellite DNA. Results of a microsatellite DNA loci amplification in two samples. Results were screened by using an automated laser fluorescent DNA sequencer. There are two alleles (dark picks) of this microsatellite locus for each individual. The picks noted with M, are the molecular weight markers

In the case of sturgeons, there is at least one microsatellite DNA (Afu-39, so called for being microsatellite number 39 used in *A. fulvescens* by May et al., 1997) which distinguishes *A. stellatus* (the species that originally produced sevruga caviar) and all its products, including caviar, from the other sturgeon species (Jenneckens et al., 2001).

However, microsatellites can be used for many other purposes (such as establishing paternity tests to differentiate populations or variants within a species) as well as the identification of caviar and distinguishing the hybrid nature of sturgeons. Thus, for example, in the USA, there are two sturgeon species (pallid and shovelnose sturgeons, *S. albus* and *S. platorhynchus*, respectively) that are normally isolated reproductively. However, the combined analysis of some mitochondrial DNA regions, on the one hand, and of certain microsatellites, on the other, has demonstrated the existence of hybrid specimens in some river zones of the Montana and Louisiana (Missouri River and Atchafalaya rivers, respectively). That is, some specimens present intermediate characteristics between the two species for these markers (Tranah et al., 2004). Therefore, the combined analysis of certain mitochondrial regions and of certain microsatellites can be useful to detect the hybrid nature of caviar-producing sturgeons. At present, together with the above-mentioned hybrid, 'bester' (HUS × RUT), hybrids are being artificially produced between several species, such as those produced by GUE × BAE and NAC × TRA, or, finally, by NAC × BAE.

18.3.1.3 RAPDs

In the case of random amplified polymorphic DNAs (RAPDs), as the name indicates, the bands that are amplified by PCR are randomly assigned nucleotide sequences. In fact, there are commercial primer collections comprising hundreds of sequence combinations. For living beings, the DNA amplification with these primers by PCR gives rise to a set of bands as a result of the high degree of repetition of the sequences taken at random in the genomes (Fig. 18.7). This also opens the possibility

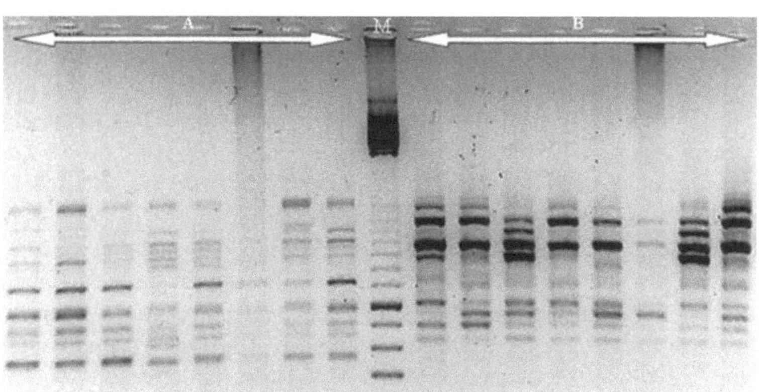

Fig. 18.7 Random amplified polymorphic DNAs (RAPDs). Electrophoresis in agarose gel of amplified fragments in different individuals (eight samples) using two different set of primers (A and B). The line M is the molecular weight marker

that specific bands may appear in certain individuals, populations, in one of the two sexes, and even in certain species. In fact, after the detection of a band specific to one of these levels, it is usually cloned, sequenced, and from this information, a specific primer-pair is designed. After the specificity of the amplification with these new primers is confirmed (especially at the sex or species level), it is usually used as a specific marker for sex or species identification—in this case, a specific marker for a sequence characterized amplified region (SCAR).

In the case of sturgeon, although various analyses using RAPDs have been made trying to distinguish between populations of different regions of some sturgeon species (*H. huso*), or attempting to differentiate problematic species (*A. persicus-A. gueldenstaedtii*), the results are not conclusive. This is partly due to the lack of adequate material. However, RAPDs are of some use as in Comincini et al. (1998) exploring the possibilities of identifying the caviar produced by different species. In this case, applying eight combinations of primers to six species of sturgeons, these authors found some species-specific bands of potential utility in caviar identification.

18.3.1.4 AFLPs

In amplified fragment length polymorphisms (AFLPs), firstly the DNA is digested by two restriction enzymes, one of them that cuts frequently (*Mse*I, 4 bp recognition sequence) and one that cuts less frequently (*Eco*RI, 6 bp recognition sequence). Then, the generated fragments are linked to end-specific adaptor molecules and amplified by means of PCR using primers complementary to each of the two adaptor sequences, except for the presence of one additional base at the 3' end, which is chosen by the user. Now, the generated products are used as template in a second amplification reaction. In this selective PCR the primers used containing two further additional bases are chosen by the user. In this case, the *Eco*RI-adaptor specific primer used bears a label to reveal the pattern of fragments amplified. Therefore, this technique is in a way a mixture of what is used to obtain microsatellites, as mentioned above, and that used to obtain restriction fragment-length polymorphisms (RFLPs), which will be discussed below. In any case, this technique gives a pattern of multi-bands for each individual, so that, in some cases, it has been found that some of them may be specific to one of the sexes or some species, and in theory they can serve our objective here (Fig. 18.8).

In the case of sturgeons, the most complete study was made by Congiu et al. (2001) with the aim of identifying the hybrids that appear naturally or are produced artificially, but the multi-band analysis enables the reliable identification of hybrids, using the parent species as a reference. This type of analysis, together with the previously mentioned analysis of certain mitochondrial regions and RAPDs, is potentially useful for the detection of caviar produced from hybrid sturgeons.

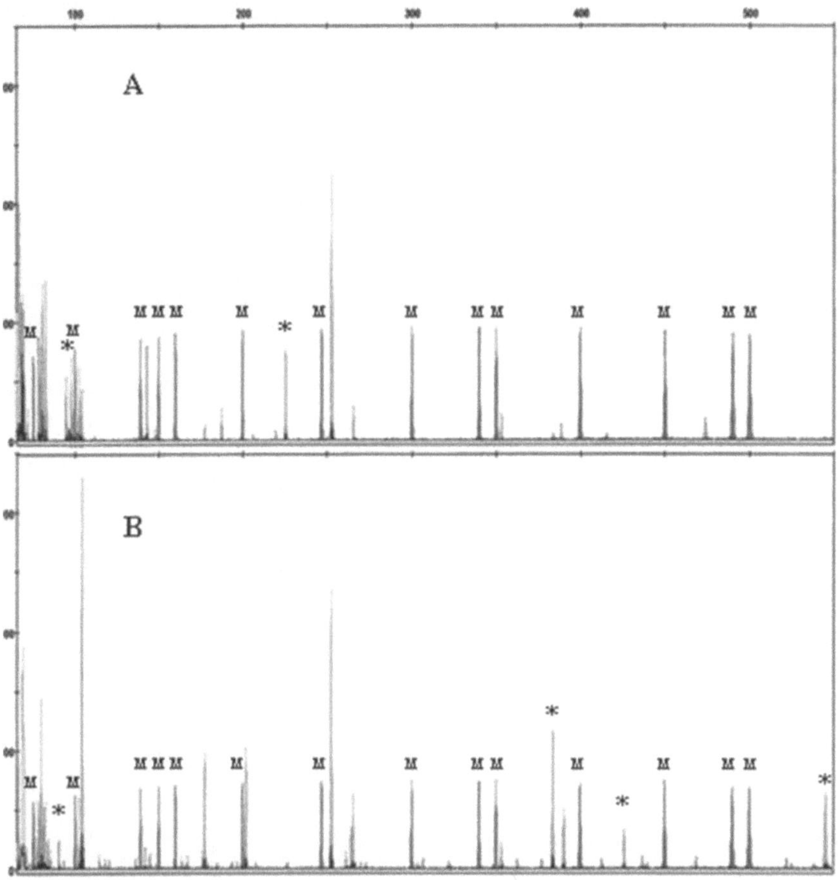

Fig. 18.8 Amplified fragment length polymorphisms (AFLPs). The figure shows the comparison of different amplified fragments of DNA in male (A) and female (B) sturgeon using a selected primers combination after having carried out the protocol described in the text. The picks noted with an asterisk are some differential amplified fragments on male and female individuals. The picks noted with M are the molecular weight markers

18.3.2 DNA Markers for Which There are No Qualitative Differences

18.3.2.1 Mitochondrial DNA

Normally, identification studies of biological material such as caviar are based on the analysis of certain mitochondrial-DNA regions. This DNA, which is normally of maternal origin for being transmitted through the cytoplasm of the ovule, is

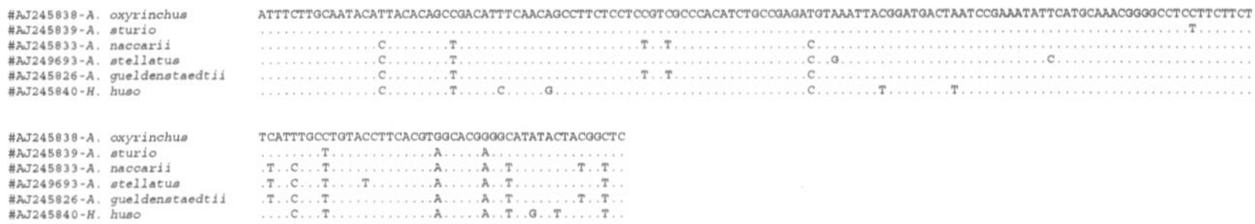

Fig. 18.9 Partial (171 bp) sequence alignment of the cytochrome *b* gene of several sturgeon species. According diagnostic specific nucleotide sites, this mitochondrial marker have been extensively used for genetic identification and for phylogenetic purposes. In the figure, the nucleotide sequence of the first species (*Acipenser oxyrinchus*) is shown while in the other five species only nucleotide differences are revealed. Dashes indicate the same nucleotide as in *A. oxyrinchus*. GenBank/EMBL accession numbers for the sequences of each species are shown

abundant and easily obtained. Moreover, it presents a faster evolutionary rate than for nuclear DNA, and thus it constitutes a good molecular marker to differentiate closely related biological groups and even species.

In the case of sturgeons and their caviar, mitochondrial DNA is easy to obtain. However, again, sturgeons are exceptional regarding this part of the genome because they have a very slow evolutionary rate, and thus there are very few interspecific nucleotide differences on which to base the distinction of a given caviar-producing species (Fig. 18.9). In addition, the use of certain mitochondrial-DNA regions, such as molecular markers to distinguish the caviar is further aggravated by the existence in sturgeons of two additional phenomena, as mentioned above: occasional or historically repeated interspecific hybridizations in the direction of a certain species, called in this case introgressions.

Despite this, most studies made to date seeking to distinguish the specific origin of caviar have been based on mitochondrial DNA. Two main techniques are used. One consists of the sequencing of mitochondrial DNA followed by PCR amplification using primers specific to each species. And the other, consists of the amplification by means of PCR with generalist primers for all the species of sturgeon, finally cleaving the amplified ones with restriction enzymes, so that each species gives a specific cleavage pattern or RFLP.

The molecular method of identifying caviar (or any biological material) with the best diagnostic power would be the total sequencing and comparison of mitochondrial (and/or nuclear) DNA of sturgeons and their caviar. However, this method is unfeasible to apply because it is time consuming and expensive even for a small molecule, such as mitochondrial DNA (less than 17 kb), and is not suitable for analysing large samples. Therefore, alternative techniques are used, based only on the sequencing of certain regions of this molecule. Furthermore, after these regions are sequenced in different species, and when diagnostic or differentiated positions are found between different sturgeon species, two strategies can be followed to avoid sequencing this region subsequently in all the samples. In the first strategy (which is called SEC-PCR) primer pairs are designed to be specific to each species and, in agreement with the information provided by the sequence, will amplify only when the caviar belongs to the species indicated and not to the rest. In the second strategy, universal primers are designed for each mitochondrial-DNA region, and these are amplified with the caviar of all the sturgeon species. Later, the amplified ones are subjected to a series of cleavages with sturgeon enzymes in such a way that the fact of whether they cleave or not and the size of the fragments depend on the particular nucleotide sequence that each species has for each region. This technique is called SEC-PCR-RFLP.

The first technique has been applied successfully to distinguish the caviar produced by the three species traditionally used for caviar (*H. huso*, *A. gueldenstaedtii* and *A. stellatus*). The application of this method by DeSalle and Birstein (1996) and Birstein et al. (1998) using specific primers for the three aforementioned species in three regions of the mitochondrial DNA (cytochrome *b*, and 16 S and 12 S ribosomal genes) enabled the determination of a caviar sample marketed in the USA as belonging to these species, for which there was up to 20% mislabelling. Fundamentally, what was detected with this technique was that several types of caviar were sold mislabelled as beluga/osetra. However, it is difficult to apply this method, considering all the sturgeon species, because specific primers must be designed for each species (a difficulty in some regions), and for the exact determination of the species, a certain number of amplification reactions is needed (at least four PCR in the case of the three species mentioned).

With regard to the second technique, the so-called PCR-RFLP, application began in 1999 by Wolf et al. (1999) for a few species and later this was extended by Ludwig et al. (2002) to 22 species of Acipenseridae. The latter study succeeded in differentiating up to 17 species, lacking only the three species of *Scaphirhynchus* and *A. gueldenstaedtii*/*A. persicus*. In fact, applying this technique, both Wolf et al. and Ludwig et al. have detected the existence of a certain percentage of incorrect labelling in European caviar trade (13% in the former case).

However, as mentioned above, the identification of sturgeons in general and their caviar in particular, using only molecular markers found in mitochondrial DNA, can be problematic with regard to the interspecific hybrid sturgeons, not only the artificial but also the natural ones. In these cases, if only mitochondrial DNA is analysed, the caviar may be assigned to a single species (the donor of the ovule and thus of the mitochondria) when in reality it is the result of hybridization of two different species, a problem that can be resolved only by studying nuclear markers.

Furthermore, in sturgeons, there are two types of hybrids. First, there are the first-generation hybrids, such as artificial hybrids called 'bester' and those produced by hybridization between species as well as those possibly produced in the wild in regions such as the Missouri River (mentioned above). All these hybrids have the mitochondria of one species and the nuclear markers as a mixture of the two.

Second, among those that can be considered as hybrid sturgeons, there are historical hybrids resulting from processes called introgression. These hybrids originated by the initial crossing of two different species and later retro-crossing (introgression) of these hybrids during many generations with only one of the species. Moreover, it often happens that the introgression during many generations is found in hybrid females and their descendants with males of the species with which the introgression takes place. This can happen when one species invades a habitat or region inhabited by another species, or when hybrid females and their offspring retro-cross with males of the resident species. At the end of the process, the resulting specimens will have genetic characteristics from the mitochondria (ovule) of the donor species and traits from the nucleus of the other species involved in the introgression.

In natural sturgeon populations, situations of this type have been found, as for example in rivers of Western Europe (in the Guadalquivir for example, specimens have been found with the mitochondrial DNA of *A. sturio* and nuclear DNA of *A. naccari*, De la Herrán et al., 2004). In the rivers of the Adriatic basin (in the Po river), specimens have the mitochondria of *A. gueldenstaedtii* and nucleus of *A. naccarii* (Ludwig et al., 2003). In the Caspian Sea, and this is more serious for the identification of the caviar of one of the most important caviar-producing species, *A. gueldenstaedtii*. In this case, the species *A. gueldenstaedtii* has three types of mitochondria; some individuals have the characteristics of the species proper, but other specimens bear the mitochondria of *A. baerii*, and there is even a third form that has the mitochondria of *A. naccarii*, which does not currently exist in the zone (Birstein et al., 1998, 2005; Jennekens et al., 2001). All the above signifies that mitochondrial markers are unreliable when used exclusively to identify caviar produced in the Caspian Sea from *A. gueldenstaedtii*, given that false negative and positive results may be found in relation to the three species mentioned. This problem is exacerbated by the fact that a fourth species has been cited, *A. persicus*, which also produces caviar (marketed especially from Iran), for which no differences have been found with respect to *A. gueldenstaedtii* in terms of mitochondrial markers. In fact, at times, doubt arises as to whether this latter species is truly a separate taxonomic entity.

All the above (slow evolution of mitochondria in sturgeons, and hybridization, and introgression) clearly indicates that for accurate identification of caviar-producing species, not only mitochondrial markers but also nuclear markers need to be analysed, as mentioned in the foregoing section.

18.4 General Conclusions

As discussed in detail in this chapter, the methods based on DNA are useful to identify caviar produced by different sturgeon species. Although there is no one universal genetic method that can resolve the problem, there are different markers and techniques to identify the origin of the caviar produced from certain species. Thus, from the nucleotide sequences of certain mitochondrial-DNA regions, species-specific primers can be designed. The alternative involves generalist primers that can be used for all the sturgeons, enabling the amplification and cleavage for specific regions. In this way, identification is possible not only of the caviar produced by the three species traditionally considered the most important (*H. huso*, *A. gueldenstaedtii* y *A. stellatus*) but also for other species. However, the exclusive use of mitochondrial DNA to identify the caviar presents various problems due to biological characteristics peculiar to sturgeons: slow rate of change, interspecific hybridization, and introgression. Consequently, the simultaneous use of nuclear markers becomes necessary, as these in some cases show more pronounced differences between sturgeon species. An analysis has been made of the potential of molecular markers, such as satellite DNAs, microsatellites, RAPDs, and AFLPs, in identifying caviar. This demonstrates both the usefulness of satellite DNA such as the so-called *Hind*III to differentiate the caviar produced by two species, *A. oxyrinchus* and *A. sturio*, from the other species, as well as the utility of microsatellite Afu-39 to identify caviar produced by *A. stellatus*. It also shows the value of RAPDs and AFLPs in identifying caviar produced from hybrid specimens. Despite all the points discussed, further research is need in testing genetic markers that may have greater potential and better applicability, until achieving a DNA barcoding for the different species and their caviar.

Acknowledgement This research has been supported by grant number AGL2003–04716 from the Ministerio de Educación y Ciencia.

References

Arnason U, Gretarsdottir S and Widegren B (1992) Mysticete (balleen whale) relationships based upon the sequence of the common cetacean DNA satellite. *Mol. Biol. Evol.* 9: 1018–1028.

Artyukhin EN (1995) On biogeography and relationships within the genus Acipenser. Sturgeon. *Quercus* 3: 6–8.

Bemis WE and Kynard B (1997) Sturgeon rivers: an introduction to Acipenseriform biogeography and life history. *Environ. Biol. Fish.* 48: 167–183.

Billard R and Lecointre G (2000) Biology and conservation of sturgeon and paddlefish. *Rev. Fish Biol. Fisheries* 10: 355–392.

Birstein VJ (1993) Sturgeon and paddlefishes: threatened fishes in need of conservation. *Conserv. Biol.* 7: 773–787.

Birstein VJ and Bemis WE (1997) How many species ate there within the genus *Acipenser*? *Environ. Boil. Fish.* 48: 157–163

Birstein VJ and DeSalle R (1998) Molecular phylogeny of Acipenserinae. *Mol. Phylogenet. Evol.* 9: 141–155.

Birstein VJ, Doukakis P, Sorkin B et al. (1998) Population aggregation analysis of three caviar-producing species of sturgeons and implications for the species identification of black caviar. *Conserv. Biol.* 12: 766–775.

Birstein VJ, Doukakis P and DeSalle R (2002) Molecular phylogeny of Acipenseridae: nonmonophyly of Scaphirhynchinae. *Copeia* 2: 287–301.

Birstein VJ, Ruban G, Ludwig A, Doukakis P and DeSalle R (2005) The enigmatic Caspian Sea Russian sturgeon: How many cryptic forms does it contain? *Syst. Biodivers.* 3: 203–218.

Charlesworth B, Siniegowski P and Stephan W (1994) The evolutionary dynamics of repetitive DNA in eukaryotes. *Nature* 371: 215–220.

Chen IC, Chapman FA, Wei CI et al. (1996) Preliminary studies on SDS-PAGE and isoelectric focusing identification of sturgeon sources of caviar. *J. Food Sci.* 61: 533–359.

Comincini S, Lanfredi M, Rossi R et al. (1998) Use of RAPD markers to determine the genetic relationships among sturgeons (Acipenseridae, Pisces). *Fish. Sci.* 64: 35–38.

Congiu L, Dupanloup I, Patarnello T et al. (2001) Identification of interspecific hybrids by amplified fragment length polymorphism: the case of sturgeon. *Mol. Ecol.* 10: 2355–2359.

Debus L, Winkler M and Billard R (2002) Structure of micropyle surface on oocytes and caviar grains in sturgeons. *Int. Rev. Hydrobiol.* 87: 585–603.

De la Herrán R, Fontana F, Lanfredi M et al. (2001) Slow rates of evolution and sequence homogenization in an ancient satellite DNA family of sturgeons. *Mol. Biol. Evol.* 18: 432–436.
De la Herrán R, Robles F, Martínez-Espín E et al. (2004) Genetic identification of western Mediterranean sturgeons and its implication for conservation. *Conserv. Genet.* 5: 545–551.
DeSalle R and Birstein VJ (1996) PCR identification of black caviar. *Nature* 381: 197–198.
Dover G (1986) Molecular drive in multigene families: how biological novelties arise, spread and are assimilated. *Trends Genet.* 2: 159–165.
Fanning TG, Seuanez HN and Forman L (1993) Satellite DNA sequences in the New World primate *Cebus capella* (Platyrrhini, Primates). *Chromosoma* 102: 306–311.
Gardiner BG (1984) Sturgeons as living fossils. In: Eldredge N and Stanley SM (eds.), *Living Fossils.* Springer-Verlag, New York, pp. 148–152.
Garrido-Ramos MA, Soriguer C, De la Herrán R et al. (1997) Morphometric and genetic analysis as proof of the existence of two sturgeon species in the Guadalquivir river. *Mar. Biol.* 129: 33–39.
Garrido-Ramos MA, De la Herrán R, Jamilena M et al. (1999) Evolution of centromeric satellite DNA and its use in phylogenetic studies of Sparidae family (pisces, Perciformes). *Mol. Phylogenet. Evol.* 12: 200–204.
Grétarsdóttir G and Arnason U (1992) Evolution of the common cetacean highly repetitive DNA component and the systematic position of *Orcaella brevirostris. J. Mol. Evol.* 34: 201–208.
IUCN 2004. 2004 IUCN Red List of Threatened Species. Downloaded on 07 March 2005. www.iucnredlist.org
Jenneckens I, Meyer JN, Hörstgen-Schwark G et al. (2001) A fixed allele at microsatellite LS-39 is characteristic fort he black caviar producer *Acipenser stellatus. J. Appl. Ichthyol.* 17: 39–42.
Keyvanfar A, Rochu D and Fine JM (1988) Comparative study of sturgeon oocyte soluble proteins by isoelectric focusing. *Comp. Biochem. Physiol. B* 90: 393–396.
Krieger J and Fuerst PA (2002) Evidence for a slowed rate of molecular evolution in the Order Acipenseriformes. *Mol. Biol. Evol.* 19: 891–897.
Krieger J, Fuerts PA and Cavender T (2000) Phylogenetic relationship of north American sturgeons (Order Acipenseriformes) based on mitochondrial DNA sequences. *Mol. Phylogenet. Evol.* 16: 64–72.
Ludwig A, Belfiore NM, Pitra C et al. (2001) Genome duplication events and functional reduction of ploidy levels in sturgeon (*Acipenser, Huso* and *Scaphirhynchus*). *Genetics* 158: 1203–1215.
Ludwig A, Debus L and Jenneckens I (2002) A molecular approach for trading control of black caviar. *Int. Rev. Hydrobiol.* 87: 661–674.
Ludwig A, Congiu L, Pitra C et al. (2003) Nonconcordant evolutionary history of maternal and paternal lineages in Adriatic sturgeon. *Mol. Ecol.* 12: 3253–3264.
May B, Krueger CC and Kincaid HL (1997) Genetic variation at microsatellite loci in sturgeon: primer sequence homology in *Acipenser* and *Scaphirhynchus. Can. J. Fish. Aquat. Sci.* 54: 1542–1547.
McDowald MR (1999) Different kinds of diadromy: different kinds of conservation problems. *ICES J. Mar. Sci.* 56: 410–413.
Pons J and Gillespie RG (2003) Common origin of the satellite DNAs of the Hawaiian spiders of the genus *Tetragnatha*: evolutionary constraints on the length and nucleotide composition of the repeats. *Gene* 313: 169–177.
Pons J, Petitpierre E and Juan C (2002) Evolutionary dynamics of satellite DNA family PIM357 in species of the genus *Pimelia* (Tenebrionidae, Coleoptera). *Mol. Biol. Evol.* 19: 1329–1340.
Raymakers C (2002) International trade in sturgeon and paddlefish species—the effect of CITES listing. *Int. Rev. Hydrobiol.* 87: 525–537.
Rehbein H (1985) Caviar: proximate composition, amino acid content and identification of fish species. Zeitschrift Lebensmittel Unters. *Forschung* 180: 457–462.
Robles F, De la Herrán R, Ludwig A et al. (2004) Evolution of ancient satellite DNAs in sturgeon genomes. *Gene* 338: 133–142.

Rochard E, Williot P, Castelnaud G et al. (1991) Eléments de systèmatique et de biologie des populations sauvages d'esturgeons. In: Williot P (ed.), *Acipenser*. Cemagref, Bordeaux, pp. 475–507.

Ruiz Rejón M, De la Herrán R, Ruiz Rejón C et al. (2000) Genetic characterization of *Acipenser sturio* L., 1758 in relation to other sturgeon species using satellite DNA. *Bol. Inst. Esp. Oceanogr.* 16: 231–236.

Tagliavini J, Conterio G, Gandolfi G et al. (1999) Mitochondrial DNA sequences of six sturgeon species and phylogenetic relationships within Acipenseridae. *J. Appl. Ichthyol.* 15: 17–22.

Tranah G, Campton DE and May B (2004) Genetic evidence for hybridisation of pallid and shovelnose sturgeon. *J. Hered.* 95: 474–480.

Ugarković Đ and Plohl M (2002) Variation in satellite DNA profiles—causes and effects. *EMBO J* 21: 5955–5959.

Wijers ER, Zijlstra C and Lenstra JA (1993) Rapid evolution of horse satellite DNA. *Genomics* 18: 113–117.

Wirth M, Kirschbaum F, Gessner J et al. (2002) Fatty acid composition in sturgeon caviar from different species: Comparing wild and farmed origins. *Int. Rev. Hydrobiol.* 87(5–6): 629–636.

Wolf C, Hübner P and Lüthy J (1999) Differentiation of sturgeon species by PCR-RFLP. *Food Res. Int.* 32: 699–705.

Chapter 19
International Trade in Caviar and Business Perspectives in Russia

A.I. Nikolaev, D.P. Andrianov, I.A. Burtsev, L.R. Kopylenko, B.N. Kotenev, and A.S. Safronov

Abstract Following a sharp reduction of sturgeon stocks up to 50-fold in the Caspian and Azov Seas, as well as in another parts of this area, caviar production and the volume of its delivery to the world market has also reduced several-fold and supplies approximately 10% of the demand. In recent years, one of the most intensively developing trends in sturgeon breeding is aquaculture, where the total volume of caviar produced on sturgeon farms in the world has reached a level of more than 13–15 t.

However, the long maturation of most species of sturgeons render it more difficult to attract investors for such business. There are several ways to solve the problem: (1) create early-ripening broods and hybrids, (2) use closed water-supply systems, and (3) use the multiple methods of obtaining roe from the same females over their lifetime. These measures reduce the term of return in total asset investments to 5–6 years and high profitability after that time, making caviar-production projects more viable and attractive.

Possible volumes of caviar production through aquaculture in Russia can be expected to reach 20 t of caviar by 2010 and up to 50 t by 2020. These achievements will be a result of artificial sturgeon reproduction for increasing natural sturgeon stocks in the sea waters of Russia.

Keywords Acipenseridae, sturgeon stocks, caviar production, aquaculture, hybridization, selective breeding, closed water-supply system

19.1 Introduction

Surviving under wild conditions, the 'living fossils', fishes of the order Acipenseriformes, (*Chondrostei*) appeared 250 million years ago, but at present have come under pressure from the advanced civilization, especially from the

A.I. Nikolaev, D.P. Andrianov, I.A. Burtsev, L.R. Kopylenko, B.N. Kotenev and A.S. Safronov
Russian Federal Research Institute of Fisheries & Oceanography (VNIRO), 17,
V. Krasnoselskaya, St. Moscow, 107140, Russia

increased human population living near the water bodies and being equipped with new fishing gear: speedy motorboats, nylon gill nets, and so on. This has resulted in legal and illegal overfishing. Damming of rivers and water pollution have also seriously damaged sturgeon environmental conditions. These circumstances became especially drastic in West Europe, Southeast Asia and North America, where several of the local sturgeon species have become gravely endangered or even disappeared at the end of the nineteenth century. It should be stressed that a similar situation has occurred for sturgeons of Eastern Europe and Northern Asia 100 years later.

Sturgeons are represented by 26 species, 11 of which live in Russia. The main stocks of sturgeons (up to 90%) have been concentrated in the Azov and Caspian Sea basins. The total catch of sturgeons in the former Soviet Union at the beginning of the 1970s reached 29,140 t (Ivanov et al., 1999; Catarci, 2004), and the production of caviar exceeded 2000 t, of which less than 10% was exported (De Meulenaer and Raymakers, 1996; Taylor, 1997).

Caviar is one of the most appreciated delicacies in many countries of the world. Its exclusive producers up to the early 1990s were the Soviet Union and Iran. By 1997, the production of sturgeons in Russia reached hardly more than 2000 t (40% of the world production), and in 2002 the catches of sturgeons fell to 700 t as accidental catches, which were mainly used for artificial reproduction and carrying out the official monitoring of the state of wild populations.

Now, caviar export from Russia has become 10- to 20-fold less. The main reason is drastic degradation of the stocks of sturgeons during the last 10–15 years, caused by a number of ecological and anthropogenic conditions. The main adverse factor is considered to be the violations of fishing rules for sturgeons developed and maintained during the Soviet period. Nowadays there is intensive and uncontrollable fishing of sturgeons, not only from the rivers, but also from the Azov and Caspian Seas.

Because of a sharp reduction of stocks (over 50 times), since 1998, all species of sturgeons have been included in Appendices I and II of the *Conventions for International Trade in Endangered Species of Wild Fauna and Flora*, dated 3 March 1973 (CITES, Raymakers and Hoover, 2002).

The scientific body on sturgeon fishes, CITES, in Russia, which involves an Advisory Board and an up-to-date centre of molecular-genetic identification, provides reliable control over the exportation of sturgeons and sturgeon products.

The introduction of international control over the export of sturgeons and their products, mainly caviar, decreased illegal exports. However, the problem of reducing illegal fishing remains to be solved: according to expert estimates, in the Caspian and Azov Sea basins, the level of illegal catch in the rivers and in the sea exceeds the legal limit eightfold, or, according to other data, 11-fold and more (Zykova et al., 2000). According to the official statistics the total world volume of legally manufactured caviar does not exceed 100 t, constituting approximately 10% of the commercial demand (according to expert estimation).

The existing situation of the natural populations of sturgeons has stimulated an active development of rearing this fish in many countries. Rearing of sturgeons originally began in the USSR, and reached a commercial scale in the production of marketable fish in the 1960s and 70s. Pond- and cage-rearing methods were mainly used. However, the continuing insufficiency in sturgeon-stocking material as well as reduced volumes of production of 'natural' caviar, caused a quick reorientation of sturgeon-breeding farms towards the creation of their own brood stocks.

Increasing retail prices for sturgeon made it possible to introduce hi-tech intensive cultivation methods. Sturgeon farms in many countries have formed their own large brood stocks, actively adopting Russian experience in marketable sturgeon growing, using both our stocking material and their own, due to favourable climate, better technical equipment and economic stability. They have advanced faster than Russia. As the demand for sturgeon meat was not very high, the majority of fish farms began to specialize in the production and processing of sturgeon caviar, using traditional technology (killing of females). By 2000, the total volume of caviar produced at sturgeon farms in the world reached 13–15 t (Beer, 2001; Jones, 2001). According to expert estimates, the volume of manufactured caviar from aquaculture can achieve some 200 t by 2020 (essential changes may be made by China, which is planning to develop aquaculture capability of up to 100 t of caviar by 2010).

At present in Russia the annual production of marketable sturgeons is over 2000 t (marketable fish not more than 3 kg in weight are mainly raised). In Russia, great quantities of illegal caviar from 'wild' fishes constrained the development of caviar production from aquaculture until recently. The price of 'wild' caviar made the farming non-competitive. However, during recent times, a substantial increase in world and domestic prices for sturgeon caviar, along with the state's measures on suppression of poaching, have predetermined a high profitability of caviar production under conditions of intensive fish-rearing.

Russian scientists of VNIRO were the first in the world to culture quickly-grown and early-ripening sturgeon breeds (bester; Burtsev et al., 2002); technologies for the maintenance and rearing of sturgeons in ponds (mono- and poly-culture) were developed; rearing of sturgeons in cages (in reservoirs-coolers of electric power stations) and in fish-breeding modules with closed water-supply systems (Andrianov et al., 2005, 2006) is followed; a technology of the intravital roe from females (Burtsev, 1969; Burtsev et al., 2007) and preparation of high-quality caviar on this basis has been developed (Kopylenko and Koryazova, 2002).

Developing a project and construction of a fish-breeding plant for caviar production on turn-key basis, as a rule, takes 10 to 12 months and the first batch of caviar can be received 4 to 5 years later. Such a complex has been built by VNIRO near Moscow (Burtsev and Nikolaev, 2004).

We estimate that the annual world demand for caviar is 1000 t, but only 10% of this demand is being satisfied by. A recent surprising demand in Russia for civil engineering designs of sturgeon-growing plants of different capacities testifies to a favourable condition for the production of caviar. It should be stressed that even the organizations funded with poaching money have expressed interest in such projects. They seem to forecast the end of illegal production!

Taking into account a high consumer demand for caviar all over the world, about it is obvious that investment is attractive for both Russian and international entrepreneurs.

19.2 Material and Methods

19.2.1 Hybridization and Selective Breeding

This report contains generalized and analysed materials which have been collected from 1996 up to the present time at sturgeon-rearing plants in the Rostov region (fish farm 'Kazachka', Fig. 19.1) and in Vologda region (fish farm at the Cherepovets thermal power station).

Cultivation at these facilities use different kinds (broods) of bester, a hybrid between beluga (*Huso huso*) and sterlet (*Acipenser ruthenus*) of several generations (F_1–F_4) as well as hybrids between the Russian sturgeon (*A. gueldenstaedtii*) and Siberian sturgeon (*A. baeri*) of the first and second generations, with initial species (Safronov and Filippova, 2001). By selective breeding and back crossings on the basis of bester, three broods were created which were subsequently registered in the State register of selection achievements of the Russian Federation as: 'Burtsevskaya' brood—bester × bester (BS, Fig. 19.2), 'Aksayskaya' brood—sterlet × bester (S.BS) and 'Vnirovskaya' brood—beluga × bester (B.BS) (Burtsev, 1997; Burtsev et al., 2002). All the three broods are recommended for commercial use. The selective

Fig. 19.1 Spring catch of pond with bester brood stock

Fig. 19.2 Dr. D.P. Andrianov (second at the left) explains an operation of VNIRO experimental closed module to staff of the Federal agency on fisheries of Russia

breeding carried out over the years involved purposeful selection of the best breeders by phenotypic characteristics, as well as by the parameters of greatest fecundity and earliest maturation rates.

19.2.2 Influence of Water Temperature on the Terms of Sturgeon Maturation

Data on growth rates, age of maturity and duration of repeated maturation of bester and other forms of sturgeons were received under conditions of pond and cage cultivation at natural temperature levels (fish farm 'Kazachka') and a raised temperature background (fish farm at the Cherepovets thermal power station), using the parameter of 'degree days'—the number of days multiplied by the daily average temperature, i.e. the total amount of heat necessary for maturation of the given sturgeon species.

19.2.3 Intravital Technique of Obtaining Ovulated Eggs

At the same facilities, the technique of repeated reception of eggs from the same female during its life cycle (Burtsev, 1969) were worked out and modernized. Thus, spawning of sturgeons was stimulated by intramuscular injections of pituitary gland extract or synthetic hormone GnRH (surfagon).

Sturgeon eggs obtained by the intravital method are shown to be in the ovulated (flowing) condition, i.e. at the V maturity stage. As their processing is aimed at obtaining a food product (caviar), the technique patented by Kopylenko and Koryazova (2002) has been used. Pasteurized samples of caviar were tested subsequently for shelf life with a full biochemical analysis of samples taken after 1–3–6–12–18–24 months of storage. At present, we have samples of caviar received by the intravital method from females grown under the pond or cage conditions but we have not investigated analogous samples from females grown in closed circuit systems.

The technique of intensive sturgeon cultivation was developed at the installations of closed water supply system (CWSS) designed and built by VNIRO in the Moscow region (it was put into operation in June, 2004, Fig. 19.2) and Tver region (working since October, 2003; Andrianov et al., 2006). At present, specifications of the growing and maturing norms of three broods of bester, maintained year-round under conditions of temperature ranging from 22–24°C are being studied.

19.3 Results

The basic problems to be overcome in sturgeon aquaculture aimed at the fastest production of maximum quantities of caviar are connected with species-specific peculiarities of the reproductive biology of sturgeons. All sturgeons under study are late-maturing fishes characterized by clearly defined seasonal type of spawning, intermittent oogenesis, determined fecundity, one-time spawning and, as a consequence, comparatively low relative fecundity, and, taking into account the absence of annual spawning, low population fecundity. It is precisely this complex set of factors, combined with the low reproductive potential of sturgeons, that makes their natural populations vulnerable both to varying ecological factors and anthropogenic load. Thus, the sturgeon as a reared fish species, seems, at first sight, unsuitable for aquaculture, especially for the purpose of caviar production.

Nevertheless, in our opinion, only two of the above-listed factors (type of oogenesis and that of spawning) are considered to be conservative reproductive parameters of sturgeons. All others, (1) late maturation, (2) seasonal fluctuations in spawning, and (3) low fecundity can be changed under aquaculture conditions.

Table 19.1 shows these labile reproductive characteristics as problems of the first order, and three approaches to their resolution are also proposed. These approaches are applicable to the same extent both to commercial aquaculture as well as to artificial reproduction on sturgeon-growing farms.

As indicated in Table 19.1, three solutions to the problems are proposed. Let us consider each of them.

Table 19.1 Problems of the first order and their solution

Problems	Solutions
Later maturation	Hybridization and selection
Seasonal fluctuations in spawning	Modelling of ambient environment
Low fecundity	Repeated obtaining of eggs

19.3.1 Hybridization and Selective Breeding

Work on breeding new hybrids is widely applied in animal husbandry and furnishes the desired result. In our case, after obtaining in 1952 (Nikoljukin and Timofeeva, 1953) an intergenetic hybrid between the beluga and sterlet and the evidence for its fertility (Burtsev, 1972), new possibilities opened up for further purposeful hybridization and selective breeding (Burtsev and Nikolaev, 2003). The 'Burtsevskaya' brood—actually bester (BS)—50% to 50% of the genomes of the beluga and starlet, is characterized by large size, high fecundity and best quality of roe, and is shown to be the universal brood suitable both for marketable fish cultivation, and for the purposes of caviar production (Fig. 19.3).

Two new hybrids (which subsequently acquired the status of broods) were bred by back crossing, their role in aquaculture being quite different. The 'Aksayskaya' brood of bester (SBS), carrying 25% of beluga genome and 75% of sterlet, is recommended by us for the purpose of caviar production as the earliest in maturation.

The second one, the 'Vnirovskaya' brood, having only 25% of sterlet genome and 75% of beluga genome, differs in its high growth rates and large size of mature fishes, but matures later. The positive property of this brood are large eggs (approaching beluga caviar in size and taste qualities), and high absolute fecundity. Bigger tanks and special devices to work with large fish are necessary for the maintenance of these brood spawners. This brood is recommended to be used in well-established fish farms working at a profitable level.

Fig. 19.3 Close to maturation bester F_4 of 'Burtsevskaya' brood female reared at the VNIRO closed module

Table 19.2 Parameters of growth of three broods of bester from larva to 1.5 years

Age (month)	Average weight of one specimen., g. Broods Aksayaskaya	Burtsevskaya	Vnirovskaya
1	1	2	3
6	100	200	300
12	1000	2000	3000
18	1500	3000	4500

Table 19.3 Age of maturation of sturgeon females at various conditions of maintenance

		Maturation age of first females (years)			
				Closed systems	
Species, brood	The total amount of heat for the maturation period (degree-days)	Natural temperature course	Warm-water plants	Forecast	Foreign experience
Bester, 'Aksayskaya' brood	28,200	6	5	3.2	2.5 (Serfling and Hamlin, 2001)
Bester, Burtsevskaya brood	42,300	9	8	4.8	
Siberian sturgeon	52,120	11	8	6.0	4.0 (Steinbach, 2005)

In our practical recommendations for the initial stage of the caviar-producing plants, our advise is to begin with cultivation of the 'Aksayskaya' brood as it will provide a way of producing commercial caviar (and hence economic efficiency) in the shortest possible time, and then gradually shift to the 'Burtsevskaya' and 'Vnirovskaya' broods of bester as they are more fecund and give higher-quality caviar.

Comparative data on the growing rates of the juveniles of three broods and the maturity-age are given in Tables 19.2 and 19.3, respectively.

In addition to the right choice for rearing (species, brood), for creating a brood stock, it is important to select breeding (productive) females, taking into account age of sexual maturation and fecundity. Among other things, for three generations of artificial selection by early ripening, the age of sexual maturity of the females of the 'Burtsevskaya' brood has, on average, decreased by 2 years (Fig. 19.4). According to the recalculation for degree-days, it took 42,300 degree-days for the first mature females of the first and second generations (F_1 and F_2), and only 32,500 degree-days for females of the third generation (F_3); the first females of the 'Aksayskaya' brood reach sexual maturity after 28,200 degree-days (Table 19.3).

19.3.2 Difference Between Generations of Bester Females ('Burtsevskaya' Brood) in Maturation Age

However, it should be stressed that selection only for early maturation, results in a reduction of size for first-time ripening females and a decrease in their fecundity.

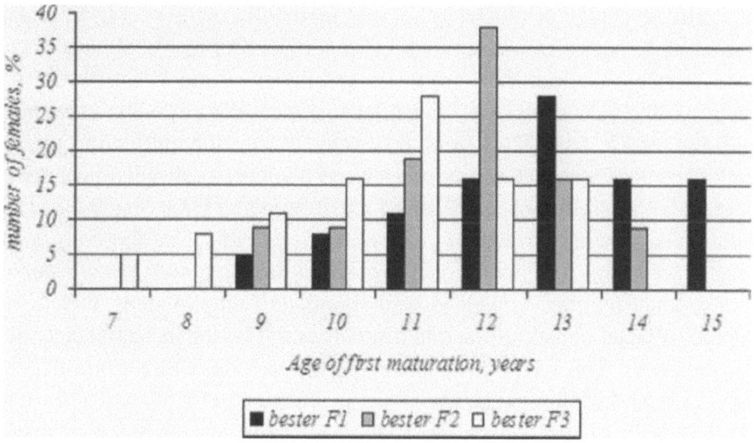

Fig. 19.4 Difference between generations of bester females ('Burtsevskaya' brood) at maturation age

Therefore, it is necessary to select early-ripening, as well as largest-sized females, these being characterized by the highest fecundity. Thus, hybridization and selection at formation of brood stocks make it possible to substantially improve such parameters as early ripening and fecundity, but do not affect seasonal changes in fish spawning.

19.3.3 Creation of a Constant Temperature Optimum

The idea given is simple and obvious, as it stems from the fact that fish are poikilotherm animals, and hence a rise in water temperature intensifies processes of growth and maturation. Such is indeed the case: it was established that the optimum temperature of life activity of most sturgeons is 20 to 24°C (Gershanovich et al., 1987). However seasonal dynamics with a prolonged winter considerably retard the highest possible growth and maturation rates.

Creation of a year-round thermal optimum accelerates growth and development rates; in this case the age of sexual maturity is essentially reduced, from 5–6 to 3 years in sterlet and the 'Aksayskaya' brood of bester, and from 8–9 to 4–5 years in the 'Burtsevskaya' brood of bester and Siberian sturgeon (Table 19.3).

However, to put this simple idea into practice is rather complicated. It is necessary to solve the problem of heating the huge quantity of running water required for the normal life activity of the fish. Such a solution seems most feasible under conditions of closed water-supply system (CWSS). In this case it is necessary to warm up only make-up water (usually 3–5% from the total volume of water in the system) and to compensate for general heat losses.

Recent studies have shown that the problem of fast creation and exploitation of brood stocks of different sturgeon species is most effectively solved in CWSS. In comparison with traditional methods this technology has many advantages.

The main advantage of closed systems is their full control and independence of the production process from natural and climatic conditions. As already noted, this offers accelerated maturation of spawners and intensive use of the brood stock, as well as providing year-round vital juveniles together with large-size stocking material and marketable fish. Thus, mortality rates during the cultivation process are minimal. This also prevents mass diseases and achieves ecological production.

There is the possibility of establishing a fish-breeding plant on the basis of polycyclic technology with permanent year-round charge of tanks and consequently, a decrease in 1.5–2 times of operational expenses (electric power, heat, water supply, etc.) per production unit, as well as year-round staff employment. CWSS reduces the volume of water consumption and discharges not less than 10 times, which also reduces expenses. The water circulation system provides utilization of fish metabolic products and ecological safety.

The leadership of our country in creating technologies for breeding marketable sturgeons, and long-term experience at managing them, has allowed us to create industrial sturgeon breeding plants for producing sturgeon caviar and allied products.

Thus, cultivation of sturgeons in CWSS solves all the three problems of sturgeon rearing:

1. Fish become mature sooner due to the constant optimum temperature.
2. Seasonal prevalence of spawning is absent as there are no seasonal differences in external conditions.
3. Population fecundity increases, as the optimum temperature makes it possible for the fish to mature every year (rather than once in 2–3 years, as at natural temperatures)

According to our plans, we received the first caviar from the 'Aksayskaya' brood of bester by the end of 2005, i.e. in the third year of cultivation in CWSS.

19.3.4 Intravital Extraction of Eggs

At present, two methods are known for egg extraction in aquaculture:

1. First, females are killed at first maturation on the IV maturity stage.
2. Second, the females are allowed to multiply as a result of intravital methods of extracting ovulated eggs.

The first method is employed by a majority of foreign companies, using a traditional salting method. An obvious shortcoming of this approach is the need for maintenance and cultivation of a large brood stock equal to the total number of all year classes prior to the first maturity year. For example, if we wish to receive caviar from 6000 females which first mature in the fifth year of life (5 kg in average weight) it is necessary to have 24,000 females (6000 individuals in each year-class) for providing continuous production in the plant for the next 4 years, and taking into account that we can separate males only at the end of the second year of life, it is necessary to have 36,000 fish(78 t in weight) (Table 19.4).

Table 19.4 Calculation of the total weight of fish kept in the system when obtaining the eggs from 6000 females using traditional method (without taking into account the natural loss during cultivation)

	First year	Second year	Third year	Fourth year	Fifth year	Total weight of fishes in the system (t)
Number of fishes (ind.)	12,000	12,000	6000	6000	6000	
Average weight (kg)	1	2	3	4	5	
Total weight (t)	12	24	18	24	30	108

Fig. 19.5 Intravital extraction of eggs from bester female by Burtsev's method

In other words, for the final goal of obtaining eggs, less than one-third of the fish was used. Furthermore, it is known that fecundity of the first-time spawning females is the lowest during their whole life cycle, and the quality of the products of young females leaves much to be desired. Thus, the given approach shows an obvious underexploitation of the sturgeon reproductive potential.

The intravital method of multiplying egg production from one female over its life cycle was offered as early as 1969 by Dr. I.A. Burtsev (1969). (Fig. 19.5). Since then, it has been somewhat modified (Podushka, 1999; Burtsev et al., 2007). This method makes it possible to receive eggs from each female up to 10 or more times. In this case the total weight of eggs produced by one female exceeds more than twice the weight of the fish itself; it is used in the majority of the Russian sturgeon-farming plants. In the CWSS under optimal environmental conditions (temperature, feed availability, oxygen, and other factors), as well as annual spawning, the productive potential of females can be realized during an even shorter period. Besides, this technology makes it possible to harvest eggs practically in any season.

In spite of all these advantages, it generates a number of problems, which are discussed in the following section.

19.4 Discussion of Results

Let us consider all of the above decisions from the point of view of new problems (real and hypothetical) created by them, and new decisions which are to be accepted in practice (Table 19.5).

Development of optimal recipes of feeds have to be different, depending on the various functional conditions of the fish (juvenile and post-juvenile development, somatic and generative growth).

19.4.1 Hybridization and Selection

The key that assures the success of a sturgeon-rearing plant focused on caviar production is the right choice of the type to be farmed.

The main criteria for the choice of object to be farmed are the following:

1. Comprehensive knowledge of its biology
2. Its appropriateness in caviar production
3. Availability of a source of replenishment of the breeding stock

From our point of view all the three requirements are met by bester. In our practical recommendations, for the start of production plants focused on caviar, we advise to begin with the cultivation of the 'Asksayskaya' brood, as it enables production in the shortest time (as early as the third year), and hence economic efficiency; gradually

Table 19.5 Problems of the second order and suggested solutions

Problems	Solutions
Significant investments and operating costs of CWSS	Use of energy-saving technologies
Deterioration of reproductive products (degeneration, change in organoleptic properties)	Development of optimal recipes of feeds which are to be different depending on various functional conditions of the fish (juvenile and post-juvenile development, somatic and generative growth)
Inevitability of hormonal stimulation of spawning	Carrying out of target studies designed to prove the absence of traces of injections in the final product (caviar)
High time expenditures for operations at intravital reception of eggs	Creation of conditions for natural spawning of sturgeons, and technical means for spawn eggs gathering
Production of caviar from ovulated eggs—roe	Use of Dr. Kopylenko's patent

farming can proceed to the 'Burtsevskaya' and 'Vnirovskaya' broods of bester as they are more fecund and their caviar quality is higher.

Selective breeding for the maintenance and improvement of bester broods and creation of new highly efficient broods of sturgeons will be continued in a number of sturgeon-breeding plants (selective-breeding centres) containing elite breeding stocks of sturgeon species and supplying caviar-producing plants factories with selected stocking material (live eggs or fry).

19.4.2 Creation of a Habitat Optimum

As already noted, the idea of modelling an optimum sturgeon habitat is the best approach in installations with closed water-supply system. However, the technology of fish cultivation in such installations involves high expense, both in capital, and operation. These facilities are industrial-type enterprises of production. The expenses are regained quickly due to the expensive final product. The average wholesale price of caviar on the world market is US $ 700 to 1000 per kg and even higher. Therefore a sturgeon-breeding plant, producing annually 10 t of caviar, pays off all its capital investments during the first year of going into operation at a designed capacity, and then it becomes super-profitable. Besides, the CWSS uses energy-saving technologies such as thermal pumps, heat recuperation, etc., which raise the profits even more.

However, the creation of a permanent optimum temperature also has its negative aspects. The fish can lose their resistance to unfavourable conditions and ecological plasticity. There is also an apprehension that the quality of eggs obtained in the CWSS can be lower than from wild fish. But all these problems can be solved.

19.4.3 Intravital Obtaining of Eggs

As evident from Table 19.5, this technique results in three inevitable consequences:

1. Necessity of hormonal stimulation
2. High labour and time consumption
3. Non-traditional raw material

It is known that in a number of countries there is a prejudice against the use of hormones in animal husbandry. Nevertheless, the exact scientific data testify that pituitary gonadotropin and a synthetic analogue gonadotropin-releasing hormone (GnRH), causing egg ovulation, have a very limited life, measured in tens of minutes (Goren et al., 1990; Zohar et al., 1990; Weil et al., 1991). These preparations have passed the necessary tests, including that of the Pharmacological Committee of Russia, so that livestock products (milk, meat and fish), obtained after hormonal application, are permitted for human consumption.

The second important problem of intravital obtaining of eggs is high labour and time consumption. Actually, our long-term experience has shown that on an average, even with good practical skills of the personnel, it takes at least half an hour for three persons to handle one fish. It involves the following steps: preparation of the fish for the operation (selection, estimate of its condition, control of the onset of egg ovulation), then the operation itself (selection of eggs, stitching, triple 'milting' of the female for the rest of eggs, rehabilitation of the fish, its transfer for fattening). If we wish to have 10 t of roe from fish of 10 kg average weight, it is necessary to use 6700 individuals, assuming that roe is 15% of the fish body weight. It takes 3400 working hours or 425 working days of one group of workers at the rate of an 8-h working day. Five such groups can perform this work in approximately 3 to 4 months.

This situation can be solved either by obtaining the roe by means of killing the fish (the shortcomings of this method have been analysed above) or by searching for methods of mass stimulation of their natural spawning with subsequent gathering of laid eggs for further processing. Such investigations are being conducted at VNIRO.

As to processing or salting of non-traditional raw material, i.e. ovulated eggs, traditional salting methods used for processing of caviar of the IV maturity stage

Fig. 19.6 Caviar of bester produced from live females

are inapplicable for eggs of the V stage. It is mainly because of the stickiness of ovulated eggs. When such eggs are washed with water, they stick together, and it is impossible to obtain the traditional dry and friable consistency of granular caviar.

The technology for removing the stickiness of sturgeon eggs for caviar production (Podushka et al., 1990) was patented in 1990. But this caviar is ranked below the traditional one because of its organoleptic characteristics, the thick envelope and its consistency which is too friable. In 2002, the superiority of the new application (Kopylenko and Koryazova, 2002) was accepted and the new normative documentation of sturgeon caviar was developed. In 2002–2003, the first experimental sturgeon batch from ovulated eggs of bester was achieved at the pond hatchery of the joint-stock company 'Kazachka' based on the new technology, which provided a high-quality product as well as microbial safety and stability of the final product for 12 months (Fig. 19.6).

Recently, experimental batches of caviar from these sturgeon species have been produced at VNIRO and investigated thoroughly for long storage. The quality of pasteurized granular caviar made from ovulated eggs does not differ from that produced with traditional technology.

19.5 Conclusions

From these facts, we can draw the following conclusion: the priority in the production of sturgeon caviar is intensive cultivation of sturgeons in installations of closed water supply. It is difficult and expensive, but highly profitable. We would like to cite further evidence of the benefit of this approach. We are sure that even low volumes of production, under conditions of the middle zone of Russia, i.e. not at all the best climatic conditions for sturgeon cultivation, are efficient and profitable for caviar production.

Possible volumes of sturgeon production in aquaculture in Russia can be estimated at 3000 to 5000 t of marketable fish and up to 20 t of caviar by 2010, and up to 15,000 t of fish and 50 t of caviar by 2020.

The realization of these projects requires state support, first of all in improving the normative legal base and providing for the development of marketable sturgeon breeding and retaining the position of Russia in the world market.

It is also necessary to develop other aspects of sturgeon breeding, aimed at increasing natural sturgeon stocks in the sea waters of Russia:

1. Maintenance of natural reproduction as a base for the conservation of stock biodiversity of sturgeon species
2. Upgrading of artificial reproduction efficiency with the help of up-to-date knowledge and technologies
3. Decrease of illegal fishery impact in the main sturgeon sea- water basins of the Russian Federation and adjacent countries

These tasks are of great strategic and geopolicy importance for Russia.

Acknowledgements We greatly appreciate the important role of our late co-author Dr. Dmitry Andrianov in popularising ideas on using closed water-supply systems for acceleration of ripening of sturgeons and creation of a VNIRO experimental closed module in a suburb of Moscow. Data of his report at the Fifth International Symposium on sturgeon in Ramsar, Iran (Andrianov et al., 2005) are used in the given paper.

References

Andrianov D.P., Burtsev I.A., Kopylenko L.R., Kotenev B.N., Nikolaev A.I. and Safronov A.S. 2005. Caviar production in aquaculture: problems and decisions. In: *Fifth International Symposium on Sturgeon*, Ramsar, Iran, Aquaculture papers, 2005, pp. 5–6.

Andrianov D.P., Petukhov A.V., Burtsev I.A. and Nikolaev A.I. 2006. Fermer's Fish Culture Komplex. Patent for useful model, with priority since 21.07.2005. Russian Federation (in Russian).

Beer, K. 2001. Commercial culture of sturgeon in North America. In: Extended Abstracts of 4th International Symposium on Sturgeon., Oshkosh, Wisconsin, USA, Coll. Aquaculture and General Biology, AQ 44.

Burtsev, I.A. 1969. Method for obtaining eggs from fish females. USSR, Personal patent on invention No 244793 , Moscow, Committee on Inventions and Discoveries, Council of Ministers of the USSR (in Russian).

Burtsev I.A. 1972. Progeny of an intergeneric hybrid of beluga and starlet. In: *Genetics, Selection, and Hybridization of Fish*, B.I. Cherfas (ed.), Academy of Sciences, Jerusalem, USSR, pp. 211–220.

Burtsev I.A. 1997. Bester in aquaculture. In: *Sturgeon Stocks and Caviar Trade Workshop*, Birstein et al. (eds.), JUCN, Gland, Switzerland and Cambridge, pp. 35–44.

Burtsev I. and Nikolaev A. 2003. Bester fish: faring well? 'Science in Russia' No. 1 Publication of the Russian Academy of Sciences, Presidium, pp. 72–77.

Burtsev I.A. and Nikolaev A.I. 2004. Innovational ways in development of sturgeon culture in Russia. 'My Moscow', *Journal of City Life*, No 8/100, 68–73 (in Russian).

Burtsev I.A., Nikolaev A.I., Krylova V.D., Filippova O.P. and Safronov A.S. 2002. First breeds of sturgeons created on a base of intergeneric hybrid between beluga and sterlet – bester. In: *Aquaculture at the Beginning of the XXI Century: Sources, State, Strategy of Development*, VNIRO Publications, Moscow, pp. 146–150 (in Russian).

Burtsev I.A., Nikolaev A.I. and Safronov A.S. 2007. Method for obtaining eggs from sturgeon females. Russian Federation. Patent on invention No 2290794 with priority since 28.04.2005 (in Russian).

Catarci, C. 2004. World markets and industry of selected commercially-exploited aquatic species with an international conservation profile. FAO Fisheries Circular No. 990, Rome, 61 pp.

De Meulenaer, T. and Raymakers C. 1996. Sturgeons of the Caspian Sea and the international trade in caviar. TRAFFIC Network Report, 71 pp.

Gershanovich A.D., Pegasov V.A. and Shatunovski M.I. 1987. *Ecology and Physiology of the Young Sturgeon*, Agropromizdat, Moscow, 215 pp (in Russian).

Goren, A., Zohar Y., Elhanati M. and Koch Y. 1990. Degradation of gonadotropin releasing hormone in the gilthead seabream, *Sparus aurata*: I. Cleavage of native salmon GnRH, mammalian LHRH and their analogs in the pituitary. *General Comparative and Endocrinology*, 79, 291–305.

Ivanov V.P., Vlasenko A.D., Khodorevskaya R.P. and Raspopov V.M. 1999. Contemporary status of Caspian sturgeon (*Acipenseridae*) stock and its conservation. *Journal of Applied Ichthyology*, 15, 103–105.

Jones, A. 2001. The Commercial farming of sturgeon in Europe. In: Extended Abstracts of 4th International Symposium on Sturgeon., Oshkosh, Wisconsin, USA, Coll. Aquaculture and General Biology, AQ 68.

Kopylenko L.R. and Koryazova I.L. 2002. Method of caviar processing in aquaculture from ovulated eggs of sturgeon fishes. Patent No. 2232523 with priority since 02.09.2002 (in Russian).

Nikoljukin N.I. and Timofeeva N.A. 1953. Hybridization beluga with sterlet. In: Reports of Academy of Science of USSR, vol. 93, No 5, pp. 899-902 (in Russian)

Podushka S.B. 1999. New method to obtain sturgeon eggs. *Journal of Applied Ichthyology*, 15, 319.

Podushka S.B., Brusovatsky R.B., Kalgina N.A., Kovda T.A. and Abdrakhmanova V.Kh. 1990. Food product from eggs of sturgeon fish. Patent SU 1824705 A1 with priority since 19.09.1990 (in Russian).

Raymakers, C. and Hoover C. 2002. Acipenseriformes: CITES implementation from Range States to consumer countries. *Journal of. Applied Ichthyology*, 18 (4–6), 629–638.

Safronov,A.A. and Filippova, O.P. 2001. Experience in growing hybrids of Russian sturgeon (*Acipenser gueldenstaedtii Brand*) and Siberian sturgeon (*A. baerii Br.*) at the warmwater Kadui Fish Farm in the Vologda region (northern Russia) In: Extended Abstracts of 4th International Symposium on Sturgeon., Oshkosh, Wisconsin, USA, Coll. Aquaculture and General Biology, AQ 49.

Serfling, S.A. and Hamlin, H. 2001. Culture of beluga-hybrid 'bester' sturgeon (*H. Huso x A. ruthenus*) in closed-cycle culture systems in Florida. In: Extended Abstracts of 4th International Symposium on Sturgeon., Oshkosh, Wisconsin, USA, Coll. Aquaculture and General Biology, AQ 51.

Steinbach P. 2005. Egg development of Siberian sturgeon *(A. baeri)* in a water recirculation system. In: *Fifth International Symposium on Sturgeon*, Ramsar, Iran, Aquaculture papers, pp. 131–133.

Taylor S. 1997. The historical development of the caviar trade and the caviar industry. In: *Sturgeon Stocks and Caviar Trade Workshop*, IUCN, Gland, Switzerland and Cambridge, UK, pp. 45–53.

Weil C., Breton B., Sambroni E., Zmora N. and Zohar Y. 1991. Bioactivity of various forms of GnRH in relation to their resistance to degradation at the pituitary level in the rainbow trout, *Oncorhynchus mykiss*. In: *Reproductive Physiology of Fish. Fish Symposium 91*, Scott, A.P., Sumpter, J.P., Kime, D.E. and Rolfe, M.S. (eds.), Sheffeld, England, pp. 51–53.

Zohar Y., Goren A., Fridkin M., Elhanati E. and Koch Y. 1990. Degradation of gonadotropin-releasing hormones in the gilthead seabream *Sparus aurata*: II. Cleavage of native salmon GnRH, mammalian LHRH and their analogs in the pituitary, kidney and liver. *General Comparative and Endocrinology*, 79, 306–331.

Zykova G.F., Zhuravleva O.L. and Krasikov E.V. 2000. Estimated landing of Russian sturgeon by unreported and poachers fisheries in the Volga River and Caspian Sea. In: *International Conference Sturgeon on the Threshold of the 21st Century*. Book of abstracts, Astrakhan, pp. 54–56.

Chapter 20
The Ecological Problems of Introduction and Reintroduction of Sturgeons

V.A. Pegasov

The intensive growth of humanity in the last 100 years has resulted in a series of ecological problems concerned with dramatic changes in and the pollution of, the environment. As a consequence of these, we see the destruction of different life systems, such as populations, communities, and ecosystems. In recent times, many developed countries are trying to restore natural land-based and water ecosystems, which are in different degrees of destruction. These operations require integrated scientific investigations of the abiotic and biotic factors for determining the conditions of living organisms. The principal aim of this work is to recover, as far as possible, the lost biodiversity in ecosystems that have been destroyed.

Scientists use the introduction and reintroduction of species as the main way of solving these problems. The first is when the destruction of a particular ecosystem leads to the total or partial extinction of aboriginal species, some of which may be endemic. In this case, theoretically, we can select related (interconnected) species which have the same mode of lifeas the vanished species. If we cannot find such a one, a new community can be created according to the environmental conditions. Sometimes, the introduction is connected with specious purposes, to improve the biodiversity and/or increase the productivity of the ecosystems. Such activities can lead to negative results and even to disaster.

In the case of fish, we can demonstrate positive, negative and catastrophic examples of their introduction.

Gambusia affinis holbrooki (Nikolsky, 1950; Rass, 1983 Karpevitch, 1975; Reshetnikov, 2002) was introduced in the 1930s in different countries (including Spain and USSR) to fight mosquitoes. This introduction saved the life of millions and did not cause serious problems in ecological communities.

The Far East salmons *Oncorhynchus keta* and *O. gorbuscha* (Karpevitch, 1975; Reshetnikov, 2002) were introduced into the rivers of Barenzov, White and Caspian seas. As a result of this, the total productivity of these ecosystems was not increased. But, these immigrants competed with local species, such as different species of

V.A. Pegasov
Academy of National Economy under the Government of the Russian Federation, Moscow, Prospect Vernadskogo, 82, Russia

R. Carmona et al. (eds.) *Biology, Conservation and Sustainable Development of Sturgeons*,

sturgeons, *Stenodus leucichtys leucichtys*, *Salmo trutta caspius*, *Salmo salar* (Nikolsky, 1950; Karpevitch, 1975; Reshetnikov, 2002). Populations of these species are decreasing or are on the verge of extinction.

The Amur River fish, *Ctenopharyngodon idella*, *Mylophayingodon piceus*, *Hypophtalmichtys molitrix* and *Aristichtys nobilis* Rass, 1983 Karpevitch, 1975) were acclimatized in the reservoirs of southern Russia. There, the capacity was raised at first, but the rate of eutrophication was reduced. Another Amur River newcomer—*Percottus glenii* (Nikolsky, 1950; Karpevitch, 1975) was introduced to the natural water system of European Russia. The result was the territorial expansion of its habitat and damage to fish farming.

Indeed the most dramatic example of mass extinction known to biology has been caused by the introduction of an alien species. It occurred in the early 1960s, when Nile perch (*Lates calcarifer*) was acclimatized in Lake Victoria (Africa) (Vinogradov, 2005) for commercial purposes. Lake Victoria is vast – the size of Switzerland. It contained a wide range of fish of the kind known generally as haplochromines. Individually, they were not spectacular in appearance: most of them were only a few centimeters long. They had diversified into 300 different species. By contrast, *Lates* can be huge, some up to two meters long, and weighing 150 kg. After 20 years of *Lates* in this lake, more than 70% of the aboriginal species of the family Cichlidae of genus *Haplochromis* have disappeared. It was a great ecological disaster, which had a strong negative effect on the native flora, fauna and human inhabitants.

There are some examples of active and passive sturgeon introduction. In the seventeenth century, *Acipenser ruthenus* was cultivated by monks of the Solovetsky monastery in the ponds of the Solovetskie isles. This species was taken from the continental population of sterlet (from river North Dvina, lake Ladoga and Onega).

In the 1930s, Russian scientists tried to introduce *Acipenser stellatus* into the Aral Sea, but those attempts failed (Tudge, 1991). On the other hand, *Acipenser nudiventris* was introduced from the Aral Sea to lake Balchasch, where it had not been found before. Ecological factors of Balchasch were apparently propitious, as this species has been maturing in this lake and spawning in the river Ily (Nikolsky, 1950; Rass, 1983 Reshetnikov, 2002).

The siberi Rass, 1983 an sturgeon—*Acipenser baerii* was introduced in the 1960s to the basins of the Caspian and Baltic Seas, and to Pskovskoe, Chudskoe, Ladoga and Seliger lakes(Karpevitch, 1975; Reshetnikov, 2002). The reason for its introduction to the Baltic Sea was the total disappearance of the Atlantic sturgeon—*Acipenser sturio* as a result of over-fishing, the construction of dams, and progressive pollution of the sea and fresh water. The Siberian sturgeon became acclimatized and even spawned, but unfortunately it has not appeared in Russian rivers, as it adapted better to conditions in Finland's rivers.

It is important to underline that there were no perceptible negative ecological effects of these introductions.

We can regard aquaculture too as an introduction.

In the 1970s, the fingerlings of *Polyodon spatula* were brought to the USSR from the USA for breeding through aquaculture, and they spread into the natural water systems also. The danger of such a situation is that the species-implanters for aquaculture can diffuse into the wild and induce destructive changes in endemic communities.

If we consider this problem using the ecological approach, we can say that the introduction of any life organism must be estimated as an intervention into the natural system with equivocal and often unforeseen effects. These effects will influence the future evolution of species and ecosystems. Nevertheless, in my opinion, introduction is not always a negative occurrence for nature. It is obvious that there is the difference for acclimatization between relatively peaceful and carnivorous species of animals. And, what is to be done, if we observe the collapse of some ecosystem, where organisms, including the endemic ones, disappear? It is natural that we should try to save such ecosystems by introducing new species, especially autochthonous ones. However, we have to do fundamental ecological research before we effect a well-founded introduction.

Reintroduction is the reinstatement of natural biodiversity. The positive result of this measure can be achieved when the abiotic and biotic factors are preferable. Some of these are: hydrological and hydrochemical regimes; the absence of contaminants in water, plants and silt such as heavy metals, radionuclides, and toxic chemical organic substances; the places for spawning and access to them; quality and quantity of food.

Several countries in the world are working at the problem of reintroduction of sturgeons. In Russia this attempt has been successful for a long time. We have methods for artificial fertilization and incubation of spawn and the growth of the different edge groups of fishes. And it has been done for most species of our native sturgeons. For example, during the last 7 years, our investigators have reintroduced the sterlet in the Moscow-river basin, including the reservoirs. This species had disappeared there for 50 years. Of course, there are many problems in introducing different species of sturgeons which have a peculiar mode of life. If these species are catadromic or semi-catadromic, we must know the particularities of their osmoregulation, because water salinity can change in these places. As for these sturgeons, there are fully adapted to the changes in salinity, because they are eurigalinical species.

Not only Russia, but the USA is also trying to restore natural fish communities. The *Acipenser oxyrhynchus* is being restored there because its numbers are few.

The problem is acute in the some countries of Western Europe as well. Such species as *A. sturio* and *Acipenser naccarii* are under threat of extinction, especially the former. We must undertake urgent measures for their restoration in the fish communities of France, Italy, Spain and Portugal.

Most of the participants of this international workshop, BIORESTURGEONS, are from these countries and I suppose there are no obstacles in reintroducing *A. naccarii* into the natural water systems of Spain and Portugal. In the first place, I am convinced that the fact of its autochthonous origin is proved. This is based on palaeontological, morphometric, cytological, biochemical and molecular-genetic data (Albuquerque, 1956; Garrido-Ramos et al., 1997; Hernando et al., 1999a; Hernando et al., 1999b; De la Herrán R. et al., 2004). In the second place, the joint dwelling of the Adriatic and Atlantic sturgeon has been attested (Soljan, 1975; Holcik et al., 1979, Hernando et al., 1999c; Garrido-Ramos M. A. et al., 1997). In the third place, the river's food base and environmental factors (where they are planning the reintroduction) correspond to the requirements of these species (Bernini and Nardi, 1990; Soriguer et al., 1999; Soriguer et al., 2001). Moreover, in Italy and Spain, there is a sufficient quantity of mature males and females and other age groups of fish. In fact, the legislative measure has been in effect in Italy for the last 10 years. The same measure can be applied in Spain (River Guadalquivir to begin with) and in Portugal (River Guadiana). All these measures will restore biodiversity, raise total productivity of freshwater ecosystems, and, in the end, provide commercial profits.

Finally, the introduction will help implement the most important ecological principle: the greater the biodiversity, the more stable and viable will be the longer and more complex food connections in any living system (communities, ecosystems and biosphere).

References

Albuquerque R.M. 1956. Peixes do Portugal e ilhas adjacentes. Chaves para a sua determinaçao. *Port. Acta Biol.* 5B, 1164 pp (p. 195).

Bernini F and Nardi P.A. 1990. Regime alimentare di *Accipenser naccarii* Bp. (Osteichthyes, Acipenseridae) nel tratto pavese dei Fiumi Po e Ticino. *Boll. Mus. red. Sci. nat. Torino.* 8 (2), 429–439.

De la Herrán R., Robles F., Martinez-Espin E., Lorente J. A., Ruiz Rejon C., Garrido-Ramos M. A. and Ruiz-Rejón M., 2004. Genetic identification of western Mediterranean sturgeons and its implications for conservation. *Conservation Genetics* 5: 545–551.

Garrido-Ramos M. A., Soriguer M. C., De la Herrán R., Jamilena V., Ruis Rejon C., Domezain A., Hernando J. A., and Ruiz-Rejón M., 1997. Morphometric and genetic analysis as proof of the existence of two sturgeon species in the Guadalquivir River. *Marine Biology* 129, 33–39.

Hernando J.A., Arlati G., Domezain A., Soriguer M.C., Poliakova-Belysceva L.A., Domezain J., Vallespín C., Bravo R. 1999a. Morphometric study of *Acipenser naccarii* (Bonaparte, 1836) in fish farm individuals. *J. Appl. Ichthyol.* 15(4–5), 46–50.

Hernando J.A., Vasil'eva E.D., Arlati G., Vasil'ev V.P., Santiago J.A., Belysceva-Poliakova L., Domezain A., Soriguer M.C. 1999b. A new proof for the historical presence of two European sturgeons in the Iberian Peninsula: *Huso huso* (Linnaeus, 1758) and *Acipenser naccarii* Bonaparte, 1836. *J. Appl. Ichthyol.* 15(4–5), 280–281.

Hernando J.A., Vasil'eva E.D., Arlati G., Vasil'ev V.P., Santiago J.A., Belysceva-Poliakova L., Domezain A., Soriguer M.C. 1999c. New evidence for a wider historical area of two European sturgeons: *Acipenser naccarii* and *Huso huso* (Acipenseridae). *J. Ichthyol.*, 39, 841–84.

Holcik, J., Kinzelback, R., Sokolov, L.I., Vasil'ev, V.P. 1989. In: J. Holcik (ed.), *The Freshwater Fishes of Europe*. Aula-Verlag, Wiesbaden, 469 pp. 367–394.

Karpevitch A. F., 1975. *The Theory and Practice of Acclimatization of Water Organisms*. Moscow, Pischepromizdat, 432 pp.

Nikolsky G. V., 1950. *The Particular Ichthyology*. Moscow, Soviet Science, 436 pp.

Rass T.S. 1983. In: Sokolov V.E. (edit.) Life of animals. Vol. 4: Fish. Pfrosveschenie, Moscow. 575 pp.

Reshetnikov Yu. S., 2002. *Atlas of Russian Freshwater Fishes*, Vols. 1 and 2. Moscow, Nauka, 378, 252 pp.

Tudge C., 1991. *Global Ecology*. Oxford University Press, New York 173 pp.

Soljan T. 1975. *Il Pesci dell'Adriatico*. Mondori. Verona. Italia. 522 pp.

Soriguer M.C., Domezain J., Domezain, A., Bernal M., Esteban C., Pumar J.C. and Hernando, J.A. 1999. An approximation of the feeding habits of *Acipenser naccarii* (Bonaparte 1836) using an artificial river. *J. Appl. Ichtyol.* 15 (4-5), p 348.

Soriguer M. C., Domezain A., Domezain J. and Hernando J.A. 2001. Trials of feeding preference in juveniles of *Acipenser naccarii* Bonaparte 1836. In 4th International Symposium on Sturgeons 8-13 July 2001, Oshkosh, Wisconsin, USA): PP94

Vinogradov A.V., 2005. The introduction—social-ecological crime. *Journal of 'Ohrana prirodi'* 2(32), 4–8.

Chapter 21
Hydrological and Production Characteristics of the Main Basins for Reproduction and Fattening of Sturgeons

B.N. Kotenev

Abstract Sturgeons are representatives of the most ancient vertebrates, the ancestors of which appeared on the Earth several hundred million years ago. Some species of the *Acipenseriformes* order, including the Adriatic sturgeon (*Acipenser naccarii*) are threatened with extinction. However, progress in science and technology, inter alia, in ichthyology and aquaculture, may enable us to conserve and restore these endangered species.

We give here a biological overview of the principal characteristics of sturgeons and their habitat. Sturgeons are mostly anadromous migrants and reveal a large capacity to grow in seawaters with extremely rich feeding stocks. In some coastal areas of the Caspian Sea the benthos biomass totals 50–500 g/m^2 and sometimes exceeds 1000 g/m^2. As to the sturgeon restocking of the Mediterranean Sea, it is very likely that this basin could provide adequate feeding stocks for the Adriatic sturgeon, though specific issues need additional studies.

All sturgeons reproduce in fresh waters. Generally, sturgeons prefer spawning grounds located at some distance from the estuary with gravel substrate and a rapid flow of water (more than 1 m/s).

There are strong doubts about the adequacy of the current hydrological regime in the Guadalquivir River for the successful natural reproduction of the Adriatic sturgeon. Therefore we would suggest that sturgeon hatcheries should be developed for restoration of the Adriatic sturgeon population in its historical area. In this respect the experience of artificial reproduction of sturgeon in Russia and other states could be quite helpful.

Keywords Acipenseridae, *A. naccarii*, reintroduction, feeding stocks, Caspian and Azov Seas, spawning conditions, productivity

---

B.N. Kotenev
Russian Federal Research Institute of Fisheries and Oceanography,
17, V. Krasnoselskaya st., Moscow, Russia

21.1 Introduction

Sturgeons are representatives of the most ancient vertebrates, the ancestors of which appeared on the Earth several hundred million years ago (Yakovlev, 1977). These species survived all the natural disasters (warming, glaciation, meteorite infalls, etc.) which descended on our planet during these years and they remained almost unchanged and fairly numerous (until recently) representatives of ichthyofauna in most basins of the Northern Hemisphere.

It was only due to the intensive development of civilization during the last century, industrial progress, population growth, development of fishing techniques and gear, development of aquatic basins and damming, as well as anthropogenic pollution that the sturgeon abundance has decreased; at present some *Acipenseriformes* species, including the Adriatic sturgeon (*Acipenser naccarii*) and Atlantic sturgeon (*Acipenser sturio*, Williot et al., 1997) are threatened with extinction. On the other hand, development of civilization, namely progress in science and technology, inter alia, in ichthyology and aquaculture, have opened up possibilities for conservation and restoration of these endangered species.

Formation of breeding stocks and successful culturing of the Adriatic sturgeon first in Italy and later in Spain (Arlati et al., 1988; Arlati and Bronzi, 1993; Bronzi et al., 1999) prompted Spanish scientists to suggest restocking of this species in its historical natural area of distribution, namely in the Mediterranean Sea (Domezain Fau, 1997).

Here, we sum up the major characteristics of the Russian sturgeon and its closest counterpart, the Adriatic sturgeon, and the peculiarities of their habitat. Vasnetsov (1947) classified most sturgeons as anadromous migrants with a large potential for growth while thriving in seawaters with extremely rich feeding stocks which are one or two orders of magnitude higher than in fresh waters.

21.2 Sturgeon Feeding Stocks in Russian Seas and in the Mediterranean Sea

The sturgeon fattening in seawaters is generally related to benthic production. Benthic biota (molluscs, worms and crustaceous) are consumed by sturgeons either directly, or indirectly through consumption of demersal fish (gobies, carps, and others) which feed on benthos (Fig. 21.1).

In several coastal areas of the Caspian Sea, the benthic biomass totals achieve 50–500 g/m^2, sometimes exceeding 1000 g/m^2 (Fig. 21.2) (Yablonskaya, 1964; Karpinskij, 2002). In the 1940s the sturgeon feeding stocks in the Caspian Sea was considerably enriched due to introduction of polychaete *Nereis diversicolor* and bivalve *Abra ovata* from the Sea of Azov (Zenkevich et al., 1945).

The benthic biomass in the Black and Azov seas exceeds that in the Caspian Sea almost twofold (Fig. 21.3); as a result, sturgeons in the former seas are characterized by higher rates of growth, compared to the Caspian sturgeons (Chugunov and Chugunova, 1964).

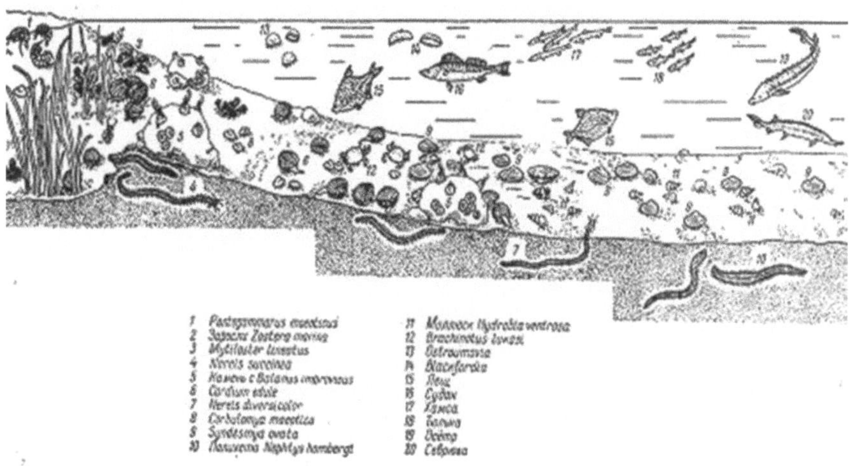

Fig. 21.1 Zonal distribution of the Sea of Azov fauna

Difference in diets between sturgeon species needs to be taken into account: the Russian sturgeon, like the Adriatic sturgeon, feeds mainly on molluscs, while stellate sturgeon (*A. stellatus*) and barbel sturgeon (*A. nudiventris*) prefer crustaceans and worms. Fish makes significant contributions to diets of each of these species when older, while Beluga (*Huso huso*) becomes an obligate predator from its early years (Zheltenkova, 1951; Yablonskaya, 1964).

From their earliest age the juvenile sturgeons in rivers feeds only on benthos: chironomid larvae, oligochaetes and gammarids. Plankton feeding is not typical of any sturgeon, though when young fish are grown in ponds with great biomass of plankton organisms, mostly cladocera of over 5 g/m^2, the larvae and young sturgeons in their first weeks consume mostly plankton. The lower jaw, barbells, and a developed sense of smell indicate that the young sturgeons are adapted to feeding on bottom organisms. The young sturgeons find their food using olfactory and tactile organs; the feeding intensity of the young fish in the daytime and at night is virtually the same. The absence of sturgeons at depths of less than 5 m is explained by the fact that they fail to compete with the bony fishes which stay in shallow places and use vision searching food (Stroganov, 1956).

In the estuary and the fore part of the Volga and the Kura Rivers the juvenile sturgeons increasingly consume gammarids, corofeids, and mysids. The young consume gammarids of 0.75–9.0 mm; prey size ranges with the size of the predator (Zheltenkova, 1964). In the pre-estuary area the young show better stomach fullness characteristics compared to those in rivers, where estimates require better physiological indicators.

At sea the number of gammarids and chironomids in the food of sturgeons declines with the growth of this fish while the amount of fish food (sculpins and herrings) and shellfish (mytilaster, cardium and monodacna) increases. The large isopod *Mesidothea entomon* makes up a major proportion of food in the East

Fig. 21.2 Distribution of benthic biomass in the Caspian Sea

Caspian Sea. The average stomach-fullness index in an adult sturgeon is about 30% or 0.3 of the total body weight, but there have been some instances when the total sturgeon stomach-fullness index reached 142.6% mostly because of mitilaster. Sturgeon stomach fullness is greater at night: 36.5% versus 24.4%.

Shorygin (1952) summarized the sturgeon feeding data for the entire Caspian Sea and produced a scheme of dynamics of the sturgeon food depending on the sturgeon age: a fish of under 50 cm feeds mainly on crustaceans, mostly gammarids, which are 88% of its food, and *Nereis*. A sturgeon of 50–100 cm takes in much more fish food (35–43%); the role of fish goes down with further growth while molluscs go up. Some 63 to 80% of food in sturgeons of over 150–230 cm long are molluscs.

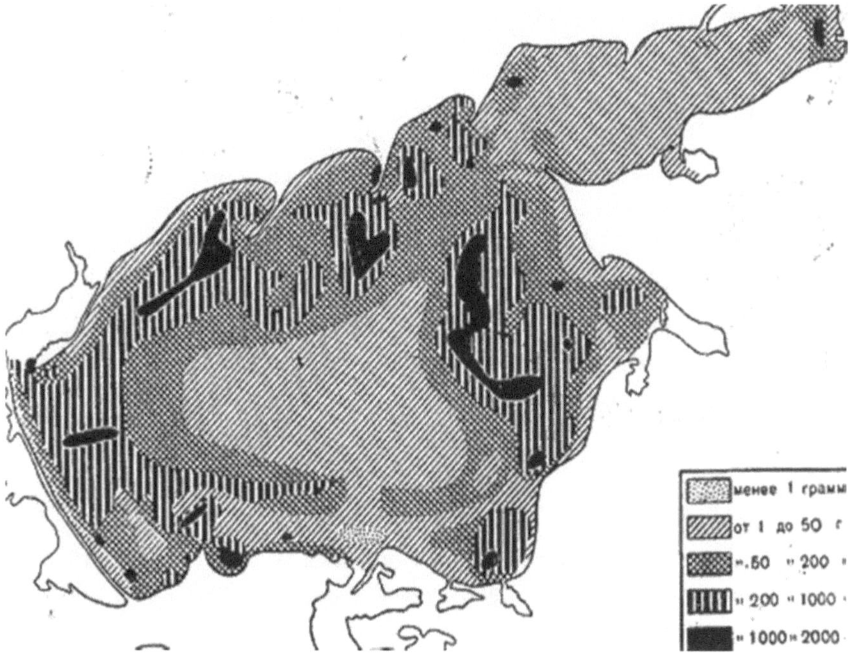

Fig. 21.3 Benthic biomass in the Sea of Azov (g/m^2)

The study of food habits of the Caspian Sea fish led Shorygin to conclude that this sea should be turned into a predominantly sturgeon basin.

At the Sea of Azov and the Black Sea the food of the Russian sturgeon includes mostly molluscs (sindesmia and corbulomia) and crustaceans, too.

Considering the Mediterranean Sea as a sturgeon habitat in connection with the issue of the species restocking, we conclude with a high probability that this basin satisfies the feeding needs of the Adriatic sturgeon, for which the growth rates and size are reported to exceed those of the Russian sturgeon. Some peculiarities of the feeding stocks and diet of the Adriatic sturgeon, however, could need additional studies during the process of the species restocking. In any case, production and the feeding stocks of the Mediterranean Sea, which supported a great abundance of sturgeons in the past, should not become limiting factors for the Adriatic sturgeon restocking process in the historical area of distribution.

21.3 Hydrological Conditions at Natural Reproduction of Sturgeons

We would call 'inadequate opportunities for natural reproduction' of sturgeons a limiting factor, which could be overcome by the modern regulated run-off of the Guadalquivir River.

Generically, all sturgeons are fresh water species. Spawning migration of mature sturgeons from the sea into rivers starts both on internal and external signals, which include such factors as the state of gonads and neurohumoral system (internal signals) and temperature and salinity gradients as well as seasonal changes in photoperiodicity and increased turbidity of water in the spawning rivers and estuaries in spring. The selection of spawning grounds is also determined by the presence of gravel substrate and a rapid flow of water (more than 1 m/s).

Survival of eggs and larvae is highly dependent on the water turbidity during floods, which allows them to escape from potential predators, i.e. almost all fish species. Efficiency of natural reproduction is directly related to the volume of the river run-off: the higher the water discharge, the higher the survival rates of young sturgeons and, consequently, the higher the number of the year-class recruitment (Raspopov and Egorova, 2001). Damming of several Russian rivers (e.g. the Don and the Kuban' rivers) has virtually brought to an end the natural reproduction of sturgeons in the Azov Sea basin.

Large-scale measures were taken on the Volga River to ensure passage for sturgeons to the upper reach of the Volgograd dam by their delivery by boats for live fish, or using a fish lift which operated successfully in 1961–1969, until the Saratov hydropower station was constructed in 1967 near Balakovo, somewhat upstream of Saratov. Totally about 546,000 parents of the spring-spawning sturgeon, several dozens of stellate sturgeon and a small number of the beluga were transplanted to the upper reach of the Volgograd dam (Shylov et al., 1971).

Some spawning of sturgeons was recorded in the remaining spawning grounds of the upper reach of that dam, and there was a downstream migration of the larvae, young fish and the spawned parent stock through the turbines. A part of the brood stock remained in the reservoir where they even repeated maturation. About 80% of the young went down from the reservoir as early as in the first year, however the young survived worse than under the old river regime because of the high degree of water transparency standing still in the reservoir and the slow speed of the current.

After the construction of the Saratov hydropower station there remained only a small stretch of about 90 km in the upper reach of the Volgograd reservoir that was fit for spawning of sturgeons. However, the passage of the parent stock to the upper reach was ended because of the low effectiveness of spawning, and the decreased number of spawners reaching the lower stretch of the Volgograd dam. At present there is only a local river starlet stock that remains in the Volgograd reservoir.

Many of the ecological factors essential for successful reproduction (e.g. opportunity to migrate for spawning to the headstream with rapid flow and gravel bottom) are diminished or completely missing in regulated rivers. Therefore we have strong doubts about the adequacy of the current hydrological regime in the Guadalquivir river for successive natural reproduction of the Adriatic sturgeon.

21.4 Necessity for Artificial Reproduction of Sturgeons

Therefore we would suggest that sturgeon hatcheries should be developed for restoration of the Adriatic sturgeon population in its historical area of distribution. In this respect the experience of sturgeon artificial reproduction in the Russian Federation and other Caspian states could be quite valuable (Gerbilskij, 1953; Kozhin, 1964). Large-scale industrial hatching of various sturgeon species in the main sturgeon basins which was developed in the former Soviet Union more than five decades ago proved to be efficient and provided for conservation of these valuable relicts even when their natural reproduction in dammed rivers was stopped (e.g. the Don and the Kuban' Rivers of the Azov Sea basin).

Italian and Spanish fish culturists have already become well acquainted with biotechnology of artificial reproduction of sturgeons in operating aquaculture developments growing the Adriatic sturgeon and other sturgeon species.

21.5 Resistance of Young Sturgeons to High Salinity

Spanish scientists have performed experiments to find out whether juvenile individuals of the Adriatic sturgeon (age 1^{+}–4^{+} with the weigh of 387–5410 g, reared in fresh water) can successfully adapt to the Mediterranean Sea environment with the water salinity of up to 33‰ (Sánchez de Lamadrid et al., 1999).

It would be reasonable to make additional experiments and determine the osmoregulatory ability of still younger fishes (age 1–3 months with a weight of 3–50 g) which are intended to be released in the lower reaches of the Guadalquivir River, under conditions of both a gradual transfer from fresh to seawater, and a direct release into salty areas of the Guadalquivir River estuary.

It is known that the small fry of Russian sturgeon (average length 6.3 cm, average weight 1.0 g) adapt to salt water more weakly than do large fry (average length more 10.2 cm and average weight more 3.8 g). A large sturgeon fry adapt to salt water more easily and faster in comparison with smaller fry (Kraiyshkina, 1983).

21.6 Biological Diversity of Sturgeons

The final issue I would like to highlight for the benefit of my European colleagues is that high productivity of sturgeons in Russian seas and in adjacent national waters is based on multispecies (six anadromous species) composition of sturgeon stocks in the Caspian and Azov basins (six anadromous species), and biological differentiation of some sturgeon species populations into seasonal races or biological groups and local populations of various spawning rivers.

Therefore, we would advise our western colleagues not to limit themselves by restocking only one native sturgeon species, but to aim at restoration of stocks for all sturgeon species which previously inhabited the Western Atlantic and the Mediterranean Sea (the Adriatic sturgeon, the Atlantic sturgeon (*A. sturio*) as well as Beluga (*H. huso*).

As a rule the species of migratory sturgeons have a complex population structure; local population reproduce in concrete river basins (Volga, Ural or Kura Rivers, etc.). In turn, local populations of sturgeon in large rivers are split into various biological (ecological) groupings, or seasonal races. The fish of different biological groups migrate to rivers at various times having different gonad conditions; prior to spawning they spend different time periods in rivers, spawn at different times at varying spawning temperatures; and the location of the spawning grounds is diverse as well.

The Volga–Caspian population of the Russian sturgeon is subdivided into four biological groups: first, the 'early summer' group, migrates to the river in early spring and spawns in spring to mid-summer, second, the 'late summer' group, migrates in mid- to late summer and spawns at end of the same summer; third the 'winter group of summer migration', migrates in summer but spends a winter in the river before spawning; and fourth, 'winter group of autumn migration', migrates in autumn and spends a winter in the river and spawns in the late spring of the next year (Gerbilskij, 1953; Barannikova, 1957; Kazanskij, 1962).

The first group begins to run regularly in the Volga estuary in late March or early April at water temperatures of 0.2–9.6°C, and goes on till the middle of May. It spawns in the mid-reaches of the Volga, downstream of Volgograd, at 15–17°C. The running brood stock has gonads at maturity stage IY.

The 'late summer sturgeon' come into the Volga somewhat later, between early May and late June. Its gonads are less developed and are diverse. It spawns later–from late May till July.

July and early August sees the run of 'winter sturgeon of summer migration' which has fat gonads at III-IY stage of maturity; these fish winter in the river. Its gonads pass to stage IY of completed maturity by early May and its spawning occurs somewhat earlier: May to early June.

The winter sturgeon of summer run enter the Volga in September–October; spawning occurs in May–June at 12–14°C.

The other species of sturgeons inhabiting large and long rivers such as the Ural and Kura in the Caspian basin, the Don and Kuban flowing into the sea of Azov, and Dnieper and Danube on the Black Sea basin have a similarly complex population structure.

During their run to the rivers the winter sturgeons have a much greater body-energy resource; prior to damming they were able to make much longer spawning migrations in terms of time (9–12 months) and range (up to 2000 km). They could reach the spawning grounds in the upper reaches of rivers after leaving the spawning sites downstream for the summer parent stock to spawn at. The spawning places and those where the young fed in rivers were thereby used by the sturgeon to the utmost, thus providing the maximum abundance of recruitment of the local populations of the species by that generation's recruits. Consequently the fishing stock of these most valued species would reach its maximum (Kazanskij et al., 1983).

During the period of the most favourable sturgeon stock condition (1960–1980) the earlier summer group of sturgeon producers were used mostly for artificial reproduction since they provided quality eggs immediately after capture and long conditioning was not needed. Given the current acute shortage of parent stock the sturgeon hatcheries are making reserves of both the summer and winter races of the Russian sturgeon, which helps maintain their natural genetic structure and biodiversity.

The consistent conservation of conditions for natural reproduction of sturgeon species both in rivers retaining the natural flow regime (Ob, Ural, Amur, etc.) and in the reaches below dams in dammed rivers (Volga, Kura, Don, etc.) serves the same purpose.

Following the construction of a series of hydropower stations on the Volga the area of sturgeon spawning grounds went down by 85%; only 15 spawning grounds of 415 ha still remain in its lower reaches.

21.7 Creation of Artificial Spawning Grounds

As a measure to enhance the natural reproduction of sturgeons, many Russian scientists have suggested establishment of artificial spawning grounds (Aliavdina, 1952; Vasnetsov, 1954). P.N. Khoroshko and A.L. Vlasenko of the former Central Sturgeon Fisheries Research Institute worked out standards and rules for designing human-made sturgeon spawning grounds to be applied by designers and construction organizations.

Seven experimental artificial spawning grounds totalling 58.4 ha were constructed in the lower reaches of the Volga between 1966 and 1986 (Vlasenko and Slivka, 1987). Such grounds, depending on the level of their position over the summer water, can belong to the class of those flooded in spring-time, which are submerged under water only during the spate, and those of the main stream which remain under water even during the low water level.

Multi-annual studies show that the intensity of the use of the constructed spawning grounds by sturgeons depends on the spring volume of the river runoff, a combination of the hydrological and thermal conditions, and the status of the spawning substrate. In different years the egg-shedding densities in the best-positioned spawning grounds were 900–4000 ind./m^2, whereas in the other ones it varied between 50 and 300 ind./m^2. The major reason for the low rate of spawning in those grounds was silting of the gravel patches, and accumulation of sand there.

The mean rate of fish production in the artificial ridges within the low reaches of the Volga which are being used was: 7 t/ha in the main stream and 12 t/ha in the spring-flooded ones. The criterion is the fishing return of sturgeons ensured by the number of eggs laid there.

The human-made grounds should be set up in the routes of the spawning run of parents, and at potential sites of their significant aggregation. The selection of sites for the construction of spawning grounds requires studying the hydrological features of the main stream processes in several river stretches for 2–3 years. The optimum speed of current at the chosen part of the river must be 1.0–1.6 m/s. The spawning substrate laid should include gravel or pebbles with a fraction diameter of 3–10 cm; the split layer should be at least 20 cm. The area of the artificially made ground per one female sturgeon must be 350 m^2, given a total area of artificial spawning ground of at least 5 ha. The chosen place must have a solid underlying ground of clay, dense loam or sands well compressed with shells.

21.8 Fishery Regulations

Depleted stocks of sturgeon species in the Azov and Caspian Seas during the last 15–20 years were a result of weakening of the state regulation of fishing activities in theses basins and increase in illegal fishing for sturgeons after the Soviet Union collapse and dramatic social and economic changes in the resultant independent states. We anticipate that this negative situation is temporary and will be changed through the cooperative efforts of states within the sturgeon area. Meanwhile European states should learn even from this negative experience of ours: they should take steps to develop rigorous regulations for the sturgeon fishery and prevent illegal fishing for sturgeons ahead of actual restocking of sturgeons in the Mediterranean Sea.

21.9 Conclusions

Under sound exploitation of these newly developed resources the fisheries could annually yield about 3000–4000 t as a rough approximation (Fig. 21.4).

Summarizing all the above, we consider restocking of natural basins with native sturgeon species quite timely and feasible. In the coming years, the achievement of this goal should receive all the attention necessary, and we wish our western colleagues success in restoring this valuable natural asset (Fig. 21.5)!

Acknowledgements I would like to express my gratitude to Dr. Alberto Domezain and all the persons of Bioresturgeon's Organization Committee for hospitality and Drs. I. Burtsev and A. Nikolaev for their great help in the preparation of the present paper.

Perspectives of sturgeons
(A. naccarii and A. sturio)
in the Guadalquivir River area
depend on the following:

- productivity of the marine area
- river discharge
- sturgeon multispecies diversity

Possible productivity about
3000-4000 t/year

Fig. 21.4 Possible productivity of sturgeons in Mediterranean Sea

Fig. 21.5 Sturgeon fry reared at the Sturgeon Hatchery for release into the Caspian Sea

References

Aliavdina L.A. 1952. Artificial spawned grounds for sturgeons at the Volga River. *J. Rybnoe khozaistvo*, No. 1, 29–31 (in Russian).

Arlati, G. and Bronzi, P. 1993. Sturgeon fish farming in Italy. *Proceedings of the International Sturgeon Symposium*, VNIRO Publication, Moscow, pp. 321–332.

Arlati, G., Bronzi, P., Colombo, L. and Giovannini, G. 1988. Induced breeding of the Italian sturgeon (*Acipenser naccarii*) reared in captivity. *Riv. Ital. Acquacol.*, 23, 94–96.

Barannikova, I.A. 1957. Biological differentiation of Volga-Caspian sturgeon stock (in connection with tasks of commercial reproduction at the Volga River delta.). Scientific reports of Leningrad State University, Ser. Biology, vol.44 No228, Ed. of Leningrad State University, USSR, pp. 54–71 (in Russian).

Bronzi, P., Rosenthal, H., Arlati, G. and Williot, P. 1999. A brief overview on the status and prospects of sturgeon farming in Western and Central Europe. *J. Appl. Ichthyol.*, 15, 224–227.

Chugunov, N.L. and Chugunova, N.I. 1964. Comparative fishing and biological characteristics of the Azov Sea sturgeons. *VNIRO Trans.*, 51, 87–182 (in Russian).

Domezain Fau, A. 1997. Recuperación del Esturión en el Guadalquivir. Riofrío, 72 pp. (in Spain).

Gerbilskij, N.L. 1953. Interspecial groups of sturgeons and importance of their studies for development of sturgeon reproduction in connection with hydrological construction. Reports of All-Union conference for the questions of fisheries in USSR (Moscow 1951), Ed. Academy of Science USSR, pp. 291–300 (in Russian).

Karpinskij, M.G. 2002. *Benthos Ecology in the Middle and Southern Caspian Sea*, VNIRO Publication, 283 pp (in Russian).

Kazanskij, B.N. 1962. Experimental analysis of seasonal reproduction of Volga River sturgeons in connection with evidence of interspecies biological differentiation. Scientific reports of Leningrad State University, Ser. Biology, vol. 47 No311, Ed. of Leningrad State University, USSR, pp, 19–45 (in Russian).

Kazanskij, B.N., Podushka, S.B. and Burenin, O.K. 1983. About significance of migrants winter-annual type for sturgeon economy. In: *The biological basis of sturgeon culture, Ed. Science*, Nauka, Moscow, pp. 42–54 (in Russian).

Kozhin, N.I. 1964. The USSR sturgeons and their reproduction. In: «Sturgeons of the Southern Seas of the Soviet Union», *VNIRO Trans.*, 52, 21–58 (in Russian).

Kraiyshkina, L.S. 1983. Functional generation of osmoregulatory system of sturgeons fry at dependence from size and age. In: *The biological basis of sturgeon culture, Ed. Science*, Moscow, pp. 158–166 (in Russian).

Raspopov, V.M. and Egorova, A.E. 2001. Natural reproduction of the Russian Sturgeon at condition of adjustable drainage of the Volga River. In: *Young Fish Ecology and Caspian Fish Reproduction Problems*, VNIRO Publication, Moscow, pp. 228–238 (in Russian).

Sánchez de Lamadrid, A., Garcia-Callego, M., Sanz, A., Munoz, J.L., Domezain, J., Soriguer, M.C., Domezain, A. and Hernando, J.A. 1999. Acclimation of the sturgeon, *Acipenser naccarii* Bonaparte 1836 to saltwater: effect of age and weight. In: *Recent Advances in Mediterranean Aquaculture Finfish Species Diversification*, Vol. 47, Centre Int. de Etudes Agronomiques Mediterraneennes, Cahiers, pp. 337–342.

Shorygin, A.A. 1952. *Nutrition and Nutritional Mutual Relations of the Caspian Sea Fishes*, Pischepromizdat, Moscow (in Russian).

Shylov, V.I., Khazov, Yu K., Ivoilova, N.K. 1971. Specific weight of sturgeons in ichthyofauna of Saratov and Volgograd water bodies, catch structure and allocation. In: Sturgeons in Volgograd and Saratov water bodies. Scientific reports of Saratov branch of GosNIORKH.vol.11, Ed. GosNIORKH, pp.5-51 (in Russian)

Stroganov, N.S. 1956. Nutrition and growth of sturgeons in aquariums and tanks (to substantiation of industrial growing method of sturgeons). 'Bulletin of MGU' No. 2 (in Russian).

Yablonskaya, E.A. 1964. Feeding Stock for Sturgeons in the Southern Seas, *VNIRO Trans.*, 54, 81–112 (in Russian).

Yakovlev, V.N. 1977. Phylogenesis of *Acipenseriformes*. In: *Phylogenesis Phylogenesis and systemathic essayes about fossil fishes and agnates. Ed. Academy of Science USSR*, pp. 116–144 (in Russian).

Vasnetsov, V.V. 1947. Fish growth as adaptation. *Byull. Mosk. Obsch. Ispyt. Prirody*, 52 (1), 22–34 (in Russian).

Vasnetsov, V.V. 1954. Artificial spawned grounds for anadromous species of fishes. *Voprosy Ichtyologii*, 2, 69–74 (in Russian).

Vlasenko, A.D. and Slivka, A.P. 1987. Biological basis of creation of an artificial spawned grounds at the Volga River low flow. In: *Reproduction of Sturgeon Fish in the Caspian and Black Seas Basins*. VNIRO Publication, Moscow, pp. 14–19 (in Russian).

Williot, P., Rochard, E., Gastelnaud, G., Rouault, T., Brun, R., Lepage, M. and Elie, P. 1997. Biological characteristics of European Atlantic sturgeon, *Acipenser sturio*, as the basis for a restoration program in France. *Environ. Biol. Fish.*, 48, 359–370.

Zenkevich L.A., Birshtein Ya.A., Karpevich A.F., 1945. First successful efforts in fauna reconstruction of the Caspian Sea. *Zool. J.*, 24, 25–31 (in Russian).

Zheltenkova, M.V. 1951. To question of sturgeon feeding at the north part of the Caspian Sea. *VNIRO Trans.*, 18, 200–210 (in Russian).

Zheltenkova, M.V. 1964. Feeding of sturgeon fishes of Southern Seas. In: Transactions of VNIRO, vol.54, Ed. "Pischevaya promyshlennost", Moscow, pp.9-49 (in Russian).

Chapter 22
Towards the Definition of Optimal Size-Weight Standards of Hatchery-Reared Sturgeon Fry for Restoration

I.A. Burtsev

Abstract The analysis of numerous data from the literature on survival rates of Russian and other sturgeon species in the natural water bodies of Russia at natural spawning and artificial reproduction, and their determining factors, is of some interest in the reintroduction of Adriatic (*Acipenser naccarii*) and Atlantic (*A. sturio*) sturgeon species into natural water reservoirs of their former habitation.

It is known that the fishing-return coefficient from hatchery sturgeon fingerlings 1-month old and standard weight 3 g consists of 1.1–1.3 to 0.6–0.9%. The author adduces arguments for defining the optimal weight standards of hatchery-reared fry which are to be within the range from 40 to 60 g, resulting in maximum fishing return that should be not less 40%.

Keywords *Acipenseridae*, hatchery-reared fry, size-weight standards, viability, fishing return

22.1 Introduction

At present, sturgeon fisheries in Russia are harvesting sturgeons from natural water reservoirs–basins of the Caspian and Azov Seas, Siberian rivers and Lake Baikal, based on the formation of their stocks due to natural and, for the most part, to artificial reproduction on the commercial scale.

The recent reduction in sturgeon stocks observed in the majority of their inhabited areas is caused by anthropogenic factors: reduction in their spawning grounds, deterioration of the hydrological conditions of the rivers as a result of construction of hydropower stations, increased water consumption, pollution of water reservoirs, and, especially, a decrease in the number of spawners caused by illegal overfishing.

I.A. Burtsev
Russian Federal Research Institute of Fisheries and Oceanography, Moscow, Russia

This Contribution was first published by the Astrakhan State Technical University of the Russian Federation in the Russian Language

The volume of natural reproduction of sturgeons in the Volga River has decreased by more than 20-fold, and in the rivers of the Azov Sea basin it is practically nil. The replenishment of sturgeon stocks in water bodies is achieved basically by the ongoing activity of sturgeon hatcheries in Russia. Their annual production in 1998–2003 was 50–70 million specimens of sturgeon fry. Cultivation and release of sturgeon fry are also carried out by the hatcheries of Azerbaijan, Iran, and Kazakhastan. Nevertheless, the commercial reproduction does not completely compensate for the observed reduction in natural reproduction and much less the intensive illegal fishery, so that the efficiency of these hatcheries' activity needs to be raised.

22.2 The General Environmental Factors for Survival of Sturgeon Progeny in Nature

Academician Derzhavin (1947) estimated the average long-term fishing-return rates of stellate sturgeon (*A. stellatus*) in the database of long-term statistical data on the number of eggs laid by stellate sturgeon in the Kura River during the period from 1901 to 1932, which amounted to 58,360.5 million eggs, and the total catch from 1916 to 1942 which equalled 6,052,000 mature fishes of appropriate year-classes which constituted 0.0104% of the total number of laid eggs.

A calculation of the fishing-return rates for the separate 5-year periods gives wide fluctuations in these values:

Years of spawning	Fishing-Return Rates (%)
1916–1922	0.0078
1923–1927	0.0482
1928–1932	0.0161
1933–1937	0.0087
1938–1941	0.0075

The same value of the total fishing-return rates was adopted for the Russian sturgeon. Later investigations (Zhuravleva and Ivanova, 2001) confirmed these data: the total fishing-return rates of Russian sturgeon generations in the unregulated Volga River in 1951–1955 amounted to 0.00855–01012% of the number of spawned eggs.

As the natural hatching yield of larvae was estimated by Derzhavin (1947) as 30% of the total number of laid eggs, the fishing-return rates of adult fishes of hatched larvae was determined by him at 0.0347%.

Derzhavin carried out his studies before the construction of the hydropower stations on all the rivers when the natural factors defining the survival of sturgeon progeny during early life stages still existed. In particular, the turbidity of water, characteristic during the spring floods of the rivers, was one such natural factor. Thus, Kozlovsky (1953) placed special emphasis upon the fact that high turbidity of waters was a favourable factor stimulating spawning of some fishes and decreasing predation of spawned eggs, larvae and fry. It is quite evident that larvae and fry are completely defenceless and they do not react to predators (Pavlov and Sbykina, 1989). It was found that the sturgeons that spawn in the early spring are more numerous as compared with other sturgeon ecological groups which spawn in the middle of summer or in autumn. Thus, suspended matter in water favours viability of sturgeon progeny.

From the environmental standpoint, high density of food organisms for fry is of great importance for its successful survival. This depends on flooding conditions of the huge plain areas of the rivers. A good correlation was established between the level and duration of spring floods and the recruitment value of sturgeon populations by generations of corresponding year-classes (Lagunova, 2001; Raspopov, 2001). In recent years the volumes and duration of floods, as well as water turbidity have considerably decreased. As a consequence, the survival of sturgeon generations after the building of the first dam on the Volga River–Kuibyshevskaya (autumn of 1955) decreased two- to threefold in comparison with earlier periods of natural run-off (Zhuravleva and Ivanova, 2001).

In the last few years due to a sharp decrease in the number of upstream migrating spawners to their spawning grounds and deterioration of ecological conditions, the amounts, and efficiency of natural reproduction of sturgeons in the Volga River reduced to a minimum and in the Don and Kuban Rivers the maintenance of sturgeons persists exclusively at the expense of hatchery cultivation, the importance of which is growing in all the reservoirs.

22.3 The History and Practice of Artificial Reproduction of Sturgeons in the Former Soviet Union and Russian Federation

Studies on artificial reproduction of sturgeons started in Russia more than 130 years ago. They began with experiments conducted by the academician Ovsyannikov (1870) in 1869 on fertilization of eggs of the Volga River sterlet (*A. ruthenus*) and its fry rearing. In 1884, Borodin (1885, 1912) was the first to perform artificial fertilization of stellate sturgeon (*A. stellatus*) eggs, and in 1901 he continued his experiments in the Kura River.

In the first half of the last century on many rivers of the European part of Russia and former Soviet Union, there were several state sturgeon breeding hatcheries which caught mature ‘fluid’ spawners on spawning grounds, fertilized and incubated eggs and released hatched larvae in the rivers. Scales of work gradually increased and by 1940–1941 the total number of released stellate sturgeon larvae to the Kura, Volga, Kuban, Don and Ural Rivers amounted to approximately 280 mln larvae (Derzhavin, 1947).

The first experimental sturgeon hatchery was built in the lower reaches of the Kura River in 1936 where methods of hormonal stimulation of brood stocks, technologies of rearing sturgeon fingerlings in ponds and tanks, cultivation of live food organisms and other techniques were developed. Results of these studies made it possible to reach a completely new level of sturgeon reproduction, i.e. to move sturgeon breeding from the upper spawning grounds to the lower reaches of the rivers, rearing of more viable fingerlings and releasing them in natural water bodies instead of larvae (Kozhin et al., 1963).

On the basis of these achievements during the post-war period the construction of commercial sturgeon hatcheries was developed in most sturgeon rivers of the Soviet Union. By the 1970s, 25 hatcheries of total capacity over 130 mln fingerlings had been built with the purpose of offsetting the damage to the reproduction of sturgeons because of the construction of hydroelectric power stations. And now recruitment of sturgeon stocks in main sturgeon water bodies of Russia is provided due to the activity of sturgeon hatcheries. The annual release of fingerlings from sturgeon hatcheries in the Caspian basin during 1998–2003 amounted to 50–70 million. Breeding and release of sturgeon fingerlings are also carried out by the hatcheries of Azerbaijan, Iran, and Kazakhastan.

22.4 The Effectiveness of Artificial Reproduction of Sturgeons

The following standards for sturgeon fingerlings have been accepted when designing and constructing sturgeon hatcheries: for Russian sturgeon and beluga the standard weight of fingerlings was established as 3.0 g, for stellate and spiny sturgeons 1.5–2.0 g (Kozhin et al., 1963; Lukjanenko et al., 1984). To substantiate the efficiency of hatchery operations, a unified commercial fishing-return coefficient was accepted as 3% (Kozhin, 1951; Bojko, 1963; Makarov, 1964). Based on long-term results of mass release of hatchery-reared fingerlings of different species and the actual return of adult fishes these coefficients of the Volga sturgeons were refined (Buhanevich et al., 1986). For the Russian sturgeon it was 2.8%, for beluga 0.42% and stellate sturgeon 1.0% of the total number of fingerlings released. However, for the Russian sturgeon of the Azov sea, the actual level of fishing return from hatchery fry released in 1956–1972 was only 0.6% (Zaidiner et al., 2000).

Despite such low survival rates of standard fry, opinions still call for increasing the scales of industrial sturgeon reproduction at the cost of construction of new hatcheries and an increase in the numbers of hatchery-reared fry up to 150 million specimens (Karpyuk et al., 2002), but this cannot be accepted as either rational or practical. In modern practice even existing hatcheries are in short supply of brood stocks, and they reduce their fry production, which results in having to create their own brood stocks at the hatcheries (Burtsev et al., 2002; Popova et al., 2002). It is obvious that industrial reproduction does not completely compensate for the existing reduction of natural reproduction and intensive illegal fishing (Levin, 2002). In this sense, the problem of consi-derably improving hatchery reproduction efficiency becomes even more urgent (Marti, 1972).

22.5 Ways of Increasing the Efficiency of Commercial Hatchery Reproduction of Sturgeons

In the initial years of production of the sturgeon hatcheries, a higher viability of the larger hatchery-produced Russian sturgeon fingerlings as compare with the little ones was observed when they were released into the Don river (Bojko and Kalinkina, 1961). Levin (2002) and Levin and Kokoza (1996) found that the absolute number of one-summer-old juveniles of Russian sturgeon in the western part of the North Caspian Sea makes up 7–10% of the number of released fry in the Lower Volga. This is explained by their being devoured by predators, their loss in water intake installations and at dredging works, starvation during their migration in the areas of poor food supply, and effects of significant thermal and salt gradients. At the same time the in the release of fry taken out of hatcheries in special boats and released directly in water areas with high food density (Polianinova, 1983) and with optimal temperature and salinity, the survival rate of the Russian sturgeon fry proved to be rather high.

It is shown to be more reasonable to increase the reproduction efficiency of hatcheries by improving the quality and viability of released fry. The scientists from the Russian research institutes CaspNIRKh (Mikhailova, 2004) and AzNIIRKh (Gorbacheva and Rekov, 1996; Gorbacheva et al., 2002) have shown the possibility of rearing larger fry in ponds using the existing extensive technology due to lower stocking density and a slightly longer term of rearing up to 7–15 g in weight, which would result in better viability (up to three- to fivefold; Levin and Kokoza, 1989; Levin, 2002).

It has been shown to be more effective to increase the reproduction efficiency of hatcheries by enhancing the quality and viability of released fry. It would be even more effective to organize an intensive fry production using a tank-rearing method, releasing much larger fry, completely excluding its extermination by predators and providing for its survival at a level not less than 50–60% (Andrianov et al., 2004).

In this respect, remarkable research was made by the American scientists (Ireland et al., 2002) who, in the period from 1990 to 2000, determined the survival level of hatchery-reared young white sturgeon (*A.transmontanus*) of 1 to 4 years old and 23 to 72 cm long, after release in the Kootenai River. It was shown that, for 3 years after being released, most of these juveniles successfully adapted to the wild environmental conditions. Average growth rates varied from 0.8 to 8.4 cm/year, and weight increases varied from 0.010 to 0.268 kg/year. During the first year the estimated average survival rate for all size groups was 60%, and in subsequent years it approached 90%.

To summarize and illustrate all the above, we prepared a diagram to represent the dynamics of survival rates of sturgeon fry (using the Russian sturgeon as an example) beginning with its release from the hatcheries to the Volga River and using known survival levels as reference points (Fig. 22.1):

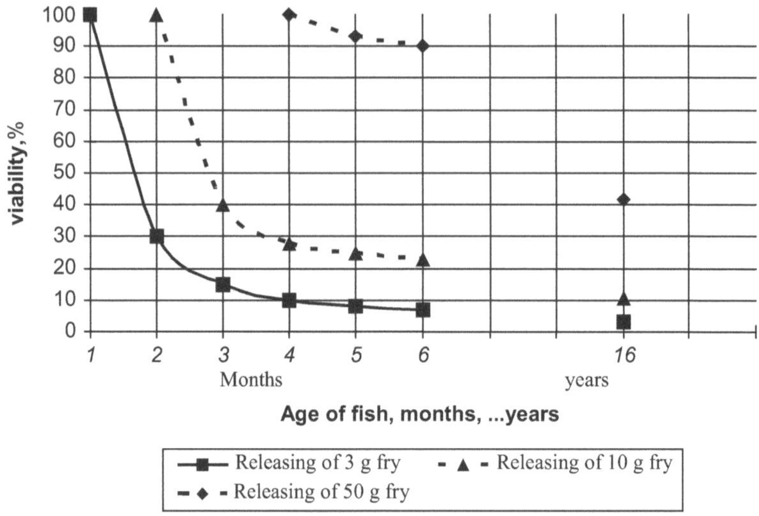

Fig. 22.1 Viability of Russian sturgeon depending on weight of released hatchery-reared fry

1. Curve 1 indicates the levels of survival rates of 1-month-old fingerlings of 3 g in weight. We have accepted the survival level during the first month as 30%, during the next 4 months 8% from the amount of released fingerlings; with a natural mortality rate of 5% during every subsequent year; the final survival rate by the age of 16 years will be 3.2%.
2. Curve 2 indicates the levels of survival rates of 2-months old fingerlings of 10 g in weight. The final survival level by the age of 16 years will be 10.0%.
3. Curve 3 concerns the release of 3-month-old fry of 50 g in weight and is constructed by the exact extrapolation of the first two curves. The final survival level by the age of 16 years will be 42%.

Releasing of larger juveniles over 50 g in weight seems to make little sense as according to the studies performed by Ireland et al. (2002), their survival rates remain on the same level.

We tried to compare the efficiency of released fry of various weight reared over a similar area of fish-breeding tanks. Figure 22.2 shows the efficiency of released fry of different weights and reared in the same area of tanks. According to the diagram, it is quite clear that the larger the fry, the less number can be reared over the same area of fish-breeding capacities. Nevertheless the efficiency of release of a lesser number of larger fry significantly increases the fishing return.

The maximum effect resulting from a release of artificially reared fry may be reached by growing fry of different weights over the same area using a step-by-step release in the river. Table 22.1 shows that if the maximum number of small-size fry is reared over the same area, some 70% of which are released, the remainder are

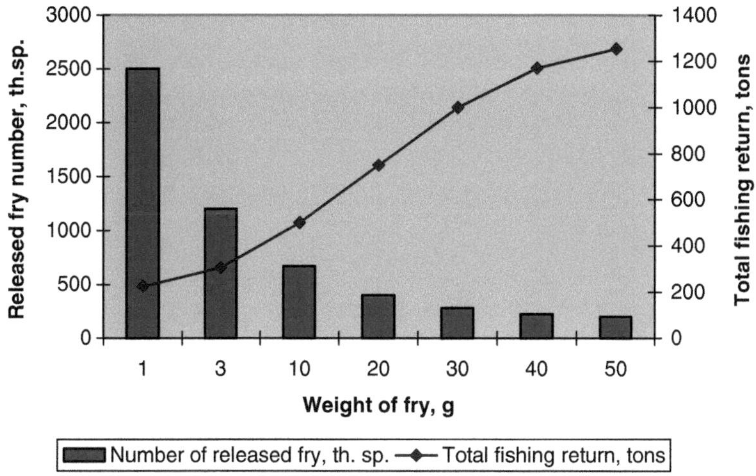

Fig. 22.2 Commercial fishing return from released sturgeon fry of different weight that were reared in the same basin area of 1000 m^2

Table 22.1 Estimated fishing return of Russian sturgeon from multigraded release of the standard and large fry reared in the same basins of 1000 m^2

Middle weight of fry (g)	Number of fry, thousands of specimens: Reared	Number of fry, thousands of specimens: Released	Fishing return coefficient, % of released fry	Fishing return: Thousands of specimens	Fishing return: Metric tons
3	1200	800	2.2	17.6	264.0
10	360	160	5.0	8.0	120.0
50	180	180	41.8	75.2	1128.0
Total	–	1140	–	100.8	1512.0

reared up to larger size and weight (10 g). Then the same operation is repeated again up to the achievement of the greatest possible density of stocking until the breeding season is over. Thus, a certain percentage of returned adult fishes will be received from different-sized juveniles released and the total effect will be maximum.

Figure 22.3 shows a relationship between the weight and absolute length of Russian sturgeon juveniles with an indication of a zone of proposed length-weight standard of juveniles reared at sturgeon hatcheries, which is optimum for their release in natural water bodies in order to achieve a rather high level of survival rates of juveniles and a maximum return of adult fishes.

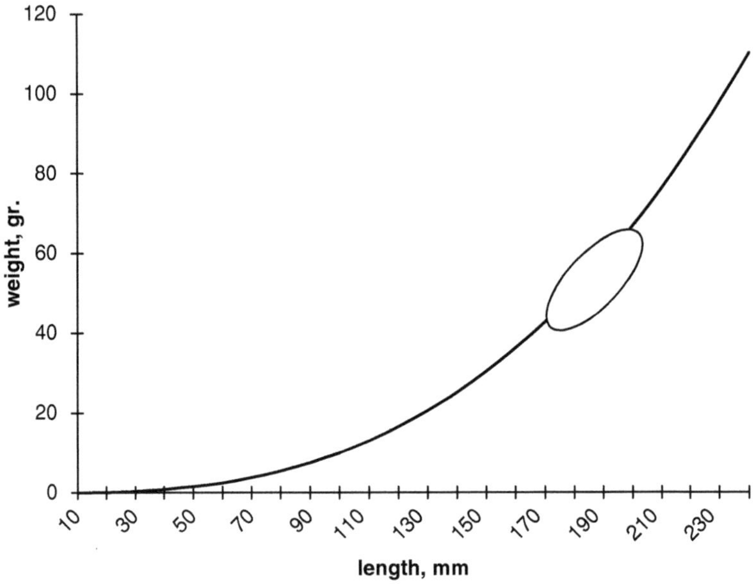

Fig. 22.3 Weight/length relationship of sturgeon fry (zone separated in oval is optimal for releasing of fry)

22.6 Conclusions

It has been shown to be more rational to increase the reproduction efficiency of hatcheries by improving the quality and viability of the fry released. It would be even more effective to organize an intensive fry production using a tank-rearing method of releasing much larger juveniles. This results in their almost complete avoidance of predators that provides their survival at a level not lower than 50–60%. In this case the use of water recirculating supply and closed systems has been shown to be highly promising, which makes it possible to receive fertilized eggs and reared fry in earlier terms and to release them in natural water bodies under optimum environmental conditions (Tyapugin and Ferafontov, 2002).

According to the concept of development of sturgeon breeding in Russia, it is recommended to reconstruct old sturgeon hatcheries for their transition to rearing and releasing bigger fry that will make it possible to solve the problem of sustainable reproduction of sturgeon stocks in the water reservoirs of Russia. Usage of this intensive biotechnology can be recommended for other countries, especially for West European coastal countries.

Acknowledgements The author expresses his thanks to VNIRO senior scientists Drs. A.S. Safronov and D.A. Vasilyev for their help in preparing and reviewing the manuscript.

References

Andrianov, D.P., Burtsev, I.A., Dokukin, M.M., Nikolaev, A.H&$$$;. and Safronov, A.S. 2004. Ways of an intensification and increase of efficiency sturgeon culture and reproduction, in *Scientific—Practical Conference, About Priority Problems Fishery Sciences in Development of Fish Branch of Russia Till 2020*, Moscow, VNIRO, pp. 118–119 (in Russian).

Bojko, E.G. 1963. Reproduction of Azov Sea sturgeons, in *Sturgeon Husbandry in Water Bodies of the USSA*, Publication of the Academy of Sciences, USSR, Moscow, pp. 160–166 (in Russian).

Bojko, E.G. and Kalinkina, E.A. 1961. About a survival sturgeon fry of different weight groups. *Rybnoe khozaistvo* 4: 18–22 (in Russian).

Borodin, N. 1885. About experiences of artificial fertilisation of sevruga eggs. The agriculture and a wood, *SPB* 148(2): 113–128 (in Russian).

Borodin, N.A. 1912. Fish culture. *SPB*, 176 p (in Russian).

Buhanevich, I.B., Dovgopol, G.F., Pavlov, A.V., Raspopov, I.M., Erman, L.A. 1986. Specification of factors of commercial fishery return of Volga River sturgeon on the basis of statistical models. In Coll: Protection and reproduction of fish stocks of Caspian sea, Moscow, VNIRO Ed., pp. 96–102 (in Russian).

Burtsev, I.A., Nikolaev, A.I., Maltsev, S.A. and Igumnova, L.V. 2002. Formation of domesticated broodstocks as a guarantee of sustainable hatchery reproduction of sturgeon for sea ranching. *J. Appl. Ichthyol.*, 18: 655–658.

Gorbacheva, L.T. and Rekov, J.I. 1996. A condition and ways of increase of efficiency of artificial reproduction sturgeon in the Azov-Don area, in *The Basic Problems of a Fisheries and Protection of Fisheries Water Bodies of the Sea of Azov*, Rostov on Don River, pp. 234–238 (in Russian).

Gorbacheva, L.T., Gorbenko, E.V., Saveljeva, E.A., Chihacheva, V.P., Burtasovskaja, L.A., Kazakova, N.M., Vorobieva, O.A, Panchenko, M.G. and Korneev, A.A. 2002. Results of cultivation Azov beluga up to the big weight at industrial cultivation, in *Modern Problems of Caspian Sea*, Materials of the International Conference Devoted to 105-anniversary CaspN IRKH, pp. 71–77 (in Russian).

Derzhavin, A.N. 1947. *Reproduction of Stocks of Sturgeon Fishes*. Academy of Sciences Azerb, SSR, Baku, 248 pp (in Russian).

Ireland, S.C., Beamesderfer, R.C.P., Paragamian, V.L., Wakkintn, V.D. and Siple, J.T. 2002. Success of hatchery-reared juvenile white sturgeon (*Acipenser transmontanus*) following release in the Kootenai River, Idaho, USA. *J. Appl. Ichthyol.*, 18: 642–650.

Karpyuk, M.I., Mazhnik, A.J. and Degtjareva, N.G. 2002. Problems of water bioresources of Caspian sea: today and tomorrow, in *Modern problems of Caspian Sea*, Materials of the International Conference Devoted to 105-anniversary, CaspNIRKH, pp. 132–136 (in Russian).

Kozhin, N.I. 1951. Coefficient of commercial fishing return. *Trans. VNIRO*, 19: 127–132 (in Russian).

Kozhin, N.I., Gerbilsky, N.L. and Kazansky, B.N. 1963. Biotechnics of sturgeon breeding and basic scheme of sturgeon hatchery, in *Sturgeon Husbandry in Water Bodies of the USSA*', Publication of the Academy of Sciences, USSR, Moscow, pp. 29–34 (in Russian).

Kozlovsky, D.A. 1953. Value of turbidity the rivers in formation of ichthyofauna and morphogenesis at fishes. *Zool. Mag.*, XXXII(6): 1052–1065 (in Russian).

Lagunova, V.S. 2001. Ecological aspects of the efficiency of young sturgeon reproduction in the Volga River under present conditions, in *Young Fish Ecology and Caspian Fish Reproduction Problems*, VNIRO Publications, Moscow, pp. 157–179 (in Russian).

Levin, A.V. 2002. About increase in efficiency of reproduction of sturgeon fishes, in *Materials of scientific conference "Problems of Reproduction, Feeding and Struggle Against Illnesses of Fishes at Cultivation in Artificial Conditions"*, Petrozavodsk University, Petrozavodsk, pp. 79–82 (in Russian).

Levin, A.A. and Kokoza, A.A. 1989. About survival rate and growth of sturgeon hatchery fry in Caspian sea, in *Morphology, Ecology and Behaviour Sturgeon*, Science, Moscow, pp. 102–112 (in Russian).

Levin, A.V. and Kokoza, A.A. 1996. A survival of sturgeon fry in experimental and natural conditions of the western part of Northern Caspian sea, in *The Condition and Prospects of Scientific—Practical Development in Mariculture Area of Russia*, VNIRO, Moscow, pp. 184–189 (in Russian).

Lukjanenko, V.I., Kasimov, R.U. and Kokoza, A.A. 1984. An age-weight standard of hatchery fry of the Kaspian Sea sturgeon. (Experimental foundation). Personal issue, Ed. Volgograd, 229 pp. (in Russian).

Makarov, E.V. 1964. An estimation of a survival sturgeonfry, reared by the Don River hatcheries. *VNIRO Trans.*, 56: 141–170 (in Russian).

Marti, J.J. 1972. Questions of development of a sturgeon economy in Caspian sea. In Coll: Scientific reports of the Central laboratory on Reproduction of Fish Stocks, Ministry of Fisheries of the USSR., Ed. of Leningrad State University, USSR, pp.124-151 (in Russian).

Mikhailova M.V., 2004. Status and perspectives of development artificial reproduction of Caspian sturgeons. In Abstracts "Sturgeon aquaculture: achievements and prospects for the development", III Int. sci. and practical conference, Astrakhan, pp.125–127 (in Russian).

Ovsyannikov, F.V. 1870. About artificial reproduction of starlets, in *Works of II Congress of Russian Scientists*, Part 2, Moscow, pp. 191–200 (in Russian).

Pavlov, D.S. and Sbykina, J.N. 1989. Some results of studying of behaviour of sturgeon fry, in *Morphology, Ecology and Behaviour of Sturgeon*, Science, Moscow, pp. 124–141 (in Russian).

Polianinova, A.A. 1983. Feeding and nutrition coverage of hatchery sturgeon fry in west region of North Caspy, in *Biological Foundation of Sturgeon Breeding*, Science, Moscow, pp. 200–216 (in Russian).

Popova, A.A. Shevchenko, L.V., Piskunova, V. N. 2002. Influence of maintenance conditions of domesticated sturgeon fevales on duration of inter-spawning cycle, In Abstracts "Sturgeon aquaculture: achievements and prospects for the development", II Int. sci. and practical conference, Astrakhan, pp.139–141 (in Russian).

Raspopov, V.M. 2001. Assessment of the natural reproduction of Russian sturgeon in the Volga River, in *Young Fish Ecology and Caspian Fish Reproduction Problems*, VNIRO Publication, Moscow, pp. 238–245 (in Russian).

Tyapugin, V.V. and Ferafontov, A.L. 2002. Experience of reception of the mature eggs of the Russian sturgeon in unusual terms in the Astrakhan region, in *Modern Problems of Caspian Sea*, Materials of the International Conference Devoted to 105-anniversaryof CaspNIRKH, pp. 331–335 (in Russian).

Zaidiner, J.I., Gribanova, S.E. and Rekov, J.I. 2000. The new data on efficiency of reproduction of the sturgeon in the Azov-Don area, in *The International Conference of Sturgeon on a Boundary of 21 Centuries*, Theses of reports, CaspNIRKH, Astrakhan, pp. 243–244 (in Russian).

Zhuravleva, O.L., Ivanova, L.A., 2001. Estimation of the reproductive capacity of Russian sturgeon in the Volga River. In Coll: Young fish ecology and Caspian fish reproduction problems, VNIRO Ed., Moscow, pp.107–113 (in Russian).

Chapter 23
Acceptability and Prerequisites for the Successful Introduction of Sturgeon Species

P. Williot, E. Rochard, and F. Kirschbaum

Abstract This study aims to synthesise the documented examples of sturgeon introductions worldwide, to pinpoint the prerequisites for success. Introductions lation and re-introduction are of concern. Our analysis enables us to classify the prerequisites for successful introduction according to the following five criteria: (1) geography, (2) habitat, (3) species-specific, (4) restocking, and (5) management. Co-managing and monitoring programmes applied to the introduction of a small number of tagged juveniles into suitable habitats of a small water system of a single country are promising conditions for success. In any case, introductions are multidisciplinary actions, which need to be mobilised in a coordinated fashion to set up clear objectives and hypotheses and take all the measures needed to verify each step of the introduction programmes.

Keywords Sturgeons, introduction, reintroduction, restocking, habitat, biology of conservation, co-management, monitoring

23.1 Introduction

The harmful consequences of species introductions are well known and well documented, so why would we challenge such a well-established notion? Due to the dramatic decline in sturgeon populations worldwide (Rochard et al., 1990; Birstein, 1993; Birstein et al., 1997) including Eurasia (Williot et al., 2002a), sturgeon

---

P. Williot
Cemagref, Estuarine Ecosystems and Diadromous Fish, Cestas cedex, France
4 rue du Pas de Madame, 33890 Audenge, France

F. Kirschbaum
Leibniz-Institute of Freshwater Ecology and Inland Fisheries, Department of Biology and Ecology of Fishes, Müggelseedamm, Berlin, Germany
Institute of Animals Sciences, Humboldt-University of Berlin, Germany

R. Carmona et al. (eds.) *Biology, Conservation and Sustainable Development of Sturgeons*,

are threatened or have been eliminated from most river drainage systems. As a result, there are projects in the field of introduction or reintroduction of species or specimens either to prevent a species from complete extinction or to remedy biodiversity issues. Many other fish species have tolerated this kind of action, especially the salmonids. So why focus on sturgeon? They are a rather ancient group of fish (~200 My; Bemis et al., 1997), and various species have colonised most temperate countries of the northern hemisphere. Two-thirds of them are migratory fish occupying very different biotopes during their life span; they reach puberty late (5–18 years for females depending on the species) with non-yearly spawning. It is only fairly recently that attention has been paid to these species and this is mainly the consequence of the decline in caviar production in the Ponto-Caspian area.

The aims of this paper are to describe briefly the different types of introduction, to report on documented attempts of sturgeon introduction, to provide an analysis and to put forward possible suggestions for successful introductions.

23.2 Terminology of the Term 'Introduction'

Many potential meanings of the term 'introduction' have been listed by the IUCN (1987, 1995). Three main terms are recognised: (1) introductions which might be accidental or intentional, (2) re-introductions and (3) restocking. Introduction means the 'dispersion of a living organism out of its historically known native range' (IUCN, 1987). One example is the escapement of siberian sturgeon (*Acipenser baerii*) from an over-flooded fish farm located close to the Gironde estuary (France) (Rochard et al., 2002). There are very few examples in the second category called 're-introduction' (Ireland et al., 2002; Runstrom et al., 2002). Most known examples deal with restocking, as illustrated by the huge programme set up by the former USSR in both the Caspian and Azov Seas (Charlon and Williot, 1978; Williot, 1984; Barannikova, 1987). The objectives were to mitigate the deleterious effects of damming most of the rivers in the Ponto-Caspian area and to increase further the production of sturgeon fishery. Other synonyms for restocking are supplementation (Hildebrand et al., 1999; Smith et al., 2002), stock enhancement (a general study of the efficiency of such a practice is presented by Stottrup and Sparrevohn (2007)), and ranching (Secor et al., 2000).

23.3 Examples of Sturgeon Introductions

A preliminary and partial synthesis of stocking Eurasian sturgeon species has already been reported (Williot et al., 2002a). The present paper aims to broaden the subject by providing more examples and in particularly giving more detailed characteristics of introductions in order to provide data for a sound synthesis (Table 23.1) integrating data on (1) type of introduction (three IUCN categories, the

Table 23.1 Different kinds of sturgeon introductions and their known characteristics

Type of introduction	Species	River drainage	Stocking characteristics	Sources	Efficiency of introduction	Comments
Accidental introduction[a]	*A. gueldenstaedtii*	Baltic and North Seas		Gessner et al. (1999)	At present no installation	
	A. baerii			Paaver (1999)		
	Others			Arndt et al. (2000, 2002)		
	A. baerii	Gironde (Fr)		Rochard et al. (2002)	At present no installation	
Introduction in foreign biotopes[b]	*A. baerii*	Baltic Sea, Ladoga & Onega lakes		Barannikova (1987)	Unsuccessful	
		Baltic Sea (Gulfs of Finland and Riga)			Easily catchable	
	A. baerii			Paaver (1999)	Unsuccessful	
	A. gueldenstaedtii				Easily catchable	
	A. ruthenus					
Supplementation, restoration	*A. gueldenstaedtii*	Caspian Sea tributaries (except Iran)	70–100 M annually	Charlon and Williot (1978)	Unable to prove efficiency of stocking (Secor et al., 2000)	Huge quantities
	A. stellatus		1 g ≤ BW[c] ≤ 5 g	Barannikova (1987)	Unsuccessful	Large freshwater and brackish water systems
Supplementation, restoration	*H. huso*		82.4 M 1989 – 55.7 M 1995	Khodorevskaya et al. (1997)	*H. huso and A. stellatus* (poaching, pollution)	Several countries
				Ivanov et al. (1999)	Unsuccessful (poaching)	Sharp decline in spawning grounds
	A. persicus	South Caspian Sea		Moghim et al. (2006)	Signs for efficiency[d] (captures)	Sharp decline in spawning grounds
	A. persicus	South Caspian Sea (Iran)	No monitoring	Abdolhay et al. (2006)	?	Sharp decline in spawning grounds
	A. stellatus					

(continued)

Table 23.1 (continued)

Type of introduction	Species	River drainage	Stocking characteristics	Sources	Efficiency of introduction	Comments
	H. huso					
	A. gueldenstaedtii	Caspian Sea (Iran)	≥ 90 M annually (Russia, Azerbaijan, Kazakhstan, Iran)	Pourkazemi (2006)	?	Sharp decline in spawning grounds
	A. stellatus					
	A. persicus					
	H. huso					
	A. nudiventris[e]					
Supplementation, restoration	*A. gueldenstaedtii*	Azov Sea	~25 M annually ≤ 1997	Chebanov and Savelyeva (1999)	Unsuccessful with regard to trend in capture	Large freshwater and brackish water systems two countries
	A. stellatus		1.5–2.5 g			Sharp decline in spawning grounds
	A. schrenckii	Amur River	1988 ≤ 900,000 (0.2–0.4 g) + 168,000 (1.2 or 25 g) ≤ 1991	Wei et al. (1997)	Number of young seemed to have increased	Large freshwater system
			1988 < 8.45 M fry < 2005, no tagging or marking, no monitoring programme	Wang and Chang (2006)	?	Boarder river
	A. sinensis	Yangtze River	1983 ≤ 2.8 M larvae + 17,000 (2–10 g) ≤ 1993	Wei et al. (1997)	?	Large freshwater and marine[f] water system
						Sharp decline in spawning grounds
	A. transmontanus	Kootenai River (upstream falls), ID, USA	1992 ≤ 2630 1–4+ ≤ 1999	Duke et al. (1999); Ireland et al. (2002)	Successful	Small freshwater system

			23 < mean TL (cm) < 58 tagged and scute removal monitoring programme		2.6% < recaptured < 23%	Small quantities of juveniles
Supplementation, restoration	*A. brevirostrum*	Savannah River and adjacent rivers, SC, USA	1985 ≤ 97,483 ≤ 1992, 19% marked a/o tagged	Smith et al. (2002)	Successful	Rather limited migration[g]
			28 < TL (cm) < 116, monitoring programme		> 10% recaptured, ~10% vagrants	
	A. oxyrhinchus	Hudson River (USA)	1996	St Pierre (1999)	Broader potential to go further	
			2400 (TL = 100 mm, 4 g), 900 yearlings (TL = 330 mm, 145 g), good growth and widespread			
		Peene River (Oder drainage, De-PL)	2006 some juveniles for telemetric studies			
		Drawa River (Oder drainage, De-PL)	2007, ca. 120 small juveniles; some for telemetric studies			
	A. sturio	Garonne and Dordogne Rivers (Fr)	1995, 5000 (~1 g) and 2000 (~6.5 g)	Williot et al. (2005)	3–4% calculated from data (Rochard et al., 1997b)	Large freshwater and, estuary and marine water systems
			Marked, monitoring similar growth	Rochard et al. (1997b) (Lochet et al., 2004)		
	A. ruthenus	Danube and its Slovakian tributaries	1989 ≤ 77,390 ≤ 2005	Holčik et al. (2006)	?	Large freshwater system

(continued)

Table 23.1 (continued)

Type of introduction	Species	River drainage	Stocking characteristics	Sources	Efficiency of introduction	Comments
			No tagging or marking			Border river
			No monitoring programme			
	A. ruthenus	Hungarian Danube	~100,000 annually	Guylas in Williot et al. (2002a)	'Successful'	Large freshwater system
			No tagging			International river
			No monitoring			
Re-introduction	*A. fulvescens*	Wolf River (upstream dams) WI, USA	≥ 1994, 30,000	Runstrom et al. (2002)	Successful first spawning in 2001 in 50 years	Small freshwater system
			Some marked fingerlings			Limited quantities
			Some juveniles and adults radio tagged			
			Good growth and survival monitoring programme			
			Co-management			
	A. naccarii	Pô river drainage	~ 0.1 M annually ≤ 2000 several months old to 3+ years	Arlati in Williot et al. (2002a)		Freshwater system

[a]Some introduction were most probably intentional by anglers and aquarists (Gessner et al., 1999).
[b]In biotopes from where the species was naturally absent.
[c]Mean body weight.
[d]The status of species is deteriorating due to increasing poaching from foreign fishermen (M. Pourkazemi, personal communication).
[e]Only by Iran and Kazakhstan which released 2–3 M fingerlings annually (Pourkazemi, 2006).
[f]The species spent its main growth period in marine waters (Wei et al., 1997).
[g]The species spent time in estuary and closed marine waters (Kynard, 1997).

first being subdivided into accidental and intentional introduction), (2) species, (3) river drainage system, (4) stocking characteristics (number, period, size, tagging or marking, monitoring programme), (5) sources, (6) efficiency of introduction with the criterion used for this statement, and (7) comments by the authors on conditions of their introductions (Table 23.1).

Documented examples of accidental introductions mainly due to escapements from aquaculture are scarce. They have occurred in northern and western Europe with the siberian sturgeon (*A. baerii*) as one of the main species (Gessner et al., 1999; Paaver, 1999; Arndt et al., 2000). At present, no reproduction has been signalled in either case by capturing fingerlings or juveniles, taking into account that there are still commercial fishermen in the area. We should mention that a female (stage III of maturation) was caught in the River Garonne (Fr) in July 2005 (P. Williot, unpublished data).

Intentional introductions into foreign biotopes consist of attempts to introduce mainly the siberian sturgeon and to a lesser extent the Russian sturgeon (*A. gueldenstaedtii*) into the large lakes close to Saint Petersburg (Russia) and the Gulfs of Finland and Riga in the Baltic Sea. All these attempts failed because the fish were easily caught by fishermen (Barannikova, 1987; Paaver, 1999).

The following category, introduction for supplementation or restocking, is the most widely represented action. This is also the most ancient practice as there were attempts in the USA in the period 1880–1920 with variable rearing success and without measurable effects and these activities were abandoned (Binkowski and Doroshov, 1985). Similar actions were started in Germany in 1874 in reproducing *A. sturio* from the Elbe river and releasing the fry (Quantz, 1903). Such trials have been carried out in a variety of important German sturgeon rivers such as the Stör and the Eider (tributaries of the Elbe River) (Spratte and Hartmann, 1992). Most of the recent examples were initiated in the second half of the twentieth century. The largest restocking programme was pioneered by the former USSR in the early 1960s in the Ponto-Caspian basin (Charlon and Williot, 1978; Williot, 1984; Barannikova, 1987). In spite of this enhancement programme (restocking, fishery regulations), the status of the sturgeon fisheries has dramatically deteriorated (Birstein, 1993; Dumont, 1995; Luk'yanenko et al., 1999; Williot et al., 2002a). In their attempts to analyse the efficiency of this programme, Secor et al. (2000) pointed out that this measure has contributed a maximum of about 30% of the adult stock of both *A. gueldenstdaedtii* and *A. stellatus*, with the beluga (*Huso huso*) being almost totally supported by restocking.

In the meantime, the Azov Sea sturgeon fishery has also collapsed (Chebanov et al., 2002; Williot et al., 2002a).

There are many reasons for the failures of restocking programmes. The sturgeon habitats are large freshwater and brackish water systems where several countries are involved and most of the spawning grounds were inaccessible. Large quantities of small untagged fingerlings were released in habitats probably not very suitable for survival.

Restocking of the Amur and Yangtze rivers did not prove efficient most probably because these are large water systems, with the Amur river being a border river and the stocking was performed with untagged fingerlings.

In the following two examples of supplementation, which concern the white sturgeon (*A. transmontanus*) in an upper section of the Kootenai river and *A. brevirostrum* in the Savannah river, stocking was considered successful. In both cases, the systems were either geographically restricted or marine migrations were fairly limited. Furthermore, the number of released fish was very low, and most or a significant number were tagged and a monitoring programme had been set up.

The last two examples of supplementation deal with *A. sturio* in France and *A. ruthenus* in central Europe. Restocking with *A. sturio* was extremely sporadic due to the dramatic decline in the number and quality of the wild-originating brood-fish (Williot et al, 2002b); in addition, it is currently impossible to carry out artificial reproduction successfully with captive fish (Williot et al., 2007). However, a moderate degree of success was recorded 2 years after the release of the single stocking with a limited number of marked fish (Lochet et al., 2004). In the case of the sterlet (*A. ruthenus*), restocking and decreasing pollution in the Hungarian part of the Danube was considered to be the reason for the improving status of the population (see Gulyas in Williot et al. 2002a).

Of the two examples of sturgeon re-introduction, the case of *A. fulvescens* in the Wolf river, as part of a co-management programme accompanied by a monitoring programme, proved to be successful, as documented by the natural spawning observed some years later (Runstrom et al., 2002). The reintroduction of *A. naccarii* into the Pô river drainage system, using some old juveniles (up to 3 years old) was apparently successful as it led to a new revival in sport fishery (G. Arlati, personal communication).

23.4 Analysis

What is the definition of 'successful introduction' in the context of our paper? Different levels of success have been mentioned: (1) the ratio of recapture of released fish, (2) statements from experts, and (3) the spawning of released fish. It is worthwhile noting that there is neither a long-term normal demographic structure available, nor a stable genetic pattern for a sustained population. This illustrates the difficulties we are facing in evaluating sturgeon introduction, where success is very difficult to achieve.

The first conclusion from the documented examples deals with geography. It highlights the fact that successful introduction is more likely in a small freshwater system with only one country involved (Table 23.2).

Apart from the geographical aspect, the primary key condition for successful introduction is that 'habitat requirements of the species are satisfied' (IUCN, 1987). This means that in case of previous decline, the causes should be known and later eliminated. However, for the most part, the causes are multiple and we are unable to rank them, as pointed out in a review of Eurasian sturgeon populations (Williot et al., 2002a). Apparently, as indicated later for caspian sturgeons (Khodorevskaya et al., 1997; Moiseeva et al., 1997; Luk'yanenko et al., 1999) or suspected for *A. sturio* in France (Williot et al., 1997, 2002b), this is probably due to the deleterious consequences of pollution on the introduction procedures.

Table 23.2 Tentative synthesis of favourable/unfavourable characteristics for the successful introduction of sturgeon

Group of characteristics	Favourable/unfavourable	Some corresponding examples from Table 24.1 positive/negative
Geography	One country/several countries, border river	Caspian and Azov Seas sturgeons, *A. chrenckii*
	Freshwater system/both freshwater and marine systems	*A. transmontanus*/Caspian sturgeons, *A. sturio*, *A. sinensis*
	Small system	*A. fulvescens*, *A. transmontanus*
Habitat	Removal of original causes of extinction	*A. ruthenus*[a] (Hungary), *A. transmontanus*[b]
	Accessible and functional spawning grounds	Caspian Iranian sturgeons
	Ample food availability	*A. sturio*
	Absence of predators	Volga sturgeons[c]
Species-specific	Origin of the fish	*A. fulvescens*, *A. transmontanus*
	Steady supply in healthy broodfish (wild or farmed)	*A. sturio*, *A. sinensis*
	Ability to produce specimen to be released on a regular basis	*A. sturio*, *A. sinensis*
	Juveniles/fingerlings	*A. naccarii*, *A. brevirostrum*, *A. transmontanus*
Restocking	Identification of released fish/absence of any	*A. shrenckii*, *A. sinensis*, *A. ruthenus*
	Small quantities of released fish/huge quantities	*A. transmontanus*, *A. sturio*, *A. oxyrinchus* (De-PL)/Caspian sturgeons
	Releasing conditions	
Management	Co-management	*A. fulvescens*
	Monitoring programme	*A. fulvescens*, *A. transmontanus*, *A. brevirostrum*, *A. sturio*
	Controlled regulation of fishery	Caspian and Azov Seas sturgeons
	Sport fishery/commercial fishery	*A. fulvescens*/Caspian sturgeons, *A. sturio*

[a]Success was due to the reduction of pollution after Gulyas in Williot et al. (2002a).
[b]Flow was re-established (Duke et al., 1999).
[c]With one former exception (see text).

Among habitat requirements, spawning grounds are of primary importance. For that reason, artificial spawning grounds were built downstream of the Fedorovsk Dam in Kuban river, southern Russia (Vlassenko, 1980). Potential spawning grounds were mapped in the Garonne and Dordogne rivers (Fr) (Jego et al., 2002) to find out if non-availability of spawning grounds could be a cause for the recent decline of *A. sturio*. In anticipation of a future introduction, potential spawning grounds were checked in the Odra river (De-PL) (Gessner and Bartel, 2000; Arndt et al., 2006). In order to obtain functional spawning grounds it is sometimes necessary to increase the flow rate during the spawning period (Duke et al., 1999) as the recruitment is correlated with flow (Kohlorst et al., 1991; Duke et al., 1999). In some cases, the estuarine habitat of juveniles was limited, as illustrated for *A. sturio* in the Gironde estuary (Fr) (Taverny et al., 2002). This coincided with food availability as polychetes proved to be the main food item (Brosse et al., 2000).

The survival of released fish was examined with regard to the release of fingerlings directly into the Volga River (Charlon and Williot, 1978). To prevent predation by *Silurus glanis* and *Stizostedion lucioperca,* released fish were transported to the Northern part of the Caspian Sea (Secor et al., 2000).

The third group of characteristics is species-specific. It is essential to know the origin of the broodfish. In most supplements, corresponding wild broodfish are used. In the case of reintroduction, it is recommended that broodfish from the original population be used, as in the case of *A. sturio* (Holcik, 2000). Recently, the situation has became more complicated in southern and northern Europe with the potential former presence of *A. naccarii* in the Iberian peninsula (Garrido-Ramos et al., 1997) and *A. oxyrinchus* in the Baltic Sea (Ludwig et al., 2002); however, in this habitat the existence of historical hybridisations between *A. sturio* male and *A. oxyrinchus* female; (Tiedemann et al., 2006) have been reported. These recent genetic data have guided the setting up of recovery programmes in Germany (see Chapter 24, Kirschbaum et al., this volume). The second step relating to species-specific characteristics is the availability of the most appropriate broodfish. In the case of a dramatic decline in stocks, the current number of captured wild broodfish is either continuously decreasing, as in the case of *A. sinensis* (Xiao Hui et al., 2006), or there are no wild brood stock available at all, as for *A. sturio* (Williot et al., 2002b). In addition, in both these examples, the quality of the wild broodfish sperm decreased considerably. It is most likely that deterioration in gonadogenesis in the Caspian sturgeon contributed to their decline (Moiseeva et al., 1997; Luk'yanenko et al., 1999). Furthermore, in the critically endangered species, in addition to the lack of wild broodfish, a lack of established functional ex situ broodstocks is apparent, as exemplified in the Chinese sturgeon (*A. sinensis*) and the common Atlantic sturgeon (*A. sturio*) (Williot et al., 2007). The rearing of larvae appears to be another possible bottleneck for *A. sturio* (Williot et al., 2005), especially if offsprings that are several months old are released to improve their survival. It is worth noting that in such situations attention has to be paid to potential imprinting.

A fourth aspect is focusing on restocking and a key factor is the possibility of identifying the fish after release and assessing the success rate of restocking. Obviously, the larger the fish, the easier the tagging. However, tagging large fish is time-consuming and therefore the number of large fish released is rather limited. The absence of convincing attempts in tagging fish has led to difficulties in assessing the effects of restocking programmes, which were tentatively performed through commercial returning rates for *A. gueldenstaedtii* and *A. stellatus* in the northern part of the Caspian Sea (Secor et al., 2000) or by genetic investigations comprising a set of microsatellites for *A. sinensis* in the Yangtze river (Zhu et al., 2002). Tagging by removing of barbells or scutes of fingerlings increases mortality, but these structures may also regenerate (Secor et al., 2000) and proved to be useful in juveniles of *A. transmontanus* (Ireland et al., 2002).

The last group of characteristics concerns management. A condition for success in any introduction is the participation of stakeholders, who should all be involved in creating the introduction programme, as illustrated by the management of *A. fulvescens* in the Wolf River (WI, USA) (Runstrom et al. 2002); these activities are called co-management. Without it there may be captures of escaped or released specimens as reported in the two former examples of accidental introduction. It should be mentioned that released fish of a protected species are more safeguarded post capture if they are tagged, as proved in the case of *A. sturio* (Rochard et al., 1997a). A complementary approach is to propose a modification in the mesh size of gillnets used by fishermen in order to limit the catchability of intentionally released sturgeon (Gessner and Arndt, 2006). The absence of a monitoring programme leads to a lack of evaluation of the introduction measures, as seen in many examples in Table 23.1. The need for a strong monitoring programme has already been pointed out (Legg and Nagy, 2006; Stottrup and Sparrevohn, 2007) and its absence may lead to unsuitable traditional management (Pullin et al., 2004). A monitoring programme should include both biological and ecological research as well as restocking actions. Hypotheses, targets, and means of testing the expected performance over time have to be clearly stated. Fishing regulations have to be applied to avoid poaching (Lepage and Rochard, 1997) or, even better, a totally 'extravagant' action on the part of the administration responsible for sturgeon protection, as demonstrated for the protected *A. sturio* in France (Guth and Laurent, 2004). This might be in part a consequence of the 'anthropogenic Allee effect' recently pointed out by Courchamp et al. (2006) who states that the rarer the species the more valuable the specimens, 'leading them into an extinction vortex'. Finally, it seems that the involvement of commercial fishermen might be considered a disadvantage for the success of any introduction.

23.5 Review and Conclusions

Introduction as a way of conserving wild populations has been accepted by the sturgeon scientific community for years (Beamesderfer and Far, 1997). It must be seen as a complex and long-term process (Peterson et al., 2006). In this perspective, climate change may impact introduction, and 'our failure to understand or predict evolutionary dynamics under climate change precludes much conservation planning'

(Hellmann and Pineda-Krch, 2007). Not only do sturgeon live long, but also very little is known of their dynamics, especially with regard to mortality rates, or their ecology, homing for example. Are these characteristics (if any) stable? We are inclined to think that mortality rates may vary in the same wide range as recruitment, i.e. 1 to 9 as recently exemplified for the shortnose sturgeon (*A. brevirostrum*) in the Hudson river (USA) (Woodland and Secor, 2007). This reinforces the absolute necessity of identifying the released fish. There is no proof for homing; however, recent genetic investigations concluded a rather high level of natal fidelity in American, Atlantic, and lake sturgeons respectively (Waldman, 2004; DeHaan et al., 2006). This supports the suggestion of setting up stream-side rearing facilities to ensure that juveniles are properly imprinted (Peterson et al., 2006). Providing guides on the building or at least of the installing of rearing facilities supposes that rearing broodfish are then obtained, but unfortunately this is not the case for all sturgeon species. Moreover, as building a successful broodstock takes a long time, it is strongly recommended that this measure should be started as soon as possible for every species concerned (Williot et al., 1997, 2002a).

As mentioned above, quantitative considerations are important in restocking. However, for a successful introduction, qualitative aspects have also to be taken into account. Genetic investigations are important, especially when conservative aquaculture is involved. Of concern is the characterisation of genetic variability of available specimens (Chebanov et al., 2002; Jager, 2004; Ludwig et al., 2004) as well as analysis of the consequences of some aquaculture practices, e.g. density, which proved to impact the variability of some loci in stellate sturgeon (Ryabova et al., 2006). It is also important that population viability analysis (PVA) studies be carried out as often as possible to help managers (both scientific and decision-making) choose the main lines for future breeding programmes (Jager et al., 2000). Here, the minimum viable population (MVP) is needed to maintain a self-sustainable population (Gilpin and Soulé, 1986). Part of these monitoring programmes consists of how to manage broodstock, and it has been pointed out that improvements are needed for some species such as *A. sturio* (see Chapter 15). It is necessary to be able to measure the consequences of rearing on survival, genetic variability, and the best mating protocols. If possible, releasing characteristics such as size, location, dispersion, etc. should be noted and rearing, survival and economics studies carried out.

In addition, it is worth recalling that any success in introduction has to pre-suppose that the earlier causes of decline have been eliminated. And as a precaution, as well as to test some methods, an experimental introduction should be carried out, as stated in the IUCN rules (1985) and as illustrated for the baltic sturgeon (Gessner et al., 2006).

It has been shown that co-managing and monitoring programmes are essential for successful introduction. These are necessary pre-conditions for the acceptability of introduction. These last two dimensions should be included in any preliminary co-managing and further monitoring programmes. Objectives and hypotheses have to be clearly established as well as the necessary tools to verify each step of the introduction process.

A final comment is needed to avoid any misinterpretation of our statements. Our objective was to point out the main criteria which are important for the introduction of sturgeon. In doing so, it is supposed that conditions are ideal, including those occasions when species are endangered and where it is often urgent to convince policy-makers to provide help in such preservation actions. Our comments should in no way be taken as criticisms of introduction programmes that were initiated under very difficult circumstances.

References

Abdolhay HA, Tahori HB. 2006. Fingerling production and release for stock enhancement of sturgeon in the Southern Caspian sea: an overview. *J. Appl. Ichthyol.* 22:125–131.

Arndt GM, Gessner J, Anders E, Spratte S, Filipiak J, Debus L, Skorra K. 2000. Predominance of exotic and introduced species among sturgeons captured from the Baltic and North Seas and their watersheds, 1981–1999. *Bol. Instit. Esp. Oceanogr.* 16:29–36.

Arndt GM, Gessner J, Raymakers C. 2002. Trends in farming, trade and occurrence of native and exotic sturgeons in natural habitats in Central and Western Europe. *J. Appl. Ichthyol.* 18:444–448.

Arndt GM, Gessner J, Bartel R. 2006. Characteristics and availability of spawning habitat for Baltic sturgeon in the Odra River. *J. Appl. Ichthyol.* 22:172–181.

Barannikova I. 1987. Review of sturgeon farming in the Soviet Union. *J. Ichthyol.* 27:62–67.

Beamesderfer RCP, Far RA. 1997. Alternatives for the protection and restoration of sturgeons and their habitat. *Environ. Biol. Fishes* 48:407–417.

Bemis WE, Findeis EK, Grande L. 1997. An overview of acipenseriforms. *Environ. Biol. Fishes* 48:25–71.

Birstein VJ. 1993. Sturgeons and paddlefish: threatened fishes in need of conservation. *Conserv. Biol.* 7:773–787.

Birstein VJ, Bemis WE, Waldman JR. 1997. The threatened status of acipenseriforms species: a summary. *Environ. Biol. Fishes* 48:427–435.

Binkowski FP, Doroshov SI (eds.), 1985. *North American Sturgeons: Biology and Aquaculture Potential.* Dr Junk Publishers, Dordrecht. 163 pp.

Brosse L, Rochard E, Dumont P, Lepage M. 2000. Premiers résultats sur l'alimentation de l'esturgeon européen, *Acipenser sturio* Linnaeus, 1758 dans l'estuaire de la Gironde et comparaison avec la macrofaune estuarienne présente. *Cybium* 24:49–61.

Charlon N, Williot P. 1978. Elevage d'esturgeons de repeuplement et de consommation en URSS. *Bull. Centr. Etud. Rech. Sci. Biarritz* 12 (1):7–156.

Chebanov MS, Savelyeva EA. 1999. New strategies for broodstock management of sturgeon in the sea of Azov basin in response to changes in patterns of spawning migrations. *J. Appl. Ichthyol.* 15:183–190.

Chebanov MS, Karnaukhov GI, Galich EV, Chmir YuN. 2002. Hatchery stock enhancement and conservation of sturgeon, with an emphasis on Azov Sea populations. *J. Appl. Ichthyol.* 18:463–469.

Courchamp F, Angulo E, Rivalan P, Hall RJ, Signoret L, Bull L, Meinard Y, 2006. Rarity value and species extinction: the anthropogenic effect. *PLoS Biol.* 4(12):e415. DOI: 10. 1371/journal.phio.0040415.

Dumont H. 1995. Ecocide in Caspian Sea. *Nature* 377:673–674.

DeHaan P.W. and S.V. Libants. 2006. Genetic population structure of remnant lake sturgeon population in the upper great lakes basin. *Trans. Am. Fish. Soc.* 135:1478–1492.

Duke S, Anders P, Ennis G, Hallock R, Hammonbd J, Ireland S, Lauffe J, Lauzier R, Lockhard L, Marotz B, Paragamian VL, Westerehof R. 1999. Recovery plan for Kootenai River white sturgeon (*Acipenser transmontanus*). *J. Appl. Ichthyol.* 15:157–163.

Garrido-Ramos MA, Soriguer MC, de la Herran RJM, Ruiz Rejon C, Domezain A, Hernando JA, Ruiz Rejon M. 1997. Morphometric and genetic analysis as proof for the existence of two sturgeon species in the Guadalquivir River. *Mar. Biol.* 129:1–7.

Gessner J, Arndt GM. 2006. Modification of gill nets to minimize by-catch of sturgeon. *J. Appl. Ichthyol.* 22:166–171.

Gessner J, Bartel R. 2000. Sturgeon spawning grounds in the Odra river tributaries: a first assessment. *Bol. Instit. Esp. Oceanogr.* 16:127–137.

Gessner J, Debus L, Filipiak J, Spratte S, Skora KE, Arndt GM. 1999. Development of sturgeon catches in German and adjacent waters since 1980. *J. Appl. Ichthyol.* 15:136–141.

Gessner J, Arndt GM, Tiedemann R, Bartel R, Kirschbaum F. 2006. Remediation measures for Baltic sturgeon: status review and perspectives. *J. Appl. Ichthyol.* 22:23–31.

Gilpin ME, Soulé ME. 1986. Minimum viable populations: processes of species extinction. In: *Conservation Biology, The Science of Scarcity and Diversity* (M.E. Soulé, ed.), Sinauer Associates Publishers, Sunderland. pp. 19–34.

Guth MO, Laurent JL. 2004. Retour d'expérience sur la capture et la vente illicite d'un esturgeon en criée aux Sables d'Olonne (Vendée). *Rapport de l'inspection générale de l'environnement No. IGE/04/034.* Ministère de l'Ecologie et du Développement Durable, Paris, 30 pp.

Hellmann JJ, Pineda-Krch M. 2007. Constraints and reinforcement on adaptation under climate change: selection of genetically correlated traits. *Biol. Conserv.* 10.1016/j.biocon.2007.03.018 137:599–609.

Hildebrand L, McLeod C, McKenzie S. 1999. Status and management of white sturgeon in the Columbia river in British Columbia, Canada: an overview. *Environ. Biol. Fishes* 15:164–172.

Holčik J. 2000. Major problems concerning the conservation and recovery of the Atlantic sturgeon *Acipenser sturio* L. 1758. *Bol. Instit. Esp. Oceanogr.* 16:139–148.

Holčik J, Klindová A, Masár J, Mészáros J. 2006. Sturgeons in the Slovakian rivers of the Danube River basin: an overview of their current status and proposal for their conservation and restoration. *J. Appl. Ichthyol.* 22:17–22.

Ireland SC, Beamesderfer RCP, Paragamian VL, Wakkinen VD, Siple JT. 2002. Success of hatchery-reared juvenile white sturgeon (*Acipenser transmontanus*) following release in Kootenai River, Idaho, USA. *J. Appl. Ichthyol.* 18:642–650.

Ivanov VP, Vlasenko AD, Khodorecskaya RP, Raspopov VM. 1999. Contemporary status of Caspian sturgeon (Acipenseridae) stock and its conservation. *J. Appl. Ichthyol.* 15:103–105.

IUCN. 1987. IUCN statement on translocation of living organisms. http://www.iucn.org/themes/SSC/publications/policy/transf.htm

IUCN. 1995. Guidelines for re-introductions. http://www.iucn.org/themes/SSC/publications/policy/reintf.htm

Jager H. 2004. Genetic and demographic implications of aquaculture in white sturgeon (*Acipenser transmontanus*) conservation. *Can. J. Fish. Aquat. Sci.* 62:1733–1745.

Jager HI, Lepla K, Chandler J, Bates P, Van Winkle W. 2000. Population viability analysis of white sturgeon and other riverine fishes. *Environ. Sci. Policy* 3:S483–S489.

Jego S, Gazeau C, Jatteau P, Elie P, Rochard E. 2002. Spawning grounds available for the European sturgeon *Acipenser sturio* L. 1758 in the Garonne-Dordogne basin. Methods used, present status and prospects. *Bull. Fr. Pêch. Piscic.* 365/366:487–505.

Khodorevskaya RP, Dovgopol GF, Zhuravleva OL, Vlassenko AD. 1997. Present status of commercial stocks of sturgeons in the Caspian sea basin. *Environ. Biol. Fishes* 48:209–219.

Kirschbaum F, Würtz S, Williot P, Tiedemann R, Arndt GM, Anders E, Bartel R, Gessner J. (this volume). Prerequisites for the restoration of the European Atlantic sturgeon, *Acipenser sturio* and the Baltic sturgeon (*A. oxyrinchus* ♀ × *A. sturio* ♂) in Germany.

Kohlorst DW, Botsford LW, Brennan JS, Caillet GM. 1991. Aspects of the structure and dynamics of an exploited central California population of white sturgeon (*Acipenser transmontanus*). In: *Acipenser* (P. Williot, ed.), Cemagref Publication, Antony, France, pp. 277–293.

Kynard B. 1997. Life history, latitudinal patterns, and status of the shortnose sturgeon, *Acipenser brevirostrum*. *Environ. Biol. Fishes* 48:319–334.

Legg CJ, Nagy L. 2006. Why most conservation monitoring is, but need not be, a waste of time. *J. Environ. Manage.* 78:194–199.

Lepage M, Rochard E. 1997. Estimation des captures accidentelles d'*Acipenser sturio* réalisées en mer. In: *Restauration de l'esturgeon européen Acipenser sturio* (Elie P, coordinator), Contrat Life rapport final du programme d'exécution. Cemagref Bordeaux étude n°24, pp. 377–381.

Lochet A, Lambert P, Lepage M, Rochard E. 2004. Growth comparison between wild and hatchery-reared juvenile European sturgeons *Acipenser sturio* (Acipenseridae) during their stay in the Gironde estuary (France). *Cybium* 28:91–98.

Ludwig A, Debus L, Lieckfeldt D, Wirgin I, Benecke N, Jenneckens I, Williot P, Waldman JR Pitra C. 2002. When the American sea sturgeon swam east. *Nature* 419:447–448.

Ludwig A, Williot P, Kirschbaum F, Lieckfeld D. 2004. Genetic variability of the Gironde population of *Acipenser sturio*. In: Gessner J & Ritterhoff J, editors. "Species differentiation and population identification in the sturgeons Acipenser sturio L. and Acipenser oxyrhinchus" *Naturschutz* 101:54–72.

Luk'yanenko VI, Vasil'ev AS, Luk'yanenko VV, Khabarov MV. 1999. On the increasing threat of extermination of the unique Caspian sturgeon populations and the urgent measures required to save them. *J. Appl. Ichthyol.* 15:99–102.

Moiseeva EB, Fedorov SI, Parfenova NA. 1997. On the pathologies of the gonad structure in female sturgeons (Acipenseridae). *J. Ichthyol.* 37:624–630.

Moghim M, Kor D, Tavakolieshkalak M, Khoshghalb MB. 2006. Stock status of Persian sturgeon (*Acipenser persicus* Borodin, 1897) along the Iranian coast of the Caspian sea. *J. Appl. Ichthyol.* 22:99–107.

Paaver T. 1999. Historic and recent records of native and exotic sturgeon species in Estonia. *J. Appl. Ichthyol.* 15:129–132.

Peterson DL, Vescei P, Jennings CA. 2007. Ecology and Biology of the lake sturgeon: a synthesis of current knowledge of a threatened North American *Acipenseridae. Rev. Fish. Biol. Fisheries.* DOI 10.1007/s1160-006-9018-6. 17:59–76

Pourkazemi M. 2006. Caspian Sea sturgeon conservation and fisheries: past, present and future. *J. Appl. Ichthyol.* 22:12–16.

Pullin AS, Knight TM, Stone DA, Charman K. 2004. Do conservation managers use scientific evidence to support their decision-making? *Biol. Conserv.* 119:245–252.

Quantz H. 1903. Störfischerei und Störzucht im Gebiet der deutschen Nordseeküste. *Mitt. Des Deutschen Seefischerei-Vereins* 19 (6):176–204.

Rochard E, Castelnaud G, Lepage M. 1990. Sturgeon (Pisces: Acipenseridae); threats and prospects. *J. Fish. Biol.* 37:123–132.

Rochard E, Lepage M, Meauzé L. 1997a. Identification and characterisation of the marine distribution of the European sturgeon *Acipenser sturio. Aquat. Living Resour.* 10:101–109.

Rochard E, Lepage M, Gazeau C, Lambert P, 1997b. Tableau de bord de la population. Estimation de l'abondance des différentes classes d'âge. In: *Restauration de l'esturgeon européen Acipenser sturio* (Elie P, coordinator), Rapport final d'exécution Contrat Life N° B4-3200/94/754, Etude Cemagref n° 24, pp. 349–374.

Rochard E, Lepage M, Williot P, Mayer N, Brosse L, Sertier M, Gonthier P, Rouault T, Gazeau C, Jatteau P. 2002. Incidental escapement of farmed Siberian sturgeon *Acipenser baerii* Brandt 1869 in last watershed of the endangered European sturgeon *Acipenser sturio* L. 1758. In: *Fourth International Symposium on Sturgeon*, Oshkosh, USA, 8–13 July 2001.

Runstrom A, Bruch RM, Reiter D, Cox D. 2002. Lake sturgeon (*Acipenser fulvescens*) on the Menominee Indian reservation: an effort toward co-management and population restoration. *J. Appl. Ichthyol.* 18:481–485.

Ryabova GD, Klimonov VO, Afanas'ev KI, Vyshkvartsev DI, Moskaleichik FF, Rubsova GA. 2006. Variation in morphometric and genetic characteristics of stellate sturgeon juveniles raised at different densities. *Russ. J. Genet.* 42:182–191.

St Pierre RA. 1999. Restoration of Atlantic sturgeon in the northeastern USA with special emphasis on culture and restocking. *J. Appl. Ichthyol.* 15:180–182.

Secor DH, Arefjev V, Nikolaev A, Sharov A. 2000. Restoration of sturgeons: lessons from the Caspian sea sturgeon ranching programme. *Fish Fisheries* 1:215–230.

Smith TIJ, McCord JW, Collins MR, Post WC. 2002. Occurrence of stocked shortnose sturgeon *Acipenser brevirostrum* in non-target rivers. *J. Appl. Ichthyol.* 18:470–474.

Spratte H, Hartmann U. 1992. Daten zur limnischen Fischfauna im Eidergebiet. *Ministerium für ernährung, Landwirtschaft, Forsten und Fischerei des landes Schleswig-Holstein und Landessportfischerverband Schleswig-Holstein e.V.*, p. 137.

Stottrup JG, Sparrevohn CR. 2007. Can stock enhancement enhance stocks? *J. Sea Res.* 57:104–113.

Taverny C, Lepage M, Piefort S, Dumont P, Rochard E. 2002. Habitat selection by juvenile European sturgeon *Acipenser sturio* in the Gironde estuary (France). *J. Appl. Ichthyol.* 18:536–541.

Tiedemannn R, Moll K, Paulus KB, Scheer M, Williot P, Bartel R, Gessner J, Kirschbaum F. 2006. Atlantic sturgeons (*Acipenser sturio, Acipenser oxyrinchus*): American females successful in Europe. *Naturwissenschaften* 94:213–217.

Vlassenko AD. 1980. Problems regarding the artificial spawning grounds of sturgeon in Kuban River. *Trudy VNIRO* CII: 2–29 (in Russian).

Waldman J. 2004. Mitochondrial DNA variation in North American Sea Sturgeon *Acipenser oxyrinchus* in relation to its historic colonization of the Baltic sea. In: Gessner J & Ritterhoff J, editors. *Bundes Naturschutz* 101:73–82.

Wang Yamin, Chang J. 2006. Status and conservation of sturgeons in Amur River, China: a review based on surveys since the year 2000. *J. Appl. Ichthyol.* 22:44–52.

Wei Qiwei, Fu'en Ke, Jueming Zhang, Ping Zhuang, Junde Luo, Rueqiong Zhou, Wenhua Yang, 1997. Biology, fisheries, and conservation of sturgeons and paddlefish in China. *Environ. Biol. Fish.* 48:241–255.

Williot P. 1984. L'expérience soviétique en matière d'exploitation des stocks d'esturgeons en mer d'Azov et mer Caspienne. *Cemagref, Etude n° 20 « Série Esturgeon n° 3*, 50 pp.

Williot P, Rochard E, Castelnaud G, Rouault T, Brun R, Lepage M, Elie P. 1997. Biological characteristics of European Atlantic sturgeon, *Acipenser sturio*, as the basis for a restoration program in France. *Environ. Biol. Fishes* 48:359–370.

Williot P, Arlati G, Chebanov M, Gulyas T, Kasimov R, Kirschbaum F, Patriche N, Pavlovskaya L, Poliakova L, Pourkazemi M, Kim Yu, Zhuang P, Zholdasova IM. 2002a. Status and management of Eurasian sturgeon: an overview. *Intern. Rev. Hydrobiol.* 87:483–506.

Williot P, Rouault T, Brun R, Pelard M, Mercier D. 2002b. Status of caught wild spawners and propagation of the endangered sturgeon *Acipenser sturio* in France: a synthesis. *Intern. Rev. Hydrobiol.* 87:515–524.

Williot P, Brun R, Rouault T, Pelard M, Mercier D. 2005. Attempts at larval rearing of the endangered western European sturgeon, *Acipenser sturio* L. (Acipenseridae), in France. *Cybium* 29:381–387.

Williot P, Rouault T, Pelard M, Mercier D, Lepage M, Davail-Cuisset B, Kirschbaum F, Ludwig A. 2007. Building a broodstock of the critically endangered sturgeon *Acipenser sturio* L.: problems associated with the adaptation of wild-caught fish to hatchery conditions. *Cybium* 31:3–11.

Williot P, Rochard E, Rouault T, Kirschbaum F. (this volume). *Acipenser sturio* recovery research actions in France.

Woodland RJ, Secor DH. 2007. Year-class strength and recovery of endangered shortnose sturgeon in the Hudson River, New York. *Trans. Am. Fish. Soc.* 136:72–81.

Xiao Hui, Liu Dengong, Tang Daming, Guo Bofu, 2006. Structural and quality changes in the spawning stock of the Chinese sturgeon *Acipenser sinensis* below the Gezhouba dam (Yangtze River). *J. Appl. Ichthyol.* 22:111–115.

Zhu B, Zhou F, Cao H, Shao Z, Zhao N, May B, Chang J. 2002. Analysis of genetic variation in the Chinese sturgeon, *Acipenser sinensis*: estimating the contribution of artificially produced larvae in a wild population. *J. Appl. Ichthyol.* 18:301–306.

Chapter 24
Prerequisites for the Restoration of the European Atlantic Sturgeon, *Acipenser sturio* and the Baltic Sturgeon (*A. oxyrinchus* × *A. sturio*) in Germany

Frank Kirschbaum, Sven Wuertz, Patrick Williot, Ralph Tiedemann, Gerd-Michael Arndt, Eckardt Anders, Angela Krüger, Ryszard Bartel, and Jörn Gessner

Abstract *Acipenser sturio* was once a prevalent fish species in all the major rivers of Northern Germany. From the end of the nineteenth century, the population sizes have decreased rapidly. The last large population was observed in the River Eider, where the last specimen was caught in 1969. Under a cooperation agreement with the French Cemagref, the Leibniz-Institute of Freshwater Ecology and Inland Fisheries in Berlin received juvenile *A. sturio* of Gironde origin in 1996 and initiated an ex-situ measure. The main scientific results found since then have been: (1) first gonad maturation occurred in 11-year-old *A. sturio* (110 to 140cm total length) kept in freshwater at 20°C under a natural photoperiod; (2) during vitellogenesis the growth factor insulin-growth-like factor (IGF-I) plays an important role as a paracrine modulator, as observed in the model species *A. ruthenus*; (3) analysis of recent and historical material revealed the presence of the *A. oxyrinchus* mitochondrial haplotype A in the Baltic. Investigation of the MHC (Major Histocompatibility Complex) nuclear

F. Kirschbaum, S. Wuertz, A. Krüger, and J. Gessner
Leibniz-Institut für Gewässerökologie und Binnenfischerei, Müggelseedamm, Berlin, Germany

F. Kirschbaum
Humboldt-Universität zu Berlin, Institut für Nutztierwissenschaften, Berlin, Germany

Patrick Williot
Cemagref (Institut de Recherche sur l'Ingénierie de l'Agriculture et de l'Environnement), Cestas, France and 4 rue du Pas de Madame, 33890 Audenge, France

Ralph Tiedemann
Universität Potsdam, Institut für Biochemie und Biologie, Evolutionsbiologie/Spezielle Zoologie, Potsdam, Germany

Gerd-Michael Arndt
Fisch und Umwelt Mecklenburg-Vorpommern e.V., Rostock, Germany

Eckardt Anders
Landesforschungsanstalt für Landwirtschaft und Fischerei Mecklenburg-Vorpommern, Institut für Fischerei, Born/Darß, Germany

Ryszard Bartel
Inland Fisheries Institute, River Fisheries Laboratory, Gdansk, Poland

R. Carmona et al. (eds.) *Biology, Conservation and Sustainable Development of Sturgeons*,

gene, however, showed that these fish carrying this haplotype represent a hybrid population (*A. oxyrinchus* × *A. sturio*). The current restoration strategy concerning the Baltic (restocking with *A. oxyrinchus*) therefore needs to be reconsidered. The mtDNA studies in addition demonstrated the genetic similarity of the Gironde and the North Sea population; (4) *A. sturio*-specific microsatellites were established for brood stock management of the German and French brood stocks; (5) Evaluation of historical spawning grounds in the River Oder drainage in collaboration with Polish scientists showed intact spawning grounds in the Drawa River.

Keywords Atlantic sturgeon, restoration, brood stock, hybridization, vitellogenesis

24.1 Introduction

Historically, *Acipenser sturio* ranged from the Black Sea via the Mediterranean and the Eastern North Atlantic to the North, Baltic and White Seas (Holcik et al., 1989). During the nineteenth and particularly the twentieth century, the stocks decreased drastically. Today only one relict population in the Gironde-Garonne-Dordogne basin in France is documented (Rochard et al., 1990; Lepage and Rochard, 1995; Williot et al., 1997, 2002).

In German waters *A. sturio* played a major role in the fisheries in former times (Benecke, 1881; Quantz, 1903; Blankenburg, 1910; Seligo, 1931; Ehrenbaum, 1936; Mohr, 1952; Kinzelbach, 1987, 1997) (Fig. 24.1).

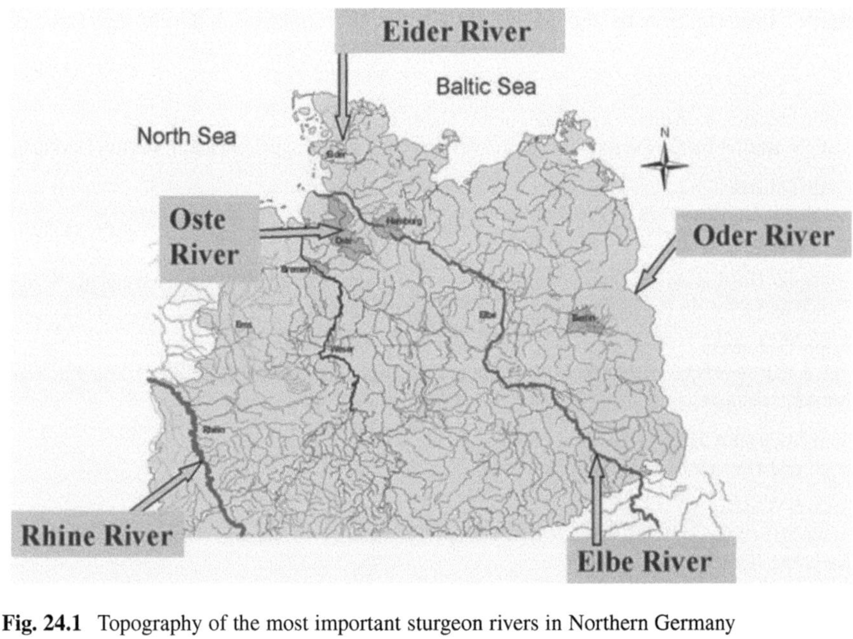

Fig. 24.1 Topography of the most important sturgeon rivers in Northern Germany

For example, between 1880 and 1890, approximately 4000 specimens were caught in the River Elbe system (Fig. 24.2) annually. However, since 1890, stocks dramatically decreased, leading to the economic insignificance of the species, indicated by insignificantly low catches within 25 years. The main reasons were overfishing, hydroconstruction, and pollution.

If fishing pressure, however, especially by the coastal marine fishing, had been the major reason for the extirpation of *A. sturio* in all large German rivers, it should have been apparent in the Eider River, too. The difference between the Eider (Fig. 24.3) and the other German rivers (Fig. 24.1) was that only small settlements and little industry were located along the river in the nineteenth century. Apart from damming the tributaries Sorge and Treene in the sixteenth and seventeenth centuries, no major hydroconstruction took place until 1890 (Fig. 24.3), when the construction of the Kiel Canal had a detrimental impact on the river habitat, eliminating approximately 35% of the catchment area from the main river (Fock and Ricklefs, 1996). The tidal zone and the sediment transport from the Wadden Sea into the river severely increased due to flow reduction, thus resulting in more frequent floods. To prevent these floods, a second, even more detrimental measure for the sturgeon population of the Eider was taken in 1934 with the erection of the dam at Nordfeld, which blocked the migration route to the spawning sites (Ehrenbaum, 1923). As a consequence of recruitment failure, the sturgeon catches declined in the 1950s to incidental catches of single individuals (Spratte, 2001). In 1953 the idea to artificially reproduce the few remaining fish arose, but was insufficiently prepared and came too late. No delivery of a ripe male or female was recorded for the subsequent years. The occasional catches were killed before reporting (Spratte and Hartmann, 1992) and the last

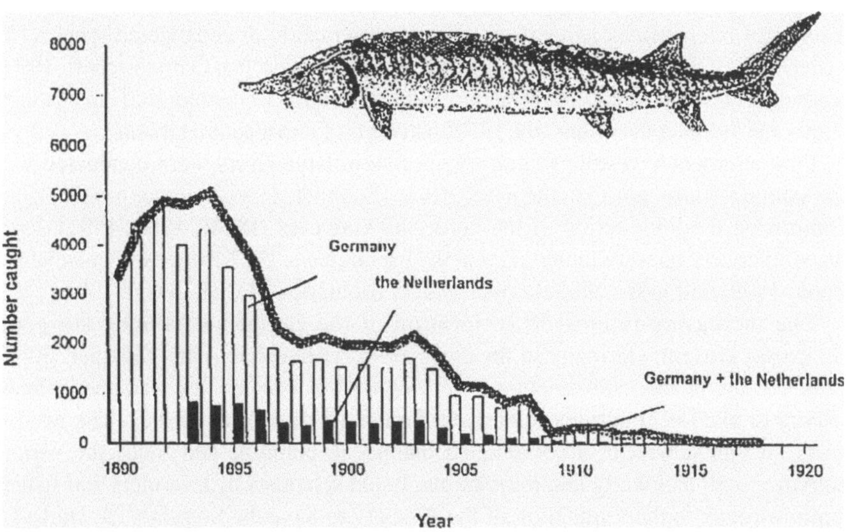

Fig. 24.2 Decline of sturgeon catches (*Acipenser sturio*) in German (light bars) and Dutch waters (black bars) between 1890 and 1920 (after Kausch, 1996, modified)

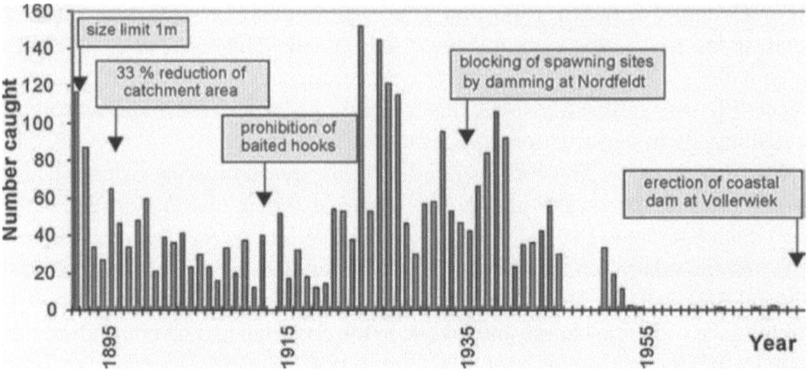

Fig. 24.3 Catches of *Acipenser sturio* in the Eider River, Northern Germany, between 1891 and 1973; arrows indicate major management efforts and impact upon the population (after Kirschbaum and Gessner, 2000)

sturgeon in the Eider River was caught in 1969. The development of the sturgeon in the Eider demonstrates that each river system has to be analysed separately to identify the underlying reasons for the decline of the population.

The last natural reproduction in German waters occurred in 1957 in the Oste river (Gaumert and Kämmereit, 1993), a tributary of the Elbe river. In the last two decades of the twentieth century, only occasional catches of *A. sturio* were reported in Northern Germany. The last large sturgeons were caught in the North Sea in 1993 (Arndt et al., 2000), in the Baltic Sea in 1996 (Paaver, 1996) and near Great Britain in 2004 (a female of 261 cm length and 120 kg).

In the 1990s the increasing interest in the conservation of genetic diversity led to a series of international agreements on the conservation of endangered species (for instance EU-Directive 42 EWG 92; Der Rat der Europäischen Gemeinschaft, 1992), leading in consequence to the protection and restoration of the required environment (Bern and Bonn conventions, the EU directive on Flora-Fauna-Habitats).

First attempts to restore migratory species in large rivers were conducted with the salmon (*Salmo salar*) in the River Rhine (Schmidt, 1996; Neumann et al., 1998) comprising the joint action of the bordering countries (IKSR, 1987, 1991). Later on, similar actions were undertaken in the tributaries of the Elbe river (Knösche and Zahn, 1999) and in the Wistula river and its tributaries (Bartel, 2001).

The increasing interest in restoration of the European Atlantic sturgeon, *A. sturio*, arose in Germany in the early 1990s (Kirschbaum and Gessner, 2000). This led to contacts with scientists working on this species, for instance in Spain (Elvira et al., 1991), but particularly in France (Williot et al., 1997). The need to work on this subject in a coordinated manner to combine and focus the various activities both nationally and internationally led scientists, fish farmers and fishery administrators to the foundation of the 'Society to Save the Sturgeon (*A. sturio* L.) e.V.', in 1994. The restoration activities on the species were performed mainly at the Leibniz-Institute of Freshwater Ecology and Inland Fisheries (IGB) in Berlin and were substantially supported by the Federal Agency for Nature Conservation

(Ministry of Environment), the Ministry of Education and Science, and the federal states of Mecklenburg-Vorpommern and Brandenburg.

24.2 Current Situation of the Preparatory Restoration Measures

24.2.1 Ex-situ Measure of A. sturio

The European sturgeon in the Gironde-Garonne-Dordogne basin in France currently has only a very low population size (actually the population is believed to be between 500 and 1500 individuals), exhibiting irregular natural spawnings (Rochard et al., 1990; Lepage and Rochard, 1995; Williot et al., 1997). Therefore, the restoration measures in France focused on research for brood stock development, which began in the early 1990s (Williot et al., 1997, 2000).

The establishment of a brood stock is also the basis of the restoration programme in Germany, as the species has practically disappeared from German waters. This ex-situ measure began in 1996 when juveniles originating from artificial reproduction in 1995 at the Cemagref in Bordeaux (Williot et al., 2000, 2007) were transferred to the IGB. The first experience with this species indicated that this fish is difficult to maintain under captive conditions (Williot et al., 1997, 2006) in contrast to many other sturgeon species.

The Cemagref and the IGB in their research cooperation therefore decided to apply different strategies during the establishment of the brood stock in order to gain as much knowledge as possible. In France, fresh, brackish, and salt-water were used for rearing and maturing the fish (Williot et al., 2002; Williot et al. this volume see Chapter 15); at the IGB only freshwater was applied. In France, for years, very low light intensities, a natural photoperiod and an annual temperature variation between 25°C and 10°C were applied whereas at the IGB the fish were raised at constant temperatures around 20°C at high light intensities and a natural photoperiod. The weighing intervals were several months in France, and 2 to 4 weeks at the IGB. In France, the fish were fed on different species of shrimp only (Williot et al., 1997) whereas at the IGB, a variety of natural food items was tested, each of which resulted in quite different specific growth rates (Kirschbaum et al., 2000; Hensel et al., 2002; Kirschbaum et al., 2006).

The feeding experiments aimed at good growth in order to reach the minimum size for first sexual maturity as early as possible. The reported minimum size for reproduction in males was 1.20 m and about 1.50 m in females (Mohr, 1952). The growth of the brood stock (Fig. 24.4) was lower than the maximum growth of this species observed in the wild (Lepage et al., 1994). Therefore, the rearing conditions at the IGB were considered suboptimal. In 2004 some fish died apparently due to intoxication (pesticides) accumulated via the frozen food. From 2004 on, the fish were fed, as in France, exclusively with two kinds of shrimp (*Palaeomon longirostris* and *Crangon crangon*), leading to increased growth again (Fig. 24.4).

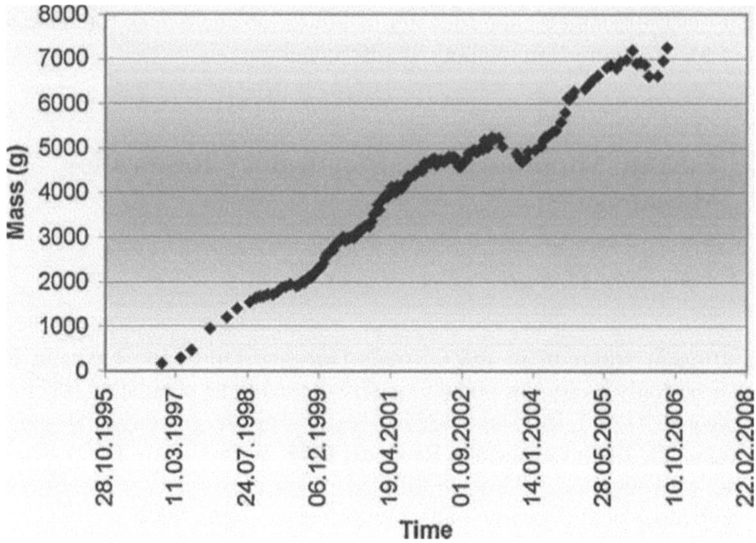

Fig. 24.4 Mean weight over time in *Acipenser sturio* brood stock ($N = 40$ in 1996 and $N = 15$ in 2006) reared on natural diets at constant temperatures (20°C) and natural photoperiod in Berlin

Table 24.1 Total length, weight, sex, and maturity stage in *Acipenser sturio* (artificially reproduced in 1995) at the end of 2005 and reared at temperatures around 20°C and ambient light conditions of Berlin; maturity stages after Bruch et al. (2001)

Tag No.	Total length (cm)	Weight (g)	Reproductive stage
Females			
81EF	136	12,595	F3–4
1906	127	9,133	F1
DDAE	112	7,211	F1
2A50	108	5,970	F1
F322	101	5,235	F1
Males			
221C	122	8,214	M2
8557	122	7,550	M2
DEOC	119	8,640	M2
8138	118	6,926	M2
6EC8	113	7,100	M2
7A59	103	6,153	M1

The 10 largest fish (of the 15 fish still alive in 2007) biopsied in 2005 included five males in advanced stages of spermatogenesis stage 2 (according to Bruch et al., 2001), four females representing stage 1 females, whereas the largest female was characterized by late vitellogenesis stage 3–4 (Bruch et al., 2001). A sixth male had been biopsied in 2004 (Table. 24.1). The first artificial reproduction of *A. sturio* might therefore be possible in 2007. Monthly blood sampling of these fish started in November 2005, allowing the analysis of steroid hormones and vitellogenin production.

24.2.2 Population Genetics and Genetic Diversity of the Brood Stocks

Historically, *A sturio* was considered to involve several geographical races (Holcik et al., 1989) separated by variable morphological features (Artyukhin and Vecsei, 1999; Holcik, 2000), and subsequently partially confirmed by genetic differences (Birstein et al., 1998). Therefore, a very important prerequisite of the restoration strategy was the genetic characterization of the European Atlantic populations using both recent (Gironde fish) and museum material (North Sea, Baltic Sea).

Ludwig et al. (2000) performed mtDNA studies that demonstrated a great genetic similarity between the *A. sturio* populations in the Gironde River and the North Sea. The Gironde fish, which are the basis of the German brood stock, therefore constitute suitable material for the restoration of the species in Germany. However, the analysis of five microsatellites showed a decrease in allele numbers between 1823 and 1992.

A set of six microsatellites and a sequence fragment of the highly variable mitochondrial control region (D-loop) were used to characterize the natural population and the brood stock on the nuclear and the mitochondrial genome level. All specimens shared the same mitochondrial haplotype, and allelic richness at the microsatellite loci was low compared to various populations of the North American sister species, *A. oxyrinchus* (Ludwig et al., 2004). Additional studies concerning the genetic diversity of the German and French brood stock integrating a higher number of *A. sturio*-specific microsatellites are currently being conducted at the University of Potsdam.

Notably, the mtDNA studies also revealed that 10 archived specimens from the Baltic and one from the Oste river (North Sea) carried the *A. oxyrinchus* haplotype A, typical for the most northern populations of *A. oxyrinchus* inhabiting the east coast of North America (Ludwig et al., 2002). This finding further supported the investigation of the differential morphology of the scutes of both Atlantic sturgeons and indicated the colonization of the Baltic by *A. sturio* about 3000 years ago, the presence of *A. oxyrinchus* in the Baltic about 1200 years ago, and after the sympatric occurrence of both species for several hundred years a dominance of the *A. oxyrinchus* mitochondrial haplotype A. The investigation of one microsatellite did not reveal any sign of hybridization in the sympatric populations of the two Atlantic sturgeons. Based on these results, an international workshop in 2002 proposed to change the restoration strategy: re-introduction of *A. oxyrinchus* into the Baltic Sea (Kirschbaum et al., 2004) and re-introduction of *A. sturio* into the North Sea tributaries, in particular the rivers Elbe and Rhine (Kirschbaum and Gessner, 2002).

However, recently Tiedemann et al. (2007) have reported extensive hybridization between both Atlantic sturgeons in the Baltic Sea based on MHCII genes.

Additional studies on this issue are in progress. In the light of these new findings the actual German restoration strategy, stocking of the Baltic with *A. oxyrinchus*, needs to be reconsidered. Hybridization would not be surprising as many sturgeon species do hybridize (Billard and Lecointre, 2001) and *A. sturio* and *A. oxyrinchus* are very similar genetically (Birstein and DeSalle, 1998), morphologically (Magnin and Beaulieau, 1963; Artuykhin and Vecsei, 1999), and at the cytogenetic level (Fontana et al., 2008).

24.2.3 Restoration in the Baltic Sea: A. oxyrinchus

Based on the genetic studies (Ludwig et al., 2002) that led to the decision in 2002 for remediation in the Baltic with *A. oxyrinchus*, the most northern populations (St. Laurence, St. John, and Hudson Rivers) were considered the most similar and therefore the most appropriate source.

Since 1998, two attempts (1998, 2001) have been successfully made to import *A. oxyrinchus* as a model for the ecological studies to be performed on *A. sturio*. These fish were used additionally for the adaptation of hatchery and rearing techniques for the remediation programme. Additional collection of material for future brood stock (Arndt and Gessner, 2005) was performed in cooperation with two research facilities in New Brunswick and Quebec. Experimental fisheries were established to collect future spawners for the brood-stock development. A total of 32 fish were successfully transferred to Germany in 2005 and 2006 (Fig. 24.5) as a basis for a future brood stock. For this purpose the fish were reared under quarantine conditions in Canada and for the first year in Germany in agreement with the ICES (International Council for the Exploration of the Sea) Code of Practice to prevent unintended transfer

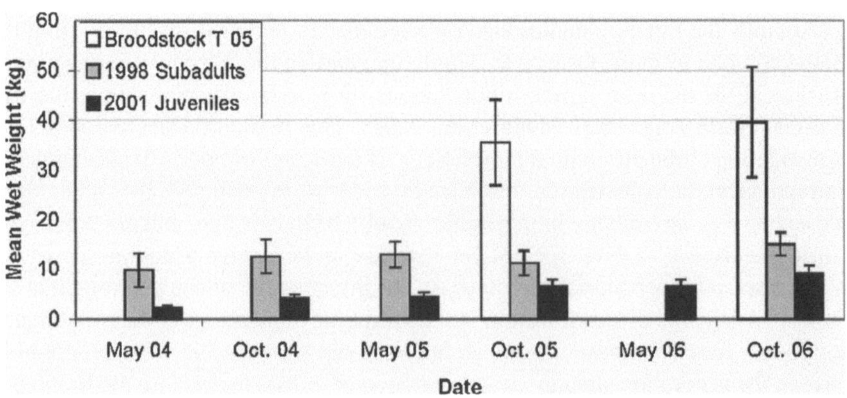

Fig. 24.5 Mean wet weight in *Acipenser oxyrinchus* of different origin and age (brood stock from wild origin, transported from Canada to Germany in 2005 and 2006; subadults from artificial reproduction in 1998 in Canada and juveniles originating from artificial reproduction in Canada in 2001) reared in raceways between 2004 und 2006

of parasites and diseases. During the quarantine, the wild caught fish were weaned to dry feed. The fish are currently reared at the Research Station for Agriculture and Fisheries of Mecklenburg-Vorpommern and at the IGB. These fish will be subjected to genetic analysis (microsatellite analysis) to select diverse material, enabling future stocking attempts with the most suitable material available.

Additional attempts at artificial reproduction were performed with wild caught brood stock in Canada for the production of stocking material. The first attempt was successful in 2002; additional attempts in 2003 and 2004 were unsuccessful due to the lack of ripe females and to technical problems in incubation. In 2005 and 2006, successful artificial reproduction trials were performed by German and Polish researchers. The larvae were transferred to Europe, the yearlings constituting the material for the first restocking trials in April 2006 (see Section 24.2.2).

Juveniles of the 2005 reproduction were considered for an initial experimental release trial in April 2006 in the Oder River. Upon release, tracking of the fish for habitat utilization and identification of potential threats were envisaged as well as the validation of a suitable telemetry technique (radio versus acoustic transmitters). These measures should be regarded as an important element of the project as there are many issues concerning the ecology and the biology of this species which can be determined only in field studies.

24.2.4 Endocrine Regulation of Gonad Maturation

In the past, the reproduction of sturgeon had been recognized as one of the major concerns for conservation measures (Williot et al., 1997, 2005). In the context of late maturity, delayed onset of vitellogenesis, and a reproductive cycle with a variable duration of several years, the underlying endocrine regulation is poorly understood. Along the hypothalamus–pituitary–gonad axis, interaction of protein hormones and steroids control the development of the gonad, mediating the environmental stimuli of external factors such as temperature and photoperiod. Recent investigations on *A. ruthenus* show that, next to the involvement of sex steroids, the insulin-growth-like factor (IGF-I) is involved in early vitellogenesis as paracrine modulator, priming the onset of puberty (Wuertz et al., 2002, 2006a–c). Application of the results to facilitate the first maturation in *A. sturio* is currently being studied.

24.2.5 Alien Species in German Waters

Recent data on sturgeon catches indicate the presence of alien species in natural open waters (Gessner et al., 1999; Arndt et al., 2000, 2002). Previous studies are demonstrating hybridizations among several sturgeon species (Birstein et al., 1998; Billard and Lecointre, 2001; Jenneckens et al., 2001) Therefore, the presence of exotic sturgeon species represents a certain risk for the restocking measures partly due to hybridization but most prominently by disease transfer (Pavlov et al., 1994). The non-native catches indicate that the environmental conditions allow the survival and growth of these species. Therefore, it can be concluded that they also indicate favourable ecological conditions for the endemic sturgeon species. No natural reproduction of non-native fish in Baltic or North Sea waters was reported until now. The removal of alien species from open waters–which requires the expertise to determine the species at hand–should be encouraged.

24.2.6 Fishing Regulations

Overfishing was an important factor for the decline of the species, and it is still considered a major threat for restoration measures in the North Sea (Nellen et al., 1994). Deriving from the reported catches of non-indigenous sturgeon species in the open waters of Central Europe, the adverse fishing techniques were identified. The results certainly are biased with regard to the feedback from anglers compared to fishermen but the general trend reflects similar findings from France (Rochard et al., 1997) and the American East Coast (Collins et al., 1996). Based on this assessment, the main risk is associated with gill netting. Therefore, development of adapted fishing techniques to reduce the probability of sturgeon bycatch in the coastal waters of the Oder River is performed in cooperation with Fisch and Umwelt MV e.V. and the Landesforschungsanstalt für Landwirtschaft und Fischerei Mecklenburg-Vorpommern. The first trials to modify the commercially utilized gill-nets indicate that sturgeon bycatch can be easily reduced to extremely low levels. The verification of the impact of such modifications upon the catch of the main target species was carried out during fishing trials in the Szczecin Lagoon, indicating that significant reduction of the catch in some target species would have to be expected (Gessner and Arndt, 2006).

24.2.7 Ecology of 'Sturgeon' Rivers Oder, Elbe and Rhine

Natural populations can be established only if the most important factors that contributed to the decline of the species are no longer present in the rivers chosen for the re-introduction measures. Management of these factors (e.g. fishing and habitat degradation) along scientifically valid guidelines in the rivers selected for restoration (the Oder, Elbe, and the Rhine) will be the main long-term challenge to meet the species requirements.

There are numerous recent studies on the fish communities of these large rivers indicating successful reproduction and recruitment of formerly threatened species (Buijse and Cazemier, 1998; Molls, 1998; Staas, 1998; Thiel et al., 1998; Bischoff and Wolter, 2001a–c; Wolter and Bischoff, 2001). These results indicate ecological conditions (Prange, 2000) that would also allow successful reproduction and recruitment of the sturgeons. Potential spawning habitats for *A. sturio* were found in the Rhine (G.W. Schmidt, personal communication) and in the tributaries of the Oder (Gessner and Bartel, 2000; Arndt and Gessner, 2006).

The assessment of critical habitat focused mainly on spawning habitat for sturgeons because it supports the most vulnerable life stages and consequently plays a key role for successful reproduction of the species, since it supports the most vulnerable life stages. Spawning grounds are river habitats with gravel substrate, sufficient water exchange in interstitial water and the absence of sedimentation (Richmond and Kynard, 1995; Elie, 1997; Sulak and Clugston, 1999). In the Oder River, the first indication of potential spawning grounds was based on historical data (Przybyl, 1976; Sych et al., 1996). Sturgeon distribution in the Oder is reported up to Raciborz (Poland). However, these sites have been inaccessible since the construction of a weir in Wroclaw (Breslau) in the nineteenth century. To assess the status of potential spawning sites, migration obstacles and water pollution, discharge dynamics, water quality, and profiles of the river were then evaluated by video cartography and water-chemistry analysis. In the middle and lower parts of the Oder river, excessive transport of fine sediments due to the construction of groyne fields can be observed throughout the river bed, thereby excluding the main river channel areas as potential spawning sites for sturgeon (suffocation of eggs). Only very limited areas with suitable substrate could be found in groyne fields and riverbank structures, where sedimentation was not detectable. In the Wartha river migration obstacles were found solely in the upper reaches of the river as well as in some of its tributaries (Notec, Obra, Prosna). At least two historical spawning sites are readily accessible in it. Here, the high nutrient load is the main reason for the limited suitability for natural reproduction. In the Prosna river, fine sands dominate the substrate throughout the investigated river stretch of 25 km along the historic spawning sites. In the Drawa River (accessible via the Oder, Wartha, and Noteć Rivers)–described as sturgeon-spawning habitat until 1939–a potentially suitable habitat was determined to be readily available with the substrate in the main channel dominated by clear cobbles and stones.

Comparable attempts to verify the status of historical spawning sites are envisaged in the Elbe river and its tributaries (Oste and Stör rivers).

Since habitat requirements for sturgeons are poorly known; a broader knowledge is indispensable for the proper characterization of potential habitats and to formulate restoration requirements. Experimental verification of the importance of substrate for the development of the early life stages revealed valuable results and are ongoing to identify the precise requirements of the species with regard to spawning and early life-stage habitat.

As mentioned above, hydroconstruction was a major cause that prevented spawning migrations and finally led to the disappearance of the species in the Eider River (see above). In the lower part of the Oder River, hydroconstruction presently imposes no major problem for the migration of the fish (Gessner and Bartel, 2000). In the Elbe River, the weir at Geesthacht about 150 km from the mouth does not prevent spawning migration of the large migratory salmonids (Fredrich et al., 1999). However, this barrier could prevent spawning migration of large *A. sturio*. Therefore, potential upstream spawning habitats of *A. sturio* may actually not be accessible for the species. However, interest in modifying this barrier has been expressed by several administrative bodies. The Rhine river (van Winden et al., 2000) can actually be used as a migration route up to the barrage of Iffetsheim, 700 km upstream from the mouth of the river. In conjunction with the availability of spawning habitats, this fact increases the potential importance of the Rhine for restoration measures both for Germany and the Netherlands (van Winden et al., 2000). Increasingly, the effectiveness of the altered sediment structure, structural diversity and the runoff dynamics need to be determined in the context of sturgeon habitat in order to assess the overall quality of the available habitat.

24.2.8 Public Awareness Campaigns

Continuation of increasing public awareness on the restoration of Atlantic sturgeons in Germany will be an important element of the project, aimed at the general public, but especially at fishermen and other users of the water resources. Exhibitions on the subject such as the one on loan from the German Marine Museum in Stralsund are a major step towards an increased public consciousness.

24.2.9 International Cooperation

The restoration of the two Atlantic sturgeon species, *A. sturio* and *A. oxyrinchus*, in Germany is a long-term project that will be successful only within the framework of close and continuous international cooperation (see for instance Gessner, 2000; Kirschbaum, 2002; Kirschbaum and Gessner, 2000, 2002). For the Baltic Sea, the project makes close cooperation with Poland an essential prerequisite. This cooperation has been fundamental to all preparations of a release since 1996. Furthermore, a project group of HELCOM Habitat was established in 1997. The project group aims at the restoration of atlantic sturgeon in the Baltic Sea by mediating the involvement of the relevant national institutions prior to the onset of the release of juveniles. In addition, it focuses on fishing information campaigns and in the middle term at an assessment of the potential of the local rivers for an expanded project. In fact, this is the first step towards an integration of a number of countries sharing the responsibility for the ecosystem, the measures to be carried out and the resource that derives from it. The starting point might be the intensive collaboration between the researchers trying to cope with the matter but in the long term an integrated management will be required to match the needs of the users of the resources. Therefore, sturgeons do have the potential of becoming powerful elements for river catchment as well as for integrated coastal-zone management. Actually there are steps to establish a European Action Plan in 2007 in the framework of the Bern Convention (Convention on the Conservation of European Wildlife and Natural Habitat).

24.3 Conclusion

Restoration measures on sturgeons need a long time. It is only after years of research that the genetic status of the Baltic sturgeon seems to emerge. The first successful artificial reproduction of *Acipenser sturio* in France in June 2007 in captivity opens the way for further progress of the restoration of the species in Europe; long-term support of these activities isneeded from the European Union.

Acknowledgements The authors thank the Federal Agency of Nature Conservation (BfN) and the Federal Ministry for Education and Science (BMBF) for their strong moral and material support (BfN: Grants Az.: Z1.3-892 11-4/96; Z 1.3-892 11-7/99; Z1.3-892 11 8/05 and BMBF: Grants FKZ 0330532, FKZ 0330718). L. Fischer's help in improving technically the manuscript is acknowledged.

References

Arndt GM, Gessner J. 2005. Aufbau eines Elterntierbestandes als Grundlage für die Wiedereinbürgerung des Störs im Ostseeeinzugsgebiet. Schriftenreihe *Fisch und Umwelt* M.-V. e.V. Selbstverl Jahresheft 2003/2004: 5–22.

Arndt GM, Gessner J, Anders E, Spratte S, Filipiak J, Debus L, Skora K. 2000. Predominance of exotic and introduced species among sturgeons captured from the Baltic and North Seas and their watersheds, 1981–1999. Symposium on Conservation of the Atlantic sturgeon *Acipenser sturio* L. 1758 in Europe, Madrid. *Bol Inst Oceanogr* 16:29–36.

Arndt GM, Gessner J, Raymakers C. 2002. Trends in farming, trade and occurrence of native and exotic sturgeons in natural habitats in Central and Western Europe. Proceedings of the 4th International Symposium on Sturgeon, Oshkosh, Wisconsin, USA. *J Appl Ichthyol* 18:444–448.

Arndt M, Gessner J, Bartel J. 2006. Characteristics and availability of spawning habitat for Baltic sturgeon in the Oder River and its tributaries. Proceedings of the 5th International Symposium on Sturgeon, Ramsar, Iran. *J Appl Ichthyol* 22 (Suppl 1):172–181.

Artyukhin EN, Vecsei P. 1999. On the status of Atlantic sturgeon: conspecificity of European *Acipenser sturio* and North American *Acipenser oxyrinchus. J Appl Ichthyol* 15:35–37.

Bartel R. 2001. The restoration of Atlantic salmon (*Salmo salar* L.) in Poland. *Arch Pol Fish* 9:219–228.

Benecke B. 1881. *Fische, Fischerei und Fischzucht in Ost- und Westpreussen.* Hartungsche Verlagsdruckerei, Koenigsberg.

Billard R, Lecointre G. 2001. Biology and conservation of sturgeon and paddlefish. *Rev Fish Biol Fish* 10:355–392.

Birstein VJ, DeSalle R. 1998. Molecular phylogeny of Acipenserinae. *Mol Phylogenet Evol* 9:141–155.

Birstein VJ, Betts J, DeSalle R. 1998. Molecular identification of *Acipenser sturio* specimens: a warning note for recovery plans. *Biol Conserv* 84:97–101.

Bischoff A, Wolter C. 2001a. Groyne-heads as potential summer habitats for juvenile rheophilic fishes in the Lower Oder, Germany. *Limnologica* 31:17–26.
Bischoff A, Wolter C. 2001b. The 0 + fish community structure in a large lowland river: first results of a study from the River Oder. *Arch Hydrobiol Suppl* 135/2. Large Rivers 12:137–151.
Bischoff A, Wolter C. 2001c. The flood of the century on the River Oder: effects on the 0 + community and implications for floodplain restoration. *Regul Rivers* 17:171–190.
Blankenburg A. 1910. Von der Störfischerei in der Elbe. *Der Fischerbote* 11:7–11.
Bruch RM, Dick TA, Choudhury A. 2001. A field guide fort the identification of stages of gonad development in lake sturgeon (*Acipenser fulvescens* Rafinesque) with notes on lake sturgeon reproductive biology and management implications. *Publ Fond Sturgeon for tomorrow*. Wisconsin Department of Natural Resources, University of Manitoba, Oshkosh, 1–40.
Buijse T, Cazemier W. 1998. Fischbestandserhebungen im Rhein im Rahmen des landesweiten Ökosystem-Monitorprogramms. *LÖBF-Mitteilungen* 23:47–56.
Collins MR, Rogers SG, Smith TIJ. 1996. Bycatch of sturgeons along the southern Atlantic coast of the USA. *N Am J Fish Manage* 16:24–29.
Der Rat der Europäischen Gemeinschaften 1992. Richtlinie 92/43/EWG Rat. 21. Mai 1992. Erhaltung der natürlichen Lebensräume sowie der wildlebenden Tiere und Pflanzen. Abl EG Nr L 206. *FFH-Richtlinie*: p. 388. 7–50.
Ehrenbaum E. 1923. Die Eider als Störfluß und die Schonung des Störs. *Fischerbote norddeutscher Fischer* 5:77–83.
Ehrenbaum E. 1936. Naturgeschichtliche und wirtschaftliche Bedeutung der Seefische Nordeuropas. In: *Handbuch der Seefischerei Nordeuropas* 2, E. Schweizerbarth'sche Verlagsbuchhandlung, Stuttgart, pp. 3–5.
Elie P. 1997. Restoration of commun sturgeon *Acipenser sturio*: final report of execution programme: operations I, II, III 1994–1997. Bordeaux, *Etude Cemagref* 24.
Elvira B, Almodovar A, Lobon-Cervia J. 1991. Sturgeon (*Acipenser sturio* L. 1758) in Spain, the population of the River Guadalquivir: a case history, a claim for a restoration programme. In: Williot P, editor. *Acipenser:* actes du premier colloque international sur l'esturgeon, Bordeaux: 03.10.-06.10.1989. Bordeaux, *Cemagref Publ*, pp. 337–347.
Fock H, Ricklefs K. 1996. Die Eider—Veränderungen seit dem Mittelalter. In: *Warnsignale aus Fluessen und Aestuaren*, Lozan JL, Kausch H (eds).. Paul Parey, Berlin, pp. 39–42.
Fontana F, Lanfredi M, Kirschbaum F, Garrido-Ramos MA, Robles F, Forlani A, Congiu L. 2008. Comparison of karyotypes of *Acipenser oxyrinchus* and *A. sturio* by chromosome banding and fluorescent *in situ* hybridisation. *Genetica* 132:281–286.
Fredrich F, Arzbach HH, Schubert HJ. 1999. Verhalten von Fischen an Querverbauungen in der Elbe und einmündenden Nebenflüssen. In: *Tagungsband Fachtagung Elbe*, Wittenberge: 04.05.-07.05.1999, Kiene S, Harms O, Büchele B (ed.). Institut für Wasserwirtschaft und Kulturtechnik (IWK), Universität Karlsruhe (TH), Kaiserstrasse 12, 76128 Karlsruhe, pp. 227–228.
Gaumert D, Kämmereit M. 1993. *Süßwasserfische in Niedersachsen*. —Hrsg.: Niedersächsisches Landesamt für Ökologie: 1–162, Hildesheim.
Gessner J. 2000. Reasons for the decline of *Acipenser sturio* L.1758 in central Europe, and attempts at its restoration. Symposium on Conservation of the Atlantic Sturgeon *Acipenser sturio* L.1758 in Europe, Madrid. *Bol Inst Esp Oceanograph* 16:117–126.
Gessner J, Arndt GM. 2006. Modifications of gillnets to minimize by-catch and its tributaries. *J Appl Ichthyol* 22 (Suppl 1):166–171.
Gessner J, Bartel R. 2000. Is there still suitable habitat for sturgeons in the Oder River tributary? Symposium on Conservation of the Atlantic Sturgeon *Acipenser sturio* L. 1758 in Europe, Madrid. *Bol Inst Esp Oceanograph* 16:127–137.
Gessner J, Debus L, Filipiak J, Spratte S, Skora KE, Arndt GM. 1999. Recent development in sturgeon catches from German and adjacent waters. *J Appl Ichthyol* 15:131–142.
Hensel E, Kirschbaum F, Williot P, Gessner J. 2002. Restoration of the European Sturgeon *Acipenser sturio* L. 1758 in Germany: effect of different food items on specific growth rates of large juvenile fish. *Int Rev Hydrobiol* 87:539–551.

Holcik, J. 2000. Major problems concerning the conservation and recovery of the Atlantic sturgeon *Acipenser sturio* L. 1758. Symposium on Conservation of the Atlantic Sturgeon *Acipenser sturio* L. 1758 in Europe, Madrid. *Bol Inst Esp Oceanograph* 16:139–148.

Holcik J, Kinzelbach R, Sokolov LI, Vasil'ev VP. 1989. *Acipenser sturio* Linnaeus, 1758. In: *The freshwater fishes of Europe, 2. Acipenseriformes*, Holcik J (ed.). AULA Verlag, Wiesbaden, pp. 367–394.

IKSR 1987. *Aktionsprogramm "Rhein"*. Internationale Kommission zum Schutz des Rheins gegen Verunreinigung, Koblenz (s. http://www.iksr.org).

IKSR 1991. *Ökologisches Gesamtkonzept für den Rhein*. Internationale Kommission zum Schutz des Rheins gegen Verunreinigung, Koblenz (s. http://www.iksr.org).

Jenneckens I, Meyer JN, Hörstgen-Schwark G, May B, Debus L, Ludwig A. 2001. First genomic marker for species identification of one Black Caviar producer: LS-39—a multipotent sturgeon microsatellite. *J Appl Ichthyol* 17:39–42.

Kausch H. 1996. Fahrwasservertiefungen ohne Grenzen? In: *Warnsignale aus Flüssen und Ästuaren*, Lozan JL, Kausch H (eds.). Paul Parey, Verlag Berlin, pp. 162–176.

Kinzelbach RK. 1987. Das ehemalige Vorkommen des Störs, *Acipenser sturio* (Linnaeus 1758), im Einzugsgebiet des Rheins (Chondrostei: Acipenseridae). *Z Angew Zool* 74:67–200.

Kinzelbach RK. 1997. The Sturgeon (*Acipenser sturio* L. 1758) in Europe. *Z Ökologie Naturschutz* 6:129–135.

Kirschbaum F. 2002. La dimension européenne de la stratégie de sauvegarde de l'esturgeon,. In: *États généraux de la Dordogne, Actes du séminaire, Quel avenir pour l'esturgeon européen?* Libourne, France, 04.-05.10.2001, EPIDOR, Castelnaud-la-Chapelle, pp. 118–132.

Kirschbaum F, Gessner J. 2000. Re-establishment programmeme for *Acipenser sturio* L. 1758: the German approach. Symposium on Conservation of the Atlantic Sturgeon *Acipenser sturio* L. 1758 in Europe, Madrid. *Bol Inst Esp Oceanograph* 16:149–156.

Kirschbaum F, Gessner J. 2002. Perspectives for the re-introduction of the European sturgeon, *Acipenser sturio* L. in the Elbe River. *Z Fischk Suppl* 1:217–232.

Kirschbaum F, Gessner J, Williot P. 2000. Restoration of *Acipenser sturio* L. 1758 in Germany, I: growth characteristics of juvenile fish reared under experimental indoor conditions. Symposium on Conservation of the Atlantic Sturgeon *Acipenser sturio* L. 1758 in Europe, Madrid. *Bol Inst Esp Oceanograph* 16:157–165.

Kirschbaum F, Ludwig A, Hensel E, Wuertz S, Kloas W, Williot P, Tiedemann R, Gessner J. 2004. Status of the protection and restoration of Atlantic sturgeon in Germany: background, actual situation, and perspectives. In: *BfN-Skripten*, 101, Gessner J, Ritterhoff J (eds.). BfN Federal Agency for Nature Conservation, Bonn, pp. 36–53.

Kirschbaum F, Hensel ECK, Williot P. 2006. Feeding experiments with the European Atlantic sturgeon, *Acipenser sturio* L. 1758 to accustom large juveniles to a new feed item and the influence of tank size and stocking density on growth. *J Appl Ichthyol* 22 (Suppl 1):307–315.

Knösche R, Zahn S. 1999. Erste Wiedereinbürgerung von Lachsen in der Mittelelbe. *Fischer und Teichwirt* 50:301–302.

Lepage M, Rochard E. 1995. Threatened fishes of the world: *Acipenser sturio* Linnaeus, 1758 (Acipenseridae). *Environ Biol Fishes* 43:28.

Lepage M, Lambert P, Rochard E. 1994. Suivi et modélisation de la croissance des formes juveniles d'sturgeon autochthones *Acipenser sturio. AGEDRA*, 18 pp.

Ludwig AN, Jenneckens I, Debus L, Ludwig A, Becker J, Kirschbaum F. 2000. Genetic analyses of archival specimens of the Atlantic sturgeon *Acipenser sturio* L. 1758. Symposium on conservation of the Atlantic sturgeon *Acipenser sturio* L. 1758 in Europe, Madrid. *Bol Inst Esp Oceanograph* 16:221–230.

Ludwig A, Debus L, Lieckfeld D, Wirgin I, Benecke N, Jenneckens I, Williot P, Waldman JR, Pitra C. 2002. When the American sea sturgeon swam east. *Nature* 419:447–448.

Ludwig A, Williot P, Kirschbaum F, Liekfeld, D. 2004. Genetic variability of the Gironde sturgeon population. In: *BfN-Skripten*, 101 Gessner J, Ritterhoff J (eds.). BfN Federal Agency for Nature Conservation, Bonn, pp. 54–72.

Magnin E, Beaulieau G. 1963. Étude morphométrique comparée de *l'Acipenser oxyrhynchus* Mitchill du Saint-Laurent et *l'Acipenser sturio* Linné de la Gironde. *Natur Can* 90:5–38.
Mohr E. 1952. *Der Stör*. Die neue Brehm-Bücherei, Geest und Portig, Leipzig.
Molls F. 1998. Die fischökologische Bedeutung der verbliebenen Altrheinarme des Niederrheins. *LÖBF-Mitt* 23:26–30.
Nellen W, Thiel R, Hölker F, Breckling P. 1994. Überlegungen zu fischereilichen Perspektiven der Elbe. *Fischer und Teichwirt* 7:265–267.
Neumann D, Ingendahl D, Molls F, Nemitz A. 1998. Lachswiedereinbürgerung in NRW. *LÖBF-Mitt* 23:20–25.
Paaver T. 1996. A common or Atlantic sturgeon, *Acipenser sturio*, was caught in the Estonian waters of the Baltic Sea. *Sturgeon Quarter* 4:7.
Pavlov DS, Savvaitova KA, Sokolov LI, Alekseev SS. 1994. *Redkie i Ischezayushchie Zhivotnye. Ryby* [Rare and vanishing animals. Fishes]. Vysshaya Shkola, Moskva.
Prange A. 2000. Die Elbe und ihre Nebenflüsse: Belastung, Trends, Bewertung, Perspektiven/die Auswertung der Forschungsergebnisse der Schadstoffforschung und der ersten Ergebnisse zur ökologischen Forschung im Einzugsgebiet der Elbe. (ATV-DVWK, Deutsche Vereinigung für Wasserwirtschaft, Abwasser und Abfall e.V., Prange A, Furrer R. editors), GKSS-Forschungszentrum, Geesthacht. GFA—Ges zur Förderung der Abwassertechnik, Hennef.
Przybyl A. 1976. Historic distribution of common sturgeon *Acipenser sturio* in Wartha River. OT: Wystepowanie jesiotra zachodniego *A. sturio* w dorzeczu Warty. *Ochr Przy R* 32:5–12.
Quantz H. 1903. Störfischerei und Störzucht im Gebiet der deutschen Nordseeküste. *Mitt des Deutschen Seefischerei-Vereins* 19:176–204.
Richmond AM, Kynard B. 1995. Ontogenetic behaviour of shortnose sturgeon, *Acipenser brevirostrum*. *Copeia* 1995:172–182.
Rochard E, Castelnaud G, Lepage M. 1990. Sturgeons (Pisces: *Acipenseridae*), threats and prospects. *J Fish Biol* 37 (Suppl A):123–132.
Rochard E, Lepage M, Meauze L. 1997. Identification et caractérisation de l'aire de répartition marine de l'esturgeon européen *Acipenser sturio* à partir de déclarations de captures (Identification and characterisation of the marine distribution of the European sturgeon *Acipenser sturio*). *Aquat Liv Res* 10:101–109.
Schmidt GW. 1996. Wiedereinbürgerung des Lachses *Salmo salar* in Nordrhein-Westfalen—Allgemeine Biologie des Lachses sowie Konzeption und Stand des Wiedereinbürgerungsprogramms unter besonderer Berücksichtigung der Sieg. Landesanstalt für Ökologie, Bodenordnung und Forsten, Landesamt für Agrarordnung NRW, *LÖBF-Schriftenr*: 11.
Seligo A. 1931. Die Seefischerei von Danzig. In: *Handbuch der Seefischerei Nordeuropas*. VIII, 7. E. Schweizerbart'sche Verlagsbuchhandlung, Stuttgart, pp. 25–28.
Spratte S. 2001. Aussterben des Störs (*Acipenser sturio* L.) in der Eider. In: *Der Stör—Fisch des Jahres* 2001, Verband Deutscher Sportfischer (ed.). Verlag M. Faste, Kassel, pp. 66–86.
Spratte S, Hartmann U. 1992. *Daten zur limnischen Fischfauna im Eidergebiet*. Ministerium für Ernährung, Landwirtschaft, Forsten und Fischerei des Landes Schleswig-Holstein und Landessportfischerverband Schleswig-Holstein e.V.
Staas S. 1998. Das Jungfischaufkommen im Rheinstrom und in künstlichen Abgrabungsseen mit Anbindung an den Rheinstrom. *LÖBF-Mitt* 23:15–19.
Sulak KJ, Clugston JP. 1999. Early life history and population structure of Gulf sturgeon, *Acipenser desotoi*, in the Suwanee River, Florida. *J Appl Ichthyol* 15:116–128.
Sych R, Bartel R, Bieniarz K, Mastynski J. 1996. Project for the restoration of migratory fish species in Poland. OT: *Projekt Restytucji Ryb Wedrownych w Polsce*. Opracowaine Zespolowe.
Tiedemann R, Moll K, Paulus KB, Scheer M, Williot P, Bartel R, Gessner J, Kirschbaum F. 2007. Atlantic sturgeons (*Acipenser sturio, A. oxyrinchus*): American females successful in Europe. *Naturwissenschaften* 94:213–217.
Thiel R, Nellen W, Ergenzinger P, Kirschbaum F, Knösche R, Larink O, Winkler HM. 1998. Ökologische Zusammenhänge zwischen Fischgemeinschafts- und Lebensraumstrukturen der Elbe. *Z Ökol Naturschutz* 7:67–70.

van Winden A, Overmars W, Bosman W, Klink A. 2000. *L'esturgeon atlantique—le retour dans le rhin*, Stichting A, Hoog K (eds.). WWF, Netherlands, 130 pp.

Williot P, Rochard E, Castelnaud G, Rouault T, Brun R, Lepage M, Elie P. 1997. Biological and ecological characteristics of European Atlantic sturgeon, *Acipenser sturio,* as foundations for a restoration programme in France. *Environ Biol Fish* 48:359–370.

Williot P, Brun R, Pelard M, Mercier D. 2000. Unusual induced maturation and spawning in an incidentally caught pair of adult of the critically endangered European sturgeon *Acipenser sturio* L. *J Appl Ichthyol* 16:279–281.

Williot P, Arlati G, Chebanov M, Gulyas T, Kasimov R, Kirschbaum F, Patriche N, Pavlovskaya L, Poliakova L, Pourkazemi M, Kim Yu, Zhuang P, Zholdasova IM. 2002. Status and management of Eurasian sturgeon: an overview. *Int Rev Hydrobiol* 87:483–506.

Williot P, Brun R, Rouault T, Pelard M, Mercier D. 2005. Attempts at larval rearing of the endangered western European sturgeon, *Acipenser sturio* L. (Acipenseridae), in France. *Cybium* 29(4):381–387.

Williot P, Rouault T, Pelard M, Mercier D, Lepage M, Davail-Cuisset B, Kirschbaum F, Ludwig A. 2007. Building a broodstock of the critically endangered sturgeon, *Acipenser sturio* L: problems associated with the adaptation of wild-caught fish to hatchery conditions. *Cybium* (p. 389) 31:3–11.

Wolter C, Bischoff A. 2001. Seasonal changes of fish diversity in the main channel of the large lowland River Oder. *Regul Rivers Res Mgmt* 17:595–608.

Wuertz S, Gessner J, Kirschbaum F, Kloas W. 2002. First identification of IGF-I mRNA in the genus *Acipenser* and its quantification in maturing gonads. In: *Proceedings of the 21st Conference of European Comparative Endocrinologists*, Bonn, 26–30 August 2002, Keller R, Dircksen H, Sedlmeier D, Vaudry H (eds.). Monduzzi Editore, Bologna, Italy, pp. 253–257.

Wuertz S, Nitsche A, Jastroch M, Gessner J, Klingenspor M, Kirschbaum F, Kloas W. 2006a. The role of IGF-I system for vitellogenesis in maturing female sterlet, *Acipenser ruthenus* Linnaeus, 1758. *Gen Comp Endocrinol* 150:140–150.

Wuertz S, Lutz I, Gessner J, Löschau P, Hogans B, Kirschbaum F, Kloas W. 2006b. The influence of rearing density as environmental stressor on cortisol response of shortnose sturgeon (*Acipenser brevirostrum*). *J Appl Ichthyol* 22 (Suppl 1):269–274.

Wuertz S, Nitsche A, Gessner J, Kirschbaum F, Kloas W. 2006c. IGF-I and its role in maturing gonads of female sterlet, *Acipenser ruthenus* Linnaeus 1758. *J Appl Ichthyol* 22 (Suppl 1):346–353.

Chapter 25
Sturgeon Recovery Plan in the Rhône River (France): Preliminary Results on Species Determination and Habitat Suitability

Laurent Brosse, Patrick Berrebi, Nathalie Desse-Berset, and Mario Lepage

Abstract Sturgeons were living in the Rhône River (France) until their complete extirpation in the mid 1970s. They disappeared without a certainty about species identity and about sympatry between *Acipenser sturio* and *Acipenser naccarii*, as occurred in the Po and Guadalquivir rivers. Preliminary studies were launched in 2005 by the *Migrateurs Rhône Méditerranée* (MRM) association to overcome this lack of knowledge and to evaluate the lower Rhône River environment suitability for sturgeon. Twelve sturgeon specimens from the Rhône River and the French Mediterranean coast were found in museums and samples for genetic analyses were taken from 10 of them (analyses in progress). Genetic analyses (cytochrome *b*) on five sturgeon bone remain (among 2500 remains available) samples more than 2000 years old found in the city of Arles, close to the Rhône River, clearly show that *A. sturio* was living in this river. Morphological comparisons between ancient bone remains and recent skeletons from *A. sturio* and *A. naccarii* are in progress. Environment studies show the existence of possible suitable habitats for sturgeon in the Rhône river, mainly spawning and feeding grounds. This should be confirmed by more precise studies. All these preliminary results are very encouraging regarding the possible restoration of sturgeon in the Rhône River.

Keywords Ancient DNA, *Acipenser naccarii*, *Acipenser sturio*, archaeozoology, Rhône River, species determination

---

L. Brosse
Association Migrateurs Rhône Méditerranée, Z.I. du port Fluvial,
Chemin des Ségonneaux, Arles, France

P. Berrebi
Institut des Sciences de l'Evolution, UMR 5554, UM2/CNRS Université Montpellier II CC 065,
Place Eugène Bataillon, Montpellier Cedex, France

N. Desse-Berset
Laboratoire d'archéozoologie, UMR 6130, CNRS-Université de Nice-Sophia Antipolis,
Valbonne, France

M. Lepage
Cemagref, UR EPBX, F-33612 Cestas Cedex, France

R. Carmona et al. (eds.) *Biology, Conservation and Sustainable Development of Sturgeons*,

25.1 Introduction

Sturgeons are among the oldest existing Osteichthyan fishes. The first sturgeon fossils date from the high Cretaceous, i.e. –100 to –65 Myr (Holcik et al., 1989; Rochard et al., 1991) and Acipenseriforms date back to the lower Jurassic (Ludwig, 2006). The Acipenseridae family contains between 24 and 26 species (according to the authors) all living in the Northern Hemisphere (Bemis et al., 1997). Sturgeons have long biological cycles with a mean age at maturity of 8 years for the males and 14 years for the females (Rochard et al., 1990). All sturgeon species around the world are threatened, or on the edge of extinction (Rochard et al., 1990; Birstein et al., 1995, 1997). According to IUCN criteria (1996), a very large majority of the species belonging to the Acipenseridae family (17/26) is classified either as threatened with extinction or as being critically endangered to various degrees, testifying to the very bad condition of sturgeon populations worldwide. The most frequently cited causes to explain the strong reductions and the disappearance of some sturgeon populations are overfishing for caviar and dam construction (Rochard et al., 1990; Castelnaud et al., 1991; Birstein et al., 1997; Hensel and Holcik, 1997; Kynard, 1997; Smith and Clugston, 1997) as well as habitat degradation and water pollution (Trouvery et al., 1984; Castelnaud et al., 1991; Birstein et al., 1997; Smith and Clugston, 1997).

Although some restoration and protection measures have been taken to preserve sturgeons, their effects are limited and sturgeons remain threatened (Beamesderfer and Farr, 1997). Thus, conservation plans for sturgeons and their habitats on a large scale are necessary to preserve or restore populations (Dadswell, 1979; Rochard et al., 1990; Beamesderfer and Farr, 1997; Bemis et al., 1997; Debus, 1997; Smith and Clugston, 1997; Bruch, 1999; Wakeford, 2001). Plans for protection or restoration of sturgeons and their habitats are running or in development. For example, there are management plans for the gulf sturgeon, *Acipenser oxyrinchus desotoï* Vladykov, 1955; (Wakeford, 2001), the white sturgeon, *Acipenser transmontanus* Richardson, 1836 (Duke et al., 1999) or the lake sturgeon, *Acipenser fulvescens* Rafinesque, 1817 (Bruch, 1999). In the Mediterranean Sea area, restoration plans for the adriatic sturgeon, *Acipenser naccarii* Bonaparte, 1836, are in progress in Italy (two European Life Nature Programmes). In Spain, sturgeon reintroduction depends on the determination of the species to which the extinct populations belonged, mainly in the Guadalquivir River. A preliminary work based on molecular determination from museum specimens described two species, the European sturgeon (*Acipenser sturio* Linnaeus, 1758) and the adriatic sturgeon (Garrido-Ramos et al., 1997). This was the beginning of a controversy between geneticists and morphologists (Ludwig and Kirschbaum, 1998; Elvira and Almodovar, 1999; Doukakis et al., 2000), some stating that only *A. sturio* had been living in Spanish rivers while others contend that it was *A. naccarii*. A recent publication of De la Herran et al. (2004) seems to end the debate by confirming the presence of both species in Spain before their extinction. These results imply an extension of the distribution area for *A. naccarii* to the whole Mediterranean Sea from eastern Italy to western Spain, including the French coasts.

Sturgeons were found in the Rhône River (Fig. 25.1) until their complete extirpation in the mid 1970s (Tabardel, 1994) due to overfishing and habitat degradation. Sturgeons disappeared from the French Mediterranean coasts without any certainty concerning the species (but in the past, it was always considered to be *A. sturio*). This poses a problem for reintroduction because two species could have been living in sympatry in the Rhône River, *A. sturio*, and *A. naccarii*. Here again, historical species determination is essential before starting any action of reintroduction in the Rhône River basin.

Since the beginning of 2005, the Migrateurs Rhône Méditerranée (MRM) association is running a reintroduction plan for sturgeons in the Rhône River. This interdisciplinary project comprises several preliminary steps. The first one is aimed at determining on the one hand which sturgeon species historically inhabited the

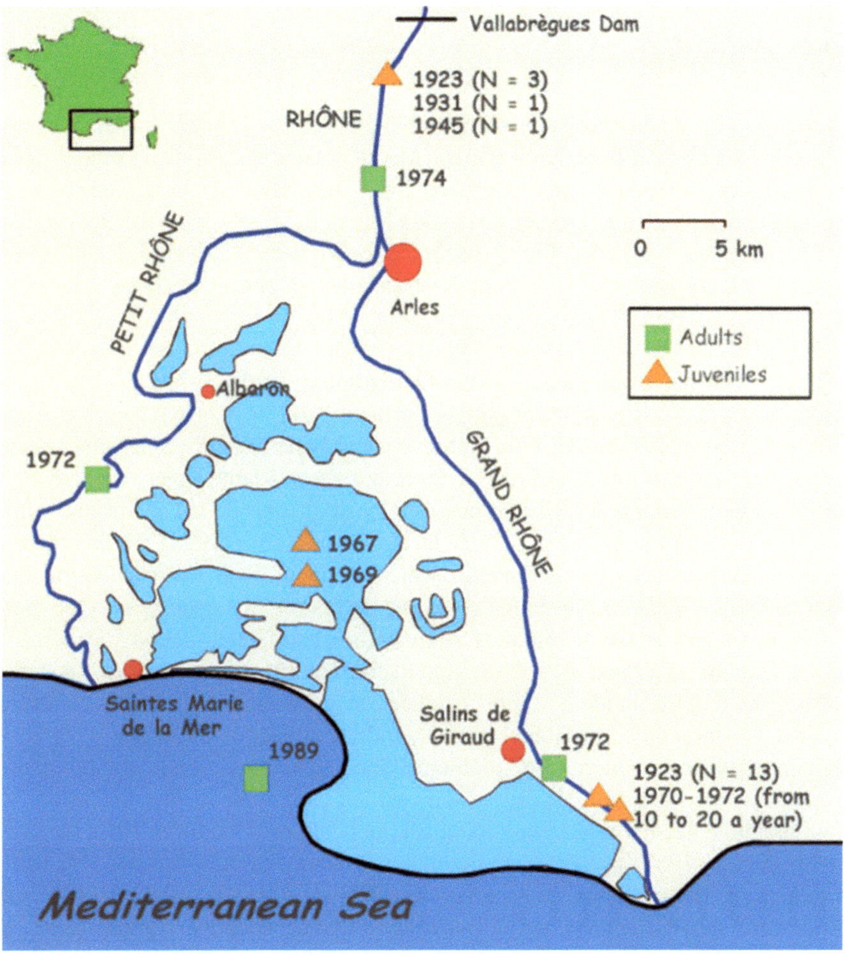

Fig. 25.1 Situation map with all recent sturgeon catches (modified from Tabardel, 1994)

Rhône River and, on the other, the degree of suitability of the environment for sturgeon reintroduction. Secondly, the potential impacts of sturgeon reintroduction on the environment will be explored. Then, depending on the results of these two points, a strategy for the next steps of the reintroduction project will be decided after concertations with sturgeon specialists in Europe.

This paper describes the studies undertaken to reach these objectives. It deals with the first step of this project, i.e. historical species determination and habitat suitability for sturgeons in the Rhône River.

25.2 Methods

25.2.1 Species Determination

25.2.1.1 Museum Specimens

Museums with a natural-history section and taxidermists located in French departments bordering the Rhône River and the Mediterranean Sea, as well as museums in Monaco and Switzerland, were contacted. They were asked for preserved specimen of sturgeons with certified origin (date and place of the catch) from the Rhône River or Mediterranean Sea. Then, for all specimens found, several measures (length, scutes number), photos and, when possible, tissue samples for genetic analyses were taken.

25.2.1.2 Archaeozoological Sources (Morphology and Genetic)

Another source of information is archaeological bone remains. The protohistoric settlement of the *Jardin d'Hiver*, in Arles (close to the Rhône River), yielded some 2500 sturgeon remains, distributed between the sixth and second century BC with a majority of them in the fourth century layer (Desse-Berset, 1994). Those sturgeon bones belong to a Greek settlement, which is previous to the installation of a Roman colony, which occurred in the second part of the first century BC. One of the priorities of any archaeozoological study is to identify the anatomical parts, to reconstitute the sizes and the masses, and to consider the minimal number of specimens. That requires modern reference collection and osteometric data. None of that existed for sturgeon before the year 1990, and the collection of an animal listed as threatened with extinction posed problems. Specimens preserved in natural-history museums were not usable, because their ossified parts were neither accessible nor measurable. Palaeontological or zoological publications (Goodrich, 1930; Gregory, 1933; Grassé, 1957) reviewing the complex anatomy of Acipenseriforms, works presenting systematic data for each species (Williot, 1991), keys based on meristic criteria for specific determination (number of dorsal, lateral or ventral scutes by example; Whitehead et al., 1984; Lelek, 1987; Holcik et al., 1989), and morphometric characteristics (percentages of lengths for each body part compared to the overall length for example) are very useful regarding the whole fish, but unfortunately are not applicable to fragmented or isolated bones to characterize archaeological bone remains.

In view of this deficiency, it was necessary to develop methods of identification by crossing the various fragmental information collected in the literature. One of the rare monographs providing a study on *A. sturio* population which is still living in the Gironde, is that of Magnin (1962). Valbonne's laboratory of Archaeozoology had only a small number of European sturgeon specimens, some fragmentary, obtained during the period 1985–1990 with the assistance of the CEMAGREF (Agricultural and environmental engineering research). Thanks to the Migrateurs Rhône Méditerranée (MRM) association and to the gifts from the CEMAGREF and from the Italian fish farm Azienda Agricola VIP, three new sturgeons could be recently obtained: one *A. sturio* of 127 cm TL, and two *A. naccarii* of 146 and 148 cm TL. These specimens will be prepared at the laboratory for the building of a species determination set between *A. sturio* and *A. naccarii* based on bone morphology.

As there was not yet assurance that morphological studies based on sturgeon bones would enable differentiation between *A. sturio* and *A. naccarii*, we decided to complement the archaeozoological study by molecular analyses. Six archaeological bones were selected from different sturgeon individuals corresponding to different chronological periods and locations (Fig. 25.2).

- Sample No. 1: JH 87 US 553 (525 to 475 BC)
- Sample No. 2: JH US 042 (400 to 375 BC)
- Sample No. 3: JH 84 US 097 (250 to 175 BC)
- Sample No. 4: JH 83 US 037 (375 to 350 BC)
- Sample No. 5: JH 86 US 376 (400 to 375 BC)
- Sample No. 6: JH 87 US 037 (375 to 350 BC)

The choice of those six bones from different periods, different parts of the fish, or different archaeological locations on the site was made to test whether all the remains of Arles are suitable for mtDNA sequencing or if only a few of them can provide good-quality DNA. After an international inquiry, searching for collaboration, the Ancient DNA Laboratory (Department of Archaeology, Simon Fraser University, Burnaby, Canada) analysed the samples.

Two pairs of primers were designed from the sturgeon sequences available in Genbank (Yang et al. unpublished data) in the cytochrome *b* region giving amplified fragment a little less than 200 bp each. DNA was extracted from archaeological samples following the protocol detailed by Yang et al. (1998) by two independent scientists of the same laboratory who performed parallel polymerase chain reactions (PCR) (Yang et al. unpublished data). The two sets of sequences obtained by the two pairs of primers were aligned with all the homologous sequences published in Genbank from the genus *Acipense*r and *Huso*, plus *Polyodon* and *Psephurus* (paddlefish) as outgroups. Then a neighbour-joining (NJ) tree was constructed and a bootstrap estimation performed. The species identification was made by sequence analogy, automatically placing the ancient DNA (aDNA) sequence in the good species cluster. The bootstrap values give an estimation of the identification liability.

Fig. 25.2 Sturgeon bone remains from the Jardin d'Hiver in Arles used for DNA analyses (Photo credit: N. Desse-Berset). No. 1: JH 87 No. 553 (ventral scute, left); total weight, 4.45 g (sample 1.65 g). No. 2: JH No. 042 (pectoral fin ray); total weight, 4.53 g (sample 1.56 g). No. 3: JH 84 No. 097 (cleithrum fragment); total weight, 3.27 g (sample 1.10 g). No. 4: JH 83 No. 037 (occipital); total weight, 4.02 g (sample 1.42 g). No. 5: JH 86 No. 376 (parasphenoid fragment); total weight, 4.69 g (sample 1.46 g). No. 6: JH 87 No. 037 (cleithrum fragment?); total weight, 1.41 g (sample 1.41 g)

Due to the fact that ancient DNA is generally highly degraded, the analysis had to satisfy all the recommendations concerning its procedure. A high level of technical precautions were taken in order to ensure that the target DNA was the only one amplified during the polymerase chain reactions (PCR), the base of any sequencing (Caramelli et al., 2003; Gilbert et al., 2005):

1. Molecular analyses were performed in a laboratory especially dedicated to ancient DNA extraction and other pre-PCR work (Wayne et al., 1999; Loreille et al., 2001)
2. Sturgeon DNA was never analysed in this laboratory before
3. For each sample, two independent DNA extractions were performed from the six fragments of bones and PCR were performed independently by two scientists
4. Similar analyses were performed, partly on the same samples, in another laboratory (Pagès et al., Online 2008), analysing a different locus, but reaching the same species identification

5. The final consensus sequences make phylogenetic sense, not resulting from contamination by exogenous DNA, since it was exactly the sequences obtained from a present living sturgeon species
6. A total of 70 PCR were performed and 46 sequencing electrophoregrams were obtained and identified as sturgeon sequences according to the Genbank data

25.2.2 Environmental Suitability

A bibliographic review was carried out to ascertain the habitat preferences of sturgeon (Beamesderfer and Farr, 1997), more precisely for the spawning grounds (Khoroshko and Vlasenko, 1970; Buckley and Kynard, 1985; McCabe and Tracy, 1994; Moser and Ross, 1995; Schaffter, 1997; Sulak and Clugston, 1999; Collins et al., 2000; Fox et al., 2000; Paragamian et al., 2001; Jego et al., 2002) and the feeding ground for juveniles (Dadswell, 1979; Levin, 1989; McCabe et al., 1993; Ruban and Konoplya, 1994; Zolotarev et al., 1996; Beamish et al., 1998; Haley, 1998; Brosse et al., 2000; Kynard et al., 2000, 2002). These two habitats are known to be critical because they are essential to the completion of the biological cycle of sturgeon. This knowledge serves as a basis for our evaluation concerning the data necessary for assessing that the quantity and quality of the habitats in the Rhône River met the sturgeon's needs.

Contacts were made with several organizations to find data on the bathymetry and sediment nature of the Rhône River in its last free 60 km downstream of the Vallabrègues dam (Fig. 25.1). These data will be used to detect a priori suitable areas for spawning grounds. The analysis of these data will enable us to find out whether it is necessary to carry out a fieldwork to gather complementary data on habitats in this area. The same approach was used concerning feeding grounds with bibliographic work on the composition and distribution of the invertebrate fauna in this area of the Rhône River.

25.3 Results and Discussion

25.3.1 Species Determination

25.3.1.1 Museum Specimens

Among the 35 museums contacted in France and the 10 in Switzerland, only six possessed sturgeons coming from the Rhône River or from the French Mediterranean coast. A total of 12 specimens were examined (Table 25.1), most of them dating from the 19th century. Five out the 12 specimens were caught in the Mediterranean Sea offshore from the city of Nice in 1876 and 1879. All the other specimens come from the Rhône River.

Table 25.1 List of the sturgeon specimens from the Rhône River or the French Mediterranean coast found in museums in France and Switzerland

Specimen number	Museum reference	Museum	TL (cm)	FL (cm)	Dorsal scutes	Lateral scutes	Ventral scutes	Catch date	Catch area	Museum Label
1	49	Nîmes	32	30	11	30	8	?	Rhône River	*Acipenser sturio* (Lin.)
2	50	Nîmes	41	34	12	35	10	?	Rhône River	*Acipenser sturio* (L.)
3	245	Nîmes	180	164	13	33	12	1871	Rhône River	*Acipenser sturio* (L.)
4	244	Nîmes	137	123	14	34	12	1928	Rhône River near Beaucaire	Sturgeon
5	000–049	Avignon	57	52	14	37	10	1938	Rhône River near Avignon	*Acipenser sturio* (L.)
6	1898–1257	Paris	41	35	12	41	9	1876	Mediterranean Sea off Nice	*Acipenser naccarii* (B.)
7	42005436/5192	Lyon	43	39	14	39	9	1879	Mediterranean Sea off Nice	No label
8	42006251/5235	Lyon	43	35	14	64	15	1879	Mediterranean Sea off Nice	*Acipenser sturio* (L.)
9	42006250/5280	Lyon	53	47	14	36	12	1879	Mediterranean Sea off Nice	*Acipenser sturio* (L.)
10	42005478/5259	Lyon	90	85	12	37	12	1879	Mediterranean Sea off Nice	No label
11	MHNM 1248	Marseille	83	77	13	34	10	1883	Rhône River	*Acipenser sturio* (L.)
12	125/100	Geneva	86	77	12	30	10	1830	Rhône River downstream Lyon	*Acipenser sturio* (L.)

[a]Species determination revised by Vasil'eva in 1997.

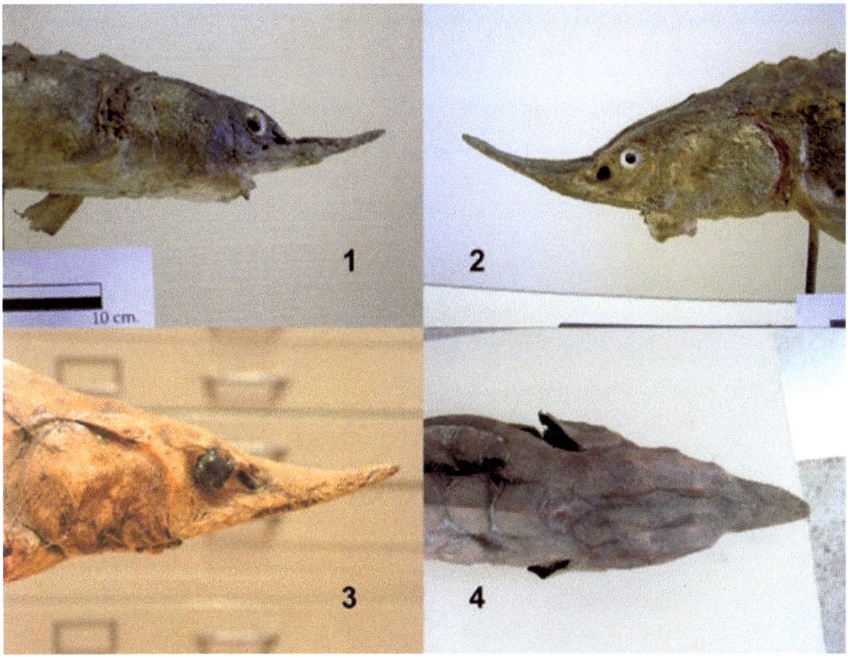

Fig. 25.3 Preserved sturgeon specimens showing snout deformation (1 and 2), body deformation (4) and missing barbels (1, 2 and 3). (Photo credit: Migrateurs Rhône Méditerranée Association.)

Only one specimen, from the National Natural History Museum in Paris, is classified as *A. naccarii* (species determination by Vasil'eva in 1997), all the others are classified as *A. sturio* or simply as sturgeon, without further specification (Table 25.1). The only specimen classified as *A. naccarii* was one of those caught at sea. The fact that only one specimen coming from the French Mediterranean coast is classified as *A. naccarii* is not enough to draw conclusions on the possible sympatry between *A. sturio* and *A. naccarii* in the Rhône River. Moreover, as most of these specimens are more than a hundred years old, they often show some degradation, such as missing barbels, or body deformation such as distorted snout shape (Fig. 25.3). These elements hamper a reliable species determination based on morphological criteria for these preserved specimens.

Among the 12 sturgeon specimens examined, samples for genetic analyses were collected from 10 of them with a total of 19 tissue samples (Table 25.2). No sturgeon specimen was found among the 42 taxidermists contacted and none of them had heard of a sturgeon specimen.

The molecular analysis of the museum specimens was programmed for the next step of this investigation. Several international laboratories will be contacted for this analysis to be performed in a specific laboratory devoted to ancient DNA analysis, using specific equipment and personal protection (Loreille et al., 2001).

Table 25.2 List of the samples collected on sturgeon specimens coming from museums. Only two sturgeons from the Nîmes museum did not allow taking samples without damaging the specimens

Genetic sample	Sturgeon specimen number	Museum reference	Sample nature
ER_1	5	000–049	Fin skin
ER_2	5	000–049	Pectoral fin skin
ER_3	5	000–049	Mouth skin
ER_4	6	1898–1257	Fin skin
ER_5	6	1898–1257	Opercule skin
ER_6	6	1898–1257	Mouth muscle
ER_7	7	42005436/5192	Fin skin
ER_8	8	42006251/5235	Fin skin
ER_9	9	42006250/5280	Fin skin
ER_10	10	42005478/5259	Fin skin
ER_11	10	42005478/5259	Anus skin
ER_12	11	MHNM 1248	Fin skin
ER_13	11	MHNM 1248	Opercule skin
ER_14	3	245	Opercule skin
ER_15	3	245	Opercule skin
ER_16	4	244	Opercule skin
ER_17	4	244	Ventral skin
ER_18	12	125/100	Skin
ER_19	12	125/100	Internal part

25.3.1.2 Archaeozoological Sources

The first archaeozoological work, based specifically on the modern specimens of *A. sturio* provided by the CEMAGREF, enabled us to carry out the morphological and osteometric study of this material (Desse-Berset, 1994). Results using these specimens showed the anatomical distribution of the remains found as follows: cranial dermal bones (4%); splanchnocranium (17%); scapular girdle (30%); dermal scutes (49%). Bones from the splanchnocranium were used for the calculation of the minimal number of specimens and for size reconstitution. Although pectoral fin rays account for only 7% of the bones determined, they proved to be rich in information for the analysis of the archaeozoological remains when coupled with the results from Magnin (1962). The table relating length/weight/age, relying on hundreds of modern individuals, provided information to build several diagrams:

1. Size distribution of the sturgeon population for protohistoric period, obtained using the relation between two measurements taken on 70 pectoral-fin rays from Arles
2. Size distribution of the current sturgeon population of the Gironde, whose age and overall length were known
3. Total length by age groups established according to the data from Magnin (1962) on sturgeons from the Gironde

The cross-referencing of all this information enabled us to draw the distribution for sturgeons found in the archaeozoological remains in various size and age groups:

1. 34% were 25 years old and more, corresponding to a TL over 180 cm
2. 35% were between 8 and 24 years old with TL from 94 cm to 2 m
3. 24% were between 4 and 7 years old, i.e. a TL from 61 to 85 cm
4. 6% were between 3 and 4 years for a TL between 47 and 61 cm
5. The youngest specimen was 2 years old and 34 cm long

The Arles sturgeon remains gave us the image of a balanced population, in spite of active and clearly not very selective fishing; all the age groups are represented. Because large specimens are numerous (about 50% were breeders), we consider that overfishing did not perturb the *A. sturio* population living in the lower Rhône River between the sixth and the second century BC (Desse-Berset, 1994).

Molecular analyses were performed using six different bones. Five of them provided usable mtDNA that permitted us to carry out 70 PCR from which 46 sequences were obtained on two 'long' and 'short' zones of the cytochrome *b* (192 and 183 bp).

These sequences were all the same with the exception of expected errors induced by DNA damage (Hofreiter et al., 2001; Gilbert et al., 2003 a,b).

The two monomorphic sequences were included in two respective NJ trees constructed with all the sturgeon homologous sequences published in Genbank data (51 and 152 published sequences, respectively). The long-sequence tree was unable to separate the archaeological taxon from *A. sturio* and *A. oxyrinchus* while the short-sequence tree neatly clustered the archaeological taxon with *A. sturio* (bootstrap value 73%), *A. oxyrinchus* being a sister species (Yang et al. unpublished data).

According to the sequences of a part of the cytochrome *b*, five specimens, represented by their bones, belonged to the species *A. sturio* with high confidence in the determination.

The morphological study made by Desse-Berset (1994) on the Arles remains based on the comparison with recent *A. sturio* skeletons from the Gironde population concluded also that these remains belong to *A. sturio*. Nevertheless, as sympatry between *A. sturio* and *A. naccarii* may have occurred, we decided to start a new morphological work comparing both *A. sturio* and *A. naccarii* skeletons with recent specimens. The three specimens were prepared at the CEPAM-CNRS Sophia-Antipolis laboratory during July 2006. The morphological work aimed at the building of specific determination keys between *A. sturio* and *A. naccarii* based on the skeletons is still in progress.

25.3.2 Environmental Suitability

Historical data indicate the existence of a functional spawning ground in the sector known as Saxy, 5 km upstream from the city of Arles. This assumption is supported by the capture, in 1974, of a 2 m long adult in this sector and the capture of 10 to

20 juveniles a year between 1970 and 1972 in the downstream part of the Grand Rhône (Tabardel, 1994). These captures occurred while the dams of Beaucaire and Vallabrègues were in service since 1970, blocking the access to the potential spawning grounds located upstream (Fig. 25.1). Sturgeon extinction in the Rhône River 30 years ago, as well as the possible evolution of the bed of this river, require an update of knowledge on the potential presence of spawning grounds between Vallabrègues and the sea. Before working in detail on a particular zone, we decided to use several criteria (depth, nature of the sediments, and current velocity) to determine the most promising sectors before going into the field from National Company of the Rhone (CNR) we obtained data on the bathymetry of the Rhône River (dating from March 2004), scale 1/5000, between Vallabrègues and its separation into two branches (Grand Rhône and Petit Rhône) as well as the longitudinal and transverse profiles of the zones appearing to us to be potential spawning grounds. The study of the bathymetric charts using a criterion based on the existence of a depression of at least 5 m depth reveals six potential spawning grounds distributed as follows (Fig. 25.4), considering that the kilometric point zero is in the city of Lyon, which is located 330 km upstream from the mouth of the Rhône River:

1. (KP) 265.3 to 265.40 the entire width, immediately downstream from the Vallabrègues dam
2. (KP) 267.4 to 267.5, the entire width, immediately downstream from the car bridge between Tarascon and Beaucaire
3. (KP) 267.7 to 267.9, the entire width, immediately downstream from the train bridge between Tarascon and Beaucaire
4. (KP) 272.9 to 273.4, right bank
5. (KP) 274.0 to 274.4, right bank
6. (KP) 276.1 to 276.3, right bank

Sediments are known to be largely dominated by gravels and pebbles in this entire sector. For the moment, we have no data on possible filling by finer sediments. The Rhône River is a swift with high current velocity generally higher than 0.5 m s^{-1}. It is thus probable that the current velocity near the bottom in the depressions described above is favourable to sturgeon spawning. The overall impression is that the zones presented above are serious candidates as spawning grounds for sturgeon. It nevertheless remains necessary to take some complementary measures on these sectors during sturgeon spawning (April to June, surface water tempterature of 18 °C) to gain a more precise perception of the suitability of these sectors as spawning grounds.

Contacts were engaged with several organizations such as universities (Lyon and Marseille) as well as engineering and design departments (ARALEP) to formulate an overall vision of the existing data on benthic fauna. The studies carried out by these organizations are restricted in time and space. In fact, there is no exhaustive cover on the composition of benthic fauna between Vallabrègues and the sea. The available data indicate that benthic fauna is primarily made up of platyhelminths, chironomids larvae or other insects, molluscs, shellfish and oligochaetes (Franquet et al., 1995; Bournaud et al., 1996; Franquet, 1999). Because of the

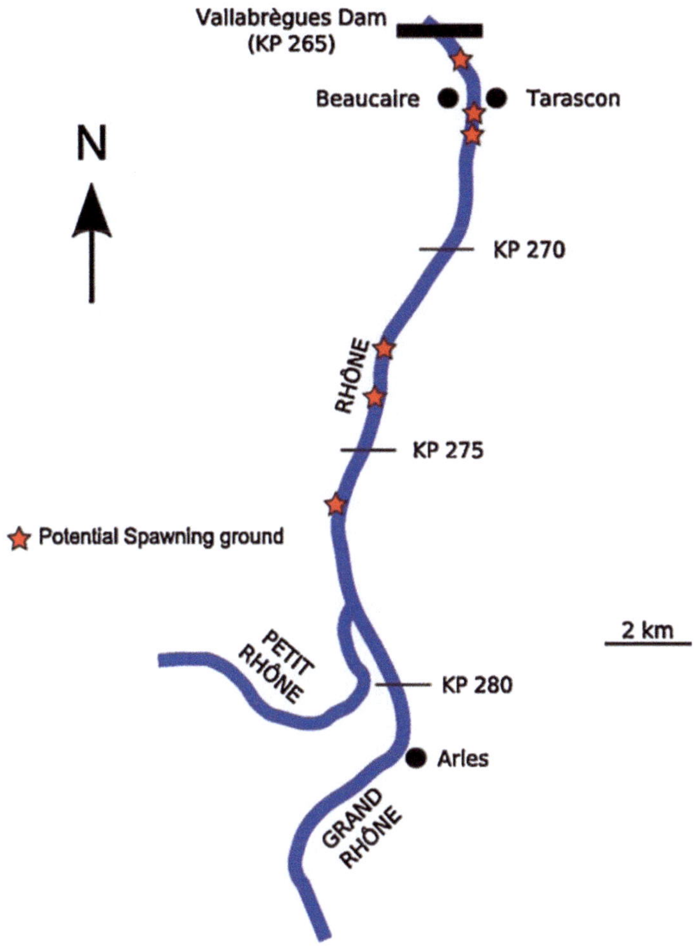

Fig. 25.4 Localization of the potential sturgeon spawning grounds identified between KP 265 and 280 using bathymetric from 2004

very limited cover of these data, it is not possible to draw conclusions on the existence and the location of feeding grounds favourable to sturgeon. However, the faunistic lists provided by Fruget et al. (1995) indicate the presence, sometimes in huge quantities, of groups (platyhelminths, chironomids, small crustacean and oligochaetes) that sturgeons feed upon. The scant data available thus constrained us to set up a study of benthic fauna for the whole of our study area (Fig. 25.1) with replicates. During spring 2006, we started the benthic-fauna study by collecting 30 samples using a Van Veen grab sampler in the last 25 km of the Grand Rhône before the sea. The analysis is in progress. More samples are planned to provide better coverage.

25.4 Conclusions

25.4.1 Species Determination

The results of the present study show that *A. sturio* was living in the Rhône River more than 2000 years ago. They are based on a first set of analyses using morphological (Desse-Berset, 1994) and genetic evidence. This is the first step and we need to perform many more analyses (morphological and genetic) on the ancient bone remains (2500 remains representing at least 100 sturgeons) and on samples coming from museums to form a more reliable perception of the sturgeon species that was historically found in the Rhône River. These analyses are in process and results will soon be available.

Several tools are available for species determination, such as morphology and genetic.

The cytochrome *b* sequence amplification using the DNA extracted from five of the six archaeological samples selected for the analysis was successful. This shows that DNA preservation in ancient bones is good enough in each stratigraphic level to allow the extraction of mtDNA sequences long enough to allow species determination. As samples cover the whole chronological periods of the settlement, these results are very promising for further trials on mtDNA and maybe on nuclear DNA based on a large selection of samples among archaeozoological remains.

Sturgeon specimens found in museums are listed mainly as *A. sturio*, except for the specimen in Paris listed as *A. naccarii*. This is not enough to say that the adriatic sturgeon was found in the Rhône River, for several reasons:

1. This specimen was caught at sea and not in the Rhône River
2. It is a single specimen that could be a wandering juvenile far away from its native watershed as it occurs with European sturgeons in the Atlantic or North Sea (Rochard et al., 1997)
3. No other fish identified as *A. naccarii*, youngs or adults, were caught at sea or in the Rhône River during the same period and the presence of a two years old fish suggest a recent reproduction.

This poses the question of the origin of this specimen in relation to some possible labelling problems in museums concerning its origin.

According to mitochondrial evidence obtained from the five specimens analysed, confirmed by a further analysis (Pagès et al., online 2008), the European sturgeon was living in the Rhône River 2500 years ago. These results are fully consistent with those obtained using morphometric analyses (Desse-Berset, 1994). These results are also consistent with the historical range area for *A. sturio*, from the Baltic Sea to the Black Sea as described in Holcik et al. (1989).

This work is based on a very limited number of samples ($N = 5$), which is not enough to reject the hypothesis of a sympatry with *A. naccarii* in the Rhône river, as in the Po River or in the Guadalquivir River according to De la Herran et al. (2004). Thus, to gain a more precise idea of this hypothesis, we need to perform additional analyses on more samples coming from archaeozoological remains and on samples coming from museums. The results from archaeozoological remains using genetic will be compared with those found using new morphological criteria based on comparative studies between *A. sturio* and *A. naccarii* skeletons (work in progress). Concerning museum specimens, results found using genetic analyses will be compared with the same species determination of specimens as labelled in the museums.

25.4.2 *Environmental Suitability*

The environment seems suitable for sturgeon spawning grounds and feeding grounds, even if our results are very limited. The Rhône River between KP 265 and 280 seems to offer six potential spawning grounds with characteristics close to those needed by *A. sturio* (Jego et al., 2002). Maybe several more suitable sites are available downstream but, for the moment, we do not have suitable data on the bathymetry to find these additional potential spawning grounds. As data on benthic fauna are very scarce, we cannot yet give a definitive opinion on this element in the Rhône River in relation with European sturgeon needs (Brosse et al., 2000). Samples collected during Spring 2006 could give a general if preliminary idea concerning the benthic fauna found in the last 25 km of the Rhône River. This is not enough to draw conclusions about environment suitability or carrying capacity concerning juvenile sturgeons for a restoration plan at this precise point. We need to collect more samples on a broader area concerning the entire river from the Vallabrègues dam to the sea (Fig. 25.1).

The next step in this project is to investigate possible ecological impacts of sturgeon reintroduction more than 30 years after it disappeared from the river. Sturgeon reintroduction in the Rhône River could result in disease problems. As an example, the introduction of stellate sturgeon, *Acipenser stellatus* Pallas, 1771, with its nematodes parasites from the Caspian sea into the Aral Sea resulted in serious losses among the population of the endemic ship sturgeon, *Acipenser nudiventris* Lovetzky, 1828, as described by Zholdasova (1997). What would be the sturgeon reintroduction impact on the food web or on the benthic fauna composition and abundance? From the fish-community point of view, what will be the impact of sturgeon reintroduction and what interactions could occur with species introduced after the sturgeon disappeared from the Rhône River such as Wels catfish (*Silurus glanis* Linnaeus, 1758)? Future work should consider the sturgeon-reintroduction project in the Rhône River from the community-ecology approach.

Acknowledgements Thanks to the 'Migrateurs Rhône Méditerranée' association who initiated the study for sturgeon restoration in the Rhône river and paved the way for this study to start. Thanks to the CEMAGREF in Bordeaux who provided us with a dead specimen of *A. sturio* and also to Sergio and Giacinto Giovannini (Azienda Agricola, VIP, Brescia, Italy) who gave us two specimens of *A. naccarii* for morphological studies. Thanks to Patrice Arcelin, Director of the archaeological excavations of the Jardin d'Hiver, Arles, for all stratigraphical data. This study was funded by the French 'Ministère de l'Écologie et du Développement Durable', the 'Groupement d'Intérêt Scientifique sur les Amphihalins', the 'Conseil Supérieur de la Pêche', the 'Agence de l'Eau Rhône-Méditerranée et Corse', the members of the 'Migrateurs Rhône Méditerranée' association, the 'Conseil Régional Provence-Alpes-Côte-d'Azur', the 'Conseil Régional Rhône Alpes', the 'Conseil Général des Bouches du Rhône', the 'Nature et Découvertes' foundation and the 'Compagnie Nationale du Rhône'. The help of the Ancient DNA Laboratory (Simon Fraser University, Burnaby, Canada) and especially of Prof. Dongya Yang, was essential for this phase of the project this is an ISEM 2007–162 publication.

References

Beamesderfer, R. C. P. and R. A. Farr. 1997. Alternatives for the protection and restoration of sturgeons and their habitat. *Environmental Biology of Fishes* **48**:407–417.

Beamish, F. W. H., L. G. N. David and A. Rossiter. 1998. Feeding ecology of juvenile Lake Sturgeon, *Acipenser fulvescens*, in northern Ontario. *Canadian Field—Naturalist* **112**:459–468.

Bemis, W. E., V. J. Birstein and J. R. Waldman. 1997. Sturgeon biodiversity and conservation: an introduction. *Environmental Biology of Fishes* **48**:13–14.

Birstein, V. J., W. E. Bemis and J. R. Waldman. 1995. The threatened status of acipenseriform fishes: a summary. In: V.J. Birstein, W.E. Bemis and J. R. Waldman (eds.), *Sturgeon Biodiversity and Conservation*, Kluwer Academic Publishers, Dordrecht, pp. 427–435.

Birstein, V. J., W. E. Bemis and J. R. Waldman. 1997. The threatened status of acipenseriform species: a summary. *Environmental Biology of Fishes* **48**:427–435.

Bournaud, M., B. Cellot, P. Richoux and A. Berrahou. 1996. Macroinvertebrate community structure and environmental characteristics along a large river: congruity of patterns for identification to species or family. *Journal of the North American Benthological Society* **15**:232–253.

Brosse, L., M. Lepage and P. Dumont. 2000. First results on the diet of the young European sturgeon, *Acipenser sturio* Linnaeus, 1758, in the Gironde estuary. *Boletín Instituto Español de Oceanografía* **16**:75–80.

Bruch, R. M. 1999. Management of lake sturgeon on the Winnebago System—long term impacts of harvest and regulations on population structure. *Journal of Applied Ichthyology-Zeitschrift Fur Angewandte Ichthyologie* **15**:142–152.

Buckley, J. and B. Kynard. 1985. Habitat use and behavior of pre-spawning and spawning shortnose sturgeon, Acipenser brevirostrum, in the Connecticut River, p. 111–117. In F. P. Binkowski and S. I. Doroshov (eds.), *North American Sturgeons: Biology and Aquaculture Potential.* Dr. W. Junk., Dordrecht, Netherlands.

Caramelli, D., C. Lalueza-Fox, C. Vernesi, M. Lari, A. Casoli, F. Mallegni, B. Chiarelli, I. Dupanloup, J. Bertranpetit, G. Barbujani and G. Bertorelle. 2003. Evidence for a genetic discontinuity between Neandertals and 24,000-year-old anatomically modern Europeans. *Proceedings of the National Academy of Sciences of the United States of America* **100**:6593–6597.

Castelnaud, G., E. Rochard, P. Jatteau and M. Lepage. 1991. Données actuelles sur la biologie d'*Acipenser sturio* dans l'estuaire de la Gironde. In: *Acipenser. Actes du Premier colloque international sur l'esturgeon*, Bordeaux, 3–6 Octobre 1989, Cemagref Editions, Antony, pp. 251–275, 520 pp.

Collins, M. R., T. I. J. Smith, W. C. Post and O. Pashuk. 2000. Habitat utilization and biological characteristics of adult Atlantic sturgeon in two South Carolina rivers. *Transactions of the American Fisheries Society* **129**:982–988.

Dadswell, M. J. 1979. Biology and population characteristic of the shortnose sturgeon, *Acipenser brevirostrum* LeSueur 1818 (Osteichthyes: Acipenseridae), in the Saint John river estuary, New Brunswick. *Canadian Journal of Zoology* **57**:2186–2210.

De la Herran, R., F. Robles, E. Martinez-Espin, J. A. Lorente, C. R. Rejon, M. A. Garrido-Ramos and M. R. Rejon. 2004. Genetic identification of western Mediterranean sturgeon and its implication of conservation. *Conservation Genetics* **5**:545–551.

Debus, L. 1997. Sturgeons of Europe and causes for their decline. In: Sturgeons Stocks and Caviar Trade Workshop, V.J. Birstein, A. Bauer and A. Kaiser-Pohlmann (eds.), IUCN, Gland., Switzerland and Cambridge, pp. 55–67.

Desse-Berset, N. 1994. Sturgeons of the Rhône during Protohistory in Arles (sixth-second century BC). *Annales du Musée Royal de l'Afrique Centrale, Sciences Zoologiques* **274**:81–90.

Doukakis, P., V. J. Birstein, R. De Salle, A. N. Ludwig, A. Ludwig, A. Machordom, A. Almodovar and B. Elvira. 2000. Failure to confirm previous identification of two putative museum specimens of the Atlantic sturgeon, *Acipenser sturio*, as the Adriatic sturgeon, *A. naccarii. Marine Biology* **136**:373–377.

Duke, S., P. Anders, G. Ennis, R. Hallock, J. Hammond, S. Ireland, J. Laufle, R. Lauzier, L. Lockhard, B. Marotz, V. L. Paragamian and R. Westerhof. 1999. Recovery plan for Kootenai River white sturgeon (*Acipenser transmontanus*). *Journal of Applied Ichthyology* **15**:157–163.

Elvira, B. and A. Almodovar. 1999. A morphological study of native sturgeon *Acipenser sturio* in Spain, and recent records of exotic Siberian sturgeon *A. baerii. Journal of Applied Ichthyology* **15**:278–279.

Fox, D. A., J. E. Hightower and F. M. Paruka. 2000. Gulf sturgeon spawning migration and habitat in the Choctawhatchee River system, Alabama-Florida. *Transactions of the American Fisheries Society* **129**:811–826.

Franquet, E. 1999. Chironomid assemblage of a Lower-Rhone dike field: relationships between substratum and biodiversity. *Hydrobiologia* **397**:121–131.

Franquet, E., B. Cellot, D. Pont and M. Bournaud. 1995. Environmental and macroinvertebrate dynamics in the lower Rhone River and a lateral dike field—a study matching 2 functioning descriptors. *Hydrobiologia* **308**:207–217.

Fruget, J. F., H. Tachet and D. Pont. 1995. Structure of macroinvertebrate community of the Rhône River delta. In: *Abstract of the First International Symposium on The Ecology of Large River*, Krem, Austria.

Garrido-Ramos, M. A., M. C. Soriguer, R. De La Herran, M. Jamilena, C. Ruiz Rejon, A. Domezain, J. A. Hernando and M. Ruiz Rejon. 1997. Morphometric and genetic analysis as proof of the existence of two sturgeon species in the Guadalquivir river. *Marine Biology* **129**:24–36.

Gilbert, M. T. P., H. J. Bandelt, M. Hofreiter and I. Barnes. 2005. Assessing ancient DNA studies. *Trends in Ecology & Evolution* **20**:541–544.

Gilbert, M. T. P., A. J. Hansen, E. Willerslev, L. Rudbeck, I. Barnes, N. Lynnerup and A. Cooper. 2003a. Characterization of genetic miscoding lesions caused by postmortem damage. *American Journal of Human Genetics* **72**:48–61.

Gilbert, M. T. P., E. Willerslev, A. J. Hansen, I. Barnes, L. Rudbeck, N. Lynnerup and A. Cooper. 2003b. Distribution patterns of postmortem damage in human mitochondrial DNA. *American Journal of Human Genetics* **72**:32–47.

Goodrich, E. S. 1930. *Studies on the Structure and Development of Vertebrates*, Macmillan, London.

Grassé, P. P. 1957. *Traité de zoologie. Anatomie, systématique, biologie*, Masson & Cie Edition, Paris.

Gregory, W. K. 1933. Fish skulls: a study of the evolution of natural mechanisms. *The American Philosophical Society* **23**:75–481.

Haley, N. 1998. A gastric lavage technique for characterizing diet of sturgeons. *North American Journal of Fisheries Management* **18**:978–981.

Hensel, K. and J. Holcik. 1997. Past and current status of sturgeons in the upper and middle Danube River. *Environmental Biology of Fishes* **48**:185–200.

Hofreiter, M., V. Jaenicke, D. Serre, A. von Haeseler and S. Paabo. 2001. DNA sequences from multiple amplifications reveal artifacts induced by cytosine deamination in ancient DNA. *Nucleic Acids Research* **29**:4793–4799.

Holcik, J., R. Kinzelbach, L. I. Sokolov and V. P. Vassilev. 1989. *The Freshwater Fishes of Europe*, Aula Verlag, Wiesbaden.

Jego, S., C. Gazeau, P. Jatteau, P. Elie and E. Rochard. 2002. Les frayères potentielles de l'esturgeon européen *Acipenser sturio* L. 1758 dans le bassin Garonne-Dordogne. Méthodes d'investigation, état actuel et perspectives. *Bulletin Français de la Pêche et de la Pisciculture* **365–366**:487–505.

Khoroshko, P. N. and A. D. Vlasenko. 1970. Artificial spawning grounds of sturgeon. *Journal of Ichthyology* **10**:286–293.

Kynard, B. 1997. Life history, latitudinal patterns, and status of the shortnose sturgeon, *Acipenser brevirostrum. Environmental Biology of Fishes* **48**:319–334.

Kynard, B., M. Horgan, M. Kieffer and D. Seibel. 2000. Habitats used by shortnose sturgeon in two Massachusetts rivers, with notes on estuarine Atlantic sturgeon: a hierarchical approach. *Transactions of the American Fisheries Society* **129**:487–503.

Kynard, B., R. Suciu and M. Horgan. 2002. Migration and habitats of diadromous Danube River sturgeons in Romania: 1998 2000. *Journal of Applied Ichthyology* **18**:529–535.

Lelek, A. 1987. The freshwater fishes of Europa. In: *Threatened Fishes of Europe*, Vol. 9, A. Verlag, Wiesbaden, pp. 42–57.

Levin, A. V. 1989. Characteristics of feeding behaviour of juvenile Russian sturgeon, *Acipenser gueldenstaedti*, in relation to food availability. *Journal of Ichtyology* **27**:41–47.

Loreille, O., L. Orlando, M. Patou-Mathis, M. Philippe, P. Taberlet and C. Hanni. 2001. Ancient DNA analysis reveals divergence of the cave bear, *Ursus spelaeus*, and brown bear, *Ursus arctos*, lineages. *Current Biology* **11**:200–203.

Ludwig, A. 2006. A sturgeon view on conservation genetics. *European Journal of Wildlife Research* **52**:3–8.

Ludwig, A. and F. Kirschbaum. 1998. Comparison of mitochondrial DNA sequences between the European sturgeon and the Adriatic sturgeon. *Journal of Fish Biology* **52**:1289–1291.

Magnin, E. 1962. Recherches sur la systématique et la biologie des Acipensérides. *Annales de la station centrale d'hydrobiologie appliquée*, Tome 9, Paris, 242 pp.

McCabe, G. T. and C. A. Tracy. 1994. Spawning and early life history of white sturgeon, *Acipenser transmontanus*, in the lower Columbia river. *Fishery Bulletin* **92**:760–772.

McCabe, G. T., R. L. Emmett and S. A. Hinton. 1993. Feeding ecology of juvenile white sturgeon (*Acipenser transmontanus*) in the lower Columbia river. *Northwest Science* **67**:170–180.

Moser, M. L. and S. W. Ross. 1995. Habitat use and movements of shortnose and Atlantic sturgeons in the lower cape-fear river, North-Carolina. *Transactions of the American Fisheries Society* **124**:225–234.

Pagès, M., N. Desse-Berset, C. Tougard, L. Brosse, C. Hänni and P. Berrebi. 2008. Historical presence of the sturgeon Acipenser sturio in the Rhône basin determined by the analysis of ancient DNA cytochrome b sequences. *Conservation Genetics*. Online First. DOI 10.1007/s10592-008-9549-6.

Paragamian, V. L., G. Kruse and V. D. Wakkinen. 2001. Spawning habitat of Kootenai river white sturgeon, post-Libby dam. *North American Journal of Fisheries Management* **21**:22–33.

Rochard, E., G. Castelnaud and M. Lepage. 1990. Sturgeon (Pisces: Acipenseridae); threats and prospects. *Journal of Fish Biology* **37**:123–132.

Rochard, E., P. Williot, G. Castelnaud and M. Lepage. 1991. Eléments de systématique et de biologie des populations sauvages d'esturgeons. In: *Acipenser. Actes du Premier colloque international sur l'esturgeon*, P. Williot (ed.), Bordeaux, 3–6 Octobre 1989. Cemagref Editions, Antony, pp. 475–507, 520 pp.

Rochard, E., M. Lepage and L. Meauze. 1997. Identification and characterisation of the marine distribution of the European sturgeon *Acipenser sturio. Aquatic Living Resources* **10**:101–109.

Ruban, G. I. and L. A. Konoplya. 1994. Diet of the Siberian sturgeon, *Acipenser baeri*, in the Indigirka and Kolyma rivers. *Journal of Ichthyology* **34**:154–156.

Schaffter, R. G. 1997. White sturgeon spawning migrations and location of spawning habitat in the Sacramento River, California. *California Fish and Game* **83**:1–20.

Smith, T. I. J. and J. P. Clugston. 1997. Status and management of Atlantic sturgeon, *Acipenser oxyrinchus*, in North America. *Environmental Biology of Fishes* **48**:335–346.

Sulak, K. J. and J. P. Clugston. 1999. Recent advances in life history of Gulf of Mexico sturgeon, *Acipenser oxyrinchus desotoi*, in the Suwannee river, Florida, USA: a synopsis. *Journal of Applied Ichthyology* **15**:116–128.

Tabardel, M. 1994. Le point sur la situation de l'esturgeon (*Acipenser sturio* L.) en Méditerranée occidentale et possibilités de réintroduction dans le Rhône. Mémoire de fin d'études ENSA, Rennes, Rennes/Arles, 57 pp.

Trouvery, M., P. Williot and G. Castelnaud. 1984. Biologie et écologie d'*Acipenser sturio*. Etude de la pêcherie. Série esturgeon n° 1 Etude n° 17, Cemagref de Bordeaux, Division. ALA/ Agedra, Bordeaux.

Wakeford, A. 2001. State of Florida conservation plan for Gulf sturgeon (*Acipenser oxyrinchus desotoi*). Technical Report 8, Florida Marine Research Institute.

Wayne, R. K., J. A. Leonard and A. Cooper. 1999. Full of sound and fury: the recent history of ancient DNA. *Annual Review of Ecology and Systematics* **30**:457–477.

Whitehead, P. J. P., M. L. Bauchot, J. C. Hureau, J. Nielsen and E. Tortonese. 1984. *Fishes of the North-eastern Atlantic and the Mediterranean*, UNESCO, Paris.

Williot, P. (ed.). 1991. *Acipenser. Actes du Premier colloque international sur l'esturgeon*, Bordeaux, 3–6 Octobre 1989. Cemagref Editions, Antony, 520 pp.

Yang, D. Y., B. Eng, J. S. Waye, J. C. Dudar and S. R. Saunders. 1998. Technical note: Improved DNA extraction from ancient bones using silica-based spin columns. *American Journal of Physical Anthropology* **105**:539–543.

Zholdasova, I. 1997. Sturgeons and the Aral Sea ecological catastrophe. In: *Sturgeon Biodiversity and Conservation*, V.J. Birstein, J.R. Waldman and W. E. Bemis (eds.), Kluwer Academic Publishing, Dordrecht, Netherlands, pp. 373–380.

Zolotarev, P., V. Shlyakhov and O. Akselev. 1996. The food supply and feeding of the Russian sturgeon *Acipenser gueldenstaedti* and the starred sturgeon *Acipenser stellatus* of the north-western part of the Black Sea under modern ecological conditions. *Journal of Ichthyology* **36**:317–322.

Chapter 26
Main Steps and Proposals for a Recovery Plan of Sturgeon in the Guadalquivir River (Spain)

Alberto Domezain

Abstract Sturgeons in the Guadalquivir area predate the appearance of humans, but there is historical evidence of use of sturgeons in this region at least since the period of the Phoenicians, until human activity exterminated these fish in the twentieth century.

During the middle of the twentieth century, the sturgeon was actively caught for caviar production. Caviar was produced from the two autochthonous species of the area, *Acipenser sturio* and *A. naccarii.*

The idea of recovering the Guadalquivir sturgeons is not new. Russian scientists who worked in the caviar factory made the request and Spanish naturalists of the period discussed the possibility in their publications. More recently, the controversy was reopened when it was demonstrated that a good part of the Guadalquivir sturgeons were the species *A. naccarii*, and, thanks to the efforts of different groups, aquaculture had provided sufficient specimens of this species with adequate quality to guarantee proper recovery.

The problem arose because the government did not act. It recognized the ecological and social value of the recovery, but it did not take action, stating that 'some scientists' did not recognize *A. naccarii* as autochthonous of the Guadalquivir. Today, nobody who acts in good faith can doubt that *A. naccarii* is autochthonous; this is no longer disputed. Now the question is how to proceed with the recovery.

Keywords sturgeons recovery plan, Guadalquivir River, Iberian Peninsula, *Acipenser naccarii*, taxonomic mistakes, sustainable development

26.1 Introduction

Sturgeons in the Guadalquivir area (S Spain) predate the appearance of humans, but there is historical evidence of the use of sturgeons in this region at least since the period of the Phoenicians, until human activity exterminated these fish in the twentieth

A. Domezain
Dpto. I + D Piscifactoría 'Sierra Nevada' S.L., 18313 Riofrío (Granada), Spain

century. The history of this extermination (the term 'extinction' would not be correct in this context as less than 50 years has passed since their presence has been reported) is the history of human development itself. While humans were less developed, their techniques for capturing fish were not efficient enough and the sturgeons survived, but with technological improvement, overfishing led to their extermination.

As in the other sturgeon zones, three basic factors caused this extermination: pollution, obstacles against free migration, and overfishing. However, in the case of the Guadalquivir, the incidence of these three causes was asymmetrical.

- Pollution hampered reproduction, but available data indicates that this cause was not excessively important, as during the time of maximum pollution, the sturgeon population remained viable.
- Obstacles to migration were concentrated in this case fundamentally in the Alcalá del Río (Seville) dam, an impasse for the sturgeons moving upstream through the Guadalquivir, but this was not decisive, either, as reproduction occurred repeatedly downstream of this dam.
- Overfishing was the direct cause of the extermination, as all the adults that returned from the sea were captured, without allowing them to reproduce and without providing the necessary repopulation.

During the middle of the twentieth century, the sturgeon was actively caught for caviar production. Caviar was produced from the two autochthonous species of the area, *Acipenser sturio* and *A. naccarii*. According to the data available, most of the specimens were of the second species (*A. naccarii*), both receiving the name 'Sollo del Guadalquivir' (Guadalquivir sturgeon). In fact, nobody cared whether it was one species or the other, but simply caught them and sold the caviar under the brand 'Ybarra'.

This fishery was the main support for professional fishermen of the zone, and it established a caviar factory in Coria del Río (Seville), which functioned with Russian technology (hiring Russian experts both to optimize the fishery as well as to prepare the caviar). The factory closed when the sturgeons disappeared.

The idea of recovering the Guadalquivir sturgeons is not new. Russian scientists who worked in the caviar factory of Ybarra made the request and Spanish naturalists of the period discussed the possibility in their publications. In fact, a small laboratory was even built for this purpose around 1950 (Laboratorio Ictiogénico de Alcalá del Río, Seville). More recently, the controversy was reopened when it was demonstrated that a good part of the Guadalquivir sturgeons were the species *A. naccarii*, and, thanks to the efforts of different groups, aquaculture had provided sufficient specimens of this species with adequate quality to guarantee proper recovery.

At the moment, Andalusia has the largest stock in the world of the species *A. naccarii*, with F1 and F2 obtained from wild parents. This stock has been submitted to strict genetic and sanitary control. Furthermore, there are other stocks in Italy, and thus recovery in Spain is possible.

In 1997, the Colegio Oficial de Biólogos de España (Official Spanish College of Biologists) presented the regional government of Andalusia with a project called

'Recovery of the Guadalquivir sturgeon' based on a previous work 'The sturgeon as a valid tool for sustainable development'. In this work, it was proposed, under the philosophy of sustainable development, to recover the autochthonous sturgeon *A. naccarii*, in the Guadalquivir river in addition to all the species that inhabited the river, and at the same time provide economic recovery of the zone.

Basically, this project continues being viable today, although it needs to be adapted and improved, but its main lines continue to be valid. In outline, the philosophy of the project is very simple, although perhaps not easy to apply, since the sturgeon, adequately managed by the fishermen themselves, must provide sufficient economic profits to stimulate them to take measures to protect the sturgeon in the river. Also, it has been proposed that the middle-term aim of the project should be not only socially and economically sustainable but also biologically, becoming independent of an outside supply of sturgeons. In short, that it should be true sustainable development.

To understand the serious problems that sturgeons currently face and the attempts at conservation in Spain, and why recovery has not begun, it is necessary to know something of their history, use, exploitation, and what is not known. Then the fundamental decision must be made:

Is it worth recovering the sturgeons of the Guadalquivir?

26.2 History of Sturgeons in the Guadalquivir

Sturgeons are part of the original fauna in the Guadalquivir river and its environs since its formation. There are certain records of sturgeon presence at least from the period of the Phoenicians, in which this fish formed part of the diet of the inhabitants of the zone. In excavations near Cadiz, abundant remains have been found dating from the Phoenician period. Afterwards, the Romans and other peoples of the area also used sturgeon, to the point that a Roman banquet was not considered of high standing without sturgeon meat (Apicius, 2007); and thus sturgeon was elevated to the stature of being stamped on coins

The history of sturgeons becomes nebulous for many centuries until around 1524 the Romantic travellers retrieved it from literary oblivion by recounting how there were great captures of *Sollo* (the vernacular name of sturgeon throughout the Iberian Peninsula) in the lower Guadalquivir (Seville, Coria del Río, Sanlúcar de Barrameda, etc.) (Fig. 26.1).

Afterwards, around 1700, the use of sturgeon meat was ceded exclusively to Cabildo de Sevilla, under the obligation that the first specimen captured had to be delivered to the court of the king.

From this period on, fewer public data are available on the consumption, until in 1932 the news was released concerning the beginning of new economic activity in the Guadalquivir, based on the sturgeon. However, for the first time, this was not based on its exquisite meat, but rather on its caviar; the first caviar factory was established in Spain, in the town of Coria del Río (Sevilla province). The facilities

Fig. 26.1 Print of the period related to a great capture of sturgeons in Cordoba

were small, under the name of Villa Pepita, situated on the right bank of the Guadalquivir.

Entrepreneurs on a national scale, the family Ybarra, took advantage of the great quantity of sturgeons being caught in the Guadalquivir, to produce caviar. The caviar gained renown in that period in Spain, but was also exported to different European countries, mainly France (Paris), where it successfully competed with prestigious Russian caviars.

As there was no tradition of using caviar in the zone, the owners contacted experts in sturgeon fishing and caviar preparation, such as T.H. Classen and T. Spiczakow, who lived many years on the Guadalquivir (especially the former), and who authored works and provided the most useful data available today on sturgeon fishing in the Iberian Peninsula (Classen, 1936, 1944, 1947; Spiczakow and Classen, 1942). These authors provided new techniques of more productive fishing, such as the bottom boulter or weighted rope which divides the river into parts and from which unbaited hooks were suspended to snag the sturgeons (Fig. 26.2), and new techniques were also introduced for preparing caviar (Fig. 26.3).

The sturgeons were caught at the mouth of the Guadalquivir (Sanlúcar de Barrameda), to Alcalá del Río, a town upstream of the city of Seville, the point beyond which the sturgeons could not pass in that period because of the inauguration of a dam for irrigation and hydroelectricity. Up to that date, sturgeons were known to travel upstream as far as the city of Córdoba, even entering the Genil River and other tributaries. Groups of professional fishermen were hired expressly for the factory Ybarra to catch sturgeon, while other fishermen caught sturgeon more by chance, these also living at the same factory. The fishermen usually transported sturgeon catch by rowboat or sailboat to the factory in Coria del Río. After

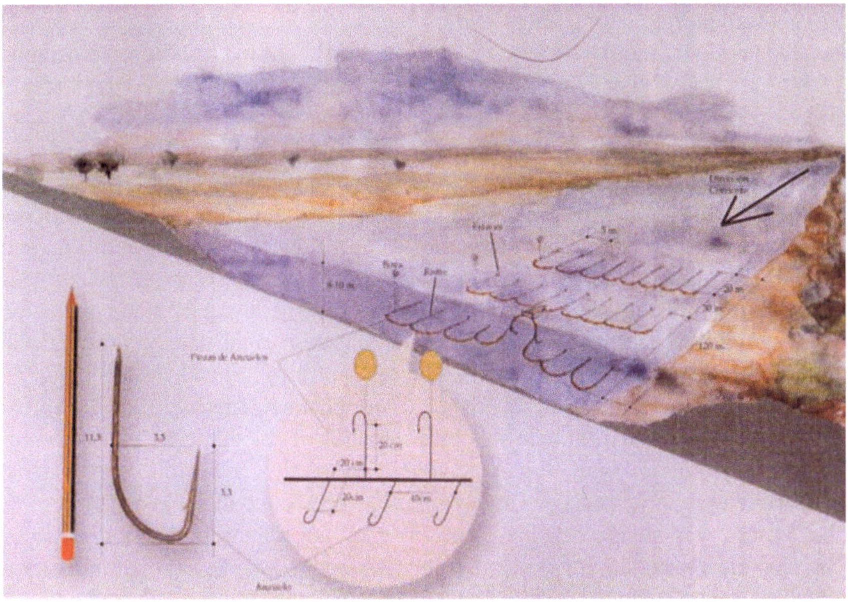

Fig. 26.2 Illustration taken from the magazine *Azotea* vol. (13–14) of the town hall of Coria del Rio (Sevilla province)

Fig. 26.3 Photos of old tins of caviar from the Guadalquivir, produced in Villa Pepita. (Photo provided by Antonio Sabater.)

a few years of functioning, the facilities were improved by the availability of a motorboat, the first available in the area for fishing (known as the 'gas boat' for the fuel it used).

This activity lasted until about 1965, when sturgeons were no longer captured in the Guadalquivir, and the factory Villa Pepita was about to close. It remained open a few years longer thanks to the activity of smoking other fish that were captured in the river.

The real reasons for the disappearance of the sturgeons from the Guadalquivir have often been confused. Elvira et al. (1991), among others, suggest that the main cause was pollution and the Alcalá del Río dam, which interfered with reproduction. These factors, despite having a degree of clear influence were not at all the main ones, as the official records show that the pollution in final years of the sturgeons was no greater than in years with good reproduction; as for the dam, the main effect was not to impede reproduction, given that downstream of the dam several spawning grounds remained (see Fig. 26.4, in which the spawning grounds are marked), and in fact there was reproduction recorded in these zones (Gutiérrez, 1962). The main cause behind the disappearance was extermination. Thus the term 'extermination' is more correct than 'extinction', since humans directly caused this disappearance, and furthermore the presence of the species has been cited in the last 50 years. The direct cause was intensive fishing, not offset by adequate repopulation measures. The dam clearly had an influence, but its main action was to facilitate the activity of the fishermen, acting as a large fish trap, in which the sturgeons entered and were held until captured by the fishermen. Evidently, the sturgeons could not climb the fish ladders, as the dimensions were too small (see Figs. 26.5 and 26.6) for the size of the sturgeons (44–46 kg average for female adults in the Guadalquivir) and remained below the dam for many days. The fishermen knew this well and took advantage of it to the point of catching all the fish.

As the catches were aimed expressly at females with eggs—that is, before reproduction—and as the effectiveness of the catches was very high, the females did not reproduce. This, added to the lack of any repopulation or the release of fry of any type, exterminated the species. Since 1974 to the present, only four autochthonous specimens have been captured in the Guadalquivir, and one Siberian sturgeon (*A. baerii*), this latter probably released from an aquarium (these fish often being sold in aquarium shops). Unfortunately, despite some attempts to keep the fish alive, the four specimens died. Three are preserved dry or in alcohol in the museum of the Biological Station of the Consejo Superior de Investigaciones Científicas (The Council of Scientific Research) in Seville while the fourth was eaten in a restaurant in Sanlúcar de Barrameda (at the mouth of the Guadalquivir, Cádiz).

In short, humans were the direct cause of the disappearance of sturgeons from the Guadalquivir river.

Nevertheless, it is only fair to observe that the specialists of the day detected the problem and proposed a solution (repopulation). Thus, Classen (1936, 1944, 1947), Vélez-Soto (1951), and Gutiérrez (1962), among others proposed repopulation, as done for many years in Russia, and fish farming was even suggested. In fact, what we consider the first experimental fish farm for the recovery of the sturgeon was installed, named the Ichthiogenic Laboratory (Laboratorio Ictiogénico) of Alcalá del Río. This laboratory was built specifically for holding and reproducing Sollo de Guadalquivir. Unfortunately, despite some captures and attempts at reproduction

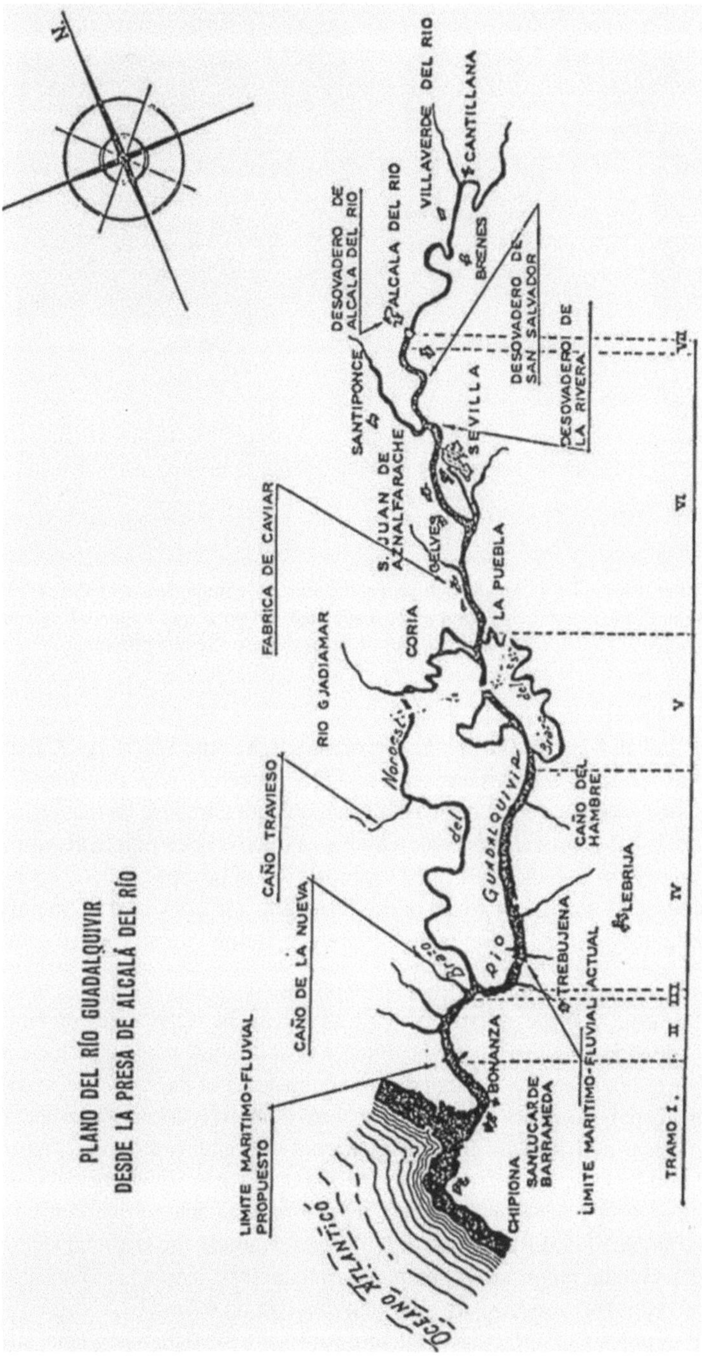

Fig. 26.4 Map of the Guadalquivir, taken from Gutierrez (1962), showing the spawning grounds of the sturgeons, among other locations

Fig. 26.5 View of the fish ladder at Alcalá dam, as can be seen by comparison with the people shown, the dimensions of the ladder being completely inadequate for sturgeons. Figure taken from the magazine *Azotea*, vol. (13–14) from the town hall of Coria del Rio (Sevilla province)

(three between 1950 and 1962, once by D. Francisco Vélez and twice by D. Juan de Aizpuru) the experiment was unsuccessful and the laboratory was abandoned.

In 1962, Gutiérrez published what can be considered the first treatise on recovering the Sollo, presenting a theoretical proposal, which was relatively complete, together with a description of how to handle artificial reproduction of the species, correcting what were considered to be the key errors of the foregoing attempts, blamed mainly on the lack of suitability of the Ichthiogenic Laboratory. Below, we reproduce some paragraphs of interest:

- From 'Conclusions': 'All the improvements implicit in the application of these reforms would provide a relative advantage. Finally, conservation would be achieved. But the number of drawbacks is so great that to eliminate all of them at once would be difficult and there is no solution to overcome them effectively other than artificial reproduction. This would offer not only the security of maintaining the species alive, but fry could also be taken to the river in the period and places where the environmental conditions, both natural and artificial, are the most suitable.'
- From the section 'Artificial Reproduction': 'The trials made for artificial reproduction in the Guadalquivir have not given satisfactory results.... Repeated attempts have been made and we are sure that this can be solved.'

  '...This involves putting all effort possible into capturing the fish, using nets, and never hooks, as the latter system injures the body during capture. Given that the fish mature during their journey upstream, the capture should be made when

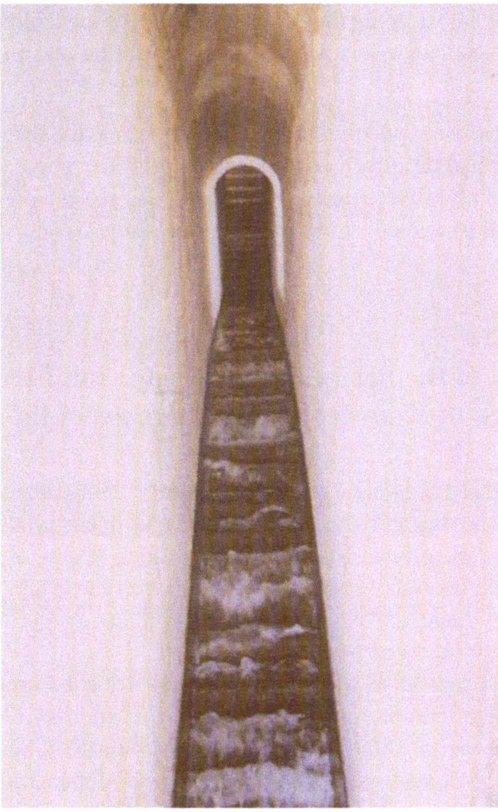

Fig. 26.6 Detail of the interior of the fish ladder. If a sturgeon succeeded in entering, it would get stuck on the steps. Figure taken from the magazine *Azotea*, vol (13–14) from the town hall Coria del Rio (Sevilla province)

the fish are almost mature, and the appropriate place in our opinion is the foot of the dam and nearby, with the advantage that these places have for their proximity to the Ichthiogenic Laboratory.'

'…Once captured, as many males and females as possible, and all together near the dam so that they are closer to maturing, should be moved to common ponds for both sexes so that from the time of their capture they are together. The ponds will be situated in such a way that the river water can enter directly, establishing a continuous current through the ponds, which will measure c. 9 m long, 3.5 m wide, and 1.5 m deep.'

'…All means possible should be taken so that the female does not lay her eggs in the pond, and therefore we propose as a solution, which we consider very good(given by Mr. Juan de Aizpuru), to seal the urogenital canal with burlap.…'

'…If any male matures before the females, the milt should be removed and kept in refrigerated thermoses, as viability has been confirmed to last almost two days under these conditions. The fecundation system to be used will be the dry method. Taking into account the current condition of the Ichthiogenic Laboratory of Alcalá del Río, we explain below the renovation and innovations that we consider necessary in case artificial reproduction is attempted.'

The treatise continues, specifying the material necessary, the type of incubators, etc. Previously, it had described (with what we consider good accuracy) the main causes of the decline of the species. For all of these reasons, we regard this to be the first serious effort to recover the Guadalquivir sturgeon and to start farming this species.

26.3 History of the Scientific Knowledge (or, More Accurately, the Lack of Knowledge) of Sturgeons of the Guadalquivir

It is well known that sturgeons were once present in the Guadalquivir river, so why has their recovery not begun if the technique and specimens needed were available?

To understand this, it is necessary to have a quick review of what happened in the scientific world related to sturgeons in the Iberian Peninsula, where the incompetence of some scientists led us not to repopulate or recover autochthonous sturgeons, though it has been possible since 1997.

Since 1818, the naturalists of the region repeatedly cited up to three species as being present in the Iberian Peninsula, including *A. naccarii* (Anonymous, 1818; Machado, 1857; Capello, 1869; Capello, 1880; Osorio, 1894; Seabra, 1911; Nobre, 1931; Nobre, 1935; Gonçalves, 1942; Helling, 1943; Albuquerque, 1956; and more recently Bauchot, 1987). These naturalists in general based their statements on direct analysis of specimens captured by fishing, mainly in Portugal. However, in 1988, a Portuguese scientist, who was not a specialist in sturgeons, published data according to which he deemed all the previously cited works to be erroneous, claiming that the only species present in the Iberian Peninsula was *A. sturio* (Almaça, 1988). This peculiar conclusion was drawn after reviewing museum specimens classified originally by the prestigious naturalists cited above as *Huso huso* and *A. naccarii*, which Almaça assigned to *A. sturio*. More incomprehensible, he changed the classification on specimens without even seeing them but rather studying the museum identification cards, since, according to him, the specimens had been lost in fires. In his opinion, these cards did not adequately demonstrate that the specimens were not *A. sturio*, and thus he classified them as *A. sturio*.

In Spain, these conclusions were immediately received by some authors who expanded them and ratified them (e.g. Elvira et al., 1991; Elvira and Almodovar, 1993).

What was the cause of this error? If we overlook the fact, absurd in itself, of disregarding the work of prestigious naturalists without even analysing the specimens that they had studied directly, we are faced with something very evident: these modern authors, although they were permitted to publish on the subject, were not specialists in sturgeons, as demonstrated by the fact that they were unable to distinguish *A. sturio* from *A. naccarii*. And what is even more striking is that they were incapable of distinguishing between the only two genera of the group, *Huso* vs. *Acipenser*, although the differences are more than apparent for anyone minimally familiar with these fish.

As a visual example, we provide images below (Fig. 26.7), beginning with a key which is easy to use even for non-experts and which is valid for differentiating between the three species that have been found in the Iberian Peninsula. This information is an extract and adaptation of those found in the most important treatises on this topic.

1 (2,3) Mouth very large, crescent shaped, and the brachiosteg membranes (of the opercula) joined together and free over the isthmus. Snout pointed, barbels laterally flattened and longer than the mouth:

....***Huso huso***

2 (3) Mouth transverse, relatively large and slightly curved at the ends; the maximum internal width of the mouth does not fit twice between the mouth and the tip of the snout; snout short, rounded and wide, with barbels closer to the tip of the snout than to the mouth. Generally more than 30 branchiospines:

....***Acipenser naccarii***

3 (2) Mouth transverse, small, straight; the maximum internal width of the mouth fits twice between the mouth and the tip of the snout; snout pointed, long, with barbels equidistant between the tip of the snout and the mouth, or closer to the mouth. Generally fewer than 30 branchiospines:

....***Acipenser sturio***

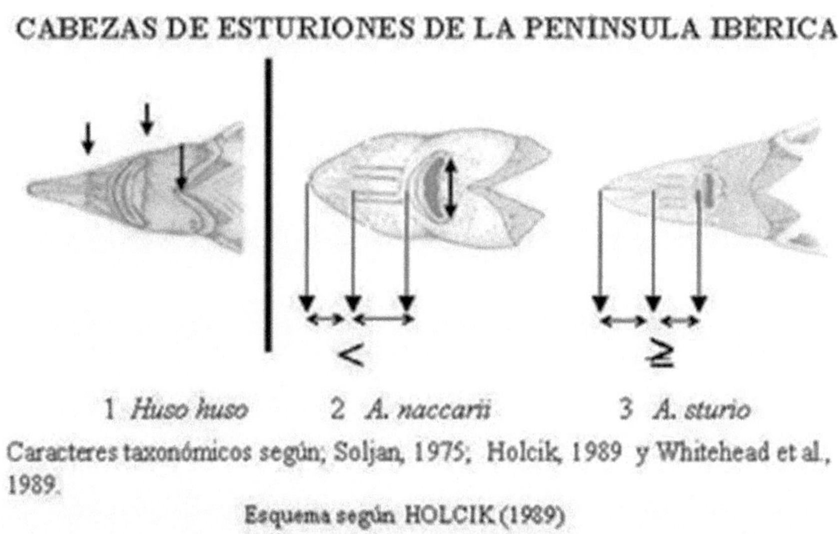

Fig. 26.7 Taxonomic key, specifically for differentiating the three species found in the Iberian Peninsula, extracted and adapted from those of the most important treatises on this topic, by Soljan, Svetovidov, and Sokolov, and collected, respectively, in Soljan (1975), Holcik (1989), and Whitehead et al. (1989); modified by Domezain et al. (2003)

The above are only some graphic examples that clearly illustrate that in the Iberian Peninsula not only is *A. sturio* autochthonous, but obviously *A. naccarii* also is, and there is evidence at least of the presence of *Huso huso* (Figs. 26.8, 26.9, 26.10, 26.11, 26.12, 26.13 and 26.14). The question again arises of how something so obvious has been denied by some scientists who were allowed to publish as though they were experts on sturgeons, when clearly they had little knowledge of the subject and, worse still, they had some interest in concealing the truth.

The graphic documents and taxonomic studies of the museum specimens, as cited above, are not the only proofs of its existence. In the scientific literature, we found data indicating the same. For example, the size of the sturgeon population, thoroughly studied and dated by Classen (1944), does not coincide at all with the rest of the known populations of *A. sturio*. In general, the populations of females in stable fish farms of *A. sturio* are between 80 and 120 kg, while in the Guadalquivir the mean size of the adult females that produced caviar was only 46 kg (in the well-known period; Fig. 26.15) and with very few specimens in the 80–120 kg range (Classen, 1944). This would indicate that the main component was of the 30–50 kg range.

Similarly, in the digitalization that we provide below, we can see that the example of the sturgeon population of the Guadalquivir provides the technical description of five specimens that the author considers morphologically representative of the population.

As can be seen in Fig. 26.16, the description by Classen is the appropriate one to apply the most usual taxonomic keys in order to differentiate *A. sturio* from *A. naccarii*, (those given by Svetovidov and Sokolov), which appear in the most important treatises on the subject, as for example by Holcik (1989) and Whitehead et al. (1989), respectively. On doing so, we find that the five specimens correspond to *A. naccarii* and not to *A. sturio*. Here are many other details, photos, data from the spawning period, etc., that coincide with the above commentary and evidence demonstrating that not only is *A. naccarii* autochthonous, but that it furthermore

Fig. 26.8 This specimen is preserved in a perfect state at the University of Oporto. According to some authors, it is *A. sturio*. However, all its taxonomic characteristics correspond to the genus *Huso*

Fig. 26.9 Detail of the branchiosteg membranes of the same specimen in Fig. 26.8. It can be seen how they join together, so that an object can pass under them without touching the isthmus. This is the main differentiating trait between the genera *Huso* and *Acipenser*. Clearly, the specimen is not *Acipenser sturio*, but of the genus *Huso*

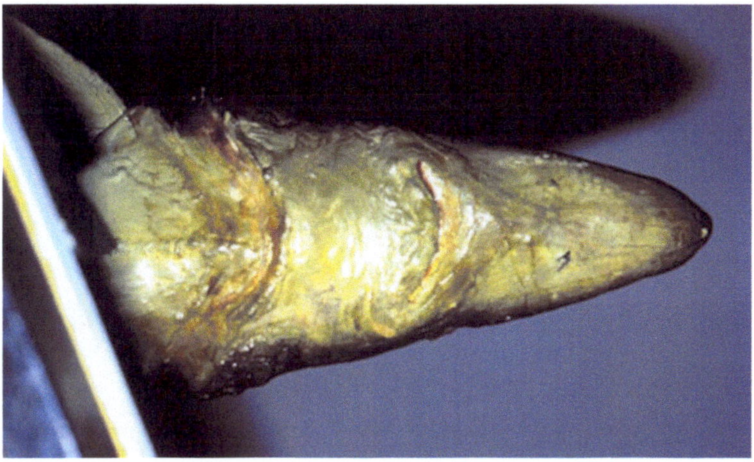

Fig. 26.10 This specimen, almost 2 m long and quite well preserved, is found in the Museum of the University of Oporto (Portugal). Incomprehensibly, it was reclassified as *A. sturio* despite having all the morphological characters of *H. huso*

was the most abundant component of the population of the area. For its aesthetic value, we provide one of these photos to which we attach scientific value in clarifying this situation (Fig. 26.17).

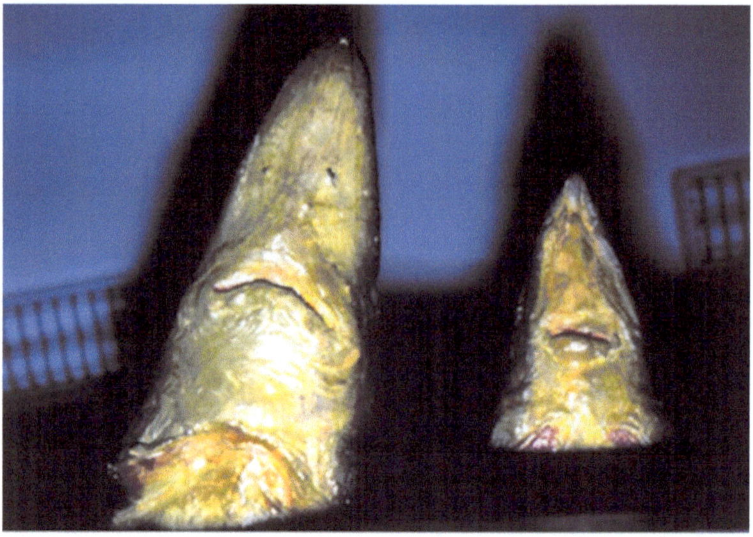

a

b

Fig. 26.11 **(a,b)** In these two photos, we can see the same specimen as in Fig. 26.10, *Huso huso*, which was incomprehensibly reclassified as *A. sturio*, together with an authentic *A. sturio*, preserved with similar techniques, in the same museum (University of Oporto, Portugal). As can be seen, the differences are more than noticeable. The difficulty is to confuse them, unless the person classifying them considers any fish with five rows of scutes to be *A. sturio*

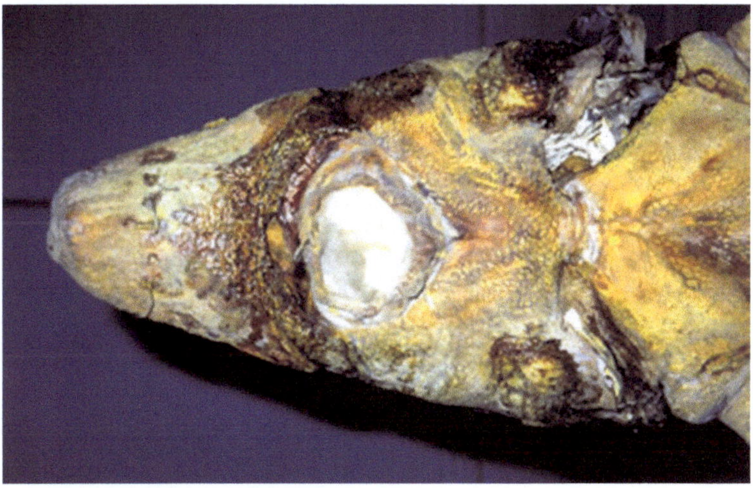

Fig. 26.12 From this specimen, also captured in an Iberian river and preserved in the University Museum of Coimbra (Portugal), is also classified as *A. sturio*. This time, at least the genus is correct (*Acipenser*), but the snout is short and wide and the barbels are closer to the tip of the snout than to the mouth, identifying it without a doubt as *A. naccarii*

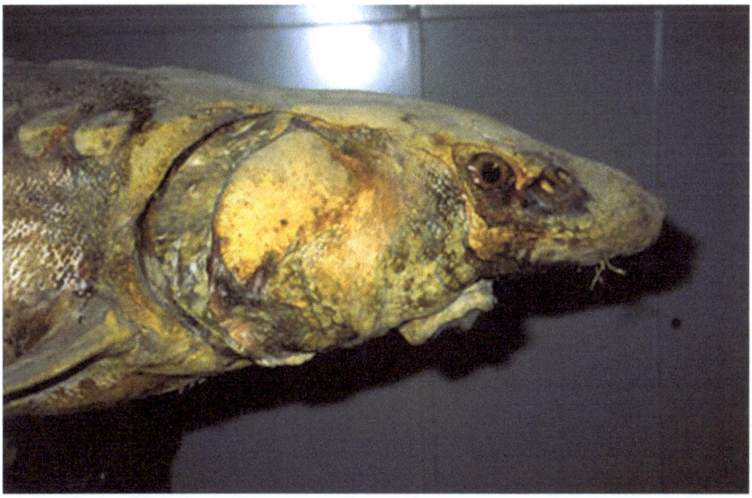

Fig. 26.13 Another specimen from an Iberian river and preserved at the University of Coimbra (Portugal) was labelled as *A. sturio*. However, the position of the barbels, closer to the tip of the snout, which is evidently short, indicates another error. This specimen is not *A. sturio*, but rather *A. naccarii*

All the above, which have been presented in different publications (e.g. Garrido-Ramos et al., 1997; Hernando et al., 1999a,b; Domezain, 2003; Domezain et al., 2003; Domezain et al., 2005) is, from a scientific standpoint, more than adequate

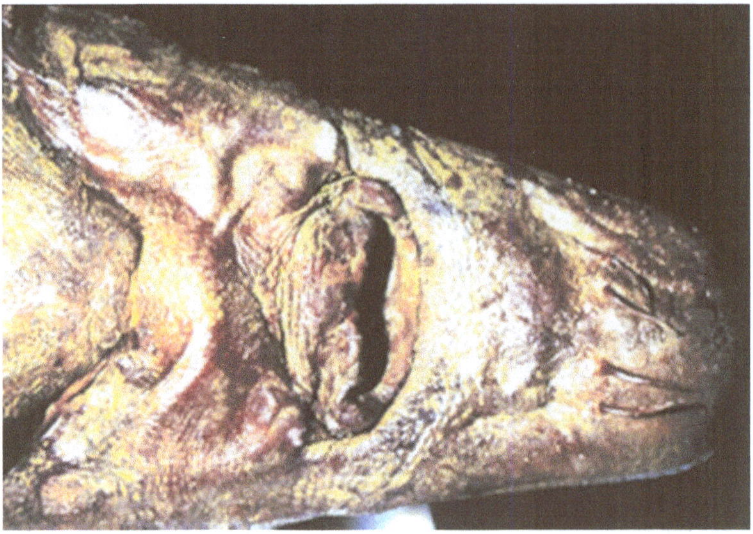

Fig. 26.14 A specimen captured in the Guadalquivir, preserved in the EBD (Estación Biológica de Doñana, Seville). The same authors identify it as *A. sturio*. Morphologically, it is evident that it is *A. naccarii*. The genetic analysis confirmed that it was not *A. sturio* (e.g. Garrido-Ramos et al., 1997; Hernando et al., 1999a,b; De la Herrán et al., 2004)

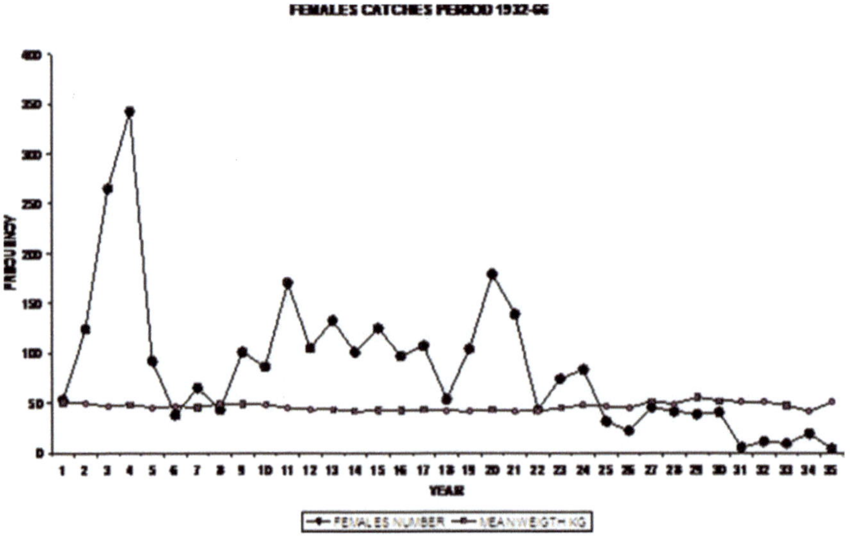

Fig. 26.15 Frequency of mean weights of reproductive females of Guadalquivir sturgeons captured between 1932 and 1966 based on data from Classen (1944)

ESTUDIO DEL ESTURIÓN DEL GUADALQUIVIR 7

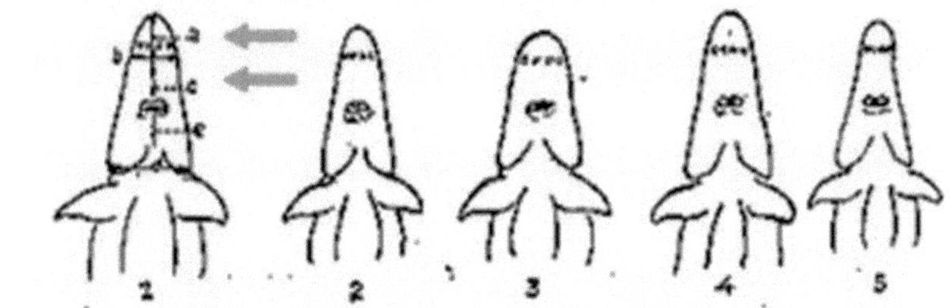

Figura 1.ª—Variación de las dimensiones cefálicas en cinco machos del mismo tamaño, pescados juntos en El Puntal (23 de febrero de 1940).

Como ilustración de la gran variabilidad de los caracteres biométricos sirve la figura 1.ª que representa esquemáticamente la vista inferior de las cabezas de cinco esturiones machos, cogidos en el mismo sitio y el mismo día, con el mismo arte y siendo todos casi del mismo tamaño. Las medidas son las siguientes:

a.—Distancia de la punta del hocico hasta los apéndices filiformes.
b.—Anchura del hocico a la altura de las barbillas.
c.—Distancia entre las barbillas y la boca.
d.—Anchura máxima de la cabeza entre las porciones distales de los opérculos.
e.—Distancia entre la boca y la "garganta" o frenulum, entre las aberturas branquiales.

TABLA I

		1	2	3	4	5
En centímetros	a	[illegible]	6	6	7	6
	b	4,6	3,5	9,3	8,5	7
	c	9,6	8	7,25	9,5	7,6
	d	17	14,5	15,5	10	13
	e	7	8	7,5	8,6	7,5

Fig. 26.16 As in the original table, it cannot be clearly read; we reproduce the distances 'a', tip of the snout-barbels, and 'c' barbels-mouth. In all cases, 'a' is less than 'c', and thus the five specimens, representative of the Guadalquivir population, are *A. naccarii*. Figure and data from Classen (1944)

to verify the autochthonous character of *A. naccarii*, but furthermore different genetic analyses were made with different techniques in several independent laboratories, and in all cases genetics confirmed the conclusions drawn from the morphological studies. That is, that *A. naccarii* is autochthonous of the Guadalquivir river, of the Iberian Peninsula and of the whole of southern Europe (e.g. Herrán, 1998; Robles, 2003; Herrán et al., 2004; Robles et al., 2005).

From the scientific perspective the subject is closed. No one has scientific grounds to doubt that *A. naccarii* is autochthonous of the Iberian Peninsula. Nevertheless, there are still people who wish to maintain the doubt. Why? The

Fig. 26.17 This photo, of a specimen captured in the Guadalquivir 40 years ago, lacks scientific value, as we do not have the exact measurements, but it appears quite evident that it is not *A. sturio*

answer to this remains difficult, because the only answers that we can find are related to economic matters, of interests created and not of scientific origin.

In short, we can say that science has supported the work of the old naturalists (such as Machado, Capello, Osorio, Seabra, Nobre, or Gonçalves) unjustly invalidated by Almaça, (1988), Elvira et al., (1991), and Elvira and Almodovar (1993), among others.

26.4 Is it Worthwhile to Recover the Sturgeons of the Guadalquivir?

Perhaps this question should have been placed at the beginning of the work, since a negative response would make the rest irrelevant. However, the response to us appears obvious, but perhaps for others it is not so clear. For all of us who have collaborated in the research that in one way or another has demonstrated that *A. naccarii* is autochthonous, the response is painfully evident: it is not only worthwhile, but it is our duty and obligation to undertake the recovery.

Most of the social groups involved in this matter agree. The European normative is very clear and poses not only the suitability but even the obligatory nature of recovering the sturgeons (or at least an attempt to do so). However, the false controversy explained above demonstrates that not everyone supports the recovery, and some sectors or government representatives show that they do not want this but rather the complete opposite, and they seek to block the project. The reason can be analysed by anyone. The clearest motive that we have been able to find is that the recovery of the sturgeon conflicts with powerful agricultural interests and fundamentally urbanization interests in the lower area of the Guadalquivir.

26.4.1 Real Possibilities of Recovering the Sturgeon Populations in the Iberian Peninsula and Specifically theGuadalquivir

Once it is known that two different species of autochthonous sturgeons coexisted in the region and that, although the opinion is not shared by all, there is evident interest in their recovery, the main issue becomes how to recover the sturgeons.

The ideal way would be to recover both species, *A. sturio* and *A. naccarii*, but in fact there is no strong motive to link the recovery of one to that of the other. Let us analyse the situation of the two species.

A. sturio is found in a truly difficult situation. Only one zone of resistance of the species is known in the natural environment, which is the area of the Garonne (France); nowhere else in the world, has the presence of this species been verified (Williot et al., 1997.). The ex situ of the species is also very complex, as it depends on two nuclei, the main one in the facilities of the CEMAGREF, near Bordeaux (France) and the second in the facilities of the IGB of Berlin (Germany). We are convinced that the good work performed by the specialists involved will guarantee that the species will not go extinct, but unfortunately, currently, it does not have enough specimens to initiate the recovery of this species outside the zones mentioned, and lamentably there are no guarantees that such stock will be forthcoming in the short term (Rochard, 2002).

In any case, anything that advances the recovery of *A. naccarii* in the free environment will be of great help when specimens of *A. sturio* become available for their recovery, as the history and biology of both species show that the two species can share their habitat without harming each other.

A. naccarii is in a substantially better situation, as in Italy, through the work of Giacinto Giovannini, who began 30 years ago to breed and raise the species with the aim of conservation of the species in one of its original zones, succeeding in its first reproduction in 1988 (Arlati et al., 1988). In addition, in general, the current situation of *ex situ* conservation is much better than for *A. sturio*, thanks to the work undertaken in the first phase in Italy (see Appendix A) and afterwards work that was continued and broadened in Spain (Domezain, 2006). In fact, today, there is more than a sufficient quantity of *A. naccarii*, with all the health and genetic guarantees to undertake recovery of the species at the very moment when politicians are deciding what to do. Today, there are only political-economic interests that limit the recovery of the sturgeon.

26.5 Recovery Plan

As we have seen, the conditions have now been satisfied for undertaking the recovery of the sturgeons, beginning with the species *A. naccarii*. This is an autochthonous species. We have enough appropriate specimens to do so. We have sufficient basic knowledge to forge ahead and the experience in other parts of the world can serve us as examples. 'Now we are able'.

With this situation, not only are we able, but it is our moral duty and the legal obligation of the different regional and national governments at least to try to recover this species. 'Now we must'.

Many species are recovered at great economic cost to the society and this is completely correct, because what is important is to recover them. If we are able, we must. But the sturgeon is different. It has the capacity to generate economic profits at the same time as it is being recovered, and this can help with the recovery of other species. That is, this could happen if a tool of real and sustainable recovery is applied. And as such, we must use it. The project that we propose (scheme in Fig. 26.18) is applicable to many waterways of the region with the appropriate adaptations, according to the peculiarities of each zone.

This is a possibile proposal, which as such seeks a possible result, not the theoretically ideal proposal that is sometimes unreachable. Nor does it attempt to be novel, nor is it an invention. We need only adapt, to our situation, the system that for more than a century has enabled Russia (in its different socio-political expressions) to maintain populations of sturgeons viable at the same time as economically profitable. We could copy and adapt what we know can function, and we could proceed to generate the greatest possible economic benefit. If we do so, in addition to having recovered the sturgeon, we will have revived the economy of the zone.

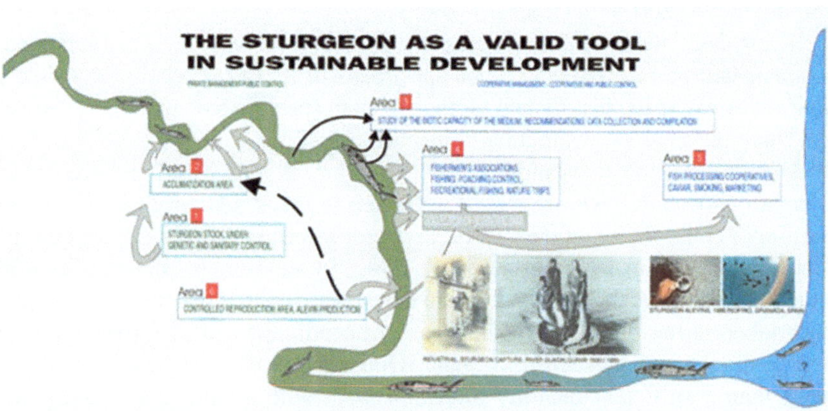

Fig. 26.18 General proposal for sturgeon recovery under the philosophy of sustainable development, adapted in Domezain et al. (1996) from Domezain et al. (1999)

On demonstrating to the social agents that by taking care of the river they earn more money, they will make an effort to protect the river, the sturgeon, and simultaneously all the other species that inhabit the river.

26.5.1 Recovery Plan for Sturgeons of the Guadalquivir, 2007

With the aforementioned premise, in 1997 the Official College of Biologists of Spain presented to the Andalusian regional government the project entitled 'Recovery of the Guadalquivir Sturgeon' (Domezain, 1997). This project, based on prior work entitled 'The sturgeon as a valid tool for sustainable development' (Domezain et al., 1996), sought to recover the autochthonous sturgeon *A. naccarii* in the Guadalquivir river, and with it all the species inhabiting the river, including *Homo sapiens* (economically) within a plan of sustainability (Domezain et al., 1999).

Today, 10 years later, the recovery has not yet begun, and we adapt this proposal to the current reality of the Guadalquivir and its environs to the possibilities, knowledge, and technology available.

The philosophy of the project is very simple, although perhaps for this reason it is difficult to apply. The aim is for those who have stripped the river, the professional fishermen, who today continue to be very numerous in the Guadalquivir and who, well organized, would be a great social force, to become the ones who protect, defend, and take care of the Guadalquivir, demanding that the authorities do the same. It is evident that if this notable social force decides to do so, the Guadalquivir would quit being a sewer for polluters.

To achieve this, the project proposes something as 'old' as making sustainable use of the economic wealth that sturgeon fishing, especially for the production of caviar, could generate. But the key is that this money should not go to any private business outside the fishermen, but directly to them, adequately organized into cooperatives and evaluating their product in situ.

26.5.2 Steps of the Recovery Plan

The steps of the project, summarized, would be the following:

Before beginning work, a team of trained technicians and scientists would be formed to act as the consultants for the government and its coordinators involved in the recovery. This team should be made up of specialized scientists, of those in charge of sturgeon stock, and of representatives of professional fishermen of the waterway (Guadalquivir in this case, but it could really be any other waterway). In addition to the technical capacity of those involved, the ability to maintain good relations within the group should be considered an indispensable condition for participation in the team, to ensure adequate cooperation.

Step 1

The project would begin with a stock of *A. naccarii* under genetic and health control. The specimens best adapted for this would be selected from a wide cohort, approximately one year of age, after olfactory and gustatory adaptation to Guadalquivir waters. The fish would be vaccinated against the most common diseases of the river, would be marked with microchips, and would be equipped with small radio-tracking and sonotelemetric equipment. Previously, and for greater security, the proper genetic marks would have been established. These operations would be performed in the original water before their transport to the Guadalquivir.

Ample cohort: As these would be specimens for aquiculture which have not undergone individual natural selection as strong as wild specimens (given that they have always had food available, certain protection against predators, etc.), it is considered advisable not to use the entire cohort, but rather a phenotypic and physiological selection.

One year of age: According to previous studies on adaptation to wild feeding (Soriguer et al., 2002; Soriguer et al., 2005), as well as resistance to predation and diseases (Domezain, 2003), the greatest facility to adapt to the physiological changes related to salinity (e.g. Martínez-Álvarez et al., 2005), as well as the decrease of residence time in the river and the greater possibility of return (Domezain, 2003; see Chapter 22), among other reasons, make it advisable to use specimens of more than a year old.

Olfactory and gustatory pre-adaptation to the Guadalquivir: To facilitate the fastest adaptation possible to wild feeding, it is advisable in a previous phase for the specimens in their habitual pond to consume feed (semi-moist compound feed) which includes, among the routine meals and fish oils, specimens of the fauna and flora that will serve as their food in the waterway where they will be released. This will not only adapt them to the smells and tastes of their new habitat but will also prepare them for non-dried food.

Vaccination: Sturgeons, like any other animal, if given the right conditions, develop immune protection against pathogens with which they will have contact over their lifetime. However, as they will be introduced into an environment which is new for these specimens in particular and which will contain pathogens different from those that existed in their previous habitat, it is highly recommended that the sturgeons be vaccinated against the main pathogens (and strains) that they will encounter in the new environment. For the strains present in the fish of the new habitat should be isolated and the appropriate autobacterins applied.

Step 2

For maximum assurance that the sturgeons used in the recovery will return to the Guadalquivir and not to another river after being taken out of the aquiculture facilities and before being released into the river, they should be kept in a temporary auxiliary facility prepared in the Guadalquivir itself. At this facility, the fish should be kept for the time that the scientists deem necessary for the double aim: that the fish develop the physico-chemical memories of the facility and that they develop the necessary ability to feed in the free environment. For this the fish should be fed with food from the river, captured/provided by the fishermen, and the fish should not be released until gastrointestinal analyses (using emetics, etc.) verify that the fish had fed correctly on the feed supplied.

Temporary auxiliary facility: This would be formed by ponds dug into the natural terrain of the Guadalquivir beside the water, which could be filled by pumping so as not have build a derivation dam. The ponds should be relatively large channels, no less than 100 m long and 10 m wide by 2 m deep. The fish kept in the ponds should not exceed 2 kg/m^2 (in this case, the most common expression, kg/m^3, is not very useful). Evidently, simple laboratories would be needed for the assays and scientific examinations, facilities for the guards, workers, generators and pumping equipment, etc. In short, it would be a simple aquiculture facility which would be used only at a certain time of year.

Physico-chemical memory: So that the sturgeons return to the Guadalquivir in the greatest possible number and decrease their dispersion in other nearby zones, the ideal situation would be for them to be born and to live in some tributary of the Guadalquivir. If this is not possible, this could be partially substituted by keeping the fish temporarily in ponds before their release into the Guadalquivir. This would be to avoid their leaving to the sea too rapidly, without acquiring the necessary physico-chemical memories that would enable their return.

Ability to feed in the free environment: Because the sturgeons to be used for the recovery project will come necessarily from aquiculture, in order to facilitate their acclimation and greater success, it would be advisable to take advantage of the residence time in the facility to feed them with the same natural feed that they would find in the natural environment, not releasing them until it is verified that they consume sufficient natural food—that is, that they know how to feed in the wild. To confirm this, gastrointestinal analyses should be performed (emetics, etc.) together with a calorimetric test.

Step 3

Although we again repeat here, before beginning the project, the scientific team should be organized and have a monitoring technician for the project. As stated above, it would be a necessary condition for the scientists and technicians to have a good relationship and common working capacity. A complete scientific programme would have been begun to study the biotic capacity of the environment and the variation in its physico-chemical characteristics, and correction measures would have to be implemented at each phase of the project. After the release of the sturgeons, at step 3, the monitoring programme would begin, with the capture of specimens, compiling of data, release, monitoring, recapture, etc.. This will provide much more information on the biology of this species and provide measurements as well as corrections that would guarantee the success of the recovery. Sonotelemetry and radiotracking would be used. This work would continue throughout the entire project and furthermore would enable the improvement of one of the steps, improving the general knowledge of sturgeons, possibly answering questions that remain. Where do sturgeons go when they are in the sea? What do they eat at the depths and types of bottom that they prefer? Do they remain gregarious? The tracking would continue both in the river as well as in the sea.

From this point on (step 3), it is fundamental that all the rest of the project not be managed but actually be carried out and controlled by the fishermen, with the proper advice from the administration and scientific-technical team. This means that each fisherman in the Guadalquivir must understand clearly that these juveniles that might be caught during migration towards the sea in reality belong to the fishermen and not to the administration nor to any outside company. Furthermore, if the fish are killed, the fisherman may earn three euros, whereas if the fish are allowed to go to the sea, when they return they will be worth some 9000 euros, which will belong to the fishermen and not to the administration or any outside company. Therefore, it is necessary that before releasing any of the specimens the cooperatives must be adequately formed and that all the fishermen on the riverbanks must participate and collaborate in the preparations and use of the fish. An adequate campaign of awareness is evidently necessary, but awareness is easier when there is a good economic prospect in the offing.

Step 4

When the mature sturgeons return to the river, they should be caught by professionals, the rightful fishermen, and be transported to ponds or manipulated (step 5). At that point, the technicians and scientists, after taking all the opportune data from the specimens captured, will decide according to the state of maturity whether the female is more appropriate for caviar (step 5) or reproduction (step 6).

In this fourth step, special emphasis should be placed on the capacity of generating sustainable economic resources using sturgeons. As explained above, the professional fishermen should be adequately organized in their professional associations in order to avoid poaching and individual abuses, which could ruin the project.

Not only the direct products of the sturgeon should be used, as explained in steps 5 and 6, but also the entire industry related to their use should be developed, including tourist, ecological, and sports aspects.

Professional fishermen: Their preparation should include the way of capturing the fish with minimum stress to the specimen, using the adequate nets (rather than hooks), sensors of net capture (to minimize the time that the sturgeon suffers in the net between the netting and the collection), ecoprobes, and other technologies that enable capture with minimum suffering for the sturgeon, as well as sedation and transport techniques necessary for steps 5 and 6.

Tourist use: Sturgeons attract great attention from the public, as shown by the many times that they appear in the press and other media, being the subject of entire documentaries, even in countries such as Spain, where no wild specimens remain. This attraction should be encouraged and focused towards the development of rural tourism with a naturalist bent, with trips around the fishery area, the spawning areas. Lodging with guided boat trips could be included, as well as swimming and diving with the sturgeons, zootherapy, etc.

Ecological use: It is necessary to make use of the approval given to the recovery of the sturgeon so that it acts as an umbrella species, in such a way that the effort dedicated to protecting the sturgeon becomes a protective service for all the species, animals and plants and the environment itself, which by themselves have little economic or social attraction but which under the aegis of the sturgeon could be recovered and their situation improved. This goes especially for species that have a life cycle directly dependent on the sturgeon, as for example the *Margaritífera* sp., as well as those that are not dependent, such as the woods on the riverbanks.

Sports use: It would be very useful to develop diving, underwater photography, etc., always guided, controlled, and appropriately undertaken from the fishing associations and the other beneficiaries of the project. Also, much interest could be developed for sports fishing, where the professional fishermen become guides who would use their boats to transport amateur fishermen to the fishing areas to be shown the practice of sturgeon fishing. The fish would be captured, photographed, and then released again. Also, data could be taken and prizes awarded for the best catches.

Step 5

The specimens caught professionally (step 4), should be mainly females, since the males could be better (when the fishing environment permits) used to attract the tourist, ecological, and sports activities mentioned in step 4. After the capture, the fish should be immediately transported live in mobile tanks to the selection zone, where the technicians and scientists of the project would take all the pertinent scientific data, including the state of maturation of the ovules, and would decide whether to dedicate the specimen to reproduction or (moving on to step 6) the preparation of caviar.

Live transport: The fishing teams should plan to use mobile tanks and ponds for immediate sedation of the fish caught (using natural substances that do not affect subsequent feeding of the fish, e.g. essence of cloves) and for moving the fish to better conditions (current speed, water oxygenation and temperature, etc.) to the selection zone.

Selection zone: Areas equipped with adequate ponds, ecoprobes, operating facilities, etc. for taking data and making the necessary decisions should be connected as directly as possible with the preparation or reproduction zones (step 6).

Preparation and value of the caviar: In the zones with the appropriate hygiene and health conditions, and equipped with the specialized techniques and personnel, the caviar can be prepared and evaluated under the best conditions and in the final form—that is, these facilities would produce caviar in the same containers and with the same labels and brands as the consumer would finally receive, so that the added value does not fall into the hands of intermediaries and packaging companies, as often happens in this sector to the detriment of its in situ value. Logically, those who make the most effort should receive the most profit, not the intermediaries.

Derivatives of sturgeon: The hot and cold smoking of sturgeon meat, marinated meat, pâtés, and the rest should be developed. That is, every effort should be made to earn the most from the sturgeon by taking advantage of its added value in the capture zone, rather than to sell the unprepared meat, which would leave profits to others. In addition to the many possible uses of the meat, it would also be helpful to evaluate other products, some of which bring high prices, such as sturgeon oil, cartilage, swim bladder (for restoration of old paintings and high cuisine), the scutes (for supposed aphrodisiacs), the skin (for leather goods), etc.

Direct marketing by the producers: To keep most of the added value in the Guadalquivir area, it is necessary for the producers (those who catch and prepare the sturgeons) also to be marketers. They must take the product as close as possible to the consumer, and this is feasible, since this project needs strong promotion and publicity, which can also be taken advantage of in this sense. In fact, it would be highly positive for tourists who visit the fisheries to become customers of these products.

Step 6

This step is divided into two phases. During the first few years after the return of the sturgeons, the main objective would be to raise the awareness of the fishermen and the riverside population and demonstrate the economic benefits for them. Therefore, in this first phase, only with the females too mature to produce good caviar, the specialized technicians would perform assisted reproduction with aquicultural techniques. After the assisted fecundation, the eggs would be transported to an appropriate facility to develop and be reintroduced into the Gualdalquivir, gradually opening the way to independence from the external supply of the fish.

In the second phase, when the cooperation of the fishermen is complete, because they see that they can really live from this industry, they will be asked to allow a certain percentage of the fish to migrate upstream for natural reproduction. Evidently, following the advice of the scientific team, suitable areas for sturgeon reproduction would be prepared upstream, and in the event that Alcalá dam could not be eliminated (elimination would be the ideal solution), elevator traps would be installed, or equivalent systems to enable the sturgeons to surmount the obstacle.

Good caviar: The technicians who prepare caviar, are well aware of the great differences of quality that can arise even in the same caviar, depending on its degree of maturation. In this project, it is especially important to be very strict on quality, and sell as caviar only what really deserves to be sold as such. This policy always gives good results, and, furthermore, when it is not a normal company but rather a cooperative that is in charge, and one formed and backed by the government, this virtue becomes an ineluctable obligation. Furthermore, this policy, necessary in itself, enables the ovules which are too mature to serve as good caviar, but viable, to be used for assisted reproduction, minimizing the objection of the fishermen to lose caviar production.

Aquicultural techniques: The use of aquicultural techniques for the reproduction and development of fry and juvenile fish would ensure the success of the project.

Independence: Although initially the project would depend totally on the aquiculture and exterior supply of sturgeons, as there is no other origin or any other real possibilities, the aim is that in the middle term the project could become completely independent of these sources and achieve real sustainability. After a few years, the Guadalquivir population should be able to become independent of any non-wild support, and only external supplies of milt would be necessary to ensure maximum genetic diversity possible, to avoid endogamy (a latent danger in all recovery programmes of animals, which by definition, when the availability of different genetic input of the species in question in nature is unavailable or very poor).

Alcalá Dam: This is the first physical obstacle that the sturgeons encounter on their return to freshwater to mate, and the fish alone cannot surmount the obstacle as designed. Only a few meters high, it is located some 10 km upstream of the city of Seville and some 100 km from the Atlantic Ocean. At the foot of the dam, the river contains freshwater, practically from the city of Seville upstream. Historically, sturgeon breeding grounds have been noticed between the city of Seville and the dam of Alcalá (see Fig. 26.4). From the standpoint of recovery of the sturgeon and of strict ecology, the only good solution would be to demolish the dam. Our desire is clearly the same, the total elimination of the dam. However, if the socio-political situation and the forces of equilibrium between conservation and agricultural and electric development make this demolition impossible, there are simple economic techniques that make the recovery of the sturgeon viable even while maintaining the dam. That is, elevator traps, or similar systems could be installed to move the sturgeons up over the dam.

Elevator traps: There are many types of fish elevators available on the market that could help the sturgeons overcome the small impasse of Alcalá dam. The fish alone cannot get over the dam, and the fish ladder is too small and badly designed for sturgeons. If the dam cannot be demolished, the solution could be a ramp elevator for the fish, adequately designed for sturgeon. If the authority responsible does not admit this solution because it allows too much water to escape continuously from the dam, an automatic elevator trap (or any similar device that exists and functions in other places) could be installed. If necessary, given the size of the fish in question, the individual specimens could even be captured and moved by hand to the higher part of the dam. This latter solution is clearly the least desirable, but it does illustrate the many possibilities at different costs. It only requires the desire to arrive at a solution.

26.6 Conclusion

In short, the aim of this project is that all the potential added value of this species should remain on the riverbanks of the Guadalquivir. On achieving this, the fishermen will become great defenders of the river, and there will be no need for the authorities to police the project but rather the fishermen will be the most vigilant agents because their labour and economy will be at stake. On protecting the sturgeon in the river, even if the fishermen did not want to (and we are certain that they will), they would be protecting not only all the other species that inhabit those waters, but also the riverbanks, etc. This is the way to protect the future. In summary, we would have an applied case of sustainable development.

Humans have exterminated the sturgeons of the Guadalquivir. Earlier, we did not know how to proceed and could not recover them. Today, we know how and we can do so. We depend only on the politicians to decide whether they are more interested in recovering the sturgeon and with it the Guadalquivir and its people, or continue using the river as a sewer to dump all the waste of our society into the sea.

We can recover the sturgeon! We must!

Acknowledgements This work was made possible thanks to the collaboration of several Andalusian and foreign research groups, the involvement of the Colegio Oficial de Biólogos de España and to the economic collaboration from many research projects from the Agencia de investigación y Desarrollo de Andalucía (Junta de Andalucía), the IFAPA (Junta de Andalucía), the Plan Andaluz de Investigación (Junta de Andalucía), the CDTI (Spanish government), the PROFIT (Spanish government), the Plan Nacional de Investigación (Spanish government), Fondos Europeos, and the company Piscifactoría de Sierra Nevada s.l.

References

Albuquerque, R.M. 1956. Peixes do Portugal e ilhas adjacentes. Chaves para a sua determinaçao. *Portg. Acta Biol.* (B) 5–1164 (p. 195).

Almaça C. 1988. On the sturgeons, *Acipenser sturio*, in the Portuguese rivers and sea. *Folia Zool.* 37: 183–191.

Anonymous. 1818. Observaçoes sobre algunas peixes do mar e rios do Algarve. *Mem. Acad. R. Sc. Lisboa* 5 (2): 1–48.

Apicius, C. 2007. Un pescado principesco. Diario de Navarra 20.03.2007. p. 73.

Arlati, G., Bronzi, P., Colombo, L. and Giovannini, G. 1988. Induzione della riproduzione nello storione italiani (*Acipenser naccarii*) allevato in cattivitá. *Riv. Ital. Acquacol.* 23: 94–96.

Bauchot, M.-L. 1987. Poissons osseux. In: W. Fischer, M.L. Bauchot and M. Schneider (eds.) *Fiches FAO d'identification des espèces pour les besoins de la pêche: Mediterranée et mer Noire, zone de pêche 37*, Revision 1, vol. 2. FAO Publ., Rome, pp. 891–1422.

Burtsev, I.A. (2008). Towards the definition of optimal size-weight standards of hatcheryreared sturgeon fry for restoration. In Carmona et al. (eds). Biology, Conservation and Sustainable Development of Sturgeons. Springer p: 359–368

Capello, F.B. 1869. Catalogo do peixes do Portugal que existem no Museo de Lisboa. *Jorn. So Math. Phys. Nat. 1ª serie*, 2: 131–193.

Capello, F.B. 1880. Catalogo do peixes do Portugal Mem. *Acad. R. Sc. Lisboa*, 6: 1–78.

Classen, T.E.A. 1936. Notas preliminares sobre la biología y el aprovechamiento del esturión en el Guadalquivir. *Pub. seccion de pesca*, Ministerio de Agricultura, Industria y Comercio. Serie 1 (2): 15–41.

Classen, T.E.A. 1944. Estudio bioestadístico del esturión o sollo del Guadalquivir (*Acipenser sturio*). *Trab. Inst. Esp. Oceanog.* 19: 1–112.

Classen, T.E.A. 1947. Notas sobre el Sollo o esturión del Guadalquivir. *Revista de Montes* 256–262.

Domezain, A. 1997. Recuperación del esturión del Guadalquivir. Proyecto presentado a la Consejería de Medio Ambiente. Colegio Oficial de Biólogos 49 + 13 pp.

Domezain, A. 2003. La acuicultura como herramienta para la recuperación de especies: el esturión autóctono de la Península Ibérica *Acipenser naccarii* Bonaparte 1836. Tesis Doctoral, Universidad de Granada, Granada, Spain.

Domezain, A. 2006. La acuicultura en las aguas continentales. En Tinaut, A. y Pascual, F. (Coord.). Enciclopedia: Proyecto Andalucía. Naturaleza. Vol XIX. Publicaciones Comunitarias, SL., Sevilla, pp. 336–350.

Domezain, A., Soriguer, M.C., Herranz, R. de la, Garrido-Ramos, M.A., Ruiz-Rejon, M. and Hernando, J.A. 1996. El esturión como una herramienta válida para el desarrollo sostenible. I Congreso Ibérico de Biólogos Ambientalistas, Lisboa.

Domezain, A., Soriguer, M.C., Ruíz Rejón, M. and Hernando, J.A. 1999. The sturgeon as a valid tool for sustainable development: application in the river Guadalquivir (Andalusia, Spain). *J. Appl. Ichthol.* 15 (4–5): 299–301.

Domezain, A., Soriguer, M.C., Domezain, J. and Hernando-Casal, J.A. 2003. Esturiones ibéricos: actualización del conocimiento y plan de recuperación. In Memoriam al Prof. Dr. Isidoro Ruiz Martínez, Edit, Universidad de Jaén, pp. 329–349.

Domezain, A., Soriguer, M.C., Domezain, J. and Hernando, J.A. 2005. *Acipenser naccarii* Bonaparte 1836: Historia de un desconocimiento. Actas del IX Congreso Nacional de Acuicultura: Conserjería de Agricultura y Pesca, Junta de Andalucía, Sevilla, pp. 37–43.

Elvira, B. and Almodovar, A. 1993. Notice about the survival of sturgeon (*Acipenser sturio* L., 1.758) in the Guadalquivir estuary (SW Spain). *Arch. Hydrobiol.* 129: 253–255.

Elvira, B., Almodovar, A. and Lobón-Cerviá, J. 1991. Sturgeon (*Acipenser sturio* L., 1758) in Spain. The population of the river Guadalquivir: a case history and claim for a restoration programme. In : P. Willot (ed.), Actes du premier colloque international sur l'sturgeon, CEMAGREF, Bordeaux, France, pp. 337–347.

Garrido-Ramos, M.A., Soriguer, M.C., Herranz, R. de la, Jamilena, M., Ruiz Rejón, C., Domezain, A., Hernando, J.A. and Ruiz Rejón, M. 1997. Morphometric and genetic analysis as proof for the existence of two sturgeon species in the Guadalquivir river. *Mar. Biol.* 129: 33–39.

Gonçalves, B.C. 1942. Colecçao ocenográfica de D. Carlos I.-Peixes Trav. *Stat. Biol. Mar. Lisbonne* 46: 1–108.

Gutierrez, F. 1962. El esturión del Río Guadalquivir. Servicio Nacional de Pesca Fluvial y Caza. Folleto informativo 5, Madrid, 73 pp.

Helling, H. 1943. Novo catalogo dos Peixes do Portugal em colecçao no Museu de Zoologia da Universidade de Coimbra. *Mem. Est. Mus. Zool. Univ. Coimbra.* 149: 1–110.

Hernando, J.A., Vasil'eva, E.D., Arlati, G., Vasil'ev, V.P., Santiago, J.A., Belysceva-Poliakova, L., Domezain, A. and Soriguer, M.C. 1999a. New evidence for a wider historical area of two European sturgeons: *Acipenser naccarii* and *Huso huso* (Acipenseridae). *J. Ichthyol.* 39 (9): 803–806.

Hernando, J.A., Vasil'eva, E.D., Arlati, G., Vasil'ev, V.P., Santiago, J.A., Belysceva-Poliakova, L., Domezain, A. and Soriguer, M.C. 1999b. A new proof for the historical presence of two European sturgeons in the Iberian Peninsula: *Huso huso* (Linnaeus, 1758) and *Acipenser naccarii* Bonaparte, 1836. *J. Appl. Ichthol.* 15 (4–5): 280–281.

Herrán, R. de la. 1998. Utilidad del ADN satélite y ADN ribosómico en la filogenia, taxonomía y diagnósis de enfermedades de peces y almejas, Doctoral thesis, University of Granada, Granada.

Herrán, R. de la, Robles, F., Martínez-Espín, E., Lorente, J.A., Ruiz-Rejón, C., Garrido-Ramos, M. and Ruiz-Rejón, M. 2004. Genetic identification of western Mediterranean sturgeons and its implication for conservation. *Conserv. Genet.* 5: 545–551.

Holcik, J. (ed.). 1989. *The freshwater fishes of Europe*. Aula-Verlag, Wiesbadem, 1/II, 469 pp.
Machado, A. 1857. Catálogo de los peces que habitan o frecuentan las costas de Cádiz y Huelva con inclusión de los del río Guadalquivir. Sevilla. *Azotea*, 8 (1991): 59–87.
Martínez-Álvarez, R.M., Sanz, A., García Gallego, M., Domezain, A., Domezain, J., Carmona, R., Ostos-Garrido, M.V. and Morales, A.E. 2005. Adaptative branchial mechanisms in the sturgeon *Acipenser naccarii* during acclimation to saltwater. *Comp. Physyol. Biochem.* 141(A): 183–190.
Nobre, A. 1931. Peixes das aguas doces de Portugal. *Bol. Min. Agric.* 13 (2): 73–112.
Nobre, A. 1935. Fauna Marinha de Portugal. Vertebrados, Porto., 579 pp.
Osorio, B. 1894. D'algunas especies a juntar ao Catalogo dos peixes do Portugal de Capello. *Jorn Sc Math phys Nat.* 3 (11): 186–188.
Robles, F. 2003. Análisis de la evolución concertada, filogenia e identificación de especimenes en el Orden Acipenseriformes. Tesis Doctoral, Universidad de Granada, Granada.
Robles, F., De la Herrán, R., Garrido-Ramos, M.A., Ruiz-Rejón, C., Martínez-Espín, E., Lorente, J.A. and Ruiz-Rejón, M. 2005. Identificación de los esturiones del Guadalquivir utilizando cinco marcadores moleculares y técnicas forenses. Actas del IX Congreso Nacional de Acuicultura. Consejería de Agricultura y Pesca, Junta de Andalucía, Sevilla, pp. 69–73.
Rochard, E. (coord.) 2002. Restauration de l'esturgeon européen *Acipenser sturio* Rapport scientifique Contrat Life n° B – 3200/460, Etude Cemagref n° 80, Groupement de Bordeaux, 224 pp.
Seabra, A.F. 1911. Catalogue systematique des vertebros du Portugal. V. Poissons. *Bull. Soc. Port. Sc. Nac.* 5: 129–224.
Soljan, T. 1975. Il pesci dell'Adriatico. Mondadori, Verona, 522 pp.
Soriguer, M.C., Domezain, A., Aragones, J. and Hernando, J.A. 2002. Feeding preference in juveniles of *Acipenser naccarii* Bonaparte 1836. *J. Appl. Ichtyol.* 18: 691–694.
Soriguer, M.C., Domezain, J., Domezain, A. and Hernando, J.A. 2005. La dieta de *Acipenser naccarii* Bonaparte 1836 en un mesocosmos cerrado. Actas del IX Congreso Nacional de Acuicultura. Consejeria de Agricultura y Pesca, Junta de Andalucía, Sevilla, pp. 79–81.
Spiczakow, T. and Classen, T. 1942. Notas sobre la cria de los alevines de esturion procedentes de huevos artificialmente fecundados. *Bol. RSEHN* XL: 455–466
Vélez-Soto, F. 1951. Observaciones sobre la pesca del esturión en el rio Guadalquivir durante el año 1.950. *Montes* 37: 33–40.
Whitehead, P.J.P., Bauchot, M.L., Hureau, J.C., Nielsen, J. and Tortonese, E. 1989. *Fishes of the North-eastern Atlantic and Mediterranean*, Vol. 1 (2nd edn.). UNESCO, París, 510 pp.
Williot, P., Rochard, E., Castelnaud, G., Ronault, Th., Brun, R., Lapage, M. and Ellie, P. 1997. Biological characteristics of European Atlantic sturgeon, *Acipenser sturio*, as the basis for restoration program in France. *Environ. Biol. Fishes* 48: 359–370.

Appendix A
Good Morning Mr. Sturgeon

Giacinto Giovannini

Good Morning Mr. Sturgeon, have a very happy and joyful day. I am Giacinto Giovannini, your beloved friend.

I like thinking about our origin as the inhabitants of planet earth. I let myself fall into the fantasy of vast time and space; of energy, mass, and atoms that always evolve in complex shapes to explain what was created in this world. Ever since the world has been in existence, life has existed in its various and complex shapes that adapted continuously to the changing conditions of the ecosystem.

Now my mind follows the trail to find an answer to reach humbly the beginnings and perceptions of the infinite.

Sometimes, I feel like a child; the rain arrives, the sun rises and I stare in amazement at the rainbow; I run to try to catch it with my tiny hands. Dreaming about the impossible is like chasing the rainbow and we feel lonely, shy and hopeless.

When I was young, imitating crepuscular poetry, I wrote my first composition.

HUMAN LIFE
Brief moment
Which gets lost
In the immensity of the universe

Later, I wrote

FROM RED TO BLACK
In the absolute darkness
A light beam
Covers the earth
And I stare silent
The life that lives
Every space, every thing
Be it black or be it pink
Has its own truth

'Azienda Agrícola V.I.P. di Giovannini Giacinto', Via Convento Aguzzano, Orzinuovi, Italy

In the bright splendour
I feel someone
That moves in the eternity
I see, enraptured
Millennia of love
And I sing a note
To the true pure reality

Many years ago emerged the roots of my personality. In the beginning, there was the love of my mother and my father, of my whole family (I am the youngest of 12 kids). I was also influenced by the education I received at school and the various realities of life.

Step by step, I learnt to love and respect every part of the universe. I learned how to do good to others and myself with a fierce willingness to act. In my case, everything was directed towards the investigation and acquisition of knowledge.

You might wonder what all this has to do with my friend the sturgeon. Well, life is a cultivated land; it transforms itself and grows if the environmental conditions are the best.

I started thinking: How can a sturgeon live without water? Definitely water is a very important element and essential to the sturgeon. We need to work with love, respect, and peace, and with human and ecological values. Only in that way can we care for and protect our most valuable and irreplaceable treasure: Life.

My father was a very good businessman and the director of an industrial site. During the Second World War, in search of greater prosperity and safety, and to facilitate the growth of a very numerous family, he took advantage of agriculture. He worked the land of a farm in the north of Italy with great ability. He produced wheat, flowers, fruits, vegetables, milk, and honey, for his beloved family. Then in 1951, he had the idea of trying aquaculture.

That decision greatly influenced my future. Three of my brothers and I started the fish farm and that took a great part of our efforts.

Breeding is a very passionate and addictive profession. There were problems of finance, personalities, and environment.

Every fish breeder has his own achievements and those are what he has managed to reach with a lot of sacrifice, which is how he becomes a master in this field.

I was little more than 30 years old when I started to dream of one day being able to breed the endemic Italian sturgeon. I pioneered it in 1977, with 100 sturgeons, captured occasionally in diverse rivers of the Po basin by experienced fishermen.

The first years of breeding were very hard, the fish were not adapting themselves to the conditions of captivity, and feeding them was a real problem, despite modern techniques.

Many sturgeons died: maybe they had not been able to adapt themselves or were stolen and eaten by ignorant humans we encounter on the streets every day.

Fifty fish survived at the end of 1982. They were a unique and important group of very strange animals.

With the help of some friends, I discovered that every fish looked very much like the species *Acipenser naccarii*, commonly known as cobice sturgeon. This meant that the other two species, the *Acipenser sturio* and the *Huso huso,* were already in great danger because they were almost extinct.

When someone is a pioneer, it means one receives little and inconstant help; as the days go by, every new thing just adds up, making the enterprise more difficult.

Finally, June 1988 marked the first reproduction; we had made it! The most important part of the dream, the achievement of the impossible.

I remember the great excitement that led to crying a whole day just because of the joy.

Today aquaculture of the sturgeon is a known reality in various countries where the meat and the precious caviar are produced. We have been responsible for beginning the breeding of sturgeons. It has transformed the 'fish for kings' into the 'king of fish' (Fig. A1).

What makes my work so unique and special is the fact of working with an almost extinct species, a group that inadvertently became the last survivors.

In other countries the sturgeon appeared without water, in museums. Because of that, in the Italian rivers the species were disappearing (Fig. A2).

Thanks to the goodwill of the Great Black, Mamma, Crazy and other fish, the effective cooperation for so long of my sons Alberto, John and Sergio, and the help of the veterinarian Dr. Giovanni Arlati and the Lombardy region, every year we produced thousands of sturgeons. Then we started the first reintroduction of sturgeons into the northern Italian rivers flowing through Lombardy. That was followed by a period of patient labour and waiting for the restocking of the environment.

Fig. A1 *Acipenser naccarii*—from the fish for kings to the king of fish

Fig. A2 Thanks to 'Mamma Storione' for the precious present

Since 2002, if I remember correctly, the provinces of Veneto, and Venice in particular, has been a great example of proper educational and ecological management.

Alessandro Manzoni, the famous writer and literary genius, once said: 'The divine grace, little by little, shows itself in our hearts, to prepare a wider way to respect and love'.

And now that is exactly what it is happening. There are hundreds of people worried about the survival of the cobice sturgeon. Meanwhile, we have identified 20 more biodiverse families.

From September 2004 to September 2007, the European community has been involved in the project LIFENATURA/IT/000126, a project that has built up a lot of expectations and tangible action toward the recovery of the cobice sturgeon species, helping various partners with contributions, which are being invested every day.

We look at the rainbow that is impossible to catch; at the same time, we are sure that at least we have offered new life and living space to the sturgeon. Thanks to actions of this type, human values keep on growing: solidarity, tolerance, respect, democracy, and peace. At the setting of a life or the sun, tired after all the work has been done, lost in our thoughts and dreams, we rest.

Good morning Mr. Sturgeon, may you have a nice and peaceful day. Thanks for opening my eyes. Thank you for giving me this great opportunity of life and love.

Index

D

E

H

I

P

R

MIX
Papier aus verantwortungsvollen Quellen
Paper from responsible sources
FSC® C105338

If you have any concerns about our products,
you can contact us on
ProductSafety@springernature.com

In case Publisher is established outside the EU,
the EU authorized representative is:
Springer Nature Customer Service Center GmbH
Europaplatz 3, 69115 Heidelberg, Germany

Printed by Libri Plureos GmbH
in Hamburg, Germany